ELECTRON
AND ION
MICROSCOPY
AND
MICROANALYSIS

OPTICAL ENGINEERING

Series Editor

Brian J. Thompson
Provost
University of Rochester
Rochester, New York

Other Volumes in Preparation

ELEGTMN AND IN MKNSCIPY AND MI6NIANAIYSIS

MUKMES MMPNKHMB

SECOND EDITION
REVISED AND EXPANDED

LAWKNCE L M M

University of Texas at El Paso
El Paso, Texas

CRC Press
Taylor & Francis Group
Boca Raton London New York

CRC Press is an imprint of the
Taylor & Francis Group, an **informa** business

CRC Press
Taylor & Francis Group
6000 Broken Sound Parkway NW, Suite 300
Boca Raton, FL 33487-2742

First issued in paperback 2019

© 1991 by Taylor & Francis Group, LLC
CRC Press is an imprint of Taylor & Francis Group, an Informa business

No claim to original U.S. Government works

ISBN-13: 978-0-8247-8556-7 (hbk)
ISBN-13: 978-0-367-40294-5 (pbk)

Visit the Taylor & Francis Web site at
http://www.taylorandfrancis.com

and the CRC Press Web site at
http://www.crcpress.com

Library of Congress Cataloging-in-Publication Data

Murr, Lawrence Eugene.
    Electron and ion microscopy and microanalysis: principles and
applications/Lawrence E. Murr. - - 2nd ed., rev. and expanded.
        p.   cm. -- (Optical engineering; 29)
    Includes bibliographical references and indexes.
    ISBN 0-8247-8556-8 (acid-free paper)
    1. Electron microscopy. 2. Field ion microscopy. 3. Microphobe
analysis. I. Title. II. Series: Optical engineering (Marcel
Dekker, Inc.); v. 29.
QH212.E4M87 1991
    502'.8'25--dc20   91-20173
                                                        CIP

I affectionately dedicate this book to my "family",
which, like the topics included, has grown over the
25 years since this effort was begun...
To: Pat; Kim and Janet; Kyle and Alison

About the Series

The series came of age with the publication of our twenty-first volume in 1989. The twenty-first volume was entitled *Laser-Induced Plasmas and Applications* and was a multi-authored work involving some twenty contributors and two editors: as such it represents one end of the spectrum of books that range from single-authored texts to multi-authored volumes. However, the philosophy of the series has remained the same: to discuss topics in optical engineering at the level that will be useful to those working in the field or attempting to design subsystems that are based on optical techniques or that have significant optical subsystems. The concept is not to provide detailed monographs on narrow subject areas but to deal with the material at a level that makes it immediately useful to the practicing scientist and engineer. These are not research monographs, although we expect that workers in optical research will find them extremely valuable.

There is no doubt that optical engineering is now established as an important discipline in its own right. The range of topics that can and should be included continues to grow. In the ''About the Series'' that I wrote for earlier volumes, I noted that the series covers ''the topics that have been part of the rapid expansion of optical engineering.'' I then followed this with a list of such topics which we have already outgrown. I will not repeat that mistake this time! Since the series now exists, the topics that are appropriate are best exemplified by the titles of the volumes listed in the front of this book. More topics and volumes are forthcoming.

Brian J. Thompson
University of Rochester
Rochester, New York

Preface

This book is part of an evolutionary process that started around 1966. It is an extension and revision of my original book *Electron Optical Applications in Materials Science* published by the McGraw-Hill Book Co. in 1970. In the preface of the original volume, I wrote "During my final year of graduate school I felt a need for a useful, comprehensive treatment of electron microscopy and its related electron optical techniques. I also felt that such a treatment should stress the uses of electron optical theory, phenomena, and devices in the study and characterization of materials. The actual writing began with a fervent desire to teach the basic theory and applications of electron optical devices and techniques, and to create an awareness of the diversity of electron optical uses for the materials scientist". In retrospect, I believe that, to a large extent, I did succeed in my original goals with *Electron Optical Applications in Materials Science*. That book received a number of favorable reviews following publication and was adopted or used by nearly two dozen universities in the United States and by universities in Taiwan, India, South America (particularly Chile and Brazil) and numerous European and other countries during the decade of the 1970s. When *Electron Optical Applications in Materials Science* went out of print in 1978, I received a great deal of encouragement from colleagues who had used the book as teachers, students, or both to consider a revision. I had thought even earlier of the alarming rate at which new concepts were being developed, and the lack of treatment in the original book. At the same time, I was concerned with the importance of the synergism rapidly being promoted and required in many characterization laboratories, not only in analytical electron microscopy, but in the combined approaches to materials characterization involving electron and ion optical techniques.

In the context of what I perceived as the success of the original text and the need to broaden the coverage with a particular aim at developing the analytical concepts underlying the synergism involving both electron and ion optical systems, which were emerging as modern methods of materials characterization, I was sensitive to criticisms leveled by some users who found the original treatment too theoretical, too general, and pitched at a relatively high (graduate) level. Indeed, during my own use of the original McGraw-Hill version through the mid 1970s, I had developed two semester-length courses: one a senior-level course involving fundamentals or elements of electron microscopy and the other a graduate-level course involving electron and ion optical applications.

The actual writing of this book began anew with a fervent desire to teach the basic theory and applications of electron and ion optical devices and techniques and to create an awareness of the diversity of electron and ion optical uses for the scientist or engineer in a wide range of disciplines. I was particularly concerned with the concepts of materials characterization, and I was not concerned that, probably like the original book, this book would be criticized as "taking a physicist's point of view too often". By virtue of the origin and evolution of this book over a period of continuous teaching of courses since 1966, this book is written for students at the senior or graduate level and for self-study by a wide range of materials scientists and engineers, chemists, physicists, metallurgists, ceramists, and others in the physical sciences and engineering who are interested or involved in the study and characterization of materials. It is a textbook designed to offer a program format and act as a reference for at least two semester or two quarter-length courses at the senior or graduate level.

The paragraphs above provided a historical perspective in the first edition of this book, and that perspective has not changed. Only the history has been extended to make this an effort which now spans 25 years.

The organization of topics begins with the basic properties of electrons and a short treatment of ions in Chapter 1. The quantum-mechanical aspects of electron optical systems, the characteristic emission properties of electrons from solids, and an explanation of electron emission and field-ion microscopy and related applications are treated in Chapter 2. Chapter 3 deals with the principles of electron and ion optics and the practical uses of an electron or ion optical system. Chapter 4 discusses a broad range of electron and ion probe microanalysis techniques and describes, somewhat synergistically, the basic analytical features of electrons and ions. Chapter 5 considers the electron and ion microscopes and

their role in the study of surfaces, and Chapter 6 deals with the theory
and applications of electron diffraction. Chapters 7 and 8 cover the principles and applications of conventional and analytical electron microscopy
as well as high-voltage transmission electron microscopy; only brief mention is made of the potential uses of transmission ion microscopy in Chapter 7. A wide range of applications is covered, which encompasses high-
resolution and *in situ* techniques. The appendices provide easy access to
practical data and aids having particular value in electron microscopy.

In this Second Edition, these topics have been expanded to include the
principles and applications of scanning tunneling microscopy and related
techniques in Chapter 2, electron reflection holography in Chapter 5, and
microprobe reflection high-energy electron diffraction in Chapter 6. Chapter 7 has been expanded to include progress and applications in lattice and
atomic imaging, computed images and computer processing of electron microscope images, transmission electron holography and energy-filtered imaging.
Specimen preparation techniques have been expanded to include cross-
sectional TEM (XTEM) and a variety of novel approaches to preparing
specimens for electron microscopy.

A list of key references, not intended to be exhaustive, but pointing
the way to historical origins or notable applications and contributions, is
included at the conclusion of each chapter along with suggestions for supplementary reading which include updated and contemporary work. These
follow problems, which are for the most part practical, and together with
the reference and supplementary readings lists are intended to aid the
reader in understanding concepts, applying particular techniques, and
solving practical research problems. A section providing solutions and/or
discussion of the problems is also appended to the book. I have given
special attention to the composition of illustrations and drawings, which
are essential to a real understanding of just how certain techniques work,
and what kinds of unique applications are possible.

For a good understanding of the bulk of this book, the reader should
have a mathematical background equivalent to sophomore-level science or
engineering courses, including differential equations and matrix algebra.
A course in solid-state physics and x-ray diffraction would also be helpful. The treatment does not assume any rigorous exposure to quantum
mechanics. In short, the serious student should have little trouble
understanding most of the essentials.

The instructor cannot possibly cover the material in this book in a
quarter-length course. At the very least, two semesters or quarters and
two separate courses would be required. This book is intended as an aid
to the development or teaching of courses in electron microscopy, electron

and ion optical applications, or electron and ion microscopy and microanalysis. It cannot be a course syllabus, and it cannot substitute for the
classroom or laboratory experience. It can only contribute to that experience. Therefore, the lack of depth at some points in the book must be
supplemented with the instructor's own notes or reference materials suggested within or at the conclusion of the individual chapters. This book
is an attempt to form a meaningful dialogue between the writer and the
reader.

Because this book was written to serve as a multi-purpose text at both
the senior and graduate levels, and as a useful reference or self-study
resource for the practicing scientist or engineer, some readers may find
its style heavy in parts. This may have occurred as a result of my own
desire to present the material in a reasonably satisfying way to the student, while still leaving room for the instructor's interpretation and
application. The instructor not only has the opportunity to supplement
for specific shortcomings, which might be perceived or real, but an obli-
gation to do so if the student is expected to achieve a full understanding.
This is not intended to be a "cookbook." To learn to "cook," the student
must get into the laboratory and operate the devices and apparatus described. This book will hopefully develop an appreciation for how things work
and how they can be applied in an efficient and useful way.

As in previous editions, I am indebted to many authors, publishers,
and industrial organizations for their kindness in supplying illustrations
and data, and for their permission to publish them. The various contributions are acknowledged in the text or in appropriate figure captions. I
am also grateful for the valuable comments and suggestions tendered by my
associates and students, as well as other reviewers during the preparation
and class testing, revisions, expansion, and updating, which has gone on
for the past 25 years. I am especially grateful to Elizabeth Fraissinet
who spent nearly a year typing the first edition of this book in camera-
ready form. Hers was a dedication beyond remuneration, and for which I
can only be eternally grateful. It provided the basis for this second
edition. Megan Harris and Faye Ekberg retyped portions of this book, and
the additions and patches which were skillfully matched with the original
camera-ready typescript to create this second edition. I am especially
grateful for their contributions.

Finally, I have dedicated this book to my family which has expanded
with these editions, and continues to provide encouragement and inspira-
tion.

LAWRENCE E. MURR

Contents

1
FUNDAMENTAL PROPERTIES OF ELECTRONS AND IONS

1.1 INTRODUCTION

In a book proposing to deal with emission or production, operations on, and detection of electrons and ions in one form or another, it would seem desirable at the outset to outline their intrinsic physical properties (and their associated historical development). In addition, it would also seem necessary to deal with the properties of electrons in atoms and solids, and then to describe the production and properties of ions. Indeed, the electron is a remarkable concept; at the risk of sounding melodramatic, it might be said that the electron represents the single most important entity in the universe.

There are two very important intrinsic features associated with electrons, namely, the fact that they are, ideally, negatively charged particles possessing a finite mass; and that an electron, or a beam of electrons, possesses a wave nature akin to that normally associated with light, x-rays, or related electromagnetic radiations.* It is this wave-particle

---

*In 1932 C. D. Anderson announced the observation of positively charged particles possessing a charge and mass identical to the negative electron. These have since come to be called positrons.

dualism that renders the electron especially suited to investigating the structure and composition of matter in an electron optical device.

The controversy over the wave-particle identity of electrons (cathode rays) reached its peak at the close of the nineteenth century and continued into the first two decades of the twentieth century. The resolution of the apparent wave-particle paradox was essentially found in quantum mechanics perhaps most notably in the form of Schrödinger's equation, the de Broglie "matter wave" concept, the Heisenberg uncertainty principle, and related contributions during the two decades after 1910. In a real sense, the quantum-mechanical treatment of electrons did not resolve the wave-particle dualism solely in terms of a simple particle or wave-train analog. Indeed, the electron must be defined in terms of its inherent features as a fundamental entity.

1.2 ELECTRON CHARGE AND MASS

The pioneering work of M. Faraday had shown, among other things, that in electrolysis a definite amount of material was deposited for every coulomb of electricity that passed through the solution. In 1881 G. J. Stoney recognized the atomic implications of Faraday's laws, and introduced the term electron to designate an elementary charge. It was not until the experiments of Sir J. J. Thomson in 1895 that the term electron was used in its present-day meaning. Since Faraday had shown that 1 g of hydrogen was liberated for each 96,500 coulombs of charge expended (based on an estimate of 10^{25} atoms of hydrogen for each gram weight of hydrogen), Stoney estimated the electronic charge to be roughly 10^{-20} coulomb (0.3×10^{-10} esu). While the method itself was entirely correct, the estimate of charge was not, since the number of hydrogen atoms estimated to compose a gram weight of the same element was in error. It was not until about 1941, in fact, that x-ray determinations showed the number of atoms in a gram atomic weight of an element (Avagadro's number) to be 6.023×10^{23} and thus allowed the correct charge to be computed on the basis originally proposed by Stoney.

The fact that electrification of a rarefied gas produces "cathode rays" whose trajectory is made visible by luminescence and fluorescence, and which are deviated by magnetic or electric fields, was already known as early as 1869. And, Sir W. Crookes about 1886 had even proposed that cathode rays were negatively electrified particles. This point was defin-

itely proven by J. Perrin in 1895 [1]. Simultaneously, in 1895, Roentgen
discovered that x-rays were emitted when cathode rays bombarded a metal
target.

Two years later, in 1897, Thomson [2] published the first accurate
measurements of the ratio of electron charge to mass (e/m), following the
experimental proposals introduced by A. Schuster [3] some 10 years before.
Thomson's value of 2.3×10^7 emu/g was considerably improved about 1900 in
similar experiments by W. Kaufmann [4] who measured $e/m = 1.8 \times 10^7$ emu/g.
This latter value is nearly in agreement with the presently accepted value.

$$\frac{e}{m} = (1.758896 \pm 0.000028) \times 10^7 \text{ emu/g}$$

The basis for experimental measurements of e/m is illustrated in the
sketch of Fig. 1.1. In this scheme, essentially employed by Kaufmann [4],
the cathode ray (electron beam) accelerated in the x direction by the anode
potential V_o enters a transverse magnetic field (acting in the z direction).
Each electron in this field is subsequently subject to a force Bev_o (where
B is the magnetic field strength, e the electron charge, v_o is the average
electron velocity), in the y direction. Consequently, the electron beam
will deviate from the central axis by an amount depending primarily on the
electron velocity and the magnitude of the magnetic field.

In effect, we say that in the field region (shown shaded in Fig. 1.1)
the electron path is circular, and defined by the radius of R. However,
on leaving the field region (at F in Fig. 1.1), the electron trajectory
remains a straight line normal to R. In the field region we then have

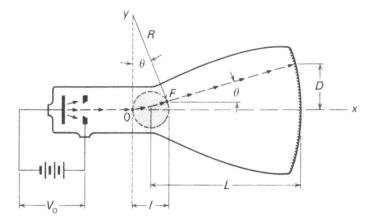

FIG. 1.1 *Magnetic deflection of electrons in cathode-
ray tube.*

$$Bev_o = \frac{mv_o^2}{R}$$

and considering

$$eV_o = \frac{mv_o^2}{2}$$

we obtain

$$R = \frac{1}{B}\sqrt{\frac{2mV_o}{e}} \qquad\qquad (1.1a)$$

Assuming the angle ϕ to be small, we can approximate

$$\theta \cong \frac{\ell}{R}$$

and further assuming L $>>$ ℓ we could consider the observed beam deflection D to be

$$D \cong L\theta$$

for small θ. Substituting for θ and R above, and rearranging, then results in the following approximate expression for e/m:

$$\frac{e}{m} = 2V_o\left(\frac{D}{\ell LB}\right)^2 \qquad\qquad (1.1b)$$

Suffice it to say that Eq. (1.1b) is an approximation that is improved by the patience of the experimentalist. Precision measurements are, however, for the most part not feasible. It was not until studies of the Zeeman effect, about 1935, that a precise value of e/m as presented was obtained. These studies also proved that electrons in a solid were identical to the cathode rays that solids emitted.

It is also instructive to note that if the magnetic field of Fig. 1.1 is replaced with an electric field by inserting square plates of a side, ℓ positioned symmetrically with respect to the central axis of the electron beam in the z plane and in the same position with respect to the cathode-ray tube shown in Fig. 1.1, then the balance of forces in the field region between the plates will become

$$eE = \frac{mv_o^2}{R}$$

where E is the electric field strength, and

$$R = \frac{2V_o}{E}$$

On invoking previous assumptions concerning the beam deflection depicted in Fig. 1.1, we then arrive at

$$D = \frac{\ell L E}{2V_o} \tag{1.2}$$

You will now observe that in contrast to deflection in a magnetic field [Eq. (1.1b)], deflection of an electron beam in an electrostatic field is independent of the charge or mass of the individual electrons. This feature, as will be demonstrated in succeeding discussions, bears considerable importance in the design of electron optical systems. Similarly, we will later see that this also has important consequences for ions as well. In fact it is this property of electrostatic fields which is required in lens-forming systems involving ions.

Experiments of the type initiated by Schuster, and refined by Thomson and Kaufmann, enabled only the ratio e/m to be determined; but until about 1911 neither e nor m were independently determined with any accuracy. However, in 1917, R. A. Millikan [5] reported on a series of now famous "oil drop" experiments that enabled the electronic charge to be determined independently as

$$e = (4.80286 \pm 0.00009) \times 10^{-10} \text{ esu}$$

or

$$e = (1.60206 \pm 0.0003) \times 10^{-19} \text{ coulomb}$$

From this value, confirmed by x-ray measurements of Avogardro's number about 1941, and application of the principle originally deduced by Stoney, the rest mass of the electron becomes

$$m_o = (9.1083 \pm 0.003) \times 10^{-28} \text{ g}$$

The concept of effective mass as opposed to the rest mass of the electron was reflected in numerous experiments to measure e/m with varying accelerating potentials for the electrons. Thus variations in e/m with electron velocity ultimately led to a proof of the relativistic expression

$$m = \frac{m_o}{\sqrt{1 - (v_o/c)^2}} \tag{1.3}$$

where m_o = rest mass (for $v_o \ll c$)

     v_o = velocity of electron

     c = velocity of light

1.3 WAVE NATURE OF ELECTRONS

The experiments with electrons, notably those of Thomson and Millikan, showed incontrovertibly that electrons were, for all practical purposes, particles. Simultaneously during the two decades following 1900, W. H. Bragg, W. Friedrich, P. Knipping, and M. von Laue were busy demonstrating the now well-known properties of x-rays. In particular, it was Friedrich and Knipping [6], at the suggestion of von Laue, whose famous diffraction experiment finally illustrated the wave-particle "corpuscle" nature of x-rays. The coincidence of L. de Broglie's doctoral thesis [7,8] in 1924 clinched the proposition of the wave nature of matter as reflected in the famous relationship

$$\lambda = \frac{h}{P} \tag{1.4}$$

where λ = associated wavelength

     h = Planck's constant

     P = particle momentum

In the case of the electron, we can now write

$$\lambda = \frac{h}{m_o v} \sqrt{1 - (v/c)^2} \tag{1.5}$$

On the basis of the de Broglie theory and on considering the previous work of C. J. Davisson and C. H. Kunsman, W. Elsasser [9] concluded that when electrons pass through a crystal, they should, like x-rays, exhibit diffraction and interference phenomena. Subsequently, in 1927, the diffraction of electrons as an experimental proof of the manifestations of their wave properties was announced independently, and nearly simultaneously, by Davisson and L. H. Germer of the Bell Telephone Laboratories; and by G. P. Thomson (the son of J. J. Thomson) and A. Reid of the University of Aberdeen. Davisson and Germer, investigating electron reflection from the surface of Ni single crystals, observed that electrons with kinetic energy

$$eV_o = mc^2 - m_o c^2 \tag{1.6}$$

were scattered at definite angles, just as in the case of x-ray diffraction [10,11]. Similarly, Thomson and Reid, on passing an electron beam through celluloid (and later thin metal foils) observed diffraction "ring" patterns on exposure of a photographic plate to the emergence side of the thin specimens, identical in all respects to those produced by x-rays (see Fig. 1.2) [12,13]. These experiments were quantitatively accounted for by admitting a wave nature of electrons, and attributing to them a wavelength described by combining Eqs. (1.5) and (1.6),

$$\frac{h}{\lambda} = \sqrt{2m_o eV_o + \frac{e^2 V_o^2}{c^2}} \tag{1.7}$$

and evaluating the constant, that is,

$$\lambda = \frac{12.27}{\sqrt{V_o(1 + 0.978 \times 10^{-6} V_o)}} \text{ Å} \tag{1.8}$$

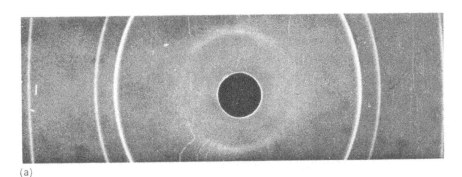

(a)

(b)

FIG. 1.2 *X-ray and electron diffraction patterns of fcc Ni. (a) Debye-Scherrer x-ray powder diffraction pattern (CuKα, λ = 1.54 Å). (b) Selected-area electron diffraction pattern (λ = 0.037 Å).*

The scattering itself was also shown to be ideally accounted for by considering the Bragg equation for x-ray diffraction

$$2\alpha \sin \theta = n\lambda \qquad (1.9)$$

where α = atomic spacing of crystal planes
$\qquad \theta$ = scattering angle
$\qquad n$ = integer denoting order of diffracted beam

R. Emden [14] showed rather conclusively in 1921 that the interpretation of light as quanta accounted for the Stefan-Boltzmann law, Planck's
law of radiation, and the undulatory Doppler effect in optics. A year later, E. Schrödinger developed an even more rigorous derivation of the optical Doppler effect solely on the basis of the quantum-corpuscular character
of light [15]. Thus the stage was set for the quantum treatment of the
electron. In 1926, Schrödinger, acting on the implications of de Broglie's
theory of a year earlier, devised a mathematical interpretation of the wave
nature of matter based on replacing the equations of motion of classical
mechanics with a matter wave equation [16].

In wave optics, the propagation of electromagnetic radiation ϕ (where
ϕ is a scalar quantity representing, for example, the electromagnetic potential) with a velocity c (the velocity of light in vacuo) is described
by

$$\nabla^2\phi - \left(\frac{n}{c}\right)^2 \frac{\partial^2\phi}{\partial t^2} = 0 \qquad (1.10)$$

where

$$\nabla^2\phi = \frac{\partial^2\phi}{\partial x^2} + \frac{\partial^2\phi}{\partial y^2} + \frac{\partial^2\phi}{\partial z^2}$$

and n is the index of refraction as given by

$$n = \frac{c}{\nu\lambda} \qquad (1.11)$$

where ν and λ are the frequency and wavelength, respectively. Substitution of Eq. (1.11) into (1.10) results in

$$\nabla^2\phi - \frac{1}{(\nu\lambda)^2} \frac{\partial^2\phi}{\partial t^2} = 0 \qquad (1.12)$$

If we now consider the de Broglie hypothesis in the form expressed by

Eq. (1.4), as did Schrödinger more than a quarter century ago, and the fact that the total electron energy is given by[*]

$$E = \frac{1}{2} mv^2 + U$$

where $mv^2/2$ = kinetic energy

U = potential energy

or

$$E = \frac{p^2}{2m} + U$$

then

$$p^2 = 2m(E - U) \qquad (1.13)$$

Finally, from Eq. (1.4),

$$\lambda = \frac{h}{\sqrt{2m(E - U)}} \qquad (1.14)$$

In the case of harmonic oscillations with a frequency ν, we can assume a solution of Eq. (1.12) of the form

$$\phi = \phi(x,y,z,t) = \psi(x,y,z)e^{2\pi i\nu t} \qquad (1.15)$$

where $\psi(x,y,z)$ is the amplitude of the oscillations of the quantity ϕ. Substitution of Eqs. (1.14) and (1.15) into (1.12) for λ and ϕ, results in

$$e^{2\pi i\nu t}\nabla^2\psi + \frac{e^{2\pi i\nu t} \cdot 2m(4\pi^2\nu^2)(E - U)\psi}{h^2\nu^2} = 0$$

whence it follows that

$$\nabla^2\psi + \frac{2m}{\hbar^2}(E - U)\psi = 0 \qquad (1.16)$$

where $\hbar = h/2\pi$. Equation (1.16) is recognized as the Schrödinger wave equation which, as it relates specifically to electrons, is completely descriptive of the dualistic nature of the electron. Indeed, Eq. (1.16) can be thought of as the mechanistic description of the electron as an entity.

---

[*]*It should be remembered that in speaking of electron energy, we must consider the fact that in the case of a completely free electron, U = 0, and consequently the total energy is simply $P^2/2m$.*

1.3.1   THE ELECTRON WAVE FUNCTION

In the treatment of electrons as discrete particles in terms of current
flow in a conductor, or particle flux through a unit area or volume of conductor, we can define the flux per unit volume as

$$\underset{\sim}{\nabla} \cdot \underset{\sim}{J} = - \frac{\partial \bar{e}}{\partial t}$$

where $\underset{\sim}{J}$[†] is the current density and $\bar{e}$ is the charge density of the electrons. The intensity of illumination of a fluorescent screen bombarded by
electrons will be proportional to the number of electrons striking the
screen per unit area of screen.

In terms of the wave concept, intensity is proportional to the square
of the wave amplitude. Thus since ψ is the representative function of
electron wave motion [Eq. (1.16)], the intensity is seen to be proportional
to $\psi^*\psi$ or simply $|\psi|^2$, since the wave function may, generally speaking, be
complex. We can then represent the total electron charge of a unit volume
as follows

$$e \iiint \psi^*\psi \; dx \; dy \; dz = e \qquad\qquad\qquad (1.17)$$

Normalization of Eq. (1.17) then results in

$$\iiint |\psi|^2 \; dx \; dy \; dz = 1 \qquad\qquad\qquad (1.18)$$

which is representative of the fact that for an electronic charge e, $e|\psi|^2$
is the charge density. Exactly akin to this interpretation, we must con-
sider the fact that the quantity $|\psi|^2 \, dx \, dy \, dz$ is effectively the probabil-
ity of finding an electron in a given volume element; and from this point
of view the quantity $\psi^*\psi$ or $|\psi|^2$ is a probability density. Continuing this
argument still further, the wave function ψ then becomes the amplitude of
the probability distribution of electrons. The former interpretation of
the wave function associated with charge density is due to Schrödinger
[17], while the latter probabilistic concept is due originally to M. Born
[18].

In the most general interpretation, ψ refers to the probability of
finding an electron in an atom, and as dictated by Eq. (1.18), the total
probability of finding an electron therein is unity. As a consequence of
the probabilistic interpretation of the wave function, it must follow that

---
[†]*Designates vectors or vector quantities.*

the position of a moving electron is also a probability, and therefore an exact location of an electron in an atom is impossible. This particular property, which in effect results as a manifestation of the wave-particle dualism and the associated probabilistic nature of the wave function, is expressed in the most general form by the Heisenberg uncertainty principle [19]. The mathematical form of this statement appears as

$$\Delta x \; \Delta P \sim h$$

or

$$\Delta P \; \Delta q = \frac{h}{4\pi} \tag{1.19}$$

where ΔP is an associated uncertainty in the determination of the electron momentum, Δq is the uncertainty in defining its position along a coordinate q, and h is Planck's constant. Equation (1.19) therefore states that with an increase in the precision of determining the electron's position, there results a concomitant decrease in the precision with which its associated momentum can be determined, and vice versa. As we shall see, the energies of electrons in atoms, or those confined to a solid, appear as discrete values. Consequently the electron orbit, or its geometric position in space, is impossible to define.

1.3.2 ELECTRON WAVE MOTION

Two important properties of the general classification of waves are that energy is transmitted from one point to another and that effectively no permanent displacement occurs as a result of the wave propagation within a time t. Let us now consider a one-dimensional harmonic electron wave to be represented in the form

$$\psi(z) = A \cos \left(|\underline{K}| \cdot z - \omega t \right) + B \sin \left(|\underline{K}| \cdot z - \omega t \right) \tag{1.20}$$

where $|\underline{K}|$ = wave number as defined by $2\pi/\lambda$

      A = constant

      B = constant

      $\omega = 2\pi\nu$

This is a solution of the time-dependent Schrödinger equation

$$\nabla^2 \psi - \frac{2m}{\hbar^2} U\psi = \frac{2m}{\hbar^2 i} \frac{\partial \psi}{\partial t} \tag{1.21}$$

in one dimension:

$$\frac{\partial^2 \psi}{\partial z^2} - \frac{2m}{\hbar^2} U\psi = \frac{2m}{\hbar^2 i} \frac{\partial \psi}{\partial t} \tag{1.22}$$

If we further suppose that due to the principle of superposition,[*] a solution exists of the form

$$\psi(z) = \psi_1 + \psi_2$$

where ψ_1 and ψ_2 are also harmonic solutions of the wave equation (1.22), both having zero phase and equal amplitudes, we can write the real part of the total wave function (considering in effect a stationary electron) as

$$\psi = \psi(z) = A[\cos(|\underset{\sim}{K}_1| \cdot z - \omega_1 t) + \cos(|\underset{\sim}{K}_2| \cdot z - \omega_2 t)]$$

or

$$\psi = 2A \cos\left[\frac{(|\underset{\sim}{K}_1| + |\underset{\sim}{K}_2|)z}{2} - \frac{(\omega_1 + \omega_2)t}{2} \right]$$
$$\left[\cdot \cos \frac{(|\underset{\sim}{K}_1| - |\underset{\sim}{K}_2|)z}{2} - \frac{(\omega_1 + \omega_2)t}{2} \right] \tag{1.23}$$

If we now consider the frequencies of the two waves to be nearly equal, the first term of Eq. (1.23) represents a modulated amplitude of the second term. The shape of this wave would appear as shown in Fig. 1.3 and possess a phase velocity

$$v_p = \frac{\omega_1 + \omega_2}{|\underset{\sim}{K}_1| + |\underset{\sim}{K}_2|} = \frac{dz}{dt} = \frac{\omega}{|\underset{\sim}{K}|} \tag{1.24}$$

where

$$\frac{\omega_1}{|\underset{\sim}{K}_1|} \neq \frac{\omega_2}{|\underset{\sim}{K}_2|}$$

the wave envelope of Fig. 1.3 travels with a group velocity

---

[*]*Compare with M. Born, "Atomic Physics," Hafner Publishing Company, Inc., New York, 1957, p. 336.*

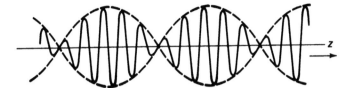

FIG. 1.3 *Electron wave packets*

$$v_g = \frac{\omega_1 - \omega_2}{|\underset{\sim}{K}_1| - |\underset{\sim}{K}_2|} = \frac{\Delta\omega}{\Delta|\underset{\sim}{K}|} \qquad (1.25)$$

In the limit as $\Delta|\underset{\sim}{K}|$ in Eq. (1.25) tends toward zero, we can write

$$v_g = \frac{d\omega}{d|\underset{\sim}{K}|} = \frac{dv}{d(1/\lambda)} = -\lambda^2 \frac{dv}{d\lambda} \qquad (1.26)$$

If we now substitute for ω in terms of Eq. (1.24) in (1.26), we obtain

$$v_g = \frac{d(|\underset{\sim}{K}|v_p)}{d|\underset{\sim}{K}|} = v_p + |\underset{\sim}{K}| \frac{dv_p}{d|\underset{\sim}{K}|} \qquad (1.27)$$

or

$$v_g = v_p - \lambda \frac{dv_p}{d\lambda} \qquad (1.28)$$

where $\lambda dv_p/d\lambda$ is the dispersion factor of the wave system. Group velocity is therefore dependent on the wave dispersion (that is, the change in velocity with frequency or wavelength); and we see that in a dispersive system, the only wave profile that could be transmitted without changing shape is a single harmonic wave train.

In the particular case of a free electron, we can also consider the total energy and momentum, given by

$$E = \frac{m_0 c^2}{\sqrt{1 - (v/c)^2}} \qquad (1.29)$$

and

$$P = \frac{m_0 v}{\sqrt{1 - (v/c)^2}} \qquad (1.30)$$

respectively.   Rewriting Eq. (1.24) in the equivalent form,

$$v_p = \frac{E}{P} \tag{1.31}$$

and substituting E and P of Eqs. (1.29) and (1.30) into Eq. (1.31) results in

$$v_p = \frac{c^2}{v} \tag{1.32}$$

where v is ideally the z component of velocity. Similarly, if we rewrite the group velocity as

$$v_g = \frac{dE}{dP} \tag{1.33}$$

we obtain on substitution of Eqs. (1.29) and (1.30)

$$v_g = v \tag{1.34}$$

with respect to the propagation of the wave train along the z direction as shown in Fig. 1.3.

We observe from Eq. (1.34) that the electron group or wave packet, whose velocity is v_g, is identical to the velocity we normally associate with the particle velocity v of the electron. Consequently we have in effect shown that Fig. 1.3 ideally depicts the electron as an entity. However, this simple wave-packet approximation is somewhat complicated by Eq. (1.32), which states that for v_g < c, the phase velocity is greater than c. This seems contrary to our understanding of particle physics, which accordingly must demand we somehow cope with a phase velocity of approximately 2c for electrons accelerated by a potential V_o of 10^5 volts.

It should be noted that if we assume a nondispersive system, then $v_g = v_p$ from Eq. (1.28). In the case of a free electron, however, this cannot occur simply on the basis of Eqs. (1.32) and (1.34). This feature therefore demands that the wave packet is continually changing, as illustrated in Fig. 1.4. Here we observe that the phase waves advance through the group profile, first increasing, then decreasing in amplitude as they give way to succeeding waves. The simplest physical example of this is the wave disturbance on the surface of a liquid. Figure 1.4 shows the successive locations of the wave-packet or group velocity v_g, and the phase-wave velocity v_p, with respect to some time t = 0, which might be thought of as corresponding to the instant a free electron having effectively zero energy experiences an accelerating potential V_o. The broaden-

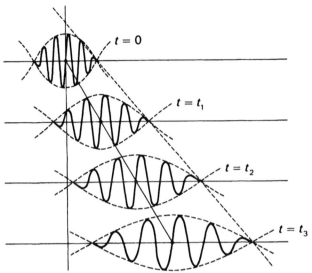

FIG. 1.4 *Electron wave dispersion. Solid connecting line relates to the group velocity, the broken line the phase velocity.*

ing of the wave packet with time is indicative of the dispersion phenomenon, and the slopes of the interconnected reference points are proportional to the group and phase velocities as indicated.

In view of the dispersion features and anomalous wave velocities associated with the electron, we are left with a feeling that the electron can't really be described, pictorially. This brings to mind the quotation of N. Bohr: "In order to obtain a consistent account of atomic phenomena, it was necessary to renounce even more the use of pictures". We are forced to continue to accept this view. However, in a vague and seemingly futile attempt to maintain some association with our physcial senses, the wave-packet analog or the conceptual description of the electron will be utilized; the electron will be symbolized as shown in Fig. 1.3.

1.4 ATOMIC ENERGY LEVELS

The general wave equation for an atom, which has associated with it N electrons, can be written in the form

$$\sum_{j=1}^{N} \left[\frac{\partial^2 \psi}{\partial x_j^2} + \frac{\partial^2 \psi}{\partial y_j^2} + \frac{\partial^2 \psi}{\partial z_j^2} + \frac{2mM}{\hbar^2(m+M)} \left(\frac{e^2 Z}{4\pi\varepsilon_0 r_j} - U_j \right) \psi \right]$$
$$+ \frac{2mM}{\hbar^2(m+M)} E\psi = 0 \quad (1.35)$$

where m = electron mass

    M = nuclear mass

    eZ = nuclear charge

    E = total atomic energy

    r_j = distance of jth electron from nucleus

and

$$U_j = \frac{1}{4\pi\varepsilon_o} \sum_{k=1}^{N} \frac{e^2}{r_{jk}} \tag{1.36}$$

where r_{jk} is the distance between the jth electron and the kth electron. In the case of the multielectron atom, Eq. (1.35) cannot be solved exactly, and solutions must be attempted through a recourse to perturbation theory, except in the simplest case of the hydrogen atom where j = 1, N = 1, and $r_{jk} \to \infty$, leaving $U_j = 0$ in Eq. (1.36). The wave equation for the hydrogen atom then becomes

$$\frac{\partial^2\psi}{\partial x_1^2} + \frac{\partial^2\psi}{\partial y_1^2} + \frac{\partial^2\psi}{\partial z_1^2} + \frac{2mM}{\hbar^2(m+M)}\left(\frac{e^2Z}{4\pi\varepsilon_o r_1} + E\right)\psi = 0 \tag{1.37}$$

The solution of Eq. (1.37) is facilitated by a change of the cartesian coordinates x_1, y_1, z_1 to the spherical coordinates r_1, θ, ϕ through general relations

$$x_1 = r_1 \sin\theta \sin\phi$$

$$y_1 = r_1 \sin\theta \cos\phi$$

$$z_1 = r_1 \cos\theta$$

and Eq. (1.22) then assumes the form

$$\frac{1}{r_1^2}\frac{\partial}{\partial r_1} r_1^2 \frac{\partial\psi}{\partial r_1} + \frac{1}{r_1^2 \sin^2\theta}\frac{\partial^2\psi}{\partial\phi^2} + \frac{1}{r_1^2 \sin^2\theta}\frac{\partial}{\partial\theta}\sin\theta\frac{\partial\psi}{\partial\theta}$$

$$+ \frac{2mM}{\hbar^2(m+M)}\left(\frac{e^2Z}{4\pi\varepsilon_o r_1} + E\right)\psi = 0 \tag{1.38}$$

The solution of Eq. (1.38), based on separation of the variables, is rather classic, and will not be pursued at length here. It should be re-called, however, that the solution of (1.38), assuming a general wave func-tion of the form

$$\psi = \psi(r_1)\psi(\theta)\psi(\theta)$$

results in the specification of quantum numbers n, ℓ, and m_ℓ: the princi-pal quantum number (n = 1,2,3,...), the azimuthal quantum number (descrip-tive of electron orbital angular momentum) specific to the atomic subshells or states ($\ell \leq n - 1$), and the magnetic quantum number m_ℓ = -ℓ...0... + ℓ), respectively; and yields an approximate expression for the total energy in the form

$$E_n = - \frac{mMe^4Z^2}{8\varepsilon_o{}^2\hbar^2(m + M)} \frac{1}{n^2} \tag{1.39}$$

Equation (1.39) is essentially coincident with the energy of atomic hydro-gen as deduced by the Bohr-Sommerfeld theory and the associated classical mechanics model of the atom. Inclusion of the relativistic variation of electron mass (assuming m + M$\cong$M) as in the work of Sommerfeld [20] results in an energy dependence on the quantum numbers n and ℓ, for example,

$$E_n(\ell) = - \frac{m_o e^4}{2(4\pi\varepsilon_o\hbar)^2} \left[\frac{1}{n^2} + \frac{\zeta}{n^4} \left(\frac{n}{\ell + 1/2} - \frac{3}{4} \right) \right] \tag{1.40}$$

where

$$\zeta = \frac{e^2}{4\pi\varepsilon_o\hbar c}$$

The fact that the energy is negative suggests that energies become increas-ingly larger with increasing n, approaching zero energy for n → ∞. This is illustrated in the energy-level diagram for the hydrogen atom as pre-sented in Fig. 1.5.

1.5 ELECTRON SPIN

The energy-level concept, although satisfactorily explaining the emission lines from various elements, fails to deal adequately with the splitting of emission lines into double lines. As a consequence of this phenomenon G. E. Uhlenbeck and S. Goudsmit about 1925 postulated that electrons, when

E_∞ ———————— 0.0 eV ———————— $n = \infty$

E_4 ———————— -0.9 eV ———————— $n = 4$

E_3 ———————— -1.8 eV ———————— $n = 3$

E_2 ———————— -3.4 eV ———————— $n = 2$

E_1 ———————— -13.6 eV ———————— $n = 1$ FIG. 1.5 *Electron-energy level diagram for simple hydrogen atom.*

viewed as stationary particles, possess a spin about some axis [21]. Short-ly thereafter, P. A. M. Dirac showed rather conclusively that spin can be predicted, on the basis on relativistic wave mechanics, to be an intrinsic property of electrons [22]. W. Pauli then, stimulated by E. C. Stoner's conclusions [23] and related contemporary spectroscopic and magnetic ob-servations, proposed that no two electrons have the same quantum numbers or occupy exactly the same quantum state [24]. Thus if we assign m_s to be the quantum number representative of electron spin, the exclusion principle implies that no two electrons can have values of n, ℓ, m_ℓ, and m_s identical.

Spin is a vector quantity, and has the dimensions of angular momentum. If we view the electron only in terms of its particle nature, spin is ac-commodated by the analog of this particle rotating about some arbitrarily defined axis, with a constant angular velocity; and with characteristic values of the spin component along such an arbitrarily prescribed direc-tion of $\hbar/2$ and $-\hbar/2$. Since spin is measured in terms of $\hbar$, we can des-cribe the characteristic value simply by +1/2 or -1/2.

In the case of multielectron atoms [actually any atom where N > 1 in Eq. (1.35)] where the quantum levels (or shells $\equiv$ K,L,M,N, etc.) degenerate into sublevels, as specified by the quantum numbers, where two electrons possess identical values of n, ℓ, and m_ℓ, we demand their spins be oppo-site. We might thus depict a quantum state containing two electrons as shown in Fig. 1.6.

We therefore observe that for each quantum or energy level we have a definite number of sublevels or quantum states designated by s, p, d, f, etc. As a consequence of the energy level degeneracies as stipulated by the inclusion of the quantum numbers ℓ, m_ℓ, and m_s into the description of

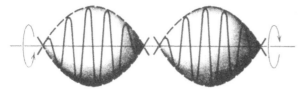

FIG. 1.6 *Electrons occupying the same energy sublevel with quantum numbers, n, ℓ, and m_ℓ, equal (s state).*

multielectron atoms, the electron energy-level diagram, as illustrated in Fig. 1.5 for a single electron without spin, can be modified to illustrate the complete electron quantum structure as shown in Fig. 1.7.

We observe from Fig. 1.7 that, for each value of n, all values of ℓ from 0 to n - 1 are possible; and for each value of ℓ, m_ℓ assumes all integral values such that $-\ell \leq 0 \leq \ell$. Thus for n = 1, ℓ = 0, and correspondingly, m_ℓ = 0. In this case only one state, the ground state, exists, and the two electrons occupying this state can be visualized as depicted in Fig. 1.6. For n = 2, ℓ can have values of 0 or 1; and correspondingly, m_ℓ takes on values of 0 for ℓ = 0, and -1, 0, and 1 when ℓ = 1. We therefore observe that eight states (including the opposing electron spins) with identical energy exist in the n = 2 quantum level. Thus, for any value of n, the degeneracy, or number of states having the same energy, is

$$2 \sum_{\ell=0}^{n-1} 2\ell + 1 = 2n^2 \qquad (1.41)$$

where the factor 2 accounts for the opposing electron spins. Equation (1.41) indicates the total number of electrons which can occupy a quantum level n.

1.6 ELECTRONS IN MOLECULES AND SOLIDS: ELECTRON ENERGY BANDS

In the simple case of the hydrogen atom we have observed, admittedly rather sketchingly, that the electron bound to such an isolated nucleus possesses certain discrete energy levels; it can be shown similarly that the associated wave functions are also unique. We now desire to construct a working model of a solid composed of atoms or molecules with overlapping energy states that ultimately must blend to stabilize the coherent properties of the solid. Let us begin with a look at what happens to the energy levels of two isolated atoms that are brought together to form a simple molecule.

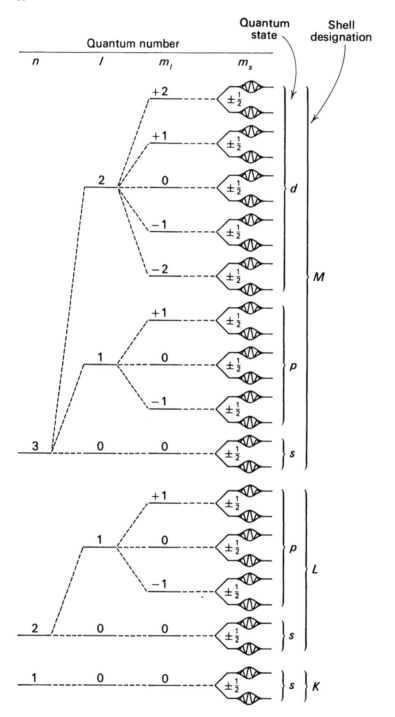

FIG. 1.7 *Complete electron energy-level diagram for n = 1,2,3.*

When two atoms are brought together, stabilization of the diatomic molecule proceeds essentially according to the balance of interacting repulsive and attractive forces, as shown in Fig. 1.8. In the same way that the coulombic forces must ultimately interact to stabilize the nuclear centers at some critical spacing r_0, the interaction of the electron energy levels produces a distinction in the energy states of the molecule as opposed to the isolated nuclei. In effect we are saying that in the case of the isolated nuclei, there exist n energy levels. However, as the atoms are brought together as shown in Fig. 1.8, the levels begin to split, in a sense, producing an energy difference that increases for each n level as the distance between the atom decreases. In the case of the diatomic molecule, the energy levels will appear as shown in Fig. 1.9. This situation ensures that the nuclei will approach each other until the coulombic repulsion between them increases to the point where the decrease in the electron energies is superseded and a bond is established between the two atoms.

The same situation as depicted in Figs. 1.8 and 1.9 will occur when bringing togther many multielectron atoms to form a solid. Ideally then, let us imagine a thin metal foil having a thickness of 10 atoms (corresponding roughly to 40 to 60 Å). Assuming that the crystal structure is simple cubic and has an interatomic spacing (lattice parameter) a, we could expect the energy levels, corresponding to the scheme outlined in Fig. 1.9, to appear as shown in Fig. 1.10. It should now become clear that the

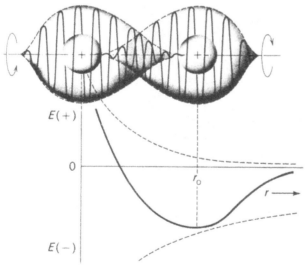

FIG. 1.8 *Stabilization of two hydrogenlike atoms in simple diatomic molecule.*

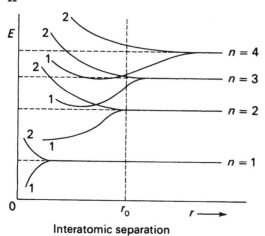

FIG. 1.9 *Energy states for electron in simple diatomic molecule. (1 and 2 refer to two states created when stabilizing molecule).*

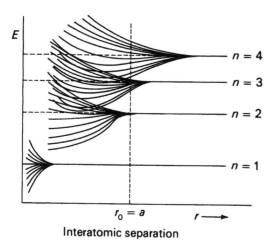

FIG. 1.10 *Energy states for ensemble of 10 atoms in metal crystal foil resolved along single line, with interatomic spacing a, and total foil thickness 10a.*

permitted electron energies are split into permitted energy bands. And, while Fig. 1.10 shows only the energy-band structure for electrons in a specified thickness of metal along a single direction in the foil, a foil containing N atoms per unit volume of solid will effectively have N energy states in each band. This particular feature, and the concept of electron energy bands in a metal, is depicted in Fig. 1.11.

Up to this point we have dealt primarily with electrons associated with atoms or molecules (and essentially solids) in the sense that they were bound to or otherwise strongly influenced by the nucleus or nuclei. This established the fact that electrons possess discrete energy levels, and correspondingly discrete wave functions. We now conclude our basic discussions of the electron by considering the energy structure and wave

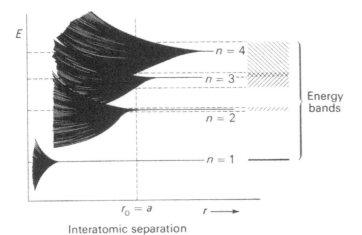

Interatomic separation

FIG. 1.11 *Electron energy-band structure of solid crystal containing N atoms separated distance a.*

properties associated with electrons in a metal crystal, especially where these electrons are considered free or weakly bound to the nuclei composing the crystal. This concept is imperative to an understanding of emission processes, discussed in Chap. 2, and to the interaction of electrons in crystals, discussed in the various portions of the text.

1.7 ELECTRON WAVES IN SOLID CRYSTALS

Let us now return to the thin metal foil having a thickness of 10 atoms. We might sketch the thin section as shown in Fig. 1.12, directing our interests specifically to the monatomic chain resolved in the z direction, and invoke the potential-well approximation shown, in line with the original approximation due to R. de L. Krönig and W. L. Penney. The one-dimensional form of the Schrödinger equation [Eq. (1.16)] for the repetitive portions of the proposed periodic potential (Fig. 1.12d) takes the following forms: In the region 0 < z < R,

$$\frac{\partial^2 \psi}{\partial z^2} + \frac{2m}{\hbar^2} E\psi = 0 \tag{1.42}$$

and in the region -b < z < 0,

$$\frac{\partial^2 \psi}{\partial z^2} + \frac{2m}{\hbar^2} (E - U_b)\psi = 0 \tag{1.43}$$

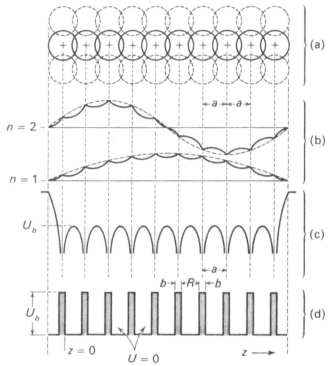

FIG. 1.12 *Thin metal foil section showing (a) idealized monatomic linear chain along z direction; (b) a-
ssociated electron waves; (c) periodic potential energy of electron in lattice; (d) square-well approximation of potential distribution.*

If we now invoke the Bloch theorem in specifying that the solution of Eqs.
(1.42) and (1.43) must be periodic with the lattice periodicity illustrated in Fig. 1.12, then for an electron traveling through the crystal with
periodically modulated amplitude as shown in Fig. 1.13, we can assume a
solution of the type

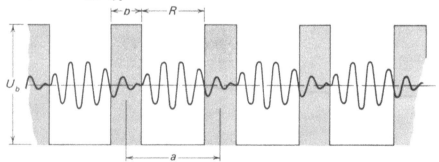

FIG. 1.13 *Electron wave modulation in periodic potential distribution
representative of linear monatomic chain.*

$$\psi = e^{i|\underset{\sim}{K}_o|z} \Omega(z) \tag{1.44}$$

where $|\underset{\sim}{K}_o|$ is the electron wave number and $\Omega(z)$ is the expression for the amplitude modulation. We can now substitute Eq. (1.44) into the forms of Schrödinger's equation given in Eqs. (1.42) and (1.43) to obtain

$$\frac{d^2\Omega}{dz^2} + 2i|\underset{\sim}{K}_o| \frac{d\Omega}{dz} - (|\underset{\sim}{K}_o|^2 - \alpha^2)\Omega = 0 \tag{1.45}$$

and

$$\frac{d^2\Omega}{dz^2} + 2i|\underset{\sim}{K}_o| \frac{d\Omega}{dz} - (|\underset{\sim}{K}_o|^2 + \beta^2)\Omega = 0 \tag{1.46}$$

for Eqs. (1.42) and (1.43), respectively; where

$$\alpha^2 = \frac{2mE}{\hbar^2} \tag{1.47}$$

and

$$\beta^2 = \frac{2m}{\hbar^2} (U_b - E) \tag{1.48}$$

The solutions of Eqs. (1.45) and (1.46), on invoking the continuity of Ω and $d\Omega/dz$ as a consequence of the periodicity of Ω in the distance R + b, and the fact that ψ and $d\psi/dz$ must be continuous at z = 0 and z = R, is probably well known in the form*

$$\frac{mR(U_b b)}{\hbar^2} \frac{\sin \alpha R}{\alpha R} + \cos \alpha R = \cos |\underset{\sim}{K}_o|R \tag{1.49}$$

where the product $U_b b$ in the limit, that is,

$$\lim_{\substack{U_b \to \infty \\ b \to 0}} [U_b b] \longrightarrow \text{finite}$$

is called the barrier strength. And, if we now define

In the event you are unfamiliar with the origin of Eq. (1.49), it is left as an exercise (Prob. 1.7) at the conclusion of the chapter.

$$P' = \frac{mR(U_b b)}{\hbar^2}$$

then Eq. (1.49) reduces to

$$P' \frac{\sin \alpha R}{\alpha R} + \cos \alpha R = \cos |\underset{\sim}{K}_o| R \qquad\qquad (1.50)$$

If we now plot Eq. (1.50) as shown in Fig. 1.14, assuming for convenience that $P' = 3\pi/2$, $\cos |\underset{\sim}{K}_o| R$ can only have real values between ± 1, which in effect means there exist only certain discrete ranges of αR that permit a wave-mechanical solution. These ranges are shown by the shaded sections in Fig. 1.14. What we have therefore shown as a consequence of the periodic assumptions in Fig. 1.12 and the application of Bloch's theorem to the appropriate forms of Schrödinger's equations, is that the electron energies in a crystal lattice are quantized and appear in bands analogous to those depicted in Fig. 1.11. Since α is related to the energy of the electron, and since the allowed ranges of energy are dependent on $\cos |K_o| R \leq 1$, we can also plot $|K_o| R$ against E to obtain Fig. 1.15. The allowed energy bands, shown shaded, result from the energy discontinuities corresponding to regular intervals of $|K_o| R$; the spaces between allowed bands are forbidden regions. Electrons having energies in the range of the allowed energies can therefore propagate from one cell to another in the periodic structure shown in Fig. 1.12. However, those electrons having energies in the forbidden regions constitute imaginary wave functions, and

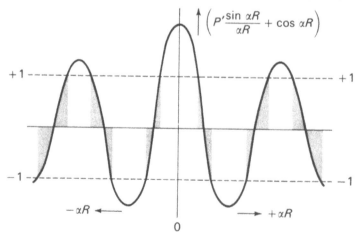

FIG. 1.14 *Graphical solution of existence of electron wave function in periodic crystal potential.*

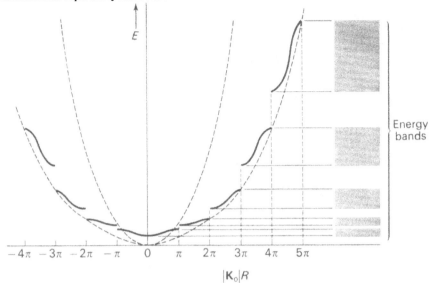

FIG. 1.15 *Plot of electron energy versus wave number for periodic solid illustrating band structure.*

therefore they do not exist. As a consequence of the discontinuities shown in the energy spectrum of Fig. 1.15, the discrete values of the wave number $|K_o|$ mark out forbidden zones in a structure known as Brillouin zones. We can, by analogy, imagine the energy discontinuities (Brillouin zones) to be represented by internal electron wave reflections, that is, a diffraction condition stipulated by the lattice structure in the same way that the physical diffraction of electrons or x-rays depends on the solid's structure.

The dashed parabolic curves of Fig. 1.15 indicate the continuous energy distribution for completely free electrons, that is, the condition that would obtain if the potential barriers of Fig. 1.11 were reduced to zero either by stipulating $b \to 0$ or $U_b = U = 0$ everywhere within the bounds of the solid, in which case R would correspond to the exterior dimensions (thickness).

1.8 IONIZATION PROCESSES AND THE NATURE OF IONS

Ions form when neutral atoms (or molecules) gain or loose electrons. This is not simply an atom in an excited state, where for example, inelastic collision of the atom causes an energy transfer to raise the energy by moving an electron from one quantum state to a higher state (Fig. 1.5). Ionization can involve an inelastic collision which ejects an electron from the atom, forming a positive ion (an atom which assumes a net positive charge, e, since one electron is lost). Obviously when this happens for a single hydrogen atom (Fig. 1.8), the nucleus is exposed, forming a proton (H^+) which has a positive charge equivalent to the electron's negative charge. The electron removed in forming a positive ion has a mass (m) which is very much less than the ion mass (M_i) (consider for example that $m_o \cong 9.11$ x 10^{-28} g for an electron and a proton mass, H^+, is about 1837 times as great!).

Ionization can occur by many other processes in addition to inelastic collisions of neutral atoms with other "particles", including x-ray and laser photons. These include gaseous and arc discharge phenomena, surface ionization (involving the probability of ionization of an atom or molecule evaporated from a surface as either a positive or negative ion), and field ionization. Secondary ionization can also occur when an ion beam interacts with a solid (target) material. We will elaborate on this concept in describing ion analysis and ion analyzers in Chapters 4 and 5. Indeed, these various ionization phenomena can be utilized in a systematic way in the operation of specific ion sources. We will deal with ion sources in Chapter 3.

In 1886, E. Goldstein [25], while experimenting with electrical discharge in gases at low pressure, observed that if a hole was placed in the negative electrode of the discharge tube, a luminous discharge extended through the hole. He concluded that this discharge was a beam or ray opposite to cathode rays (electrons), and he originally called them Kanalstrahlen. In 1898, W. Wien [26] showed that these rays (of ions) could be deflected by a magnetic field, and in a series of experiments carried out between about 1907 and 1913, J. J. Thomson [27,28] was able to resolve an ion beam into a series of parabolas, each characteristic of ions of one mass. He did this by using a simultaneous arrangement of electric and magnetic fields, giving rise to the inception of mass spectral analysis. The radii of curvature defining the parabolas of specific ion mass are essentially given by Eq. (1.1a) for magnetic field effects, and its counterpart for electric field effects:

$$R = \sqrt{\frac{2M_i V_o}{eB^2}} = \left(\frac{2V_o}{E} \right) \tag{1.51}$$

where M_i is the ion mass, and e is its fundamental charge. You should note that, as pointed out in connection with Eq. (1.2) for electrons, deflection of an ion beam in an electrostatic field is independent of the ion mass. As a consequence of this, the electrostatic system must be curved in order to achieve a radial path.

You should recognize the clear connection between electrons and ions which is implicit in the electrical discharge experiments of E. Goldstein cited above. It is this connection and indeed synergism in the context of modern analytical electron microscopy and microanalysis which is the theme upon which the bulk of this book will be based. It is this synergism of electron and ion beam analysis techniques which is emerging as the contemporary basis for modern microanalysis.

As we shall see in succeeding chapters of this book, ions and ion beams can be used to image and analyze details of matter in systems which are identical to, or similar to, those which commonly employ electrons. There are, as a result of the much greater masses for ions, inherent advantages and disadvantages in such ion-beam applications. You can readily appreciate the features of ideal resolution even for simple protons (H^+ ions) as compared to electrons (you should confirm this by working Prob. 1.14). However it must be recognized that ion collisions with matter can produce secondary and Auger electrons as well as secondary ions from the collision volume, and this can have important consequences which are different from those associated with electron beam interactions. In addition ion beam transmission through solid material can involve molecular charge transfer and give rise to other features in transmission imaging processes which do not occur for electron beam transmission.

PROBLEMS

1.1 A beam of electrons equivalent to 0.1 mA impinges on a thin foil of metal while being accelerated by a potential difference of 50 kV. How many electrons considered as particles strike the foil per unit time? Assuming the speed of the electrons to be reduced 20 percent on passing through the foil, calculate the heat developed per unit time in calories.

1.2 Show that the momentum of electrons accelerated through a potential V_0 is given by

$$P = \sqrt{2m_0 e V_0} \sqrt{1 + \frac{e V_0}{2m_0 c^2}}$$

1.3 Calculate the diffraction angle θ for the first two orders of the {111} planes in a face-centered cubic crystalline film using 100 kV electrons.

1.4 Consider electrons in a metal foil to be completely free everywhere inside the foil so that for a thickness of foil t along the z direction, $U(z) = 0$ for $0 < z < t$, and $U(z) = \infty$ at the surfaces. Show that the electron energies are quantized, that is, that $E \propto n^2$ for $n = 1, 2, 3$. Show also on the basis of the probabilistic nature of the electron [Eq. (1.18)] that the wave function is given by

$$\psi(z) = \sqrt{\frac{2}{t}} \sin \sqrt{\frac{2mE}{\hbar^2}} \, z$$

Plot the wave functions and probability distributions in the one-dimensional confinement (along the z direction) for $n = 1, 2, 3$.

1.5 Consider an electron in a solid to be bound to the nuclei composing the solid, but essentially free within an atomic unit cell of the solid. If the potential in a unit cell having dimensions $x = y = z = a$ is $U = 0$ everywhere inside, and $U = \infty$ outside, we have an analog of an electron in a box. Show that the energy of the electron in such a "box" is given by

$$E_{n_x n_y n_z} = \frac{\hbar^2}{8ma^2} (n_x^2 + n_y^2 + n_z^2)$$

where n_x, n_y, and n_z are integers

1.6 The emission lines from hydrogen (known as the Balmer series) correspond to electron energy transitions from a general state n to the quantum state $n = 2$. Calculate the wavelengths of the four emission lines of this series that appear in the visible region of the spectrum, that is, 3500 to 7000 Å.

1.7 Consider Eq. (1.45) in the range $0 < z < R$, and Eq. (1.46) in the range $-b < z < 0$ to have general solutions of the form, respectively:

$$\Omega = A e^{-i(|\underset{\sim}{K}_0| - \alpha) z} + B e^{-i(|\underset{\sim}{K}_0| + \alpha) z}; 0 < z < R$$

$$\Omega = C e^{-i(|\underset{\sim}{K}_0| - \beta) z} + D e^{-i(|\underset{\sim}{K}_0| + \beta) z}; -b < z < 0$$

If we now invoke the conditions that Ω is continuous at $z = 0$, and that $\partial\Omega/\partial z$ is continuous at $z = 0$, and in addition invoke the periodicity of Ω and $\partial\Omega/\partial z$, we will develop four equations with four un-known coefficients A,B,C,D. Show that the simultaneous solution of these equations in determinant form, stipulating that the coefficients of A,B,C,D must be zero, results in

$$\frac{\beta^2 - \alpha^2}{2\alpha\beta} \sinh \beta b \sin \alpha R + \cosh \beta b \cos \alpha R = \cos |\underline{K}_o| (R + b)$$

Show further that this equation can be written in the form of Eq. (1.49) by invoking the limit property on $[U_b b]$, the barrier strength, as discussed on page 25.

1.8   Consider the electron energy to be given by

$$E = \frac{p^2}{2m} + U$$

or

$$E = \frac{\hbar^2 |\underline{K}|^2}{2m} + U$$

and to have a general wave function

$$\psi = A \cos 2\pi(z/\lambda - \nu t) + B \sin 2\pi(z/\lambda - \nu t)$$

If A = 1 and B = i, derive the expression of Eq. (1.22).

1.9   The ability of electrons or the electron wave function to penetrate the potential barriers (shown in Fig. 1.12d) is sometimes treated as a transmission problem called tunneling. If the probability of electron transmission through a potential barrier is given by the ratio of electrons penetrating the barrier to those impinging on it in the form

$$T = \frac{4}{4 \cosh^2 |\beta| b + \left(\frac{\beta^2 - \alpha^2}{\alpha\beta}\right)^2 \sinh^2 |\beta| b}$$

calculate the probability of penetration of a single barrier of thickness 4 Å and height U_b = 10 eV for electrons of energy 8 eV in terms of the notation of Fig. 1.12d. How many electrons might be expected to penetrate the 10 - barrier analog of a thin film as in Fig. 1.12d if the impinging density on the entrance surface consists of roughly 10^4 electrons per second?

1.10   Electrons striking a color-film emulsion cause a blue-green activation. Estimate the limits of electric potential through which these electrons were accelerated.

1.11 Electron momentum expectation (momentum probability distribution) is defined by

$$\langle P \rangle = \int \psi^* P_x \psi \; dx$$

in the x direction. If P_x is the momentum operator defined by

$$P_x = \frac{h}{2\pi i} \frac{\partial}{\partial x}$$

and the wave function in the time-dependent form is

$$\psi = \psi(x,t) = Ce^{i\gamma(x-\omega t/\gamma)}$$

where C is a constant, derive Eq. (1.19) if $a/2\pi n = \Delta q$, $q = x$.

1.12 Write out the general form of Schrödinger's equation for a Li atom.

1.13 A helium ion (He^+) is accelerated through a potential difference of 1 kV. What will be its velocity? If this ion passes through a radial magnetic field; what will be the radius of its path if the magnetic field is 50 gauss? (Note: the mass of the helium ion is equal to 4 atomic mass units; 1 amu = 1.6598×10^{-24} g)

1.14 Calculate the de Broglie wavelength associated with a "proton microscope" operated at an accelerating voltage of 30 kV. At what voltage must such a system be operated to achieve the wavelength associated with 100 keV electrons? Discuss the practicality or impracticality of such a system.

REFERENCES

1. J. Perrin, *Compt. Rend.* (1895).

2. J. J. Thomson, *Phil. Mag., 48:* 547(1899).

3. A. Shuster, *Proc. Roy. Soc. (London) 42:* 371(1887).

4. W. Kaufmann, *Gottingen Mach., 2:* 143(1901).

5. R. A. Millikan, *Phys. Rev., 2:* 109(1913).

6. W. Friedrich, P. Knipping, and M. von Laue, *Ann. Physik, 41:* 971(1913).

7. L. de Broglie, doctoral dissertation, Paris, 1924.

8. L. de Broglie, *Phil. Mag., 47:* 446(1924).

9. W. Elsasser, *Naturwiss., 13:* 711(1925).

10. W. H. Bragg and W. L. Bragg, *Proc. Roy. Soc. (London), 88:* 428(1913).

11. C. J. Davisson and L. H. Germer, *Phys. Rev., 30:* 705(1927).

12. P. P. Ewald, *Krystalle und Roentgenstrahlen,* Julius Springer, Berlin, 1923.

13. G. P. Thomson and A. Reid, *Nature, 120:* 802(1927).

14. R. Emden, *Physik. Zeit., 22:* 513(1921).

15. E. Schrödinger, *Physik. Zeit., 23:* 301(1922).

16. E. Schrödinger, *Ann. Physik, 19:* 489(1926).

17. E. Schrödinger, *Ann. Physik, 18:* 109(1926).

18. M. Born, *Z. Physik, 37:* 863(1926).

19. W. Heisenberg, *Z. Physik, 43:* 172(1927).

20. A. Sommerfeld, *Ann. Physik, 51:* 1(1916).

21. G. E. Uhlenbeck and S. Goudsmit, *Nature, 117:* 264(1926).

22. P. A. M. Dirac, *Proc. Roy. Soc. (London), A117:* 610; *118:* 351(1928).

23. E. C. Stoner, *Phil Mag., 48:* 719(1924).

24. W. Pauli, *Z. Physik, 31:* 765(1925).

25. E. Goldstein, *Berl. Ber. 39:* 691(1886).

26. W. Wein, *Verhandl. Physik. Ges., 17* (1898).

27. J. J. Thomson, *Phil. Mag., 13:* 561(1907).

28. J. J. Thomson, *Rays of Positive Electricity and Their Application to Chemical Analysis,* Longmans, Green and Co., London, 1913.

SUGGESTED SUPPLEMENTARY READING

Blakemore, J. S., *Solid-State Physics,* J. V. Saunders Co., New York, 1984.

Brown, Ian G. (ed.), *The Physics and Technology of Ion Sources,* J. Wiley-Interscience, New York, 1989.

Crowther, J. A., *Ions, Electrons, and Ionizing Radiation,* Longmans, Green, and Company, New York, 1927 (interesting reading in spite of its age).

DeLogan, D., *Solid-State Devices: A Quantum Physics Approach,* Springer-Verlag, New York, 1987.

Eisberg, R. M. and Resnick, R., *Quantum Physics of Atoms, Molecules, Solids, Nuclei, and Particles,* J. Wiley and Sons, Inc., New York, 1974.

Jammer, M., *The Conceptual Development of Quantum Mechanics,* Dover Publications, Inc., New York, 1966.

McDowell, C. A. (ed.), *Mass Spectrometry,* McGraw-Hill Book Company, New York, 1963.

Mehra, J., and Rechenberg, H., *The Historical Development of Quantum Theory,* Springer-Verlag, New York, 1987.

Murr, L. E., *Solid-State Electronics,* Marcel Dekker, New York, 1978.

Pauling, L., *The Chemical Bond: A Brief Introduction to Modern Structural Chemistry,* Cornell University Press, Ithaca, New York, 1967.

Rosenberg, H. M., *The Solid State: An Introduction to the Physics of Crystals for Students of Physics, Materials Science, and Engineering,* Clarendon Press, Oxford, 1978.

Schommers, W. (ed.), *Quantum Theory and Pictures of Reality,* Springer-Verlag, New York, 1989.

2
ELECTRON EMISSION
AND EMISSION
AND IONIZATION MICROSCOPY

2.1 INTRODUCTION

From the concept of energy-band properties associated with electrons in
solids as introduced in Sec. 1.6, and the fact that metallic solids can be
thought of as ion-core aggregates, we might imagine a metallic solid as a
macroscopic entity possessing, on the whole, the properties of the compositional units. In effect we are saying that the valence electrons associated with any one ion core are all equally influenced by all neighboring
nuclei. Consequently the electron wave functions associated with neighboring nuclei overlap, and the valence electrons in a metallic solid can
be assumed to be free to pass from one nuclear potential to another. The
valence electrons[*] therefore can be thought of as free electrons, and
their three-dimensional nature within the confinement of a solid can be
effectively modeled in terms of gas particles in a box. It is in fact the
random gaslike motion of electrons in a metal that accounts for its resistivity, which is a consequence of nuclear vibrations induced by electron

---

[*]*It should be recalled that the valence band in a metal overlaps the bottom
portion of the conduction band. The free electrons are thus restricted to
the region bounded by the band overlap at $T \cong 0°K$.*

collisions and their concomitant loss in kinetic energy as heat to the
lattice.

As shown schematically in Fig. 1.12, the surface potential acts as a
steep barrier preventing the electrons from leaving the metal. Figure 2.1
depicts the bonding character that might be imagined for a single layer of

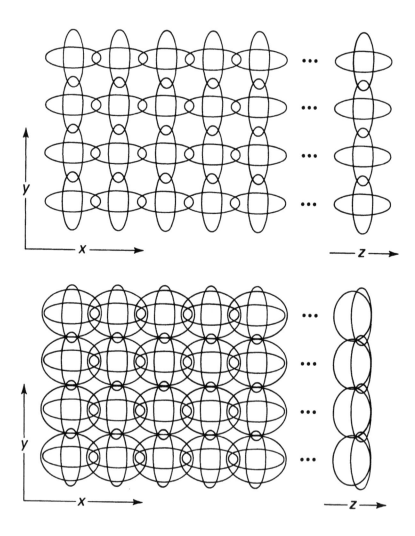

FIG. 2.1 *Simple arrangement for nuclear bonding in interior (top) and surface (bottom) of metallic solid. Necessity to accommodate extra nuclear
bond in surface layer can be thought of as an increase in effective potential as sketched in Fig. 1.12c.*

metal ions within a solid, and shows for comparison a possible arrangement of the surface layer. If electron motion in a solid is treated as a diffusion-type process, the surface constriction is obvious. In order to escape from a metal, the electrons must somehow gain sufficient energy to penetrate the barrier. It is this process of energizing the valence electrons in a metal that leads to emission. The principal modes of electron emission are outlined schematically in Fig. 2.2. Here we observe that electron energy is enhanced by heating the solid, with the result that the increased ion-core vibrations induce an increase in electron collisions and thereby impart to the electrons added kinetic energy (1) by the absorption of light near the surface of energy $h\nu$, (2) by the bombardment of the metal surface with energetic electrons that in turn impart additional kinetic energy to those electrons near the surface, allowing them to escape; and (3) by the suppression of the effective barrier height through the action of an electrostatic field. In the case of Fig. 2.2a to c, representative of photo, thermionic, and secondary emission of electrons, respectively, we are to imagine the surface potential barrier as a dam holding back the electron

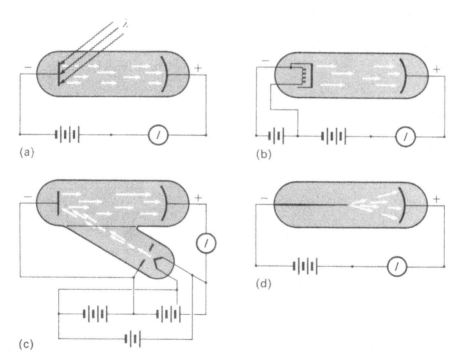

FIG. 2.2 *Emission modes of electrons. (a) Photoemission; (b) thermionic emission; (c) secondary emission; (d) field emission.*

sea, providing a mechanism whereby the electrons are "pumped" over the dam.
In the case of field emission, sketched in Fig. 2.2d, the level of the dam
is lowered by the field action, and the electrons effectively tunnel through
the dam at the level of the electron sea.

2.2 ENERGY DISTRIBUTION OF ELECTRONS IN A METAL

In order for electrons in a metal to escape by any one of the emission processes outlined in Fig. 2.2, they must either possess an intrinsic energy
equal to or in excess of the surface-barrier potential, or they must be
energized to that level. Chapter 1 has already emphasized that the energies of electrons confined within the bounds of a solid are discrete, and
that they occupy definite energy levels or bands. Using this analog, the
energy levels of a metal, as viewed for the hypothetically simplified one-
dimensional case, can be depicted as in Fig. 2.3. Here the electrons are
viewed as partially filling the available energy levels (the overlap of
the valence band on the conduction band) that extend to the surface; the
net deficiency, that is, the net energy needed to push the electrons out
onto the surface, is represented by E_W, the work function. And, if $\langle E \rangle$ is
taken to represent the maximum kinetic energy of the free electrons, the
work function can be written simply as

$$E_W = E_B - \langle E \rangle \tag{2.1}$$

It should of course be immediately obvious that Fig. 2.3 is an ideal

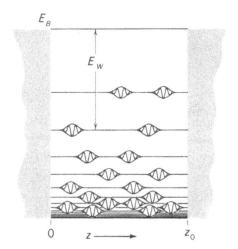

FIG. 2.3 *Electron energy levels in
one-dimensional metal.*

representation, and that in reality the extent to which the energy levels are filled, and consequently the value of <E>, will depend on the temperature of the lattice and, more specifically, on the local lattice vibrations. Thus, at temperatures above absolute zero, there will exist an uncertainty in the electron energy-level filling, which will conceivably be temperature dependent. <E> will then represent a probable energy-level filling, or the energy of the topmost electrons as depicted in Fig. 2.3.

In the case of a real metal, the hypothetical analog depicted in Fig. 2.3 must be extended to three dimensions, and as a consequence of the fact that z_0 (Fig. 2.3) as well as x_0 and y_0 are large compared with the periodicity of the lattice in a three-dimensional solid, the energy bands become, for the most part, fused into a continuous energy spectrum. In such a case we are forced to describe an energy distribution for these free (valence) electrons; and we must therefore be concerned about the density of electrons in a range, say E and E + dE, or the number of levels in a range of energies, dE at some energy E.

On considering the quantum nature of electrons, and the statistical nature of their distribution in terms of a gas confined within the bounds of a solid, E. Fermi [1] and, independently, P. A. M. Dirac [2] developed an energy-density function that, in the specific case of metals, was subsequently applied by A. Sommerfeld [3] in the form

$$N(E)dE = z(E)F(E)dE \tag{2.2}$$

Here N(E)dE is representative of the actual number of electrons in a given energy range, which is equivalent to the product of the density of states (in three dimensions),

$$z(E)dE = \frac{\pi}{2}\left(\frac{8ma^2}{h^2}\right)^{3/2} E^{1/2}\, dE \tag{2.3}$$

and the Fermi-Dirac distribution function

$$F(E) = \left[e^{(E-E_F)/kT} + 1\right]^{-1} \tag{2.4}$$

The energy term E_F in Eq. (2.4) is the Fermi energy; it is equivalent to the average or topmost electron energy <E> at the absolute zero of temperature (T = 0°K). At T = 0°K, the number of electrons in the range dE is obtained from

$$N = \int_0^{<E> = E_F} N(E)dE$$

and if we define N_o as the number of electrons per unit volume of metal (N/a^3), then

$$N_o = \frac{\pi}{2} \left(\frac{8m}{h^2}\right)^{3/2} \int_0^{<(E) \ = \ E_F} E^{1/2} dE \tag{2.5}$$

since at $T = 0$, $F(E) = 1$. Integration of Eq. (2.5) then results in

$$E_F = \frac{h^2}{2m} \left(\frac{3N_o}{8\pi}\right)^{2/3} \tag{2.6}$$

which, at $T = 0°K$, represents the topmost energy of the free electrons in the valence band as depicted (one dimensionally) in Fig. 2.3.

At $T = 0°K$, it is impossible for electrons to escape from a metal unless they are energized from without by photon or high-energy electron absorption, since the maximum energy they possess is given by Eq. (2.6). However, at $T > 0°K$, the energy distribution begins to change as illustrated in Fig. 2.4. At temperatures in excess of $T_o(T_o \gg 0°K)$, a portion of the distribution (shown shaded in Fig. 2.4) begins to "spill" out onto the surface of the solid, and electrons are subsequently lost in the emission process. Rewriting Eq. (2.1) in terms of the Fermi distribution (Fig. 2.4) then defines the work function as

$$E_W = E_B - E_F \tag{2.7}$$

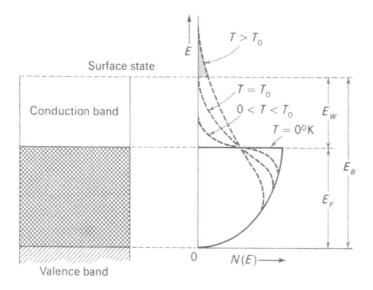

FIG. 2.4 *Energy distribution in metallic solid.*

The work function of a metal at $0°K$ is the minimum energy that must be imparted to the most energetic valence electrons (those with energy E_F) to
enable them to move into the unfilled levels of the conduction band and
finally to escape from the physical interior of the metal to the surface
exterior.

2.3 NATURE OF THE METAL WORK FUNCTION E_W

Although the work functions for most metals can be experimentally determined by simply measuring the threshold wavelength λ_o at which photoemission commences (Fig. 2.2a), that is,

$$E_W = \frac{hc}{\lambda_o}$$

the values so measured are in reality only the apparent work functions.
(You are urged to work Prob. 2.2 in order to demonstrate this feature in
addition to acquainting yourself with the concepts outlined in Fig. 2.4.)
This feature is a result of the fact that in treating the concept of a work
function, we have, to this point, considered an idealized situation where
the emission surface was considered to be clean and continuous, the emission process was crystallographically isotropic, and the potential barrier
was a step function. In reality, none of these assumptions is strictly
valid, and the idealized energy scheme outlined in Fig. 2.4 is only a
crude approximation insofar as the emission surface is concerned since we
have assumed the only necessary condition for electron emission to be that
$E > E_B$. Although this condition is necessary, it is not necessarily sufficient since those electrons that are emitted must propagate in a direction normal to the emitting surface.

Looking in retrospect at the bonding mode associated with the surface
ion cores sketched in Fig. 2.1, we can consider the atomic planes to represent a {100} plane for a cubic lattice structure, and the z direction shown
to be a <100> zone axis. (The reader unfamiliar with crystallographic notation should refer to Appendix A.) If we now consider a close packing of
the ion cores, say in a {111} plane arrangement, with the <111> zone axis
representing the normal direction of propagation of electrons, a difference
in the work function is to be expected since the potential resulting from
the overlap of incomplete bonds on the surface will be decidedly different
from the previous case. Thus the electrons experience different surface
forces depending on the particular crystallographic orientation exposed.

Even if a simple potential barrier is considered, not all electrons will be transmitted through the barrier. Ideally the transmission probability will be represented in the form shown in Prob. 1.7. Not only will electrons be reflected from the surface potential barrier, but reflection may occur just outside the surface as a consequence of the space charge produced by electrons emitted into the region adjacent to the hot metal, in the case of thermionic emission, or because of the emitting surface in general.

The apparent work function of a metal is also changed by the action of the electric field applied between the cathode (the emitting surface) and the anode (see Fig. 2.4). And, by considering the more realistic nature of the potential function associated with the surface of a metal, it can also be shown that the electrons that escape from within experience an image force equivalent to that which would be experienced if the conducting surface were replaced by a charge + e located an equal distance from the surface boundary within the solid as the electron is from without. The complete potential at the surface must therefore include the image-force potential in the form

$$E_1 = - \frac{e^2}{4\pi\epsilon_o 4z} \tag{2.8}$$

where z is the distance of the electron from the surface, and ϵ_o is the permittivity of free space (and has a value of 8.854×10^{-12} farad/m). The potential energy of an electron (in the absence of an applied field) is then given by $E_B + E_I$, and represented schematically in the corresponding curve of Fig. 2.5.

If we now consider the effects of the applied field E_z (in volts/m), the potential energy of the electrons a distance z from the surface is shown by the curve of Fig. 2.5, and written in complete form:

$$U = E_B - \frac{e^2}{4\pi\epsilon_o 4z} - ze E_z \qquad E_z > 0 \tag{2.9}$$

The important feature of this phenomenon is that the work function as expressed in Eq. (2.7) is reduced by the application of an electric field of appreciable magnitude, as depicted in Fig. 2.5. This reduction, commonly known as the Schottky effect, can be obtained by differentiating

$$U' = - \frac{e^2}{4\pi\epsilon_o 4z} - ze E_z$$

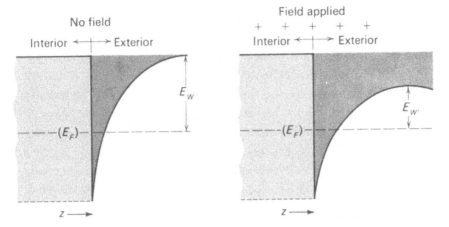

FIG. 2.5 *Potential energy of electron near surface of a metal.*

with respect to z, and setting the derivative equal to zero to obtain

$$U'_{max} = - e\sqrt{\frac{eE_z}{4\pi\varepsilon_0}}$$

(2.10)

The modified work function is then written in the form

$$E'_W = E_W - e\sqrt{\frac{eE_z}{4\pi\varepsilon_0}}$$

(2.11)

where it is observed that for extremely high fields, the effective work function E'_W will approach zero (see Prob. 2.3). The application of an electric field to an emitting surface thus acts in a fashion to lower the barrier level or, by analogy, to effectively raise the Fermi level.

2.4 PHOTOEMISSION ELECTRON MICROSCOPY

It has long been known that images of polished (solid) surfaces could be obtained using photoelectrons emitted from a sample area irradiated with ultraviolet light or some other energetic source which could free electrons from spatially unique areas, and provide an electron emission image having intensity variations due to work-function differences as a result of chemical or crystallographic differences. The first photoelectron microscope was demonstrated by Bruche in 1933 [4], and resolutions of surface detail have exceeded those for light microscopes by nearly an order of magnitude. The work-function contrast obtained in photoelectron or photoemission electron microscopy is ideally suited to the study of metals and alloys as described in the applications reviewed by Wegmann [5], and the more contemporary applications outlined at the First International Conference on Electron Emission Microscopy held in 1979 [6] (see also D. V. Edmonds and R. W. K. Honeycombe, *Metal Science,* Sept., 1978, p. 399).

Figure 2.6(a) illustrates the basic operating features of modern photoelectron microscope systems which can use coherent (laser) and other intense light sources and electrostatic lens systems [6,7]. The electrostatic lens system shown in Fig. 2.6 (a) will be described in detail in Chap. 3 along with the more popular electromagnetic lens system. You will note in this impending development that electrostatic lenses have limitations for electrons accelerated through a potential in excess of about 40 kV, but for the lower voltage emission microscopes, electrostatic lenses are very efficient. In addition, they are much simpler to fabricate in comparison to the electromagnetic lens designs requiring special magnetic pole pieces to produce sufficient magnetic field intensities for the lens-forming action at high electron accelerating voltages.

Figure 2.6(b) illustrates an example of the image resolution and contrast which can be obtained with the photoelectron microscope. You might compare this image quality with the thermionic electron emission images to be described in the following section along with those to be described for secondary electron emission images in the contemporary scanning electron microscope in Chap. 5.

Although an advantage of some importance in photoelectron microscopy involves the ability to examine uncoated biological specimens at ambient temperatures as well as other solid surfaces (including minerals) the emission characteristics are enhanced by heating the sample. As a consequence, heating-stage experiments can be conducted in which low-temperature ($\lesssim 800°C$) transformations and related phenomena can be directly observed. Actually at elevated temperatures the emission profile becomes increasingly dominated by thermionically emitted electrons, and this phenomena will be treated in more detail in the following section.

The resolution for a UV photoelectron microscope has recently been calculated by Gertrude Rempfer and O. H. Griffith (Ultramicroscopy, 27, 273, 1989) to be 50 Å. The reader might also peruse the article, "Photoelectron Imaging: Photoelectron Microscopy and Related Techniques" by O. H. Griffith and G. F. Rempfer in Advances in Optical and Electron Microscopy (A. Barer and V. E. Cosslett, eds.) vol. 10, Academic Press, London, 1987, p. 269.

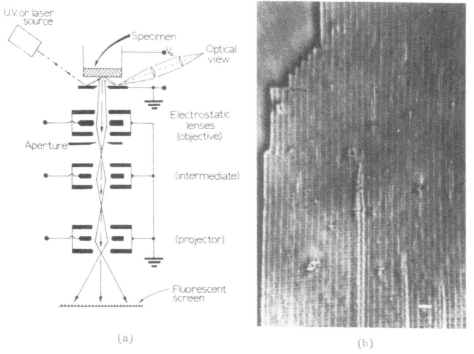

FIG. 2.6 *Photoemission microscope schematic (a) and example of image con-*
trast and resolution (b). Photoelectrons excited on the specimen surface
by energetic incident photons (of energy E = hν) are accelerated into a
three-lens electrostatic electron microscope to produce an image of a gold-
coated grating replica (2,160 lines/mm) in (b). [(b) is courtesy of Dr.
Gail A. Massey; after ref. [7]].

2.5 THERMIONIC EMISSION AND THERMIONIC ELECTRON EMISSION MICROSCOPY

2.5.1 *DERIVATION OF THE EMISSION EQUATION*

In the previous discussions of the more important properties of the Fermi
energy and the work function, it was observed that at sufficiently high
temperatures, that is, at $T > T_0$ with reference to Fig. 2.4, the electron
energy distribution tails out beyond the surface with the result that the
electrons with energies distributed within this tail are emitted. Since
the tail is itself a portion of a probability distribution, it is logical
that a description of emission-current density from an area of surface
would itself be probabilistic in nature. And, as indicated in Sec. 2.3,
it is insufficient that the electrons possess the finite energy for over-
coming the potential barrier at the surface. Those electrons that are
emitted must also possess a momentum, directed normal to the emission sur-

face, having a minimum value of the form

$$P_{z_0} = \sqrt{2mE_B} \tag{2.12}$$

where z is the direction of propagation normal to the emission surface. Consequently we require some statement of the velocity distribution in the range v_z and $v_z + dv_z$. Since the electron energy in the interior distribution is given by

$$E = \frac{1}{2}m(v_x^2 + v_y^2 + v_z^2)$$

the probable number of electrons in the velocity ranges $v_x + dv_x, v_y + dv_y$, and $v_z + dv_z$ is

$$dN_0 = \frac{2m^3}{h^3} F(E)dv_x dv_y dv_z \tag{2.13}$$

The number of electrons arriving at a unit area of the surface in a unit time (the emission-current density) is then expressed by

$$J = eT(v_z) \int v_z dN_0 \tag{2.14}$$

where $T(v_z)$ is the probability that the electrons with a velocity v_z will be transmitted beyond the surface potential barrier (the surface). $T(z)$ here can be assumed to have the general form shown in Prob. 1.7. Evaluating Eq. (2.14) as

$$J = \frac{2m^3}{h^3} eT(v_z) \int_{-\infty}^{\infty}\int_{-\infty}^{\infty}\int_{\sqrt{\frac{2E_B}{m}}}^{\infty} e^{E_F/kT} e^{-mv_x^2/2kT} e^{-mv_y^2/2kT} e^{-mv_z^2/2kT} v_z$$

$$\cdot \, dv_x dv_y dv_z$$

since in a practical range of emission surface temperatures

$$E - E_F = \left[\frac{m}{2}(v_x^2 + v_y^2 + v_z^2) - E_F\right] \gg kT$$

or

$$e^{(E-E_F)/kT} + 1 \cong e^{(E-E_F)/kT}$$

we obtain

$$J = \frac{4\pi mk^2T^2 eT(v_z)}{h^3} e^{-E_w'/kT} \tag{2.15}$$

where

$$A_o \equiv \frac{4\pi m k^2 e}{h^3} = 120 \ amp/(cm^2)(°K^2)$$

Equation (2.15) is a modified form of the classically derived emission equation of O. W. Richardson [5], and stems primarily from the later treatments of S. Dushman [9], who verified Eq. (2.15) experimentally. As a consequence of the work of these early investigators, Eq. (2.15) is commonly called the Richardson-Dushman equation when electric field effects are negligible.

2.5.2 *THE THERMIONIC ELECTRON EMISSION MICROSCOPE*

Historically, the electron emission microscopes are the earliest of the electron microscopes. The distinguishing features of the emission microscopes, however, are that the emission source is also the specimen, and that, because electrons leaving the specimen are accelerated, the lens arrangement in a magnetic instrument is not that of a true magnetic electron microscope. The first thermionic electron microscope was, in fact, an electrostatic device [10] consisting of a single cathode "lens" operated at a maximum of only a few thousand volts. A formal development of electron lenses, electrostatic and magnetic, is presented in Chap. 3. The reader completely ignorant of lens action is therefore at this point required to accept "on faith" the ability to form an emission image using magnetic-deflection lenses. Electrons emitted from the specimen surface were then simply projected onto a fluorescent screen for direct observation of the emission profile. Figure 2.7 illustrates schematically this early instrument.

Further development of the early thermionic emission microscope by

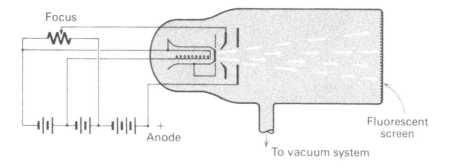

FIG. 2.7 *Early (electrostatic) thermionic emission microscope.*

M. Knoll, F. G. Houtermans, and W. Schulze [11] led to the double magnetic-lens design that forms the basis of conventional thermionic emission
instruments. These instruments, as developed originally, were the forerun-
ners of modern electron microscopes. Their range of application was limited, however, by the inherent features of the emission process, the active
nature of the specimen, the imaging and resolution characteristics, and by
the fact that only metals and alloys possessing relatively low work functions could be studied reliably.

In Fig. 2.8, the contemporary thermionic emission microscope is shown
schematically. Such instruments are capable of operation at potentials up
to 50 kV and are to be found in a number of laboratories in both the United
States and Europe. It is to be noted from Fig. 2.8 that the cathode or
specimen assembly forms the integral part of the instrument design. This
arrangement is demountable, and is fabricated in such a way that the speci-
men-cathode section is completely interchangeable. In this way emission
from numerous metals and alloy samples can be consecutively studied.

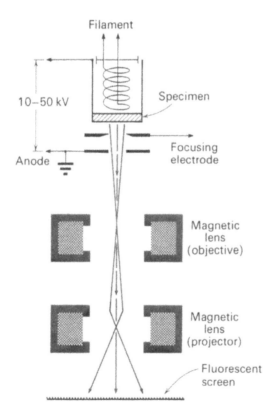

FIG. 2.8 *Magnetic thermionic electron emission microscope (schematic).*

Image contrast in the thermionic emission microscope Thermionic emission specimens, for the most part, are polycrystalline metals or alloys. As a consequence of this feature, and the fact that the effective work function of a metallic emission surface varies with the crystallographic orientation, the emission current arising from the associated polycrystals will be different. In addition, included phases or large precipitates, when exposed on the surface of the cathode specimen, will also emit an electron current having a density differentiated from the background matrix by an amount governed by the difference in the associated work functions. This "patch effect" is therefore primarily responsible for the contrast arising from the emission-specimen surface.

We might visualize this process as illustrated in the idealized sketches of Fig. 2.9. Here we imagine a polycrystalline metal surface having several low-index crystallographic-zone axes exposed as shown. If the corresponding work functions for these emission patches are characterized by

$$E_{W[100]} < E_{W[111]}$$

then the emission-current density will differ from one area to the other by an amount

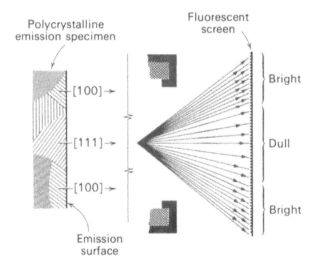

FIG. 2.9 *Image contrast for polycrystalline thermionic emission specimen.*

$$\Delta J = A \left[\frac{e^{E_{W[111]}} - e^{E_{W[100]}}}{e^{E_{W[111]}} \cdot e^{E_{W[100]}}} \right]$$

where A is a constant; and we assume that field effects are negligible at
a constant emission temperature T°K. The experimental determination of
variations in the work function for a difference in crystallographic orientation has been accomplished for a number of metals by the method of field-
electron emission microscopy (discussed in Sec. 2.6), and specific values
are listed in Table 2.1. In this connection it will be instructive to work
Prob. 2.4, which deals directly with emission-current-density calculations.

The concept of image contrast as outlined above is indeed idealistic,
but nevertheless phenomenologically accurate enough to get the point across.
In the thermionic electron emission microscope, the actual emission process
is somewhat more complicated, and the imaging qualities of a specimen surface tend to be strongly dependent on the residual gases present, and the
degree of surface reaction or adsorption. A great deal of the success in
imaging many materials, particularly those with very high work functions,
in the thermionic emission microscope involves the activation of the surface by controlled adsorption. This process has the effect of reducing
the work function to the point where emission will occur or be appreciably
enhanced at relatively low temperatures.

TABLE 2.1 *Single crystal surface work functions*[†]

Metal	Crystal structure	Crystallographic-zone axis	Work function, eV
Ag	fcc	[111]	4.75
		[100]	4.81
Cu	fcc	[111]	4.89
		[100]	5.94
W	bcc	[100]	4.90
		[112]	4.90
		[310]	4.35
		[110]	6.00
Mo	bcc	[100]	4.40
		[110]	4.90
		[111]	4.10

[†]*From L.P. Smith, Thermionic Emission, in E.U. Condon and H. Odishaw (eds.),*
Handbook of Physics, sec. 8, p. 76, McGraw-Hill Book Company, New York,
1958; and M. H. Richman, ASM Trans. Quart., 60:719 (1967).

Work function modification; thermionic emission activation We have observed in previous discussions that the work function is simply the result of the surface potential barrier, and that this potential barrier is influenced, to a great extent, by the surface crystallography and the characteristic equilibrium of the unbonded surface atoms. Bearing these features in mind, it is a simple matter to visualize a concomitant alteration in the potential barrier and the work function with any change in the surface structure. In particular it can be observed that the adsorption of a surface layer would have the ultimate effect of satisfying the bond characteristics of the surface, thereby altering the work function. The reaction of the surface atoms with a residual gas, for example, oxygen, would also tend to create a new surface structure with the result that the potential barrier and the work function would be significantly altered from that of the clean, unmodified surface.

The adsorption of a surface layer either reactive (for example, oxygen) or nonreactive (nitrogen) also influences the potential barrier by originating a surface-charge layer, which in effect functions somewhat analogous to a localized space-charge layer. This particular phenomenon accounts for the major portion of work-function modification.

We can visualize the situation ideally as follows: Suppose in a thermionic emission instrument a residual vapor atom drifts into the immediate vicinity of the emission surface. If this atom had an easily removable electron (the outer electron of the atom will tend to be captured by the metal surface if the work function exceeds the ionization potential V_i of the atom, thereby establishing a lower energy configuration), then acceptance of such an electron by the emission surface will leave the adjacent atom positively ionized, and the surface in the vicinity of the electron impingement will be negatively charged. The resulting local electrostatic attraction will cause the positive ion to adhere or adsorb to the surface. Figure 2.10 illustrates this process, and the adsorption of an ionized monolayer.

The adsorption of a monolayer of positive ions as depicted in Fig. 2.10c effectively creates an electric field at the surface that acts in a fashion similar to the external field as described by Eq. (2.11). In other words, the work function is reduced by an amount

$$\frac{n_A e^2 \delta}{\varepsilon_o}$$

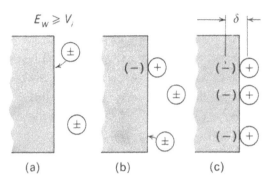

FIG. 2.10 *Ionization and ion adsorption on emission surface. (a) Neutral atoms in vicinity of surface; (b) ionization and attraction of residual-vapor atoms; (c) adsorption of ionized monolayer.*

or

$$E_W' = E_W - \frac{n_{A(+)} e^2 \delta}{\varepsilon_0} \tag{2.16}$$

where $n_{A(+)}$ is the number of positive ions (of charge e) per unit area of surface, and δ is the effective distance between charge centers (see Fig. 2. 10c). For a monatomic layer of positive ions, the change in the work function ($E_W' - E_W$) can amount to several eV (see Prob. 2.5). An adverse effect on the work function occurs for the adsorption of a negatively charged ion layer (for example, oxygen) on the emission surface because the space charge is now repulsive. Ideally, we then obtain

$$E_W' = E_W + \frac{n_{A(-)} e^2 \delta}{\varepsilon_0} \tag{2.17}$$

For the most part, work-function reduction occurs by the ionization and adsorption of heavy-metal atoms, particularly the alkaline earth and alkali elements, while the ionization of residual gases such as oxygen induces an increase in the work function. The emission current, while it depends most strongly on the effective work function of the emission surface, can also be influenced markedly by enhanced electron reflection from the adsorbed layer, or from a new surface species resulting from a reaction of the adsorbed layer with the substrate matrix (as in the formation of an oxide layer). In this way transmission (or tunneling) through the surface potential is decreased since

$$T(z) = 1 - R(z)$$

where $R(z)$, the reflection of electrons in the z direction normal to the emission surface, increases.

By inducing the formation of a controlled monolayer of positive ions on an emission surface, the work function over the entire specimen can be effectively lowered. This process, as it applies specifically to the imaging of metallic surfaces in the thermionic electron emission microscope, is called activation; and the ionizing elements are referred to as activators. The primary function of the activation process is therefore the enhancement of emission at temperatures considerably below the melting temperature of a specimen (see Appendix B).

Activation of an emission specimen is usually accomplished by the vapor deposition of approximately a monolayer of an alkali earth (Ba or Cs) onto the heated surface in situ (with the activator source located within the thermionic emission microscope). It can also be accomplished in most instances by "painting" the specimen surface with an alkali earth carbonate or formate outside the emission microscope, with the surface at effectively room temperature. However, the latter method has a number of undesirable features, such as the trapping of impurities, the inhomogeneity of the painted layer, etc., which promote spurious image effects under observation. In Table 2.2, the ionization potentials for a number of the alkali and alkaline earth elements are presented, and these are to be compared with the effective work functions for various metallic substrate materials of interest in thermionic electron emission microscopy to be found in Table 2.1.

TABLE 2.2 *Ionization potentials for several activators*

Element	Ionization potential, eV
Li	5.40
Ba	5.19
K	4.34
Rb	4.16
Cs	3.87

While an activator functions primarily to enhance image contrast by the reduction of the work function on the formation of a monolayer (depicted ideally in Fig. 2.10c), contrast is also enhanced by the preferential adsorption of ionized activator on certain crystallographic surface orientations. This is due again to the fact that the work function differs from one orientation to another, and the adsorption process, contingent on the activator ionization as described earlier, differs from one crystallographic area to another.

2.5.3 *METALLURGICAL APPLICATIONS OF THERMIONIC ELECTRON EMISSION*
       MICROSCOPY

While the thermionic emission microscope is not completely limited to the
study of metals and alloys, you may already have formed the idea that grain
growth, recrystallization, phase transformations, and related metallurgical
processes are particularly amenable to fairly detailed investigation in
this instrument since the temperatures required for the production of significant emission current and image contrast are also conducive to these
processes. Some of the first work of this type was carried out by W. G.
Burgers and J. J. A. Ploos van Amstel [12-14] about 5 years after the introduction of the magnetic instrument [11]. While a mounting of interest
has not generally been attached to the use of the thermionic emission microscope, a considerable number of important investigations have been conducted
over the past 20 years. In addition the instrument itself has undergone
numerous design changes to bring it to the appearance shown in Fig. 2.11.

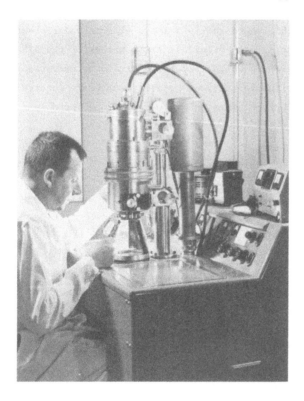

FIG. 2.11 *Thermionic electron emission microscope (From S. R.*
Rouze and W. L. Grube, Proceedings of the International
Microscopy Symposium, vol. 5, *Academic Press, New York, 1960).*

Commercially available units have been manufactured in Europe for several decades.

Recrystallization and grain growth While many earlier investigations of crystal structure and emission characteristics centered around the pure metals such as Fe, Ni, Cu, W, Ti, etc., [12-16] much of present-day thermionic emission microscopy is concerned with problems involving grain growth or recrystallization kinetics in alloys, particularly steels. Major research on these topics has been conducted in the U.S., notably at the Research Laboratories, General Motors Corp., Warren, Mich.; and at the Ford Research Laboratories, Dearborn, Mich. Commercial systems have also been manufactured in Europe (e.g. Metioscope KE3 manufactured by Balzers Aktiengesellschaft).

Figure 2.12 illustrates the typical grain growth in a modified AISI 3312 steel observed in the thermionic emission microscope employing an accelerating potential of 30 kV, with the specimen surface temperature main-

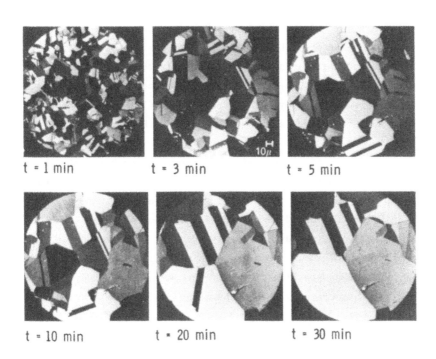

t = 1 min   t = 3 min   t = 5 min

t = 10 min   t = 20 min   t = 30 min

FIG. 2.12 *Grain growth in austenite-thermionic emission. (From W. L. Grube and S. R. Rouze, Thermionic Emission Microscopy - Applications, in* High-Temperature-High Resolution Metallography, *Gordon and Breach, New York, 1967).*

tained at 1340 K. It is easily recognized from Fig. 2.12 that growth rates
are particularly amenable from such observations since with high speed cinematography the movement of grain boundaries and the nucleation and propagation of annealing twins can be followed in considerable detail. Figure
2.13 illustrates a number of nucleation and growth characteristics of annealing twins. [See M. A. Meyers and L. E. Murr, Acta Met., 26:951(1978).]

In many respects, thermionic emission microscope studies of grain-
growth kinetics [17] and equilibration processes, and the formation of twin
boundaries, etc., present only a quantitative view of the actual physical
processes (such as variations in relative interfacial energy with composition [18]) occurring at the surface of a metallurgical specimen. In the
study of grain structure, and particularly grain or interface geometry, it
must be remembered that the emission image of a crystalline boundary is
simply a projection of a surface intersection line. Consequently we are
unable to extract from such images any information concerning the three-
dimensional structure of the polycrystalline solid; and the angles associated with the intersection of grain boundaries or grain boundaries with
twin boundaries are therefore not the true dihedral angles. [This feature
is encountered in any surface observation whether optical or electron optical. True dihedral angles can only be accurately resolved at grain-
boundary intersections, etc., by the observation of metal-foil sections by
transmission electron microscopy. This treatment is left for development
in Sec. 7.4.1; and the reader is also referred to L. E. Murr, *Acta Met.*,
16: 1127 (1968).]

Image resolution in the thermionic emission microscope is not, in
general, in excess of that attainable by optical microscopy, and the magnifications employed in most metallurgical studies such as those typically
illustrated in Figs. 2.12 and 2.13 are usually only of the order of a few
hundred times. Ultimate resolution roughly an order of magnitude of that
for light-optical microscopy is ideally possible as discussed in Sec. 3.3.2.
Nevertheless, the polycrystalline contrast enhancement and the dynamic
viewing capability associated with the thermionic emission microscope certainly make it a useful approach to many materials problems.

Phase transformations and related phemomena We have observed that the use
of high-speed cinematography in conjunction with the thermionic emission
microscope lends a dynamic and fruitful approach to the study of grain
growth. It is certainly only a slight transition to related investigations
of phase transformation in metal and alloy systems. Just as grain-growth

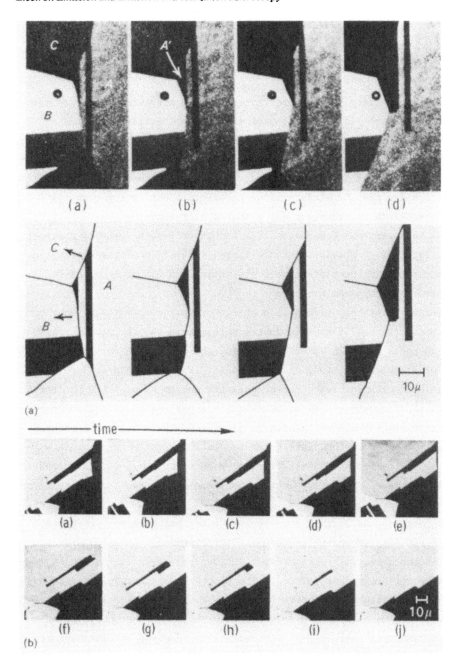

IG. 2.13 *Thermionic electron emission micrographs showing (a) broadening*
f annealing twin A' at grain corner, resulting in generation of incoherent
boundary [*elapsed time from (a) to (d) is 30 sec.*]; *(b) twin-band annihi-*
lation as result of incoherent boundary migration. [*From W. L. Grube and*
S. R. Rouze, Can. Met. Quart., 2: 31 (1965).]

rates, etc., may be accurately followed by direct observations in the thermionic emission microscope, the rate of formation of a new phase, as well as its rate of propagation, can also be followed with an equivalent accuracy.

A large portion of contemporary use of the thermionic emission micro-scope involves the investigations of transformation kinetics in Fe and various steels. G. W. Rathenau and G. Baas [19] were the first to utilize this approach in the study of pearlite formation of a high-purity eutectoid carbon steel, while Heidenreich [20] almost simulataneously used thermionic emission microscopy to study similar transformations in low-carbon steel. More recent work has involved the gamma-alpha transformation in Fe [21,22] and the various phase transformations and phase growth rates in austenite [23]. Figure 2.14 provides an excellent illustration of the observation and measurement of pearlite-nodule growth in an SAE 1095 steel at 948°K in the thermionic emission microscope.

The fact that any perceptible change in surface crystallography or surface structure will cause a contrast change in the thermionic emission image suggests a number of other uses for the thermionic emission techni-que: for example, the study of deformation [12] and fracture, and related materials studies involving surface changes due to slip or the nucleation and growth of cracks. Various other surface phenomena might also be amen-able to study by thermionic emission microscopy, but it must be pointed out again that in some cases greater resolution might be attainable with an optical microscope. In this regard, modern hot-stage metallographs are in some instances more convenient to use, especially for certain transfor-mation studies.

An attempt has been made in the preceding paragraphs to illustrate some of the simpler design features of the thermionic emission microscope, in addition to several relevant applications in the investigation of ma-terials structure and properties. The approach was certainly not rigorous, nor the descriptions conclusive. For a more comprehensive treatment of the many techniques employed in thermionic emission microscopy, consult the contributions that accompany reference 23, and the relevant suggested references at the conclusion of Chap. 2.

2.6 FIELD EMISSION AND FIELD EMISSION MICROSCOPY

It is possible to remove electrons from a metal surface, without the ne-cessity of heating the metal, by the application of an electric potential

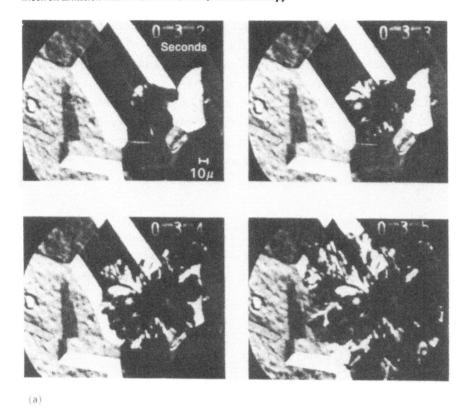

(a)

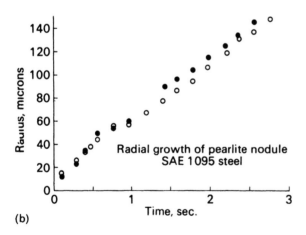

(b)

FIG. 2.14 *Pearlite-nodule growth in SAE 1095 steel at 948°K. (a) Three-second thermionic emission sequence; (b) radial-growth measurements.* [*From W. L. Grube and S. R. Rouze,* Can. Met. Quart., *2: 31 (1965).*]

between the metal cathode and anode which results in a field intensity on the metal surface, $E_z > 10^7$ volts/cm. In the presence of an applied electrostatic field of this magnitude, electrons escape from the cold cathode by tunneling through the surface potential barrier ideally depicted in Fig. 2.15. In the case of high field emission, the actual reduction in work function (Schottky effect) becomes relatively unimportant, and hot field emission, for the most part, depends on E_W to the extent that it determines the area under the surface potential curve (Fig. 2.15) as

$$A = \frac{E_W^{3/2}}{2E_z} \qquad (2.18)$$

Since the shape of the surface potential barrier under the high-field condition is no longer a simple rectangular function with a finite width b_0, the probability of electron penetration (tunneling) in the z-direction is given by

$$T(z) = \zeta \exp\left(-\frac{\sqrt{m}}{h}\int_0^{z_0}\sqrt{U-E}\,dz\right) \qquad (2.19)$$

with reference to the high-field barrier case shown in Fig. 2.15 and where ζ is a constant. The field emission current density is then found by considering Eq. (2.18) to be multiplied by the number of electrons leaving the surface per unit area per unit time; and by rigorously evaluating Eq. (2.19). R. H. Fowler and L. W. Nordheim on performing such a derivation,

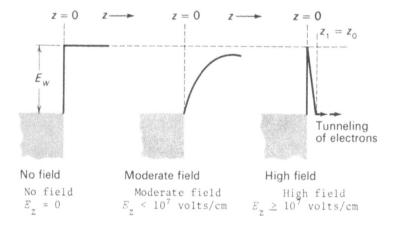

FIG. 2.15 *Surface potential barrier for cold metal with no field applied, hot metal with moderate field applied, and cold metal with very high field applied.*

obtained the equation [24]

$$J = \frac{6.2 \times 10^{-6} \sqrt{E_F/E_W} E_z^2 \exp\left(-\dfrac{6.8 \times 10^9 \ E_W^{3/2}}{E_z}\right)}{E_F + E_W} \tag{2.20}$$

which has fittingly become known as the Fowler-Nordheim equation or the F-N equation for field emission.

2.6.1 THE FIELD-ELECTRON EMISSION MICROSCOPE

While the field-electron emission and the more important field-ion emission microscope discussed later are not strictly electron optical devices as one would consider the thermionic emission or transmission type of electron microscopes to be, nevertheless these instruments do represent a simple "self-imaging" optical system of very extensive use in materials science and metal surface physics. In effect the major importance of field emission of electrons is in the utilization of these basic principles in the field emission microscope.

The field-electron emission microscope was invented by E. W. Müller at the Siemens Research Laboratory in Berlin about 1936 [25]. In its simplest form the device consists principally of an etched wire (the pointed surface of which forms the specimen surface of interest) enclosed in a highly evacuated glass tube, with a conductive-phosphor screen applied to the inner glass wall opposite the wire specimen. Figure 2.16 illustrates the instrument schematically. Wires of a desired material are etched chemically or electrochemically to a fine point, and rounded smoothly within the instrument by outgassing at temperatures near the melting point (see Appendix B).

For a very sharp specimen, characterized by a tip radius of roughly 10^{-5} to 10^{-4} cm, the field strength at the surface is given approximately by

$$E_z = \frac{V_o}{5r} \tag{2.21}$$

where E_z is the field strength normal to the surface, resolved parallel to the wire specimen axis in the z direction; V_o is the voltage applied between the wire-cathode emitter and the conductive screen (Phosphor coatings backed with a vapor-deposited thin film of Al or suitable condensed conductor. Phosphors consist of ZnS or $AgSiO_4$ as examples.) (see Fig. 2.16);

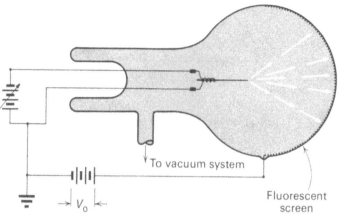

FIG. 2.16 *Field-electron emission microscope (schematic).*

and r is the tip radius. With a smooth tip, the field intensity is essentially uniform over the surface, and the emitted electrons are accelerated
radially, initially following the flux lines as indicated in Fig. 2.16.
Since the lines of electric flux enter perpendicular to the tip surface,
the resulting work-function images, corresponding to the electron emission
intensity from various crystallographic regions over the surface, are
highly magnified as illustrated in Fig. 2.16. A good approximation of the
actual magnification attained is given by

$$M \cong \frac{d}{r}$$
(2.22)

where, with reference to Fig. 2.17, d is the tip to screen distances, and
r is the tip radius.

As pointed out above, and considering the prior discussions of metal
work functions, the tip image formed on the fluorescent screen is essentially a magnified map of work-function variations over the emission surface. The resolution of discrete surface structures giving rise to the
intensity profiles is, however, limited by the velocity component normal
to the tip radius that causes a distortion with radial displacement in the
image (because of the statistical spread in the electron velocity components normal to, and creating, the fluorescent image) as well as a diffraction effect that distorts the electrons at the tip surface as a result of
their relatively long wavelength. The ultimate resolution capability of
the field-electron emission microscope (FEM) is therefore limited to about
20 Å. This is developed in the treatment of resolution in electron optical

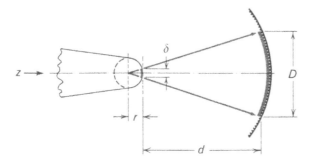

FIG. 2.17 *Radial magnification in field-*
electron emission microscope.

systems in Sec. 3.3.2. But, although individual atoms cannot be identified
in the FEM image, useful information concerning the migration of crystallographic zones, surface diffusion phenomena, adsorption, and work-function
variations can be obtained. In addition, since regions of electric field
enhancement resulting from a protuberance (in the form of a large adsorbate
molecule, for example) will cause a distortion of the flux lines emanating
from the tip surface (resulting in a concomitant increase in the electron
emission from this immediate area), it is sometimes possible to produce a
clear image of such a surface artifact.

2.6.2 APPLICATIONS OF FIELD-ELECTRON EMISSION MICROSCOPY

Work-function measurements Although the FEM in the strictest sense is not
generally employed in the measurement of effective thermionic work functions, its basic design (see Fig. 2.16) is ideal for the fairly accurate
measurement of E_W at reduced or high field intensities. Where reduced
fields and emitter heating are employed, Eq. (2.15) is applicable, but for
high fields and negligible emitter temperatures, the work function is best
obtained from the Fowler-Nordheim equation [Eq. (2.20)]. The arrangement
is, however, particularly useful for the measurement of $E_{W[HKL]}$, the work
function corresponding to a particular crystallographic-zone axis [HKL].
The thermionic mode of determining E_W is to utilize field strengths considerably less than 10^7 volts/cm ($\sim 10^3$ to 10^4 volts/cm) and to plot the
logarithm of emission current, which can be conveniently measured in the
FEM, against variations in the field strength for a fixed temperature.
This is shown in Fig. 2.18 a. The original (nonlinear) portion of the
curve is due to the space charge and Schottky effect, and the linear

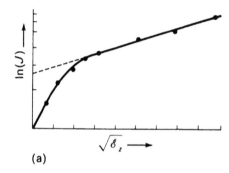
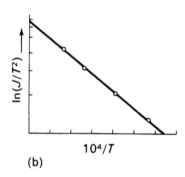

(a)   (b)

FIG. 2.18 *Measurement of E_W (effective work function). (a) Schottky effect.*
(b) Thermionic emission.

portion of the curve, extrapolated to zero field strength, therefore corresponds to the zero field emission current. This procedure is then repeated for a range of temperatures, and the corresponding logarithm of the
zero field emission current is divided by T^2, and plotted against $1/T$ as
shown in Fig. 2.18b. The slope of the curve of Fig. 2.18b is therefore
the apparent work function for the particular crystallographic orientation.
It is not necessarily the true work function since some dependence of E_W on
temperature and the surface transmission probability [as expressed in Eq.
(2.16)] is still somewhat uncertain. Several contemporary measurements of
$E_{W[HKL]}$ for a number of metals have been previously presented in Table 2.1.

The average work function of a metal is obtained in a similar manner
to that outlined above, with the exception that the specimen (again in the
form of a wire) is not so finely polished to a sharp tip. The exposed surface therefore presents numerous exposed crystallographic orientations (for
the ordinary polycrystalline wire specimen) similar to the situation presented in the thermionic emission microscope. The older, more common method for measuring average work functions consists of surrounding the emitter wire with a cylindrical anode. This facilitates the estimation of the
emission area, and enhances the total emission current flowing. The work
function, as measured from curves such as those already outlined in Fig.
2.18, is represented as

$$E_W = \sum f_g E_W$$

from f_g is the fraction of surface area having a zone axis g = [HKL], and E_W is the corresponding work function for this crystallographic patch. In the high-field mode, the work function can be obtained even more straight-forwardly from the Fowler-Nordheim equation in the form

$$I = \alpha V_o^2 \exp \left(-6.8 \times 10^7 \, \frac{\zeta E_W^{3/2}}{V_o} \right) \qquad (2.23)$$

where I = total current measured in amperes

    α = constant for particular emitter metal [compare to Eq. (2.20)]

    ζ = field-voltage proportionality factor

    V_o = applied potential in volts

The corresponding average work function for a high-field emission experiment is then given by

$$\zeta E_W^{3/2} = \sum f_g \zeta_g E_{Wg}^{3/2}$$

In some instances, a relationship of the form

$$E_W = E_{W_s} \left(\frac{V_o}{V_{o_s}} \right)^{2/3} \qquad (2.24)$$

can be employed in the determination of the work function of an adsorbed or vapor-deposited layer on an emitter of known work function, where V_{o_s} and V_o correspond to voltages required to produce an equivalent emission current without and with the deposit, respectively. Table 2.3 lists the work function for a number of metals averaged for a number of experiments using thermionic and high-field operating modes.

Studies of surface adsorption and oxidation in the FEM The presence on a surface of an adsorbate, as we have already observed in the discussion associated with Fig. 2.10, has a marked effect on the work function. This phenomenon is primarily the result of the change in the surface bond character as the adsorbate links with the local surface potential. In the field emission microscope, the presence of such an adsorbate, accompanied by the associated image intensity profile in the vicinity of the emission disturbance, can be resolved to a scale of roughly 20 to 30 Å. Thus the chemisorption of atoms and molecules on a metal surface can be followed with considerable precision as a result of the enhanced emission observed in the fluorescent image. Typical of this effect is the resolution of the

TABLE 2.3 *Average work functions*

Metal	E_W, eV
Ba	2.49
Cs	1.81
Cr	4.60
Co	4.41
Cu	5.30
Hf	3.53
α-Fe	4.48
β-Fe	4.21
Mg	3.60
Mo	4.25
Ni	4.61
Nb	4.04
Os	4.70
Pd	5.00
Pt	5.35
Re	5.10
Rh	4.58
Ag	4.78
Ta	4.19
Th	3.35
W	4.50
Zr	4.50

quadrupole emission distribution of adsorbed phthalocyanine molecules on the surface of a tungsten emitter as shown in Fig. 2.19. The larger symmetrically bright regions observed in the image of Fig. 2.19 are the result of the crystalline symmetry of the emitter surface.

The crystallography of the emission pattern can be fairly rigorously identified, and in this way the adsorption onto preferred zones <HKL> can be followed fairly reasonably by variations in the emission intensities at identifiable zones. Figure 2.20 illustrates the atomic arrangement ideally associated with several common emitter crystallographies. The coincidences for oxidation to occur preferentially can thus be followed in this way, and such phenomena as the onset of surface corrosion, or the reactions with impurities, etc., to form surface precipitates or new phases can be rather generally investigated. A fairly rigorous treatment of many of these applications of FEM, developed in the two decades following its inception in 1936, is found in a review by R. H. Good and E. W. Müller [26].

Emitter surface migration and the determination of solid surface tension
It is possible, with the FEM, to obtain fairly reliable data concerning the surface migration of agglomerated impurities [26], and of related

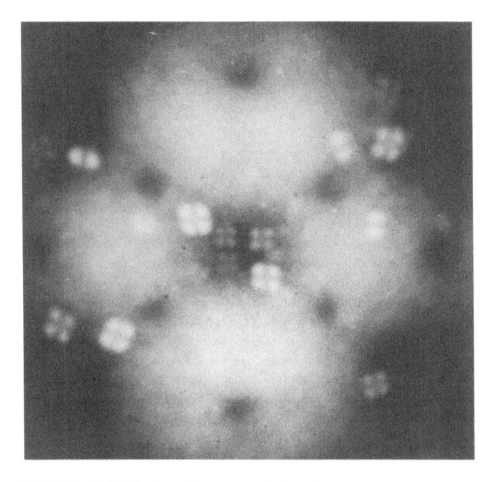

FIG. 2.19 *Field-electron microscope emission image of tungsten tip decorated with adsorbed phthalocyanine molecules.* [*Courtesy E. W. Müller, (deceased) and* Physics Teacher, 4: 53 (1966).]

transport phenomena, and to subsequently determine values for the activation energy and surface diffusion constants [27]. In recent work (within the past decade) several modifications in the field applications, such as pulsing, have allowed improved values of these constants to be obtained; and the surface tensions for tungsten, tantalum, and molybdenum have been measured directly from the effect of the electrostatic surface stress on the surface migration rate [28].

Under normal operation in the FEM, the heating of a single crystal emitter will, even under a moderate field, allow migration of those atoms at the tip to proceed (via surface diffusion) toward the shank of the

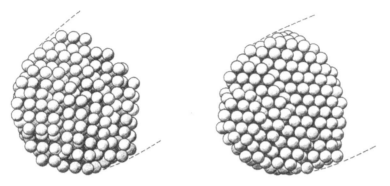

FIG. 2.20 *Packing arrangement of atoms accommodating pseudohemishperical emitter tips.*

specimen. In this way the transport of atoms, adsorbates, etc., can be followed by noting the change in the displacement of the corresponding emission rings in the fluorescent image. The electrostatic forces on the tip will nonetheless tend to reduce the surface energy by an amount proportional to the square of the applied electric-field strength. As a consequence, the balance of surface tension forces against the electrostatic forces at the emitter surface causes the migration to stop. If the electrostatic force exceeds the surface tension, the surface migration will be toward the tip, while general zone recession toward the shank (away from the tip) will occur for low fields. Using this argument, the surface tension of the emitter material is found to be

$$\gamma = \frac{r[E_z(\min)]^2}{8\pi} \tag{2.25}$$

where γ = surface tension in ergs/cm^2

   r = average emitter tip radius

$E_z(\min)$ = field intensity at which emitter surface recession (at a fixed temperature) ceases

A value of γ = 2900 ergs/cm^2 was obtained for a single-crystal tungsten emitter at 2000°K [28] which agrees quite well with the more conventional liquid-drop technique [29]. Surface energy values are also available for various crystallographies for W, Ta, and Mo in M. H. Richman, *ASM Trans. Quart.*, 60:719 (1967).]

2.6.3 THE FIELD-ION MICROSCOPE

Nearly two decades after the inception of the field-electron emission microscope, E. W. Müller also invented the field-ion emission microscope

[30, 31]. Many of the basic developments of the field-ion microscopes are described in a review by E. W. Müller in L. Marton (ed.), *Advances in Electronics and Electron Physics*, vol. 13, p. 83, Academic Press Inc., New York, 1960). The field-ion emission microscope (FIM), basically the same in design as the FEM, differs primarily in the fact that the image formed of the emitter surface occurs by the ionization of gas atoms near the sites of protruding surface atoms, and the subsequent electrostatic repulsion of these ions, which finally strike the imaging (fluorescent) screen. The field-ion microscope, as a result of the larger imaging ion mass that reduces local diffraction effects to a negligible level, and the associated uniformity of the tangential velocity component, has an optimum resolution given approximately by

$$\zeta_{opt} = \sqrt{\frac{6 \times 10^6 \ Tr}{E_z}} \tag{2.26}$$

where T is the associated temperature of the emitter surface having a radius of curvature r, subject to a field strength E_z. This means that individual atomic spacings on the order of a few angstroms can be clearly distinguished in the FIM image. Figure 2.21 illustrates the basic design features of the field-ion microscope.

Strictly speaking, the field-ion emission microscope is preferably referred to as simply the field-ion microscope, since the inclusion of the word emission might tend to convey the idea that the usual image of the emitter surface results when atoms actually leave their sites and, on striking the fluorescent screen, produce an image. The surface atoms can in fact be induced to leave the surface by a process called field evaporation or field desorption [32], which is normally employed in promoting a "smooth" emitter surface, or by the sequential exposure of deeper-lying atomic layers. The normal mode of imaging an emitter surface, however, involves the use of a suitable "imaging gas." It is this gas, or rather the ionized gas atoms, which forms a useful fluorescent emission pattern.

We can observe from Fig. 2.21 that the field-ion microscope differs from the field-electron emission microscope by the reversal of the emitter anode polarities. In the field-ion microscope, the emitter specimen is positively charged. While, like the field-electron emission instrument, an ultrahigh vacuum is initially attained, the field-ion microscope produces a useful image only after the admission of an imaging gas, usually He, at a pressure of roughly 10^{-3} torr. With an electric field of suffi-

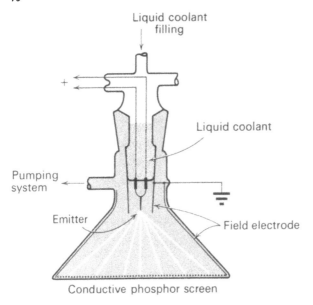

FIG. 2.21 *Field-ion microscope (schematic).*

cient strength, the gas atoms become slightly polarized (similar to the
scheme illustrated in Fig. 2.10a), and their arrival at the emitter-tip
surface is also considerably accelerated as compared with the arrival expectation on the basis of simple gas kinetics. These gas atoms, on collision with the surface, lose a portion of their kinetic energy; their re-
bound is usually insufficiently energetic to allow them to escape the field
region of the emitter surface. We can follow the course of one such gas-
atom collision in Fig. 2.22. Here a polarized atom impinges on an emitter
surface with a concomitant loss of energy to the lattice. The rebounding
atom fails to free itself from the influence of the surface field, and it
again falls onto the surface, and again loses a small portion of its kinetic energy. This process is repeated on the order of several hundred times
[31], with the atom effectively "hopping" over the specimen tip surface in
Fig. 2.21. After having lost a significant portion of its kinetic energy,
the atom gives up an electron; the transfer occurs preferentially by tunneling into the metal at a region of local field enhancement just above a
protruding surface atom. It has in fact been found that ionization occurs
in a disklike zone just above the protruding atom [33] (dashes above the
protruding atom in Fig. 2.22). The gas ion formed in this zone is then
accelerated by the high field it immediately experiences, and travels

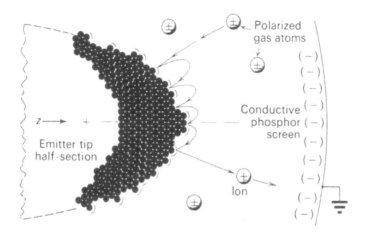

FIG. 2.22 *Collision of polarized gas atom and ion emission from high-field region of a protruding emitter-surface atom.*

radially along the associated flux line to the conductive phosphor screen. A fluorescent image dot is formed, representing the atom protruding from the surface. For an emitter tip (depicted ideally in Figs. 2.21 and 2.19), the atoms at the ledges of each layer will produce a distinct image dot. A specimen tip of about 10^3 Å radius will actually have some 10^5 such protruding surface atoms. The projection of these surface atoms via the fluxline associated gas ions onto the fluorescent screen thus results in an image of resolvable atomic surface detail. (The radial projection of the emitter surface topography results in a crystallographic projection essentially characterized by the stereographic projections of Appendix A.4.) Figure 2.23 shows a typical field-ion image of a tungsten single crystal with the same [110] orientation associated with the field-electron emission image of Fig. 2.18. Figure 2.24 shows the same [110] orientation for tungsten imaged at liquid nitrogen temperature. The generally sharper image points in Fig. 2.23 result from the lower temperature reducing the tangential velocity component; implicit in Eq. (2.26).

A radially symmetric image of the emitter surface in the FIM depends fairly critically on the smoothness of the hemispherical end form. To achieve this end form, the field experienced by a chemically prepared

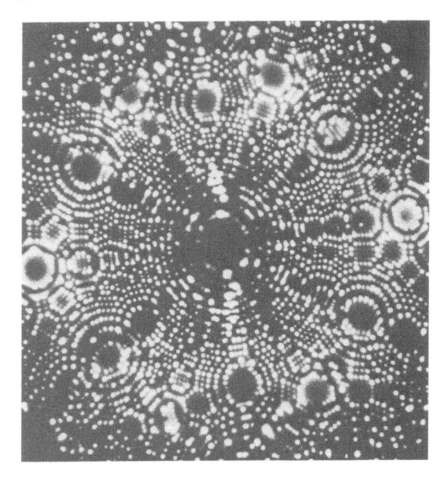

FIG. 2.23 *Field-ion microscope image of tungsten in same* [110]
orientation as Fig. 2.18. Rings characterizing crystallography
zones arise at net-plane ledges and atomic protrusions in the end
form, as depicted ideally in Figs. 2.20 and 2.22. (Liquid hydro-
gen coolant) (Courtesy O. Nishikawa and E. W. Müller.)

emitter tip is raised gradually. As the field strength builds up on the
protruding atoms and other surface irregularities, they will field evaporate [32] as ions, leaving an electron or two behind. This technique of tip
polishing is effectively self regulating since the protruding atoms and
sharp edges evaporate preferentially as a result of their locally enhanced
electric fields. Field evaporation is carried out directly within the FIM,

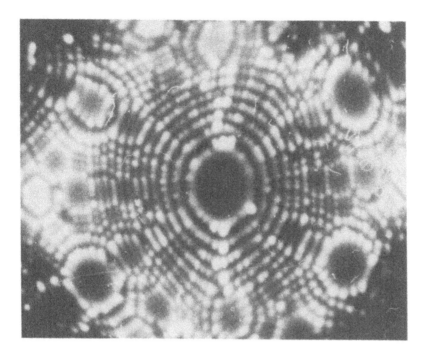

FIG. 2.24 *Field-ion image of tungsten* [110] *orientation observed at T ≃ 78 K (liquid nitrogen coolant); for comparison with Fig. 2.23. (Courtesy O.T. Inal).*

and in the presence of the imaging gas. The evaporation field, that is, the electric force required to overcome the binding energy at the surface, can be calculated from the image force theory of field evaporation [32,34]. Evaporation fields for a number of metals listed in Table 2.4 are doubly charged (2+) with the exception of Cu and Zn, which are singly charged. In actual practice, as pointed out earlier, controlled field evaporation of a smooth emitter surface can also be used to obtain images of deeper-lying structures. This feature obviously has many important applications in the study of surface continuity and the interior structure of the solid emitter metal (see Appendix B).

The evaporation field represents in some instances a limitation to the imaging of a particular emitter metal with a desired imaging gas. That is, if the field necessary to produce an emission pattern by the ionization of a particular imaging gas exceeds the evaporation field, the surface will simply field evaporate, and a useful image will not occur. In order to overcome this problem, an imaging gas having a sufficiently low ionization potential should be employed. Table 2.5 illustrates the relevant data for a number of possible imaging gases. We can observe by comparing Table 2.5 with the metals listed in Table 2.4, that all the metals listed are capable of being imaged with hydrogen at an emitter temperature of 20°K. Table 2.5 also lists the optimum resolution attainable in the image as calculated from Eq. (2.26) using a tip radius of 500 Å. However, the use of hydrogen as the imaging gas, while it does promote ionization at sufficiently lower fields, produces an emitter surface image with somewhat less resolution than He, Ne, or Ar. Since He is generally preferred for a number of reasons (resolution being chief among them), field-ion microscopes can be fitted with a suitable image-intensification system (see Sec. 3.5) to reduce the necessary imaging field, and to eliminate lengthy exposure times for photographing the fluorescent net plane pattern. The use of hydrogen to promote field emission has also been of considerable importance since it has been shown that reduction in evaporation fields up to 40 percent occurs [35,36].

TABLE 2.4 *Evaporation fields for metal ions at 0°K[†]*

Metal	E_e,MV/cm
Ir(2+)	541
W(2+)	530
Au(2+)	505
Pt(2+)	488
Re(2+)	460
Ta(2+)	450
Cu(1+)	410
Mo(2+)	388
Ni(2+)	385
Zn(1+)	383
Fe(2+)	352
Zr(2+)	345
Nb(2+)	342

TABLE 2.5 *FIM image gas data[†]*

Gas	Ionization potential, eV	Emitter temp., °K	Image field, MV/cm	δ_{op},Å (r = 500 Å)
He	*24.6*	*20*	*450*	*1.2*
Ne	*21.6*	*20*	*370*	*1.3*
A	*15.7*	*80*	*230*	*3.2*
H_2	*15.6*	*20*	*228*	*1.6*
N_2	*15.5*	*80*	*226*	*3.3*
Kr	*14.0*	*80*	*194*	*3.5*
O_2	*12.5*	*80*	*164*	*3.8*
Xe	*12.1*	*80*	*156*	*4.0*

[†]*Data from E. W. Müller, Science , 149:591 1965 by the American Association for the Advancement of Science. This reference also contains a general survey of field-ion microscopy.*

2.6.4 *APPLICATIONS OF FIELD-ION MICROSCOPY*

It is perhaps intuitively obvious that most of the phenomena readily observed in the FEM are more highly resolvable and better examined in the FIM. In actual practice, the FIM can be operated in the electron emission mode by simply heating the emitter, and reversing the emitter-screen polarity. For many investigations, an image observed in the FEM mode can be very quickly switched to the FIM mode with the admission of an imaging gas to the vacuum chamber. A comparison of Figs. 2.19 and 2.23 demonstrates this feature rather convincingly; and it is also apparent that FIM is by far the more useful technique for the detailed study of atomic surface structure. [A simple FIM design is described in L. E. Murr and O. T. Inal, *J. Vac. Sci. Technol.*, *8*: 426(1971)].

Point defects in the FIM image With the ability to resolve individual atomic details, the field-ion microscope is ideal for the direct observation of vacant or interstitial atom sites. Combined with the technique of field evaporation, the densities of vacancy or vacancy clusters and interstitials in a volume of emitter metal can be directly measured [37] and compared with theoretical estimates based on a probability of the form

$$n \cong N e^{-E_v/kT}$$

where n is the approximate number of vacancies (or interstitials) in an atomic volume of N atoms with an activation energy E_v at a temperature T, and k is Boltzmann's constant. The direct counting of vacancies in platinum quenched from near the melting point has led to the calculation of an energy of formation, E_v, of 1.15 eV, [38] and observations have also been made of radiation damage to the atomic arrangement of an emitter tip by noting dark areas or bright spots created in the emission pattern on the formation of vacancy clusters or interstitial atoms [39]. Similar observations can also be made of impurity atoms; adsorbed gas atoms such as oxygen or nitrogen; and the details of corrosion, oxidation, etc., can be studied with considerable clarity [40].

Figure 2.25 is an illustration not only of a unique application of field-ion microscopy in the detailed analysis of vacancies and vacancy clusters, but also of the systematic evaporation of successive atomic surface layers in order to measure the bulk concentration of point defects and point-defect clusters. The experiments illustrated by Fig. 2.25 were the first direct (visual) evidence of shock-induced vacancy concentrations [41,42]. In these experiments, 7.5 x 10^{-3} cm diameter molybdenum wires were sandwiched between molybdenum foils and subjected to an explosive shock wave propagating perpendicular to the wire axes, having a peak pressure of 15 GPa, and a pulse duration of 2 μs. The wires were recovered after the shock event and electroetched to emission end-forms as described in Appendix B, and quantitative estimates made of the vacancy (cluster)

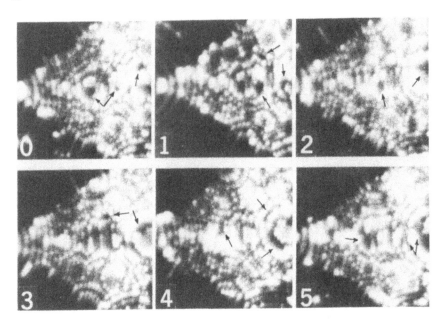

FIG. 2.25 *Field-evaporation sequence showing successive atom layers removed from the* [011] - [121] - [112] *triangle (see Appendix A stereographic projections). Zero indicates the start of the sequence with layers field evaporated denoted by successive number. Arrows show prominent vacancies and vacancy clusters. (From L. E. Murr, O. T. Inal, and A. A. Morales, ref. 41). (Liquid nitrogen coolant).*

concentrations by counting and measuring the zones revealed in field-evaporation sequences shown typically in Fig. 2.25.

Direct observation of crystal perfection and imperfection Perhaps the most valuable contribution of field-ion microscopy lies in its ability to resolve the imperfections inherent in most crystal lattices. The high-resolution capability of the FIM, combined with the layer-by-layer field evaporation of surface atoms allows details of discolorations [43,42], grain boundaries, [44] phase boundaries [45], and related crystal interfaces [46] to be followed into the emitter material. Various deformation modes have also been observed by the pulsing of emitter tips with a high field to induce "fatigue" and associated slip and failure [47]. Figure 2.26 illustrates a dislocation in a <111> zone of an Ir [001] emitter surface while Fig. 2.27 shows a twin band resolved on a Co emitter surface. (See also L. E. Murr, *Interfacial Phenomena in Metals and Alloys*, Addison-Wesley Publishing Co., Reading Massachusetts, 1975.)

The metallurgical problems readily amenable to study by field-ion

FIG. 2.26 *Field-ion micrograph of Ir end form containing dislocation (arrow). (Courtesy E. W. Müller)*

microscopy should be immediately obvious to the materials scientist. Aside from those fundamental studies indicated above, the FIM is potentially quite useful in studies of alloying (with sufficient dilutions), phase transformations, and a host of related phenomena. And, by ingeniously manipulating the emitter environment, the imaging gas, and the electric field, a count-

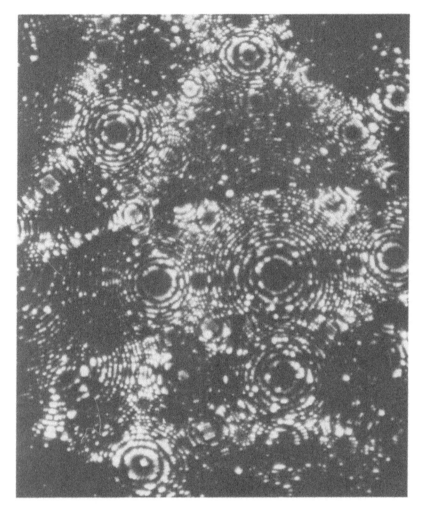

FIG. 2.27 *Field-ion image of twin boundaries in Co end form.*
(Courtesy O. Nishikawa and E. W. Müller.)

less number of important areas are open to research. It should be cautioned, however, that the interpretation of image detail in many instances presents a quite formidable task; and it should be borne in mind that the
emitter surface conditions as they exist in the high field of the FIM are
not necessarily representative of solid-state features normally dealt with
in the realization that the mechanical stress exerted by the imaging field
E_z on the surface is of the order of

$$\sigma \cong \frac{E_z^2}{8\pi}$$

or the surface energy γ, divided by the tip radius r [compare with Eq. (2.25)], and it is readily shown that stresses on the order of 50 times the technical yield stress can occur routinely during imaging of an emitter surface.

Field-ion microscope designs have afforded some interesting in-situ experiments. For example, Inal and Murr [48] have irradiated Mo end-forms with a focused laser beam in order to simulate shock-wave phenomena. Figure 2.28 illustrates the physical arrangement of the laser optical system associated with the field-ion microscope. In this FIM arrangement the emission end-form and imaging screen are at right angles to the basic unit shown schematically in Fig. 2.21. The laser beam is focused through a window in the FIM using a lens at "L" in Fig. 2.28. A focusing camera at "C"

FIG. 2.28 *Metal field-ion microscope design for in-situ laser beam focusing experiments (After Inal and Murr in ref. 48).*

insures the alignment of the incident laser beam, and the images are recorded from the fiber-optic channel-plate image intensifier mounted on the
initial phosphor screen utilizing the recording camera designated "R".

Similar in-situ experiments have involved arrangements to vapor deposit thin layers of metal upon emission end forms in order to systematically
study epitaxial and pseudomorphic growth, and related phenomena. Figure
2.29 illustrates experiments of this type which have been described in detail by Murr, et al [49]. In Fig. 2.29(a) an initially perfect segment of
a W end form is altered by vapor deposition of Pt in Fig. 2.29(b). You
should notice that the image points (atomic positions) in the (121) plane
of Fig. 2.29(a) are exaggerated somewhat by the high-field conditions,
causing the images to be distorted into egg-shaped regimes. This field-
enhanced image exaggeration also occurs for Pt atoms and atom clusters
which have been deposited onto ledges and terraces of the W end-form surface as shown in Fig. 2.29(b).

Recent interest has arisen in connection with studies involving the
behavior of amorphous metals and alloys. Very subtle property changes occurring for such materials are associated with the onset of crystallization
which is characterized by short-range ordering. For incipient phenomena of
this type and for examining the amorphous structure in general, field-ion
microscopy can provide fundamental microstructural information at the atomic level. For example, it is instructive to compare the field-ion microscope image of an amorphous Fe-B alloy in Fig. 2.30(a) with the very regular crystal structures shown in Figs. 2.24 and 2.26. Figure 2.30(b) shows
for comparison the lack of crystal structure implicit in an electron diffraction diffuse-ring pattern while Fig. 2.30(c) and (d) compare the simple representations of the crystalline and amorphous structures for an ordered and disordered arrangement of a complex alloy system. You might
glance in advance at the ring-type electron diffraction patterns for fine,
polycrystalline materials shown in Chapter 6 for comparison with Fig. 2.30
(b). It should also be noted that the imaging voltage is not the same for
the iron and boron atoms in Fig. 2.30(a), and as a consequence the image
features are primarily those for iron.

FIM image interpretation In addition to the reservations discussed above
regarding conclusions based on field-ion microscope images, it should be
realized that image interpretation in general is a difficult job. In many
experiments, the use of image intensification becomes absolutely essential
for the production of recognizable surface net-plane patterns, and where

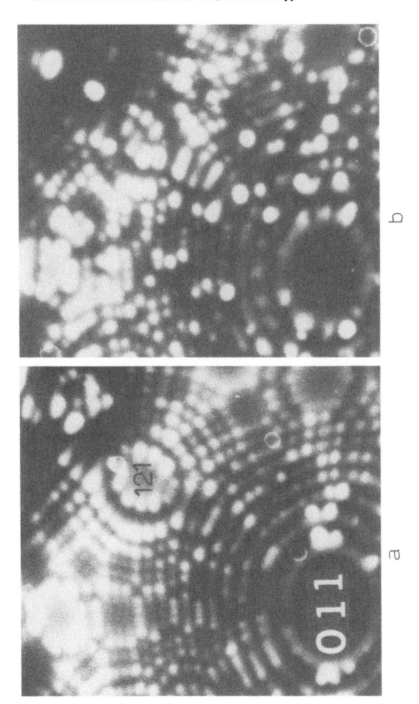

FIG. 2.29 *Vapor deposition of thin Pt layer upon a W field-emission end form surface in the FIM. (a) initial, perfect end form, (b) same surface region after in-situ vapor deposition of Pt. (liquid nitrogen coolant; He imaging gas).*

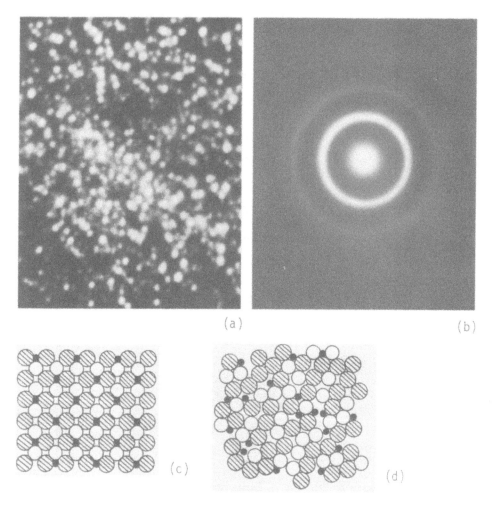

(a)   (b)

(c)   (d)

FIG. 2.30   *FIM image of amorphous* $Fe_{80}B_{20}$ *(a), Diffuse elec-*
tron diffraction pattern corresponding to the glassy metal
in (a); (c) and (d) compare a complex order-disorder regime
(crystalline and amorphous structures respectively).   (a)
is after Murr, et al. ref 50).

the close-packed atoms composing individual net planes are to be imaged, a
rigid regulation of imaging gas and electric field intensity must be imposed. As noted earlier, the net-plane-image patterns result from the high-
field regions accommodating the net-plane ledges and the various atomic
protrusions within the net planes. As a consequence, only a portion of the
total surface atomic density is imaged, thereby complicating the interpretation of emission intensities and related features.

Some advances in the interpretation of field-ion images, particularly the identification and analysis of dislocations, have been made by computer simulation of emitter image points, crystallographic poles in the image, and gross topological features [51-54] by considering a shell Model [55] using a point lattice of known coordinates corresponding to an fcc, bcc, or hcp system. The details of these methods are outlined in J. J. Hren and S. Ranganathan, "Field-Ion Microscopy," Plenum Press, Plenum Publishing Corporation, New York, 1968. In addition, the reader unfamiliar with stereographic projections might consult O. Johari and G. Thomas, *The Stereographic Projection and Applications,* J. Wiley and Sons, Inc., New York, 1969. By assuming that only protruding atoms contribute significantly to the FIM image, it has been shown that an orthographic projection (see Appendix A.4) of atoms within a critical distance (net-plane ledge heights; see Fig. 2.22) at which they will contribute to the net-plane images as simulated by the computer, reproduces FIM images essentially point for point.

If the emitter-end form is not smooth, if asperities and artifacts occur on the surface, or if the end-form is asymmetrical, there may be overlapping image features by intersections of electric field lines or the high-field conditions created at surface asperities may create image flaring or streaks. Because such features may be crystallographic as a result of preferential etching, cleavage, or related phenomena, it might be tempting to ascribe such features to the presence of crystal defects, etc. Such image features are usually artifactual, and result from surface irregularities described above [56,57].

2.7 SCANNING TUNNELING MICROSCOPY

We have seen that in the field-ion microscope, ions are "emitted" from the high-field region corresponding to an atomic protrusion on an emission end-form surface by electron tunneling from the imaging gas atoms (Fig. 2.22). In the 1960's R. D. Young, a student of E. W. Mueller working at the National Bureau of Standards, carried out the first successful metal-vacuum-metal tunneling experiments, and attempted to develop tunneling microscopy as a means for revealing surface topography at the atomic level [58]. Using a tungsten field-ion emitter, Young, et al. [59] obtained tunneling I-V (current-voltage) characteristics between the emitter (end form) and a thin metal foil; and in 1972 built a *topografiner* which allowed three-dimensional motion of the emitter over a surface. This permitted a

surface mapping to be made of a diffraction grating with a resolution of
about 30 Å in the vertical direction, and about a factor of 10^2 poorer in
the horizontal plane of the grating. The concept involved keeping the
tunneling current constant and measuring the displacement: a quantum
mechanical profilometer as illustrated schematically in Fig. 2.31.

In 1978, Heinrich Rohrer and Gerd Binnig, working at the IBM Zurich
Research Laboratory, began the development of a vacuum tunneling device
which was eventually able to control the vacuum gap to an accuracy of a
single Angstrom unit, using a piezo drive mechanism and dramatically reducing the noise. Eventually, by adjusting the emission end form in this
device so that the distance to the surface was constantly adjusted (δ_z in
Fig. 2.31) to achieve a constant current, the tip precisely traced out the
hills and valleys conforming to actual atomic positions along the displacement path (x in Fig. 2.31) of the end form. Repeated side-by-side
(moving the probe in the y-direction), these linear displacements could
build up a detailed "picture" of the atomic landscape and established the
invention of a new surface microscope: the scanning tunneling microscope
or STM [60]. This work had actually originated with efforts to utilize
Josephson tunneling junctions [61] in superconducting devices, and the
measurement of the tunneling currents associated with the superconductor-
insulator (vacuum)-superconductor configuration.

While the original description of the STM did not attract much atten-

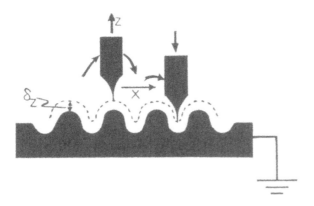

FIG. 2.31 *A field-emission end form is moved along a grating surface
so that electron tunneling current remains constant, tracing the profile
of the hills and valleys of the grating.*

tion because of poor image quality associated with a relatively large radius tungsten emission end form, a major breakthrough came in 1983 when the first real-space atomic configuration of the 7x7 unit cells on the [111] silicon surface were observed [62]. This led to a flurry of activity world wide and set in motion a new thrust in high-resolution surface microscopy with applications to a wide range of materials and materials landscapes. Binnig and Rohrer were awarded the Nobel Prize in Physics for their STM invention in 1986. The prize was shared with E. Ruska for the electron microscope.

2.7.1 *THE SCANNING TUNNELING MICROSCOPE (STM)*

The basic principle of the STM, shown schematically in Fig. 2.32, manifests itself in the exponential variation in tunneling current with the distance between the emission tip and the surface being scanned: The smaller the distance the greater the tunneling current, J_t. Expressed in terms of the STM parameters shown in Fig. 2.32(a) we can write (in a form similar to Eqs. (2.19) and (2.20):

$$J_t = CV \exp\left[-A\sqrt{E_w}\,\delta_z\right] \tag{2.27}$$

where C and A are constants and V is the voltage applied to the tip. The probe distance, δ_z (approximately 10 Å or one nanometer), is therefore varied by displacements $\Delta\delta_z$ (up and down) which are monitored by piezo drive P_z, while piezo drives P_x and P_y scan the probe over the surface of the sample. The tunneling current is usually on the order of a nanoampere as the emission tip follows a contour of constant charge density near the Fermi level in this constant impedance scanning mode to reveal the surface topography. This is called the constant height or current imaging mode of STM. Some researchers also refer to this mode as tunneling barrier height imaging (TBI).

Figure 2.32(b) shows a schematic drawing of the STM illustrated conceptually in Fig. 2.32(a). The fine adjust screw is normally turned by a motor which initially brings the emission tip [Fig. 2.32(c)] toward the sample surface with a z-drive feedback circuit in operation. As soon as a tunneling current is detected, the z-drive can be retracted and the motor switched off automatically [63,64]. This allows the probe tip [Fig. 2.32(c)] to be brought within the working range (δ_z) of the sample surface without touching it.

To achieve atomic resolution as illustrated in the topographic image

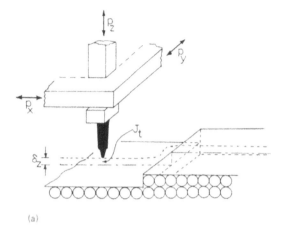

(a)

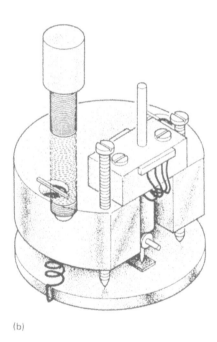

(b)

FIG. 2.32   (a) Simple schematic diagram showing the principal components
of the scanning tunneling microscope.   The emission tip or STM probe is
positioned by piezodrives, P_z, P_x and P_y to map the speciman surface
topography.   (b) Schematic drawing of the STM (Courtesy John Clarke [64].
(c) SEM image of a tungsten STM probe tip being positioned over a graphite
sample (Courtesy T. Hasegawa, Hitachi Central Research Laboratory), (d)
Atomically resolved [111] silicon 7x7 structure [66].

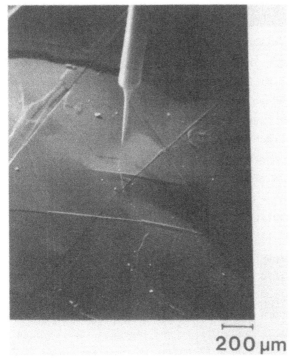

200 µm

(c)

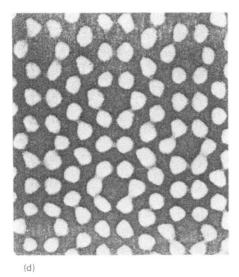

(d)

in Fig. 2.32(d), the probe tip [Fig. 2.32(c)] must have an extremely sharp curvature so that the tunneling current flows only to the single most protruding atom or atoms (as shown schematically in the emitter tip half section view of Fig. 2.22)[65]. This can be achieved by field evaporation of the end form, and field-ion microscope characterization of the probe tip has become the preferred procedure for successful, high-resolution STM [67]. Sakurai, et al. [68,69] have in fact combined a scanning tunneling microscope with a room temperature field-ion microscope so that in-situ inspection and fabrication of the STM probe tip [Fig. 2.32(c)] can be carried out at the atomic scale almost simultaneiously with a particular STM surface imaging experiment.

The success of the STM device [Fig. 2.32(b)] in imaging a specimen surface with atomic resolution is also very much dependent upon almost perfect damping of vibrational noise (mechanical, thermal, and acoustic). In addition, rather sophisticated (and highly sensitive) electronic feedback circuits are required to control the vertical motion ($\Delta\delta_z$) of the probe tip. The entire microscope is often able to be held in one's hand, and can be attached to a suitable control unit and immersed in a cryogenic liquid. In addition the STM can examine virtually any surface, even liquids, so long as there is suitable electrical conductivity. When examining nonconducting specimens, including many biological materials of low conductivity, successful imaging requires the application of a metallic or other suitable conducting coating as in the observation of specimens in the scanning electron probes and microscopes described in Chapters 4 and 5. Other similar techniques such as replicating a surface by the application of conducting films, etc. can also allow for surface imaging of nonconductors, but not necessarily at the atomic level (See Chap. 5).

2.7.2 THE ATOMIC FORCE MICROSCOPE (AFM)

The atomic force microscope, like the STM, scans a surface by moving a probe tip along a line (x) which can be displaced in the y direction to scan the surface topography. But unlike the STM, the AFM probe tip is in actual contact with the surface and no current is measured. This is truly a nanomechanical profilometer and was invented by G. Binnig and Christopher Gerber in colloboration with C. F. Quate [70,71]. As illustrated in Fig. 2.33, the original AFM used a diamond microstylus mounted on a tiny gold-foil cantilever spring. A small force, typically 10^{-8} N (or 10^{-6} g), keeps the stylus in contact with the surface without damaging it. A laser beam

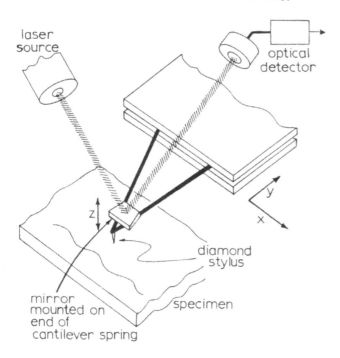

FIG. 2.33 *Optical-lever atomic force microscope schematic. Vertical displacements (z) are detected by minute changes in the position of the mirror which rides on the diamond stylus much the same way as bar codes are read in a grocery store checkout.*

is reflected from a tiny mirror mounted on the cantilever holding the stylus, and a detector senses minute displacements which are amplified to map out surface structure. Vertical displacements as small as 0.1 Å can be detected so that atomic structures similar to those shown in Fig. 2.32(d) can also be imaged in the topographical displays obtained by the AFM. Because this is a purely nanomechanical surface mapping, the specimen can be non-conducting and have virtually any structure and composition.

While the STM is phenomenologically an "electron" microscope, the laser-sensing or optical lever AFM (Fig. 2.33) is a purely mechanical device. As mechanical devices go, the AFM may be able to manipulate, position, and reposition individual atoms and molecules on a surface to perform atomic-scale engineering. Some of these features have already been achieved in laboratory experiments [72]. Similar applications may also be possible with the STM as well [73,74].

2.7.3 APPLICATIONS OF STM AND AFM

While the scanning tunneling microscope and the optical-lever atomic force
microscope can provide topographic mappings forming images of atomic surface structure, a relaxation of this resolving power can also permit unprecedented mapping of machine-finished surfaces [75]. Similar evaluations
of manufacturing techniques could push quality control to the nonoregime.
Atomic and molecular coatings on surfaces can be differentiated down to
vertical resolutions approaching 0.01 Å, and these features can have important applications in the manufacture of compact disks and thin film magnetic heads in vertical recording applications. Nanometer lithography is
of course a logical extension of the high resolution probing of STM and
AFM. Ringger, et al. [76] have succeeded in producing linear structures on
a glassy Pd-Si alloy surface, but the actual mechanism of STM "writing" is
not well understood.

Since, as we noted in connection with Eq. (2.27), the STM tunneling
current is directly proportional to the density of states at the Fermi
level (implicit in $\sqrt{E_w}$), the STM can image both the atomic lattice and the
real space charge-density wave (CDW) superstructure on a crystal surface
of a so-called low-dimensional metal. These are also often referred to as
quasi-one dimensional and quasi-two dimensional conductors or CDW materials,
[77]. Figure 2.34 shows a projected view of a STM-TBI image of a single

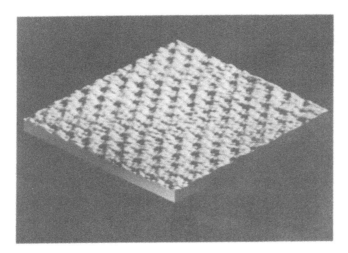

FIG. 2.34 *Local charge-density-wave (CDW) structure in 1T-TaS$_2$ single
crystal surface determined by scanning tunneling microscopy. Courtesy
John Clarke [64].*

crystal surface of IT-TaS$_2$ in the T phase, which shows a CDW phase slipped one lattice spacing when viewed at a glancing angle along the lattice image [64].

Small, two-dimensional islands of gold formed by in-situ evaporation onto single-crystal graphite cleaved in vacuum have been imaged with the STM as illustrated in Fig. 2.35. In striking contrast to the bulk situation, the lattice is found to be rectangular rather than close packed [78]. The islands observed in Fig. 2.35 contain ordered regions of roughly 50 Å in rectangular lattices. It might be instructive to compare Fig. 2.29 with Fig. 2.35.

Si pn junctions fabricated by photoresist masked As$^+$ implantation have been observed using a technique called current imaging tunneling spectroscopy (CITS) in a scanning tunneling microscope as illustrated in Fig. 2.36. Using the CITS, a specific bias was chosen to define n-type or p-type areas according to whether or not tunneling current flowed. The pn junctions could be identified from the current image at this bias and in the STM topographic image. Figure 2.36 shows the effect of sample bias which can effectively create bright and dark image reversals [79]. The junction

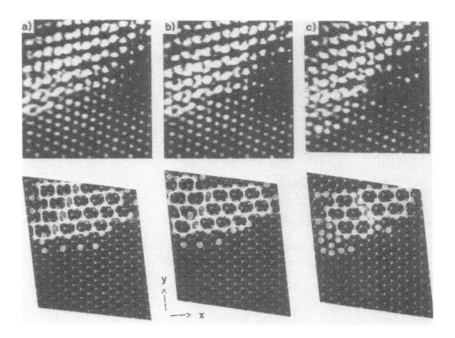

FIG. 2.35 *STM image of the local atomic structure of two-dimensional gold islands on graphite surface. Courtesy John Clarke [78].*

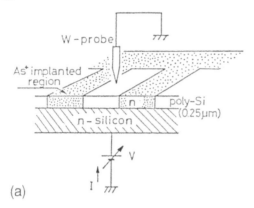

(a)

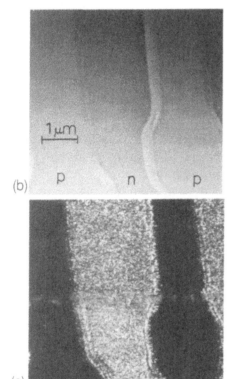

FIG. 2.36 (a) Schematic diagram for STM imaging of Si pn junctions.
(b) STM image of silicon surface with p and n-type areas at a tunnel
(bias) voltage of -2.4 volts. (c) Same image area as (b) taken at +2
volts sample bias. Courtesy Dr. T. Hasegawa, Hitachi Central Research
Laboratory [79].

structure associated with volume expansion caused by the implantation and annealing is also apparent in the topographic image of Fig. 2.36(b).

The direct observation of pn junctions is of course of great interest in ultra large scale integration (ULSI) or related down-scaled devices because it is necessary to measure the practical profile of ultrashallow junctions. These shallow junctions are essential in constructing micro-devices with high voltage and high speed operation. Auger electron spectroscopy (AES) and scanning ion mass spectrometry (SIMS) (See Chapter 4) are not adequate for these kinds of pn junction studies because they lack lateral resolution.

PROBLEMS

2.1 Calculate the average velocity of electrons that escape from a Ge crystal irradiated with an ultraviolet beam having a wavelength of $0.1 \, \mu m$, if the work function for Ge is taken to be 4.62 eV. (Assume all electrons to be emitted normal to the surface at $T = 0°K$ in the z direction, and that the experimental arrangement is essentially that of Fig. 2.2a.).

2.2 Complete the following table of constants by calculating λ_o or E_W as required. Note that all the metals listed have a cubic lattice with the lattice parameter indicated. The quantity λ_o is the threshold wavelength at which photoemission commences.

Metal	Structure	Lattice parameter, Å	λ_o, Å	E_W, eV
Cr	bcc	2.885	—	4.60
Ni	fcc	3.524	—	4.61
Ag	fcc	4.086	—	—

Assuming all valence electrons to be "free," calculate the barrier potential E_B for these metals at $T = 0°K$.

2.3 In a simple thermionic emission experiment, the voltage applied between a Cu emitting cathode and the anode is 1 kV. If the cathode and anode are considered to be square plates spaced 0.1 m apart, calculate the thermal energy required for emission if the apparent work function is taken to be 5.30 eV. What is the voltage required for cathode emission at $0°K$?

2.4 A Cu specimen is imaged in a thermionic electron emission microscope using an accelerating potential of 30 kV, with an anode-to-cathode spacing of 2 cm. The Cu emission sample is polycrystalline with an initial texturing that results in 80 percent of the grain orientations normal to [111], and the remaining grains oriented normal to [100]. With the emission temperature constant at 700°C, recrystallization occurs with the result that the field of view now consists of an equal number of grains of each orientation and of the same average size. Calculate the net change in the emission current density for a unit area of specimen that accompanies this recrystallization.

2.5 A monatomic layer of Cs is adsorbed on a single crystal of Ag in the thermionic emission microscope having the emission direction parallel to [001]. Calculate the reduction in the Ag effective work function in electron volts assuming the Cs ions, imagined as hard spheres, pack in a square array on the surface. (The mean ionic radius for Cs is taken to be 1.69 Å in four-coordination; and the interlayer separation δ is assumed to be twice the ionic radius.)

2.6 Generally, specific crystallographic orientations are not distinguished one from another in a thermionic emission microscope image, even though the contrast differs from one orientation to another by virtue of the change of E_W with crystallography. Suppose, however, a specimen to be examined by thermionic emission microscopy is first examined by x-ray diffraction methods and found to have a strongly preferred (110) grain surface texture. If the work function of this material is known for the (100), (110), (111), and (112) orientations, suggest ways for tentatively identifying the grain surface orientations in the thermionic emission image. Could the grain surface orientations be determined absolutely from crystallographic information in the emission image? If so, how?

2.7 Describe how the image magnification in a thermionic electron emission microscope can be accurately calibrated.

2.8 A field-electron emission specimen is electrolytically etched to a fine point, and observed in the scanning electron microscope as shown in Fig. P2.8(a). The tip radius, measured directly from this electron micrograph, was found to be 0.2 μm. If the tip-to-screen distance in the FEM is 0.085 m, calculate the image magnification to be

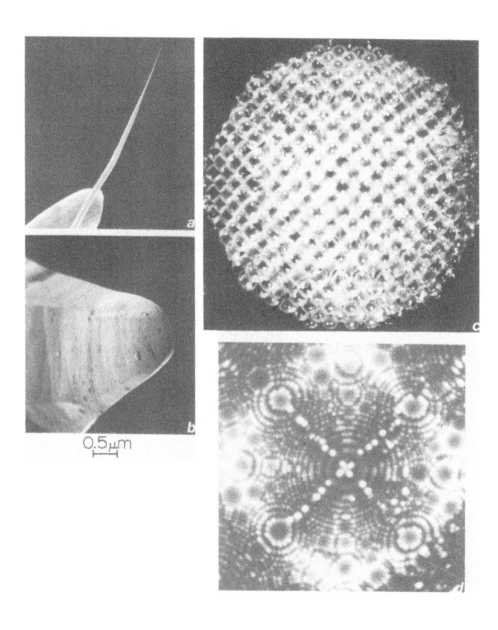

0.5 μm

FIG. P2.8

expected in the FEM operated at 10^4 volts. What is the corresponding field intensity developed on the tip surface? In Fig. P2.8(b) is shown an enlarged view of an iridium end form oriented along the [001] direction. In Fig. P2.8(c) is shown a ball model of this iridium end-form, while Fig. P2.8(d) shows an iridium surface image observed in the FIM. What is the magnification of the image in Fig. P2.8(d)? Show the indices of the prominent crystallographic planes evident in both Fig. P2.8(c) and (d). You must recognize that the FIM image illustrates the emergence on the surface of a unit cell face for iridium. Why is this image distorted and why is the face-centered atom not apparent?

2.9   If the polished wire specimen in Prob. 2.8 is a single-crystal whis-ker of tungsten, with the long axis (the growth axis) parallel to [112], calculate the corresponding emission current. Recall that tungsten is bcc with a = 3.165 Å.

2.10 A similar tungsten emitter as used in Prob. 2.9 is imaged in a field-ion microscope using He as an imaging gas at 10°K. If the tip radius is 1200 Å, and the tip-to-screen distance is 10 cm, find the magnifi-cation of the resulting surface in the image and the expected resolu-tion. Would the individual atoms in the central [112] zone be ob-served? What is the approximate emission current to be expected during the observation of the image?

2.11 A portion of a telemetry network for a space probe is designed for a special application that requires the use of wire-wound Ta resistors having an ambient temperature resistance on the earth of 1K ohms each. A materials laboratory conducting tests on these elements (after a reentry circuit failure) deduces that an increase in temperature due to reentry heating caused a resistance change governed on the order of

$$R = R_o \left[1 + 4 \text{ (percent vacancy concentration)} \right]$$

where R_o is the ambient temperature resistance. Calculate the resis-tance of an element at 2300°K and consider the vacancy activation energy in Ta to be the same as that in Pt. Explain how one might prove that an approximate percent of vacancies did in fact exist at the temperature specified.

2.12 Assume that in a hypothetical FIM experiment, failure (in the classical sense that cracks are observed to form) occurs on a Pt emitter surface imaged with Ne at 20°K. If it is estimated that the field stress just at the onset of failure is equivalent to 6 percent of the modulus of elasticity in tension, calculate the surface energy if the tip radius is 400 Å. Assuming the temperature coefficient for the surface energy of Pt to be -1.1 ergs/(cm^2)(°C), calculate the equivalent surface energy at 1000°C. At what percent of the evaporation field did the tip failure commence?

FIG. P2.14 *(Courtesy of O. Nishikawa.)*

2.13   It has long been suggested that the fatigue of metals is associated
       with the creation of vacancies in the vicinity of the surface, and
       their migration to the surface followed by their coalescence to form
       voids and microcracks. Suggest an experiment to test the accuracy
       of this philosophy using the field-ion microscope, but not involving
       the field pulsing of the emitter in situ.

2.14   In the field-ion micrograph of a Pt emitter (Fig. P2.14), identify
       the net-plane zone crystallographies indicated at A,B,C,D,E,F,G,H,I,
       J, and K. Identify the defects at 1,2,3,4,5,6,7, and 8 as vacancies,
       interstitials, or dislocations. The emitter tip axis is parallel to
       [001]. (Consult Appendix A.4.)

2.15   In the atomic force microscope illustrated schematically in Fig. 2.33,
       assume the laser beam used in the detector system is a ruby laser.
       Estimate the pressure (expressed in grams) applied by the laser beam
       to the cantilever spring, and compare this with the pressure of 10^{-6} g
       required to keep the stylus in contact with the surface without
       damaging it. Does this pressure contribute to the stylus pressure?

REFERENCES

1. E. Fermi, *Z. Physik, 36:* 902(1926).

2. P. A. M. Dirac, *Proc. Roy. Soc. (London), 112:* 661(1926).

3. A. Sommerfeld and H. Bethe, in *Handbuch der Physik,* 2 Aufl., vol. 24,
    p. 333, Springer-Verlag, OH6, Berlin, 1933.

4. E. Bruche, *Z. Physik, 86:* 448(1933).

5. L. Wegmann, *Res./Development,* July: 20(1970).

6. *Proceedings of the First International Conference on Electron Emission
    Microscopy,* Tübingen, W. Germany, Sept. (1979).

7. M. D. Jones, G. A. Massey, D. L. Habliston, and O. H. Griffith, *ibid.*
    p. 177.

8. O. W. Richardson, *Proc. Cambridge Phil. Soc., 11:* 286(1901).

9. S. Dushman, *Rev. Mod. Phys., 2:* 381(1930).

10. E. Brüche and H. Johannson, *Ann. Physik, 15:* 145(1932).

11. M. Knoll, F. G. Houtermans, and W. Schulze, *Z. Physik, 78:* 318(1932).

12. W. G. Burgers and J. J. A. Ploos van Amstel, *Physica, 4:* 5(1937).

13. W. G. Burgers and J. J. A. Ploos van Amstel, *Physica, 5:* 305(1938).

14. W. G. Burgers and J. J. A. Ploos van Amstel, *Nature, 41:* 330(1938).

15. R. D. Heidenreich, *J. Appl. Phys., 26:* 757(1955).

16. T. Fukutomi and A. Tanaka, *Japan J. Appl. Phys., 32:* 235(1963).

17. J. Nutting and S. R. Rouze, *Proceedings Fifth International Congress for Electron Microscopy,* vol. 2, CC-7, Academic Press Inc., New York, 1962.

18. N. A. Gjostein et al., *Acta Met., 14:* 1637(1966).

19. G. W. Rathenau and G. Baas, *Acta Met., 2:* 875(1954).

20. R. D. Heidenreich, *J. Appl. Phys., 26:* 879(1955).

21. E. Eichen and J. Spretnak, *Trans. ASM, 51:* 454(1959).

22. R. P. Zerwekh and C. M. Wayman, *Acta Met., 13:* 99(1965).

23. W. L. Grube and S. R. Rouze, Thermionic Emission Microscopy - Applications, in *High-Temperature-High-Resolution Metallography,* Gordon and Breach, Science Publishers, Inc., New York, 1967.

24. R. H. Fowler and L. W. Nordheim, *Proc. Roy. Soc. (London), A111:* 173 (1928).

25. E. W. Müller, *Z. Physik, 106:* 541(1937).

26. R. H. Good and E. W. Müller, in S. Flugge (ed.), *Handbuch der Physik,* vol. 21, pp. 176-231, Springer-Verlag OH6, Berlin, 1956.

27. E. W. Müller, *Z. Physik, 126:* 642(1949).

28. J. P. Barbour et al., *Phys. Rev., 117:* 1452(1960).

29. L. E. Murr, *Interfacial Phenomena in Metals and Alloys,* Addison-Wesley, Reading, Mass., 1975.

30. E. W. Müller, *Z. Physik, 131:* 136(1951).

31. E. W. Müller and K. Bahadur, *Phys. Rev., 102:* 624(1956).

32. E. W. Müller, *Phys. Rev., 102:* 618(1956).

33. T. T. Tsong and E. W. Müller, *J. Chem. Phys., 41:* 3279(1964).

34. R. Gomer and L. W. Swanson, *J. Chem. Phys., 38:* 1613(1963).

35. E. W. Müller et al., *J. Appl. Phys., 36:* 2496(1965).

36. E. W. Müller, *Science, 149:* 3684(1965).

37. E. W. Müller, *J. Appl. Phys., 28:* 6(1957).

38. E. W. Müller, *Z. Physik, 156:* 399 (1959).

39. E. W. Müller, *J. Phys. Soc. (Japan), 18 (suppl.2):* 1(1963).

40. J. F. Mulson and E. W. Müller, *J. Chem. Phys., 38:* 2615(1963).

41. L. E. Murr, O. T. Inal, and A. A. Morales, *Appl. Phys. Letters, 28:* 432(1976).

42. L. E. Murr, O. T. Inal, and A. A. Morales, *Acta Met., 24:* 261(1976).

43. E. W. Müller, *Proc. Fourth Intern. Symp. Reactivity Solids (Amsterdam), 0:* 682(1961).

44. D. G. Brandon et al., *Acta Met., 12:* 813(1964).

45. T. T. Tsong and E. W. Müller, *J. Appl. Phys., 38:* 545(1967).

46. K. D. Rendulic and E. W. Müller, *J. Appl. Phys., 37:* 2593 (1966).

47. E. W. Müller, W. T. Pimbley, and J. F. Mulson, in G. M. Rassweiter and W. L. Grube (eds.), *Internal Stresses and Fatigue in Metals,* p. 189, Elsevier Publishing Company, Amsterdam, 1959.

48. O. T. Inal and L. E. Murr, *J. Appl. Phys., 49:* 2427(1978).

49. L. E. Murr, O. T. Inal, and H. P. Singh, *Thin Solid Films, 9:* 241 (1972).

50. L. E. Murr, O. T. Inal, and S-H. Wang, *Mater. Sci. Engr., 49:* 57 (1981).

51. R. C. Sanwald, S. Ranganathan, and J. J. Hren, *Appl. Phys. Letters, 9:* 393(1966).

52. R. C. Sanwald and J. J. Hren, *Surface Sci., 7:* 197(1967).

53. A. J. W. Moore and S. Ranganathan, *Phil. Mag., 16:* 723(1967).

54. A. J. W. Moore, *Phil. Mag., 16:* 739(1967).

55. A. J. W. Moore, *J. Chem. Phys. Solids, 23:* 907(1962).

56. L. E. Murr and O. T. Inal, *Phys. Stat. Sol. (a), 4:* 159(1971).

57. O. T. Inal and L. E. Murr, *Phys. Stat. Sol. (a), 23:* K1(1974).

58. R. D. Young, *Physics Today, 24:* 42(1971).

59. R. D. Young, J. Ward, and F. Scive., *Rev. Sci. Instrum., 43:* 999(1972).

60. G. Binnig, H. Rohrer, Ch. Gerber, and E. Weibel, *Phys. Rev. Lett., 49:* 57(1982).

61. B. D. Josephson, *Phys. Lett., 1:* 251(1962); see also P. W. Anderson, "How Josephson Discovered His Effect", *Phys. Today, 23:* 23(1970).

62. G. Binnig, H. Rohrer, Ch. Gerber, and E. Weibel, *Phys. Rev. Lett., 50:* 120(1983).

63. B. Drake, R. Sonnenfeld, J. Schneir, P. K. Hansma, G. Slough, and R. V. Coleman, *Rev. Sci. Instrum., 57:* 441(1986).

64. R. E. Thomson, U. Walter, E. Ganz, J. Clarke, and A. Zettl, *Phys. Rev. B., 38*(15): 10734(1988).

65. J. Tersoff and D. R. Hamann, *Phys. Rev. Lett., 50:* 1998(1983).

66. R. S. Becker, J. A. Golouchenko, and B. Swartzentruber, *Phys. Rev. Lett., 54:* 2678(1985).

67. Y. Kuk and P. Silverman, *Rev. Sci. Instrum., 60:* 165(1989).

68. T. Sakurai, et al., *J. Vac. Sci. Technol., A7* (1989).

69. T. Sakurai, A. Sakai, and H. W. Pickering, *Atom-Probe Field Ion Microscopy and Its Applications,* Advances in Electronics and Electron Physics Suppl. 20, Academic Press, N.Y., 1989, Ch. 7-8.

70. G. Binnig, C. Gerber, E. Stoll, T. R. Albrecht, and C. F. Quate, *Surface Sci., 189-190:* 1(1987).

71. G. K. Binnig, *Phys. Ser., T19A:* 53(1987).

72. S. Alexander, et al., *J. Appl. Phys., 65*(1): 164(1989).

73. C. F. Quate, *Physics Today, 39(8):* 26(1986).

74. D. M. Eigler and E. K. Schweizer, *Nature, 344:* 524(1990).

75. N. Garcia, et al., *Metrologia, 21:* 135(1985).

76. R. Ringger, et al., *Appl. Phys. Lett., 46:* 832(1985).

77. G. Gruner and A. Zettl, *Phys. Rep., 117:* 119(1985).

78. E. Ganz, K. Sattler, and J. Clarke, *Phys. Rev. Lett., 60*(18): 1856 (1988).

79. S. Hosaka, S. Hosaki, K. Takata, K. Horiuchi, and N. Natsuaki, *Appl. Phys. Lett., 53:* 487(1988).

SUGGESTED SUPPLEMENTARY READING

Bruche, E., Electron-Optical Images Obtained with Photoelectrons, *z. Physik, 86:* 448-450(1933) (an interesting, historical treatment).

Byrne, J. G., *Recovery, Recrystallization and Grain Growth,* The Macmillan Company, New York, 1964

Gomer, R., *Field Emission and Field Ionization,* Harvard University Press, Cambridge, Mass., 1961.

Good, R. H., and E. W. Müller, Field Emission Microscopy, in S. Flugge (eds), *Handbuch der Physik,* vol. 21, pp. 105-131, Springer-Verlag OHG, Berlin, 1956.

Grube, W. L., and S. R. Rouze, Thermionic Emission Microscopy - Applications, in *High-Temperature-High-Resolution Metallography,* Gordon and Breach, Science Publishers, Inc., New York, 1967.

Hren, J. J., and S. Ranganathan, *Field-Ion Microscopy,* Plenum Press, Plenum Publishing Corporation, New York 1968.

Johari, O., and G. Thomas, *The Stereographic Projection and Applications,* J. Wiley and Sons, Inc., New York, 1969.

Modinus, A., *Field, Thermionic, and Electron Emission, Spectroscopy,* Plenum Publishing Corporation, New York, 1984.

Müller, E. W., and T. T. Tsong, *Field Ion Microscopy,* Elsevier Publishing Company, Amsterdam, Distributed by American Elsevier Publishing Company of New York, 1969.

Murr, L. E., *Interfacial Phenomena in Metals and Alloys,* Addison-Wesley Publishing Co., Reading, Massachusetts, 1975. (Reprinted by Tech Books, Herndon, Virginia, (1990)).

Sakurai, T., A. Sakai, and H. W. Pickering, *Atom-Probe Field Ion Microscopy and Its Applications,* Advances in Electronics and Electron Physics, Supplement 20, Chap. 7-B, Academic Press, New York, 1989.

Wagner, R., *Field-Ion Microscopy in Materials Science,* Springer-Verlag, New York, 1982.

The reader should peruse any current issue of *Surface Science* for examples of contemporary FEM, FIM and STM applications, etc. See also: G. Erlich, *Physics Today: 36(6),* 44(1981), and C. F. Quate, Vacuum Tunneling: A New Technique for Microscopy, *Physics Today, 39(8):* 26(1986).

3
ELECTRON AND ION OPTICS
AND OPTICAL SYSTEMS

3.1 INTRODUCTION

In the previous chapter, you have been exposed to various fundamental aspects of electron and ion optics and electron and ion optical systems. For the most part, however, the exposure has neglected the details of the optical system and more attention has been paid to the implications of basic emission processes and the utilization of such processes in the solution of materials science problems amenable to the use of emission microscopy techniques. We are now interested in extending these techniques to the more conventional electron and ion optical systems, particularly those involving some form, or deviation from, the electron microscope, and electron and ion analyzing systems.

Our approach in this chapter will be directed first toward the acquisition of a general background in the basic theoretical concepts underlying the design of electron and ion lenses; and second, toward the development of some knowledge regarding lens aberrations and the resolution limitations of electron optical systems. The theoretical treatment will, perhaps of necessity, be nonrigorous; it may simply serve as a review for the reader who has already had some formal exposure to electron and ion optics.

However, some attention will be given to the development of the aberration formulas for magnetic-lens systems, as well as an introductory treatment of instrument resolution. (The reader who by nature or necessity feels a need to supplement this treatment with a rigorous (mathematical) development of electron or ion optics is referred to appropriate text suggestions in this chapter.) In a sense, resolution will often form the basis for the acceptance or rejection of an electron optical system or analytical technique as it applies to the solution of a specific problem in materials science or engineering.

3.2 FUNDAMENTALS OF ELECTRON AND ION OPTICS

E. Bruche and O. Scherzer [1] were among the first to apply Hamilton's analogy between light optics and dynamics to electrons, where it turns out that the paths of electrons in an electric or magnetic field are identical to the ray paths we normally associate with light, where glass lenses are the refractive medium. In the case of an electric or magnetic field, however, the refractive index at any point depends on the corresponding field strengths. We shall initially consider the case of an electric potential difference as shown in Fig. 3.1. An electron or ion beam passing from a region of low potential V_1 to higher potential V_2 is, on acceleration, observed to undergo refraction as defined by Snell's law in the form

$$\frac{\sin r}{\sin i} = \sqrt{\frac{V_1}{V_2}} \tag{3.1}$$

The electron beam on passing through a region of potential difference with this condition reversed, that is, $V_1 > V_2$ as shown in Fig. 3.1, experiences a retardation, making the angle of refraction greater than the angle of incidence. Where i is very large, the beam undergoes reflection as expressed by

$$\text{Refraction:} \quad \frac{\sin r}{\sin i} = \sqrt{\frac{V_1}{V_2}} \quad i < \sin^{-1}\sqrt{\frac{V_1}{V_2}}$$

$$\text{Reflection:} \quad r' = i \quad i > \sin^{-1}\sqrt{\frac{V_1}{V_2}} \tag{3.2}$$

where r' is the angle of reflection from the plane of the potential zone shown in Fig. 3.1. An electrostatic lens for $V_1 < V_2$ is thus observed to

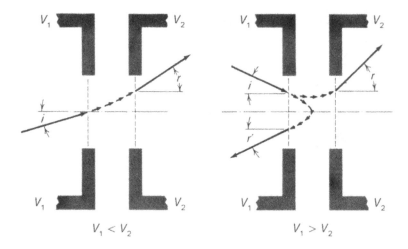

$V_1 < V_2$ $V_1 > V_2$

FIG. 3.1 *Refraction and reflection of electron or ion beam on encountering region of potential difference.*

act in an identical fashion to glass lenses with respect to the focusing action on a divergent electron or ion beam (illustrated in Fig. 3.2).

The deflection of an electron or ion beam in a uniform magnetic field has been treated somewhat generally in Chap. 1 along with uniform electrostatic deflection. In Sec. 1.2, on comparing Eqs. (1.1) and (1.2), it was noted that one distinguishing feature of the action of a magnetic field on electrons is that any deflection the electron experiences is proportional to its charge and mass. And, as noted in Sec. 1.2, a magnetic field exerts a force on a moving electron in a direction normal to both the field and the propagation direction of the electron (that is, the beam axis). This also applies for an ion beam. Thus a magnetic field acting in a direction parallel to an electron or ion beam will not affect it, while a field normal to the beam will cause it to describe a circle with a radius given by

$$r_o = \frac{1}{B} \sqrt{\frac{2mV_o}{e}} = \frac{1}{B} \sqrt{\frac{2M_iV_o}{e}}$$

r_o is in centimeters for V_o, the accelerating potential, in volts; and B is the magnetic field strength in gauss. In effect, an electron or ion in a uniform magnetic field will describe a helical path, with the radial extent limited by r_o.

An electromagnetic lens, similar in principle to that shown for the electrostatic case in Fig. 3.2 can, as a consequence of the simple feature outlined above, be fashioned so that a flux distribution identical to that

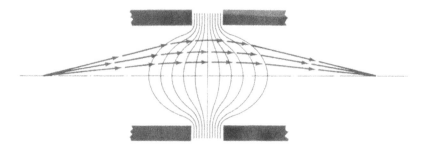

FIG. 3.2 *Cylindrical electrostatic lens action. The lens lines shown are electrostatic action lines which are always perpendicular to the electric field (or flux) lines.*

established in the potential regions of Fig. 3.1 is created. Figure 3.3 illustrates this scheme, which involves the creation of a short electromagnetic electron lens by enclosing a wire coil with a slotted soft-iron shell.

It is perhaps unnecessary to point out, with reference to Figs. 3.2 and 3.3, that the action of these simple electric or magnetic lenses is to converge a divergent stream of electrons or ions from a point of source onto another point on the central beam axis. In other words, electrons or ions traveling through such field or flux distributions obey all the fundamental laws of classical geometrical optics. We might now look in a little more detail at the basic nature of ion or electron lenses, with a view toward understanding their ultimate incorporation into an optical system.

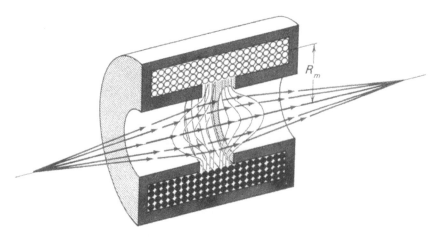

FIG. 3.3 *Cylindrical-type electromagnetic lens action (cut-away section). The lens lines shown are action lines which are perpendicular to the magnetic field (or flux) lines. R_m is the mean lens radius.*

3.2.1 THEORY OF ELECTROSTATIC LENSES

Electrostatic lenses were essentially the first types of electron lenses to be employed in an electron optical system (for example, the early thermionic emission microscope illustrated schematically in Fig. 2.6). Their usefulness in preference to magnetic lenses has, however, declined almost with the inception and development of electromagnetic lenses; today very few electron optical analytical instruments utilize electrostatic lenses as a major portion of the beam or image-forming system. While electron beams are usually focused with magnetic lenses, ion beams must be focused with electrostatic lenses because magnetic field focusing is mass independent and the magnetic fields attainable in rotationally symmetric magnetic lenses are orders of magnitude too low for effective ion beam focusing [Eq. (1.51)].

There are two electrostatic lenses of general interest. These include the so-called cathode lens and the unipotential lens. The cathode lens is primarily the basis of the image-forming system of emission schemes such as that illustrated in Fig. 2.6. This lens functions with the cathode serving as one electrode of the lens, with the subsequent aperature system at opposite polarity. Figure 3.4 illustrates two configurations of axially symmetric apertures that produce a fixed focal length and variable electron focal length. The beam-forming system of any electron or ion optical system employs, to a certain degree, some basic features of the cathode lens. Where special electron gun assemblies are concerned, certain design features involve the direct functions of the beam-focusing schemes illustrated in Fig. 3.4

The electrostatic unipotential electron lens is perhaps the most useful for incorporation into a general electron or ion optical system since it is essentially analogous in function to a single converging glass lens in a light-optical system. By this we mean that in the unipotential lens, the image and object regions of the lens are at the same potential, with the consequence that the refractive index is constant. In addition, the object-side focal length f_o, and the image-side focal length f_i, are equal as illustrated in Fig. 3.5. Where the focal length f is related to the object-image geometry (as shown in Figure 3.4) in the form (classical thin-lens notation)

$$\frac{1}{f} = \frac{1}{p} + \frac{1}{q}$$

(3.3)

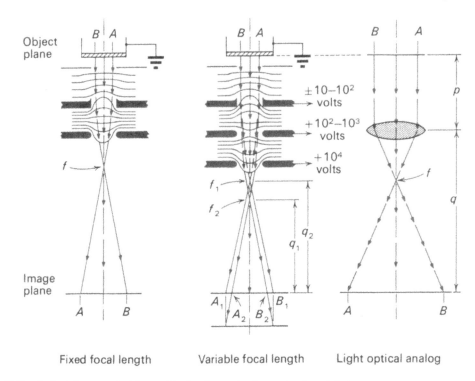

FIG. 3.4 *Electrostatic cathode-lens designs (schematic).*

this property of refractive power for the unipotential lens is expressed
by approximately

$$\frac{1}{f} = \frac{3}{16} \int_{z_o}^{z_i} \left(\frac{V_c}{V_o}\right)^2 dz \qquad (3.4)$$

Thus the unipotential lens strength depends on the ratio of the potentials
applied to the electrodes as shown in Fig. 3.5.

It can now be observed directly from Eq. (3.4) that in electron optical systems where the electron velocity is very large, the necessary lens
strength to focus such a powerful beam is only attained by the application
of large potentials to the lens electrodes. This feature therefore presents an obvious design problem from the standpoint of lens stability and
breakdown. In ion optical systems, however, the beam is not accelerated
at a high potential, and the field strength required for particular lens

---

*For a detailed treatment see Gertrude F. Rempfer, "Unipotential Electro-
static Lenses: Paraxial Properties And Observations Of Focal Length And
Focal Point", J. Appl. Phys., 57(1), 2385(1985).

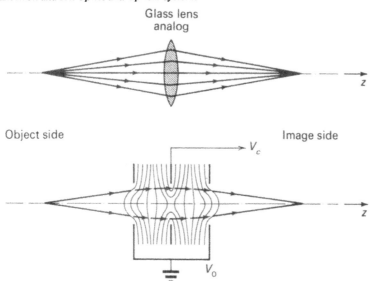

FIG. 3.5 *Schematic representation of electrostatic unipotential lens.*

action is orders of magnitude lower for ions than for electrons [compare Eqs. (1.2) and (1.51)]. In addition, electrostatic lenses do not suffer hysterisis effects or impart beam or image rotation, and as a consequence, electrostatic imaging systems have certain stability advantages over electromagnetic imaging systems.

3.2.2 THEORY OF ELECTROMAGNETIC LENSES

While electrostatic lenses vary considerably in mode of operation with simple design alterations, they are fundamentally the same regardless of such modifications. In effect, all electromagnetic lenses, where they are in fact lenses, are analogous to the unipotential electrostatic lens, which is fundamentally analogous to a glass converging lens in a light-optical system. In this sense, the field strength on the lens axis (the central electron beam axis) varies from zero on the object side to some maximum and back to zero again on the image side of the lens. Figure 3.3 schematically illustrates the conventional electromagnetic lens features.

Since, as in the case of the electrostatic unipotential lens, the object and image focal lengths of the electromagnetic lens are identical, the refractive power of the electromagnetic lens is given by

$$\frac{1}{f} = \frac{0.022}{V_o} \int_{z_o}^{z_i} H^2 \, dz \qquad (3.5)$$

where V_0 is the potential through which the electrons or ions converging on the lens have been accelerated and H is the magnetic field strength on the z axis in gauss. The field strength is related to the physical design of the lens coil by

$$\frac{4\pi NI}{10} = \int_{z_0 = -\infty}^{z_i = \infty} H \, dz \qquad (3.6)$$

from which we can observe that the lens power [Eq. (3.5)] is proportional not only to the number of turns (N) of conductor, and the current flow (I), but also to the extent of the field region. We desire that the lens be de-signed so that the field can be concentrated into as small a region as fea-sible, that is, we desire a thin-lens approximation. This is achieved by providing the lens coil system with an annularly slotted high-susceptibi-lity encasing (Fig. 3.3), such as soft iron, employing a central high-sus-ceptibility pole piece; with the ultimate goal of achieving a concentration of the magnetic field in a short region (small volume) along the axis.

A unique feature of magnetic lenses, not encountered in electrostatic or glass-covering analogs, is that the image produced by focusing an elec-tron or ion beam through an electromagnetic lens is not only normally in-verted but also rotated, with respect to some object-reference. This occurs because the rotationally symmetric magnetic flux lines act on the beam as it passes through the thin lens which causes an effective rotation related to the magnetic field strength as originally shown in Eq. (1.1). The important implications of this feature, while not immediately apparent, are considerable in the operation of electron microscopes of the transmis-sion type, especially where on going from the transmission mode to the se-lected-area diffraction mode, a portion of the lens system is deleted or its strength significantly altered so as to cause a rotation of the select-ed-area electron diffraction pattern with respect to the corresponding image. While this feature will be discussed with regard to transmission electron microscope image interpretation in Chap. 6, we shall very briefly deal with the phenomenology at this point.

Let us "focus" our attention onto the simple double-lens system de-picted schematically in Fig. 3.6a. Here the final image undergoes a 360° rotation (double inversion) in addition to the lens rotation. If the cor-responding intrinsic lens-rotation angles are denoted $\Phi_1 + \Phi_2$, as shown, then the image rotation with respect to the object is $\Phi_1 + \Phi_2$ for rotation of the same sense or $\Phi_1 - \Phi_2$ for rotations of opposite sense (clockwise or

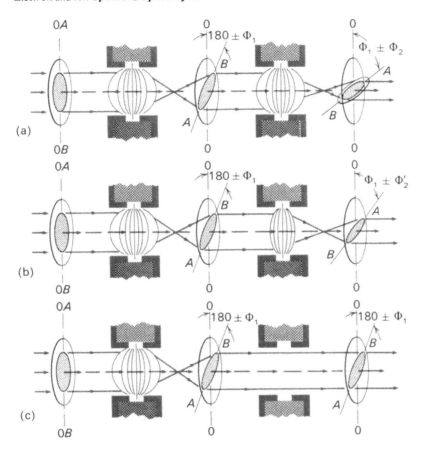

FIG. 3.6 *Image rotation in simple electromagnetic electron lens system.*

counter-clockwise). For a reduction in the strength of lens 2 of Fig. 3.6b
relative to Fig. 3.6a, we can denote the corresponding image rotation as
Φ_2', where

$$\Phi_1 + \Phi_2 > \Phi_1 + \Phi_2'$$

provided the sense of rotation is the same in lenses 1 and 2. Correspondingly, if the strength of lens 2 is reduced to a zero level, the image rotation is ultimately influenced by the single inversion so that the absolute image rotation becomes $180° \pm \Phi_1$ depending on the sense of the intrinsic lens rotation, Φ_1. This situation is depicted in Fig. 3.6c.

Since the intrinsic electromagnetic lens rotation (Φ) experienced by
the electron beam depends on the field strength, it is perhaps obvious,

with regard to Eq. (3.6), that the relative image-rotation angle as depicted in Fig. 3.6 will be proportional to the individual lens currents. This
feature leads to a direct means for calibrating image rotation in electron
microscopes (discussed in Chap. 6).

3.2.3 ELECTRON MIRRORS

As illustrated in Fig. 3.1, electrons entering a potential zone within
axially symmetric electrodes can be reflected for conditions expressed
ideally in Eq. (3.2). A simple axially symmetric configuration of electrodes can be fashioned into the analog of a convex or concave light mirror
by making one electrode so negative that a zero-equipotential surface
crosses the optic axis. Electrons incident on the axis are therefore
slowed down as they penetrate the field lines. As they reach an effective
zero equipotential, they lose their incident velocity and reverse their direction of motion. Conversely, as depicted in Fig. 3.7, electrons entering
the mirror field at some angle to the optic axis retain a portion of their
tangential velocity and describe a parabolic reflection path; the reversal
is not limited to a single reference surface as in the case of a common
glass mirror.

   You are no doubt already prompted to question the possibility of producing mirror action from virtually any electrostatic lens arrangement,
since the only real criterion is that within the potential field a region
exists corresponding to a negative potential with respect to a zero reference. And of course this condition has already, in perhaps a subtle way,
been expressed in Eq. (3.2). Yet, while mirror analogs do exist for these
electrostatic configurations, the behavior of electron mirrors is considerably more complex than the light-optical analog because the reflection surface is not a single "flat" surface for incident electron "rays" at some
angle. θ relative to the mirror field. As a consequence, the conventional
ray equations are not applicable in tracing a paraxial ray through an electron mirror. The virtual image formed by the mirrored electrons from some
object plane will also, as a consequence of the variation in the incident
angle θ (Fig. 3.7), not necessarily be sharp since electrons passing a
point on the object plane or a virtual object plane at an angle θ relative
to the electrostatic mirror field are returned to this reference plane via
the parabolic path described. If the point of reversal of the incident
ray is measured at a distance D_m from this plane, and the time required
for the electron passing a point on the plane to reverse direction and

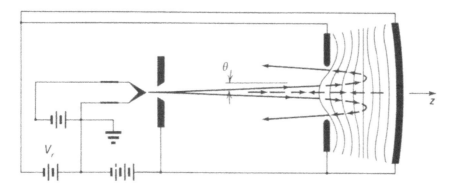

FIG. 3.7 *Electron mirror (schematic). The action lines shown are perpendicular to the electric field lines.*

and return to a displaced point on the same reference plane is denoted simply as 2t, then we can write

$$D_m = t \cot \theta \sqrt{\frac{eV_r}{2m}} \qquad (3.7)$$

where V_r is the retarding potential, or the electron velocity at the effective object plane. Thus, while the object plane can exist, in the conventional sense, at some distance from the mirror surface, it can also assume the virtual image plane of the biased electrode. This scheme is actually employed in the imaging of surface potential distributions by making the material of interest serve as the "mirror" electrode. This feature of electron mirror microscopy will be taken up later in Sec. 5.6. In this particular mode of operation, Eq. (3.7) effectively describes the displacement of the reflection field surface from the specimen surface serving as the biased electrode of the electron mirror.

The treatment of electron mirror theory in some depth was established as early as 1935 in the work of F. H. Nicoll [2] and A. Recknagel [3]. Many developments such as the calculation of the electron mirror refractive power are found in the classic treatment of electron optics by Zworykin et al [4]. This reference is especially recommended to the reader who has a need for or interest in the more subtle details of electron lens and electron mirror theory, especially where mathematical rigor is desired.

While we have not included ions in this treatment, you should appreciate the difficulties encountered as a result of the ion mass, which is apparent on substituting M_i for m in Eq. (3.7). This system would suffer the same problem as electrostatic lenses in electron lens systems, electrical breakdown because of the high-voltage requirements.

3.3 THE ELECTRON OPTICAL SYSTEM

We have been progressively exposed to electron optical systems throughout Chap. 2 and Sec. 3.2. What we wish to describe now is in effect an electron optical "bench" arrangement, that is, an arrangement whereby the electron beam is generated and then focused, and the focused beam is used in the formation of an image, or in the excitation of secondary radiations from a specimen material. We shall be particularly concerned here not only with the gross features of this electron optical arrangement but also with the more subtle operation of the composing lenses, and the intrinsic limitations and related phenomena associated with the individual lens functions.

We shall begin with an electron optical system analogous to the light-optical microscope or related system. Consequently we shall initially deal with the basic features of the electron microscope, outlining the various lens actions. Deviations from this basic column design in the fabrication of related electron optical instrumentation for specific analytical purposes will then be briefly elaborated on if these applications are specifically treated in the following chapters.

3.3.1 THE ELECTRON MICROSCOPE

Because electrostatic lenses are not utilized in most contemporary electron optical devices, they will not be treated further except for the special case of the electron mirror microscope (see Sec. 5.6). For the most part, contemporary electron microscope designs employ magnetic lenses exclusively; and, as pointed out in Sec. 3.2, the electromagnetic lens action is analogous to that of a glass light-optical lens. As a consequence, the obvious design of an electron microscope is effectively analogous to that of a light microscope as depicted schematically in Fig. 3.8. The arrangement and ray paths in these systems is identical.

The electron microscope consists initially of a source of electrons, or electron gun. In conventional instruments, the gun design may vary considerably, but for the most part it consists of using either a heated tungsten filament roughly 0.1 mm in diameter, LaB_6, or a field-emission source. The important consideration in electron gun design is the attainment of a maximum electron emission over a concentrated area, or the

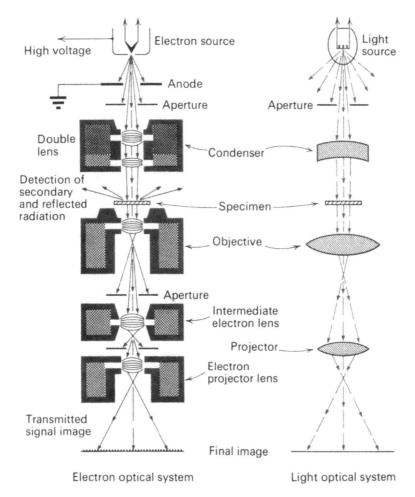

FIG. 3.8 *Schematic comparison of basic light-optical and electron optical microscope systems.*

concentration of emitted electrons into an intense initial beam. The electron gun, as an entity, is composed of the filament or cathode and the subsequent accelerating and beam-forming electrodes. The thermionic and field emission characteristics of the filament or high-field source are effectively described by the same processes discussed in Chap. 2; it is these emission properties that render these designs most suitable in electron optical systems. The inference here is that electrons emitted thermionically or by field emission processes possess much greater velocity stability, thereby producing a greater degree of homogeneity of the velocity

profile in the initial electron beam. This feature in turn results in
trivial disturbances in the final image or terminal signal.

Figure 3.9 illustrates the relative appearance of the more tradition-
al thermionic electron gun designs, which can change, depending on specific applications, in the placement and geometry of the grid cap and anode.
The filament is usually operated about 10^2 to 10^3 volts less negative than
the grid cap, with the anode at ground potential. This arrangement improves the stability of the emission stream and, because of the bias, aids
in the concentration of the electron beam. In effect, an electrostatic
lensing action is established in the gun section as shown in Fig. 3.9; and
the specific mode of action can, necessarily, be altered by a change in
geometry, etc., as indicated previously. Field emission sources offer
higher brightnesses which are especially important in certain electron
"probe" applications. Modern source designs involve LaB_6 high-brightness
thermionic emission sources and tungsten field emission sources. These
features will be discussed in later sections.

The condenser lens The condenser lens simply serves to regulate the intensity of the electron beam in the optical system. In the specific mode
of electron microscope operation, the condenser-lens system, which may
effectively contain a double-lens configuration - and does in conventional
electron microscopes - also serves to converge the beam onto the specimen
object of interest. In operation in the transmission mode, the condenser-
lens section of the electron microscope also contains a feature for beam

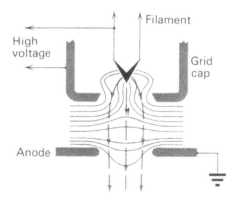

FIG. 3.9 *Basic thermionic emission*
electron gun design.

tilt to accommodate dark-field illumination of the object. A fine focus of the electron beam by the condenser lens, with a concomitant regulation of beam diameter using suitable apertures, is mostly responsible for contrast in the images obtained by specimen transmission in the electron microscope.

The lens action in the condenser and subsequent lenses shown in Fig. 3.8 is enhanced by the concentration of field lines by a highly magnetizable pole piece having a desirable slot arrangement and possessing a bore of several millimeters in diameter. Normally this pole piece is magnetically saturated, and the effective focal length is determined by an expression of the form

$$f_c = \frac{\zeta_c V_o}{N_c^2 I_c^2} \qquad (3.8)$$

where ζ_c = condenser-lens form factor (geometric parameter)

$\quad N_c$ = number turns of conductor in condenser coil

$\quad V_o$ = accelerating potential of electron beam in volts

$\quad I_c$ = condenser current in amperes

A condition of intensity crossover is attained by the focus of the condenser lens on the object (in the case of conventional or transmission electron microscopy) either by increasing or decreasing the lens current, thus causing a variation in the convergence angle of illumination at the center of the object (specimen). This condition obtains, according to Eq. (3.8), by a concomitant change in focal length with lens current (see Prob. 3.2). Consequently a variation in object intensity occurs on one side or another of crossover. Figure 3.10 illustrates graphically the illuminating-intensity on the object with condenser focal length.

The objective lens The objective lens in the electron optical system - particularly the transmission electron microscope mode - performs the same function associated with a glass objective lens in a light-optical system, namely, the focus of object detail. The electromagnetic electron objective lens, like the analogous light lens, performs the most important function insofar as the imaging and resolution of object detail is concerned: focusing the electron beam to a final area of resolution as in the secondary emission or reflection modes of operation of the electron optical system (see Chap. 4). The objective electron lens differs from preceding condenser lenses and subsequent intermediate-projector lenses primarily in terms of the more constricted field parameters necessitated by

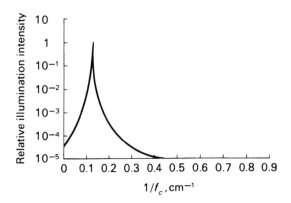

FIG. 3.10 *Illumination-intensity variation in object with condenser-lens focal length.*

a greater definition (shortening) of focal length through a concentration of magnetic field strength on the axis of the system. In this respect, the focal length is defined in an equation of the same form as Eq. (3.8), that is,

$$f_{ob} = \frac{\zeta_{ob} V_o}{(NI)_{ob}^{2}}$$

(3.9)

where ζ_{ob} = objective-lens form factor

   N = number of turns in lens coil

   I = objective-lens current

   V_o = accelerating potential

In order to obtain a greater magnetic field concentration, a more constrictive pole-piece geometry is combined with greater fabrication precision in the objective-lens design. Thus the objective lens is sometimes larger and more heavily encased with a high-permeability magnetic conductor than are the other component lenses of the electron optical system. Figure 3.11 illustrates schematically the usual design features of an objective lens.

It is obvious from Eq. (3.9) that in order to satisfy the demand that f_{ob}, the objective-lens focal length, be small for a high-performance objective function, $(NI)_{ob}^{2}$ must be large, especially where V_o is very large, thereby necessitating the physical enlargement of this lens. With the advances made recently in the design of superconductive electromagnetic lens systems [5,6], it is not really necessary to overdesign in terms of large

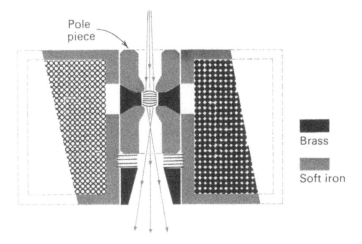

FIG. 3.11 *Objective-lens design (schematic).*

N or mass of coil casing. However, such lens systems suffer the necessity
to supercool, usually by immersion of the entire system in liquid He or hydrogen. While such superconductive lens systems are at present mostly
experimental, they do offer numerous advantages in spite of the supercooling problem. Among these is the greater local vacuum attained by the self-
pumping action of the inner column walls, and the reduction in intrinsic
contamination (discussed in greater detail in Chap. 7). Supercooling also
limits the problem of energy (mainly thermal) dissipation, since the power
dissipated as heat loss in the lens is given ideally by

$$W = \frac{2\pi (NI)^2 R_m \rho}{A} \tag{3.10}$$

where R_m = mean lens-coil radius (see Fig. 3.3)

ρ = coil-winding resistivity

A = mean cross-sectional area of lens coil (shown in Fig. 3.11)

NI = product of lens-coil turns and lens current

Intermediate-projector lenses The intermediate lens functions in most conventional electron microscopes as the magnification control, and final
image magnification is normally calibrated directly from the intermediate-
lens current or some normalization of this current (treated in more detail
in Chap. 5). It should be remembered also that total image magnification
is determined by the product of objective and intermediate-projector magnifications (see Prob. 3.1), that is, $M = M_{ob}M_I M_p$ (where M_{ob} is usually

fixed for the focus of the object detail since the object distance in the objective lens remains fixed by the geometry and spacing of the specimen holder with respect to the objective pole piece).

3.3.2 *LENS ABERRATIONS AND OBJECT RESOLUTION*

We have assumed in the foregoing discussion that the image formed (or the ultimate resolution of the object signal) was dependent only on the final magnification since the paraxial electron rays were assumed constant and energetically homogeneous; the focal length of the lens was considered descriptive of crossover and image sharpness. These assumptions in a real electron optical system are not justified since not only are the lens fields somewhat imperfectly formed by mechanical flaws in the lens design and pole-piece fabrication, but the mutual repulsion of electrons at constricted points (for example, lens apertures, etc.) along the optical axis, particularly focal points, and the variation in electron energy at various points in the beam give rise to image distortions and contribute generally to loss of contrast and sharpness.

The lens aberrations primarily responsible for deviations in electron ray intersections and concomitant loss in image clarity may be classified as geometrical aberrations, chromatic aberrations, and field-effect aberrations including the space charge of the electrons. For the most part, geometrical aberrations are all defects that cause deviations in the axial imaging electron beam on an ideal image reference surface - the gaussian image plane. These aberrations can be regarded simply as a function of the magnetic refracting field geometry of the particular electromagnetic lens in question, and in an idealistic sense they can be computed from solutions of the paraxial electron ray equation (the equation of motion of an electron in an electron optical or electromagnetic lens system). The total image aberration is then simply defined as the deviation of the actual image from the ideal or gaussian image, and is assigned a magnitude based on the radius vector of the actual intersection point of the electron ray from the ideal image intersection point on the gaussian image plane. Consequently the total aberration is simply the vector sum of the individual aberrations; and where resolution is concerned, the disk of least confusion becomes the total aberration, dependent to a large degree on the half-angle subtended from the image by the lens aperture and designated α in Fig. 3.12.

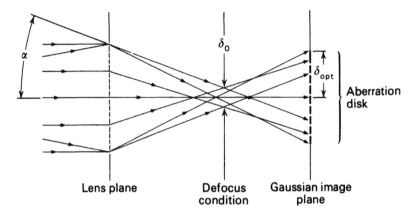

FIG. 3.12 *Aberration disk and disk of least confusion,* δ_o, *associated with electron lens.*

Coma[†] In the consideration of focal length for an electromagnetic lens, we assumed, because we were only concerned momentarily with a single ray path, that the angle of incidence of the ray to the lens field was of no consequence. However, while the angles of incidence of component rays of an electron beam are not very different, the focal length is nonetheless changed for rays imaged radially from the central beam axis. Thus the image of a point off the axis appears cometlike; and rays from the object plane to some arbitrary aperture plane before the lens field form concentric circles that because of coma appear as eccentric circles in the gaussian image plane. Any mechanical misalignment of lenses in the electron optical system contributes to the severity of the coma aberration. However, coma can be eliminated almost entirely in an electromagnetic lens by the establishment of field conditions giving rise to unity magnification.

Curvature of field If the rays adjoining the principal beams from several object points off the electron optical axis converge on the object either ahead or behind the image plane, an effective curvature of field occurs. Quite naturally, giving the object itself an equivalent-compensating curvature can result in a complete reduction of the field-curvature aberration, while some compensation is attained more realistically by properly shaping the electromagnetic lens field.

[†]*Detailed derivations of numerical estimates of aberrations and aberration coefficients are to be found in chap. 16 of reference 4.*

Astigmatism Astigmatism occurs when a longer focal length exists in one plane direction than another normal direction in the same plane. In effect, a noncircular pole-piece bore or a noncircular beam-limiting aperture causes an elliptical distortion in the image plane on focusing the electron beam, which shifts direction by $\pi/2$ with a change in focus from one side of the lens crossover to another. This is especially troublesome in the condenser- and objective-lens systems, and is further complicated by the associated field-curvature aberration. This aberration is correctable by inserting stigmators in the appropriate lens systems to compensate for the noncircularity of the image-beam profile on the image plane. The stigmator, containing symmetrical arrangements of tiny ferromagnets or suitable permanently magnetized components, acts to circularize the image.

Lens distortions Lens-distortion observations result as a deviation in the geometrically correct image; and such distortion ideally varies with the cube of the distance of the object point from the electron optic axis. For an electromagnetic lens, a square image is therefore characterized by a barrel or pincushion distortion, with the distorted image also rotated about the image plane (Fig. 3.13d). Normally, the correction of coma in an electromagnetic lens concurrently eliminates the lens distortions as well.

Spherical aberration Because of the nature of electron lens-field geometries, and the coincidence of the component electron rays, electrons originating at different points in the object from the optic axis are focused at slightly different points along the optic axis immediately before the image plane. The effect (shown in Figs. 3.12 and 3.13e) is that a characteristic unsharpness results in the image on the optic axis, depending, to a large degree, on the aperture angle α. Consequently the aberration is also referred to as the aperture defect. Spherical aberration is by far the most important of the intrinsic lens aberrations from the standpoint of the materials scientist interested in analyzing a certain object feature, since it is this aberration that limits the resolution of an electron optical system such as the electron microscope or electron microprobe or related secondary emission instruments. And, since this aberration relates directly to resolution, we shall treat its numerical significance in more detail.

    Since the image-forming lens or the critical beam-forming lens in an electron microscope or microprobe system is the objective lens, the spher-

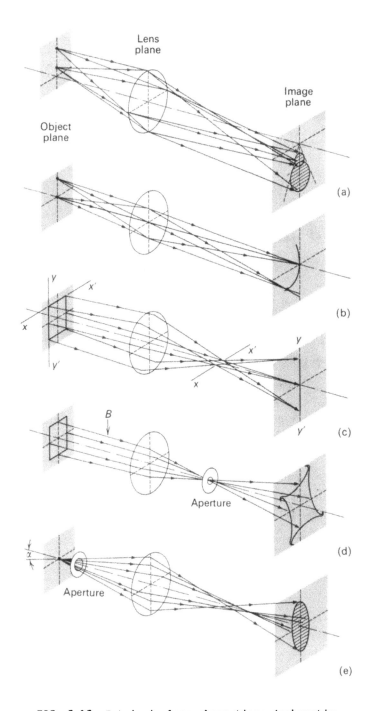

FIG. 3.13 *Intrinsic lens aberrations (schematic representations). (a) Coma. (b) Curvature of field. (c) Astigmatism. (d) Distortion (pincushion distortion with superposed rotation clockwise; barrel distortion with image bowed outward results in aperature at B). (e) Spherical aberration (aperature defect).*

ical aberration of this lens is quite naturally the "focal point" of our discussion. We can describe the disk of confusion caused by spherical aberration as

$$\delta_{Sp} = 2C_s \alpha^3 \qquad (3.11)$$

where C_s, the spherical aberration coefficient, is given by

$$C_s = \gamma_0 \left[\frac{V_o}{(NI)^2} \right]^2 \qquad (3.12)$$

which is observed to be proportional to the square of the focal length as given by Eq. (3.9), and α is the aperture angle (Fig. 3.12). We can there-fore write

$$C_s = \frac{\gamma_0 f_{ob} V_o}{(NI)^2} \qquad (3.13)$$

where γ_0 is a constant ranging from roughly 100 for strong lenses to 150 for weak lenses, respectively. Spherical aberration is therefore depen-dent on the changes in focal length with accelerating voltage and strongly dependent on the aperture angle α. The correction of spherical aberration thus rests in the design of lenses with special field distributions for allowing smaller aperture angles to be attained with the simultaneous re-duction in C_s, possibly by a design departure aimed at producing less sym-metrical lens fields. In making such compensations, other aberrations are often also simultaneously eliminated. Several contemporary attempts to overcome spherical aberration in electron lenses are enumerated in a re-cent article by A. Septier [7].

Chromatic aberration While the aberrations treated to this point have originated as an intrinsic function of the electron lens action, there are several additional electron optical aberrations that are, in a sense, in-herent features of the electron beam. Chromatic aberration as it pertains to the objective lens is a condition in the image that detracts from its contrast, and is due to velocity differences in the electron rays combined with fluctuations in the focal length due to current instabilities in the lens coils. The physical significance of the chromatic variation in beam focus is illustrated schematically in Fig. 3.14a. We observe here that for large variations in the focal length of say the objective lens f_{ob},

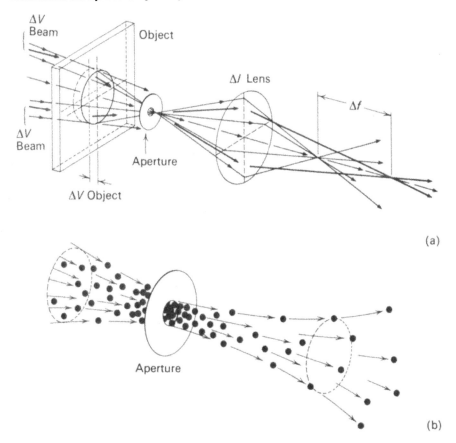

FIG. 3.14 *Chromatic image and space-charge aberrations. (a) Chromatic aberration. (b) Exaggerated space-charge effect.*

both the image position characterized by a change in the radial and rotational position and that characterized by magnification are distorted. Chromatic aberration, as it depends on voltage differences or electron energy loss ΔV, and lens-current fluctuations ΔI, is expressed by

$$\delta_{Cr} = 2C_c\alpha \sqrt{\left(\frac{\Delta V}{V_o}\right)^2 + \left(\frac{2\Delta I}{I}\right)^2} \qquad (3.14)$$

where C_c, the chromatic aberration coefficient associated with the objective lens, is given by

$$C_c = \gamma_o'f_{ob} \qquad (3.15)$$

where γ_o' is a constant varying from about 0.5 to 1.0 for strong- or weak-lens action, respectively.

It should be pointed out that the variation in the effective electron energy, ΔV of Eq. (3.14), need not necessarily occur primarily by instabilities in the high-voltage supply, but can result as a direct energy loss when electrons traverse a sufficiently thick specimen material before encountering the objective-lens system (see Prob. 4.12). Such energy losses, when detectably large, can be used as a measure of the object detail, for example, composition, etc. This energy loss can be detected as an energy spectrum and the technique, referred to as electron energy loss spectroscopy (EELS), will be treated in detail in Chap. 7. Obviously, the fluctuations occurring in the lens coils become simply a problem of electronic regulation, as do fluctuations in the cathode or anode potentials. To this extent, chromatic aberration is correctable. However, energy losses resulting from inelastic scattering in the object cannot be dealt with to the same extent, and this feature is only overcome by operation of the system at very high voltages, a topic discussed in Chap. 8. There is also a small additional energy spread in the thermionic source.

Space-charge effect   At the focal points along the electron optic axis, the concentration of electrons into small volumes produces a strong mutual repulsive action and a concomitant tendency of the beam to "spread" from the point of constriction (focus) as shown in Fig. 3.14b. This produces an effective reduction in the associated accelerating potential of the electrons, and they lose velocity. Consequently very intense beams generated in a moderate potential difference V_0 add to the overall electron energy loss and ultimately contribute to loss of clarity in the image due to measurable aberrations. This problem is somewhat less, again, at very high voltages and where lower beam currents are employed with an associated low beam intensity.

Object resolution and image clarity   Having now attained some basic feel, as it were, for the distortions experienced by an image in an electron optical system emphasizing the microscope mode, we might now look at the ultimate resolution that can be expected for the imaging of an object area in an electron optical system. We must realize at the outset that the resolution, defined as the minimum diameter of the disk of confusion prior to the gaussian image plane, possesses in addition to aberrations a limiting error due to the probabilitic nature of the electron as described in

Chap. 1. That is, as a consequence of the uncertainty principle as expressed in Eq. (1.19), the exact image of displacement of electrons diffracted from the object areas will be subject to an uncertainty of discrete (line) displacements in the object of

$$\Delta x \equiv \delta_{LL} = \frac{\lambda}{2 \sin \alpha} \qquad (3.16)$$

δ_{LL} here is simply the aperture diffraction limitation of Airy disk radius and represents the line-to-line resolution for planes of atoms in a crystalline specimen, with the atom planes parallel to the beam direction. This equation arises simply by replacing the Bragg angle, θ, in Eq. (1.9) with the aperture angle, α, and the crystal plane spacing with δ_{LL}. (See Appendix A for details of crystal structures). Equation (3.16) therefore stipulates that even for negligible lens aberrations, the minimum resolution attainable, simply as an inherent property of electrons, is wavelength dependent. If we assume an aperture angle on the order of 0.01 to 0.001 rad, Eq. (3.16) is written in approximate form as

$$\delta_{LL} = \frac{0.5\lambda}{\alpha} \qquad (3.17)$$

In an electron microscope where $\alpha = 0.003$, and for 100-kV operation, δ_{LL} becomes roughly 7 Å. However, improved objective lenses, especially in electron microscopes, now allow for considerably larger aperture angles; and this feature, coupled with modern high voltage acceleration, can allow ideally for resolutions near 1 Å.

Where lens aberrations are included in real electron optical systems, the ultimate resolution is given for the most part by considering, in addition to the diffraction uncertainty [Eq. (3.17)], chromatic and spherical aberrations. The combination of error disk radii in the image plane is then found from[*]

$$\delta_{\substack{opt \\ line}} = \sqrt{\delta_{LL}{}^2 + \left(\frac{\delta_{Sp}}{2}\right)^2 + \left(\frac{\delta_{Cr}}{2}\right)^2} \qquad (3.18)$$

---

[*]*Contemporary objective-lens designs in electron microscopes and microprobes effectively eliminate coma, field curvature, astigmatism, and image distortion; space-charge effects are also normally small in electron microscopy.*

Problems 3.6 to 3.8 illustrate the significance of the resolution limit for
the case of real electron optical systems.

If we consider the resolution limit of two points in the gaussian
image plane to be radiating incoherently with a Lambert distribution, as in
the case treated by O. Scherzer [8][*], the optimum image aperture is given by

$$\alpha = 1.414 \left(\frac{\lambda}{C_s} \right)^{1/4} \qquad (3.19)$$

and the resolution limit for these two points is

$$\delta_{PP} = \frac{0.61\lambda}{\alpha} \qquad (3.20)$$

where again we could include spherical and chromatic aberrations to obtain

$$\delta_{\substack{opt \\ point}} = \sqrt{\delta_{pp}{}^2 + \left(\frac{\delta_{Sp}}{2} \right)^2 + \left(\frac{\delta_{Cr}}{2} \right)^2} \qquad (3.21)$$

The conditions stipulated above are not really found in transmission electron microscopy because imaging is, in the limit, accomplished by contrast
of the points in question with the background. H. Niehrs [9] has made some
calculations related to contrast and resolution to be expected for imaging
individual atoms by transmission electron microscopy; he estimated a recognizable contrast to exist for carbon atoms separated a distance of 3 Å. As
stated previously, resolution can be experimentally attained that approaches
1 Å for some optimum accelerating voltage > 100 kV, since at very high
voltages, contrast reaches a plateau because of the relativistic mass effects.

Theoretical resolution limits δ_{opt} for an electromagnetic objective
lens with a spherical aberration have been predicted by E. Ruska [10] to be
accelerating-voltage dependent to the extent indicated in Fig. 3.15. It
therefore appears quite possible where contrast is not necessarily a limiting factor, to image individual atoms in a crystalline lattice. In this
respect resolution seems to be characterized quite generally by the same
notations, where in fact resolutions of atomic planes or lines are not necessarily those of distinguishing two image points. At the present time,
line-to-line resolutions (crystal plane spacings) on the order of a few

---

[*]A geometrical optics approach to resolution is given in a recent article
by G. F. Rempfer and M. S. Mauck, "A Closer Look At The Effects Of Lens
Observations And Object Size On The Intensity Distribution And Resolution
In Electron Optics", J. Appl. Phys., 63(7): 2187(1988).

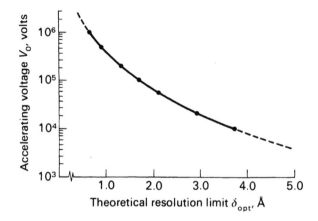

FIG. 3.15 *Theoretical resolution limit as function of accelerating potential in electron microscope. (Data from E. Ruska, "Fifth International Congress for Electron Microscopy", paper A-1, vol. 1, Academic Press, New York, 1962.*

angstroms or less have been routinely attained in commercial electron microscopes as illustrated in Fig. 3.16, while image resolution by the various secondary emission modes has been generally limited to several orders of magnitude greater. However, refinements in the electron optical systems of instruments such as the scanning electron microscope and scanning transmission electron microscope, for example, combined with a reduction in image distortion due to contamination, have led to resolutions less than 10 Å in commercially designed instruments and even less for specialized research tools.* Normally, as shown in Fig. 3.16, line resolutions resulting from lattice diffraction phenomena are about a factor 4 greater than point-to-point resolution.

The sharpness or clarity of an electron image may also depend to a large extent on the degree of specimen purity maintained during observation. Thus, to a considerable degree, the contrast loss resulting from the buildup of contamination on the object area may become a limiting factor in the object resolution. This particular phenomenon is increasingly overcome with the utilization of contemporary vacuum technology, and the incorporation of anticontamination features such as localized supercooling of the immediate specimen (object) area.

---

*In the scanning electron microscope, chromatic aberrations are negligible and only δ_{pp} and δ_{Sp} influence resolution in a manner identical to that for transmission electron microscopy.

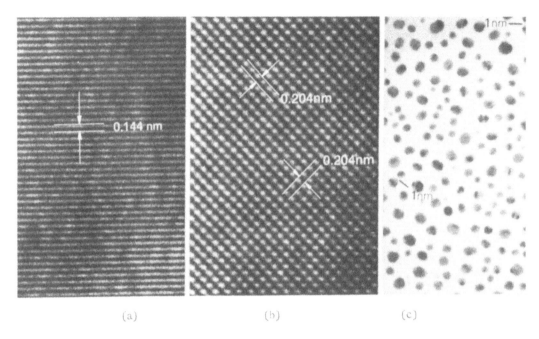

FIG. 3.16 *Line and point resolution examples in the transmission electron microscope. (a) 1.44 Å line resolution in gold film [(002) planes], (b) 2.04 Å resolution for (022) planes in gold film, (c) Gold particles with 10 Å recognizable separations. (Electron micrographs taken in the Hitachi H-600 electron microscope operated at 100 kV; Courtesy of Hitachi, Ltd, Tokyo, Japan).*

We have, in the discussions above, treated resolution on the gaussian image plane. But where we are concerned with ultimate resolution, or the disk of least confusion, we are in reality talking about the defocused condition (illustrated in Fig. 3.12) that lies just before the gaussian image plane. Perhaps it has already occurred to you that by simply reducing the objective-lens strength slightly in the case of image resolution in an electron microscope, the resolution should improve simply because the disk of error is smaller at the defocus plane than at the gaussian image plane. Similarly, when we wish to focus the electron beam onto a specimen using the objective lens (as in the case of an electron microprobe described in Chap. 4), the minimum constriction of the beam can be described as the disk of least confusion or defocus plane (Fig. 3.12). We find that in fact the diameter of the disk of least confusion is characterized by a slightly modified version of Eq. (3.21), which we shall write fully as

$$\delta_o = \sqrt{4\delta_{pp}^2 + \left(\frac{C_s \alpha^3}{2}\right)^2 + \left(C_c \alpha \frac{\Delta V}{V_o}\right)^2} \qquad (3.21a)$$

where δ_{pp} = 0.61 λ/α

$\quad\quad$ C_s = spherical aberration coefficient [Eq. (3.13)]

$\quad\quad$ C_c = chromatic aberration coefficient [Eq. (3.15)]

$\quad\quad$ ΔV = voltage change for accelerating potential V_o (including energy loss)

$\quad\quad$ α = objective aperture angle

assuming the lens current fluctuations are negligible, that is, that $\Delta I/I \simeq 0$. Equation (3.21a) again assumes that the electron density at the disk of least confusion is low enough so that space-charge effects can be ignored. This is not always true, and in most high-resolution work in electron microscopy the use of Eq. (3.21) or (3.21a) is contingent on the use of very low beam intensities. In the case of electron microprobes where sufficient intensity for detectable excitation must be focused onto the specimen, the current density is also a limiting factor that contributes to "beam resolution"; we shall treat this correction in Chap. 4, along with special gun design problems.

The implications of Eq. (3.21a) are simply that for optimum resolution, say of atomic detail, it is necessary to work in an out-of-focus condition in electron microscopy. This technique, the embodiment of phase-contrast microscopy, will be discussed in more detail in Chap. 7 with its appropriate association with transmission electron microscopy, and high-resolution electron microscopy.

Depth of focus of the image in electron microscopes The fact that electron optical systems are mostly characterized by small aperture angles leads to a decisive advantage where image focus is concerned. This feature, as illustrated in Fig. 3.17, for electron transmission, results in a large depth of focus (field) since a sharp image occurs within a finite distance above and below the object plane as given by

$$D_f = \frac{\delta_f}{\tan\alpha} \tag{3.22}$$

where for α (the aperture angle) small, and δ_f the most effective electron beam-spot size

$$D_f \simeq \frac{\delta_f}{\alpha} \tag{3.23}$$

and this dimension is maintained in the final image plane or viewing screen.

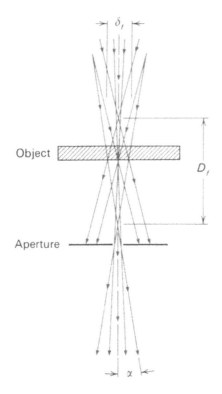

FIG. 3.17 *Depth of image focus*

Since this feature is an associated feature, as it were, of the objective lens, depth of focus is also maintained in the associated imaging of secondary emission electrons, etc. This property of the electron optical system thus forms an important function in the scanning electron microscope to be described in Chap. 5.

3.4 THE ION OPTICAL SYSTEM

Like the electron optical system treated above, the ion optical system consists basically of a source and a system of lenses to focus and control the beam action along a specified beam path. As noted previously, the ion beam is not effectively controlled by magnetic lenses because of the large mass requiring prohibitively large magnetic field strengths. Consequently, ion optical systems can be conceptually identical to many electron optical systems, with electrostatic lenses replacing magnetic lenses. Electrostatic lenses are actually simpler mechanically, and consume no net power. However they are, generally, of poorer optical quality than magnetic lenses. Magnetic fields are nonetheless effective and essential for mass separation, and the formation of discrete beams of a specific ion specie can be achieved by mass separation implicit in Eq. (1.51).

3.4.1 *ION SOURCES*

While electron sources produce only a single species, namely electrons, from a heated or high-field source, the primary species in ion sources can in principle be chosen from any of the 90 naturally occurring elements. These ions can be positive or negative. There are various types of ion sources, with the most popular involving electron bombardment or plasma discharge utilizing the Penning principle. The electron bombardment source was introduced in 1921 by A. J. Demptster for ion mass spectrometry. By the 1950s it was the most universally used ion source. Figure 3.18 illustrates schematically the operation of an electron bombardment source. In this arrangement, electrons streaming radially from the thermionic emitter bombard gas atoms introduced in a gas stream shown, producing (positive) ions which are accelerated through the source opening. Since the accelerating electrode is negatively charged, electrons are repelled, and are not mixed with the ion beam. In addition, the ion chamber itself may be biased so that electrons are confined to the anode region of the electron bombardment cylinder.

If the thermionic emitter of Fig. 3.18 is replaced by a single wire electrode and a high voltage applied between it and the anode cylinder, the introduction of a gas at reduced pressure will create a discharge "burn" or plasma. This can even be constricted by a magnetic field surrounding the source, and the ions can be extracted by the application of a suitable field (or voltage) to the accelerator plate. Positive or negative ions in the plasma can be extracted depending on the polarity of the extraction field, and the biasing relative to the plasma generator. A much simpler arrangement can consist of electrode or discharge plates between which an ionized discharge occurs.

In the past few years, more specialized ionization sources have come into use. Cesium ion sources have become important especially in secondary ion applications and analyzers because primary Cs^+ ions cause high negative secondary ion yields from certain elements [11,12]. Figure 3.19 illustrates a simple surface emission source. In the cesium source Cs vapor diffuses through a porous tungsten plug maintained at about 1400 K. The Cs leaves the opposite face of the plug as Cs^+ ions which are produced by thermal surface ionization, and accelerated to form the primary ion beam.

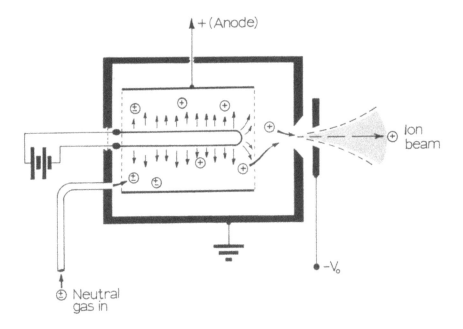

FIG. 3.18 *Ion source design based upon thermionic electron bombardment of a neutral gas.*

The surface ionization source was discovered around 1907 and is based on the fact that when an atom or molecule is evaporated from a surface it has a probability of being ionized, given approximately by the equation [13]

$$\left(\frac{n^+}{n^o}\right) \simeq \exp\left[\frac{e(E_W - E_i)}{kT}\right] \tag{3.24}$$

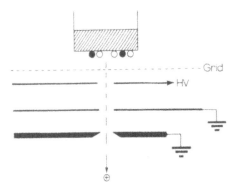

FIG. 3.19 *Schematic view of simple surface emission source.*

where n^+/n^o is the ratio of the charged (positive) to neutral constituents, e is the electronic charge, E_W is the work function of the surface or filament material, E_i is the ionization potential of the evaporated specie, k is the Boltzman's constant, and T is the absolute temperature. You might refer back to Table 2.2 and recognize that for the cesium, the ionization potential, E_i is relatively low, however, on a surface it reduces the work function. The work function is usually lower than the ionization potential. An expression similar to Eq. (3.24) can be written to express the approximate probability of an atom or molecule being evaporated from a surface as a negative ion

$$\left(\frac{n^-}{n_o} \right) \cong \exp \left[\frac{e(A - E_W)}{kT} \right] \tag{3.25}$$

where n^-/n^o is the ratio of the charged (negative) to neutral constituents and A is the electron affinity for the evaporating species.

The ions formed in most sources can differ in energy spread, current output, and brightness. Impurity ions and energetic neutrals which can originate during or after acceleration through charge exchange processes also exist in the ion beam. The primary beam can, however, be "purified" with a magnetic sector field which functions as a mass separator or mass filter.

Field emission sources which are characterized by the field ionization processes described in Chapter 2 (2.5) are also prominent especially in fine-focus (microprobe) applications requiring high brightness [14]. These sources, while having essentially very simple features, sometimes suffer instabilities as a result of poor beam confinement. Liquid metal field-ionization sources can also be used to overcome limitations on ion emission current density which results from the kinetic supply of ionizable gaseous particles (as depicted in Fig. 2.21). If a liquid film reservoir of the ionizable material is formed on the high field region of the emitter, the ionizable particle supply is enhanced by diffusion or viscous flow [15]. Such sources show great promise in providing high beam intensity with excellent spatial resolution.

3.4.2 *ION LENSES, FOCUSING SYSTEMS, FILTERS, AND PRISM OPTICS*

The unipotential electrostatic lens shown in Fig. 3.5 is used almost universally in ion optical systems. It is, in effect, the ion lens just as the cylindrical electromagnetic lens is the electron lens. The electro-static lens is an axial short-focus lens just as the electromagnetic lens is an axial short-focus lens, that is, it can accomplish linear ion beam focus along the optic axis of the ion optical system.

Divergent beam focusing in electric and magnetic fields Curved field fo-cusing of divergent ion beams can be accomplished according to Eq. (1.51) in both electric and magnetic fields respectively. In the former system mass is not a parameter and, as shown in Fig. 3.20, variations in the ac-celerating voltage allow for the selection and focus of ions or other charged particles of a given velocity, energy, or momentum (a momentum filter) since

$$R = \left(\frac{2 \, V_o}{E} \right)$$

It should also be recognized that if charged particles entering the curved electric field of Fig. 3.20 have different kinetic energies, then the beam divergence can result by differences in energy, ΔE or ΔV. Conse-quently, the divergence becomes $E_o \pm \Delta E$ ($V_o \pm \Delta V$) and there will be corres-ponding radii $\pm \Delta R$. So if the ion or charged-particle source is moved clo-ser to the electric field sector in Fig. 3.20, and an aperture is placed just outside the field sector in the exit side, some of the divergent par-ticles will be filtered out. The effectiveness of this filtering will de-pend upon the size or width of the aperture. This width will establish an energy "window" which will define the spread (ΔE) of the beam admitted to a detector or some other part of the system, and we will illustrate this fea-

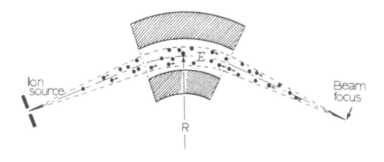

FIG. 3.20 *Velocity focusing of divergent ion beam in a curved electric field.*

ture in Chap. 4 when we discuss energy analysis systems, or particle-energy spectrometers.

As shown in Fig. 3.21, a divergent ion beam of constant mass-to-charge ratio (M_i/e) entering a wedge-shaped magnetic field at or near right angles to the field, having a specific reference geometry as shown, can be refocused according to [16]

$$\ell_f = R(\csc\theta + \cot\theta) \tag{3.26}$$

where

$$R = \sqrt{\frac{2\,M_i}{e}} \sqrt{\frac{V_o}{B^2}} \tag{3.27}$$

or

$$R = \frac{K'}{B} \sqrt{V_o}$$

where K' can be $(2M_i/e)^{\frac{1}{2}}$ or $(2m/e)^{\frac{1}{2}}$ for ions or electrons respectively, and must be constant. It has been common practice to have $\theta = 60°$ or $90°$ in Fig. 3.21.

The conditions implicit in Eq. (3.27) and illustrated in Fig. 3.21 also obviously apply to electrons (of mass m) as well as ions (of mass M_i). If the ion or electron energy is changed by an amount ΔV, the radius of the beam trajectory will correspondingly change

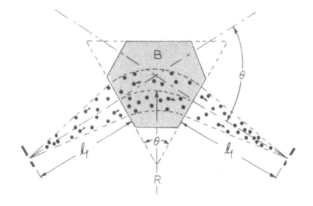

FIG. 3.21 *Direction focusing of divergent ion beam of constant (M_i/e).*

$$R \pm \Delta R = \frac{K'}{B} \sqrt{V_o \pm \Delta V} \qquad\qquad (3.28)$$

Alternatively

$$R \pm \Delta R = \frac{2}{evB} (E_o \pm \Delta E) \qquad\qquad (3.29)$$

Focusing magnetic prisms; mass and velocity filters You should observe
from Fig. 3.21 that if the divergent ion beams at the source differ in mass
(or charge) the point of beam focus will be different because the corresponding beam radii [implicit in Eq. (3.27)], will be different. This feature gives rise to mass filtering or mass spectrometry, and the magnetic
field region acts as a magnetic prism spectrometer as shown in Fig. 3.22.
If an aperture is placed along the M_2 beam, this focusing arrangement becomes an effective mass filter, blocking other ion beams which differ
in mass from the M_2 species. On the other hand, this phenomenon affords
a means to achieve a mass analysis of a complex ion beam. Recording the
relative intensity of each focused mass beam can also allow for a quantification of the total beam composition. This feature is the basis for mass
analyzers and mass spectrometer systems which will be treated in Chapters
4 and 5.

   If a homogeneous magnetic field B acts over a distance ℓ as shown in
Fig. 1.1, electrons or ions entering are deflected as shown. If an electric field, E, is applied at right angles to the magnetic field in Fig. 1.1
(crossed fields) the deflections will be in opposite directions. By properly adjusting the fields, the deflections of electrons or ions having a
velocity $v_o = E/B$ will cancel, and these charged particles will pass
through undeflected. By placing apertures in the undeflected path direc-

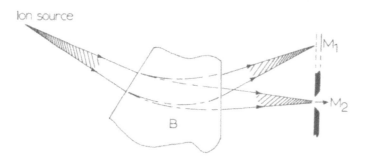

FIG. 3.22 *Ion-focusing magnetic prism*

tion, this device acts as a velocity filter for both ions and electrons. This particular system is commonly referred to as a Wein filter.

The ion probe By combining the source and beam focusing features illustrated in Figs. 3.19-3.22, it is possible to fabricate a variety of ion probe systems. If a probe is required which conforms to a very specific ion, then it is necessary to filter the ion source using the focusing magnetic prism of Fig. 3.22. This can be followed by an electrostatic unipotential lens to bring the specific ion beam to a focus at some desired focal length. Figure 3.23 illustrates this concept for a simple ion probe system. This system can in effect be operated in reverse, with a complex ion beam entering the electrostatic lens and being focused into the magnetic prism which then acts as a mass analyzer, focusing specific ion masses as shown in Fig. 3.22. To improve resolution or beam control the system shown in Fig. 3.23 can be combined with additional electric or magnetic ion

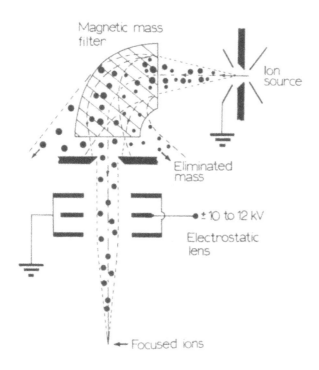

FIG. 3.23 *Simple ion probe system with mass filter for primary beam purification.*

focusing or spectrometer systems described depending upon the specific application.

We will deal with specific applications of ion probes and ion mass analyzers (spectrometer) in succeeding chapters. Of particular significance will be the contemporary synergism of electron and ion optical systems. In addition, we will describe time-dependent mass analyzers and quadrupole mass analysis in connection with specific applications in Chapter 4. In the time-of-flight mass spectrometers, time is the analytical parameter as compared to the spatial parameter in the electric and magnetic field sector analyzers.

For applications involving ion or electron beam probes, the probe size and the intensity will be significant parameters since they determine resolution and the ability to see or distinguish object features. By considering the geometry of the probe to be described by a half-angle, α, the overall probe *radius* at beam focus (as shown in Fig. 3.23) can be determined by combining Eqs. (3.19) and (3.20) to obtain

$$\delta(\text{probe}) = 0.43 \, \lambda^{3/4} \, C_s^{1/4} \tag{3.30}$$

if chromatic aberration is neglected or corrected. However the minimum size or ideal probe radius must also consider the brightness and probe current at beam focus. As a consequence, we must consider [17]

$$\delta_{\text{opt}} = \delta(\text{probe}) \left[1 + \frac{I_o}{\pi^2 \beta V_o (0.61\lambda)^2} \right]^{3/8} \tag{3.31}$$

where I_o is the probe current at the focused spot (ions or electrons per second), β is the specific brightness in ions or electrons $s^{-1}A^{-2}Sr^{-1}V_o^{-1}$, and V_o is the source or accelerating voltage.

When chromatic aberration distorts the formation of an optimum probe radius as a result of source energy fluctuations or of ion or electron energy fluctuations, Eq. (3.31) is written alternatively as

$$\delta_{\text{opt}} = \sqrt{1.1 \left[\lambda C_c \left(\frac{\Delta V}{V_o} \right) \right]} \left[1 + \frac{I_o}{\pi^2 \beta V_o (0.61\lambda)^2} \right]^{1/4} \tag{3.32}$$

From this expression and Eq. (3.31) we see that it is possible to calculate the source brightness needed to achieve a certain probe resolution if I_o,

$C_{\bar{s}}$, C_c, and λ are fixed. In addition, the specific source brightness, β, is given by

$$\beta = \frac{I}{\pi^2 \alpha^2 \delta^2 V_o} \tag{3.33}$$

where I is the primary current, α is the half-angle of emission, δ is the source radius, and V_o is the source voltage. Equation (3.33) is specific to ion beam formation in the context of ion microscopy described by Escovitz, et al [17].

For a two-lens system for probe formation where the divergent source beam is initially focused by an electrostatic lens (for an ion optical system) and then finally focused by a second lens to form the probe, the beam spot size at the final focal point is given by [18][†]

$$\left(\frac{d}{2}\right)^2 = M^2 \left[\rho^2 + \left(.25 C_{s1} \alpha^3 \right)^2 + \left(0.5 C_{c1} \frac{\Delta V}{V} \alpha \right)^2 \right]$$

$$+ \left[0.25 C_{s2} \left(\frac{\alpha}{M}\right)^3 \right]^2 + \left[0.5 C_{c2} \left(\frac{\Delta V}{V}\right) \left(\frac{\alpha}{M}\right) \right]^2 \tag{3.34}$$

where d is the beam diameter, C_{s1}, C_{c1} are the spherical and chromatic aberrations associated with the first lens, C_{s2} and C_{c2} are the corresponding aberrations for the second lens, M is the overall system magnification, ρ is the virtual source radius [δ in Eq. (3.33) above], and α is the half-angle at the source; defined by the beam-limiting aperture. In Fig. 3.23, this angle is half the angle which describes the beam divergence between the source and the magnetic mass filter. You should realize that for the system shown in Fig. 3.23, the second and third bracketed terms in Eq. (3.34) would be eliminated. Equation (3.34) also applies specifically to fine-focus ion beams employing a field ionization source. It can, however, apply more generally as is implicit in its derivation [18].

3.5 SIGNAL DETECTION, AMPLIFICATION, AND INTENSIFICATION

There are several popular charged-particle detector systems in addition to photographic recording, which should strike you as the most universal for divergent-beam recording. These include simple electrical detectors involving ion neutralization by an electrical current in a collector plate,

---

[†]*This formula appears to be a good approximation if one lens is dominant over the other, but is not reasonably valid if the two lenses have equal magnification.*

or electron impingement (where both effect the current flow in opposite ways), scintillation detectors where an ion or electron beam strikes a scintillator where the light pulse generated by the collision is amplified with a photomultiplier system, or the electron multiplier detector illustrated in Fig. 3.24. In this latter system, the particle beam or signal impinges upon an initial electrically biased dynode plate which creates secondary electron emission. These secondary electrons are, as a consequence of the biasing potential shown, accelerated to the next dynode causing additional emission. This process is repeated until a large (amplified) electron current results for each incident particle (electron or ion). Amplification can be controlled to a large extent by the potential, V, utilized in the dynode biasing as well as the number of stages.

In order to eliminate resolution limitations imposed by beam intensity, in addition to lowering the specimen heating and contamination, as well as the enhancement of image brightness, image intensification as an electron or ion optical support system has become increasingly popular in many research systems including field-ion and conventional electron transmission microscopy. True image intensification, as opposed to simple image reproduction, is accomplished by the successive reinforcement of electron intensity through accelerating potential stages. Figure 3.25 illustrates both the utilization of image intensification as well as a schematic representation of the device phenomenology, which is similar in principle to

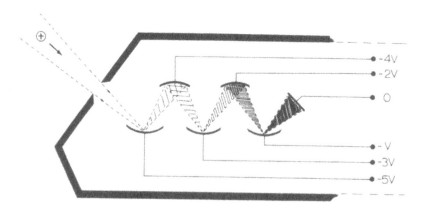

FIG. 3.24 *Signal detection and amplification by secondary electron multiplication.*

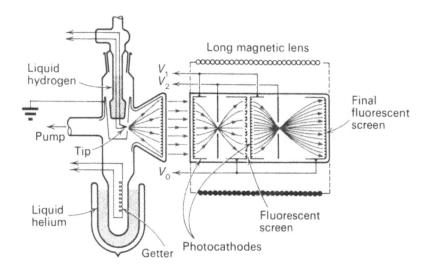

FIG. 3.25 *Image intensification in field-ion microscopy.*

the scheme for electron multiplication shown in Fig. 3.24. The biasing
voltages are denoted by V_1, V_2, etc. in Fig. 3.25.

While systems such as that shown in Fig. 3.25 function by direct intensity modification, many similar schemes devised over the past two decades,
and used in electron microscopes, function by direct beam amplification,
with the amplified signal intensity channeled into an electronic scanning
system for direct video display.

In such systems, final image contrast is ultimately influenced by the
signal-to-noise properties in the associated circuitry. However, in spite
of certain inherent circuit-noise problems, this mode of image reproduction
offers numerous advantages. For example, let us consider the basic electron microscope system employing image intensification as depicted schematically in Fig. 3.26. Here the image primary signal (the transmitted
electron beam) impinges on the intensifier, is modified, and the information, as electrical signal data, can then be channeled to a video display
unit. The signal information can also be fed to integration circuits
where the unwanted contrast features arising by electron transmission
through the object can be rejected as a noise component of the signal.
The modified signal can then be fed into computer-linked systems for the
analysis of image-contrast profiles or related features such as the detailed analysis of discrete object detail in the form of precipitates or
similar distinguishable defects.

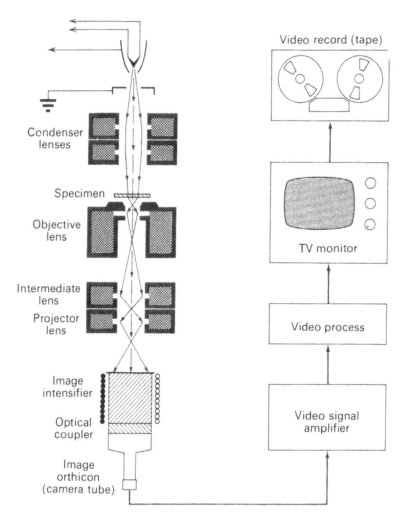

FIG. 3.26 *Image intensification in transmission electron microscopy.*

More compact image intensifiers have been in use for more than a decade, especially in field-ion microscopy which involve thin sandwich devices referred to as channel-plate image intensifiers. These devices eliminate the magnetic focusing illustrated in Fig. 3.25 because the initial
layer which faces the image screen is a thin photoconducting material. An
inner layer composed of a very thin conducting film is mated to an outer
luminescent layer. When a voltage is applied across this sandwich, light
from the initial image screen impinges upon the photoconductor side and is

converted to an electric current which produces an illuminated zone in the outer luminescent layer. The voltage applied amplifies this signal in proportion to the potential difference. A fiber optic array is normally an integral part of this sandwich arrangement to facilitate light transmission from the initially weak image or as a final viewing screen.

PROBLEMS

3.1 For a thin-lens approximation (for example, Fig. 3.4), the overall magnification in an electron optical column is

$$M = M_1 M_2 M_3 \cdots M_n$$

where M_n is the magnification associated with each successive lens. If we solve for the total differential magnification, we can write

$$\frac{dM}{M} = \frac{dM_1}{M_1} + \frac{dM_2}{M_2} + \frac{dM_3}{M_3} + \cdots + \frac{dM_n}{M_n}$$

where

$$M_n = \frac{f_n}{P_n - f_n}$$

Show that for a simple electromagnetic-lens system, such as that depicted in Fig. 3.6, the differential magnification is given by

$$\frac{dM}{M} = \left[\frac{M_1 + 1}{f_1} + \frac{(M_1 + M)^2 M_2}{f_2} \right] df_1 + \left(\frac{M_2 + 1}{f_2} \right) df_2$$

Recall that $M_n = q_n/P_n$; and assume P_1 constant. (Hint: Since the fluctuations in the second lens depend on fluctuations in the first, we can consider $dP_2 = -dq_1$.) If the focal lengths of the lenses are found to be $f_1 = 3$ mm and $f_2 = 10$ mm, and the corresponding object distances are $P_1 = 2.94$ mm and $P_2 = 9.9$ mm, find the magnification M of the final image. Evaluate dM/M for this system if $df_1 = 10^{-2}$ and $df_2 = 0.3\ df_1$.

3.2 Derive Eq. (3.8) from Eq. (3.5) assuming from basic field theory that

$$H = \frac{2\pi N I d_f^2}{5(z^2 + d_f^2)^{3/2}}$$

where d_f is the effective lens-field diameter. Express the form factor as a function of d_f (assume that $f_c \gg d_f$).

3.3   Referring to Fig. 3.10, if the accelerating potential V_0 is assumed to be 5×10^4 volts, and the condenser-lens form factor ζ_c = unity, calculate the corresponding increase in the coil current to maintain a relative crossover intensity of unity at an accelerating potential of 125 kV. What provisions in lens design must be altered in order to produce reasonable crossover response at very high voltages using conservative values of lens current? What alterations in the dimensions of the optical column might be involved in accommodating the very high accelerating potentials ($V_0 \gg 100$ kV)?

3.4   Assume that an objective electron lens design necessitates attaining a field of 15,000 gauss for a gap length in the pole piece of 0.4 cm. If 2×10^4 turns of Cu conductor compose the lens coil, which has a cross-sectional area of $10^2 \mathrm{cm}^2$ with a mean coil radius of 8 cm, find the current required, and the power dissipated in the lens. Calculate the equivalent lens resistance and the voltage applied to the lens-coil terminals.

3.5   For an effective intermediate lens-image distance of 0.65 cm, calculate the total image magnification in an electron microscope operating at 100 kV, with a fixed objective magnification of 15 if the effective form factor is ζ = 100 (volt cm)(amp-turn)$^{-1}$. The intermediate lens contains 10^4 turns of Cu wire and the intermediate-lens current is 200 milliamperes. (Refer to Prob. 3.1 and assume a thin- or short-lens approximation to be valid.) The projector magnification is 6.

3.6   Assuming an aperture angle of 0.001 rad, calculate the optimum line-to-line resolution attainable in an electron microscope operating at 0.5×10^6 volts, with an objective focal length of 1.5 mm, lens constants of γ_0 = 100 and γ_0' = 1,20,000 turns of Cu conductor composing the lens-coil winding, and focus attained at a lens current of 0.15 amp. Assume as in most conventional electron microscopes, that the voltage and current stabilities are at least 1 part in 10^5. Calculate the percentage error in neglecting lens aberrations in this particular case.

3.7   Calculate the change in the spherical aberration coefficient C_s of Prob. 3.6, assuming the parameters of focus remain unchanged for an increase in accelerating potential to 10^6 volts. Is the resolution (with reference to Prob. 3.6) increased or decreased and by how much? Calculate the point resolution for this case.

3.8 Calculate the depth of field associated with the system of Prob. 3.7. What would happen in an image formed of electrons emitted from a surface profile greater than D_f, say in a scanning electron microscope?

3.9 In the thermionic electron microscope, the energy of the emitted electrons just at the surface determines the effective resolution of the surface detail; the instrument potential simply serving to accelerate the "imaging" electrons at the surface. Surface resolution is therefore essentially diffraction dependent if lens aberrations are neglected. Assuming an aperture angle of 0.005 rad, calculate the approximate resolution (assume an effective surface electron energy of 1 eV at high temperature). Note that this is only an approximation, which serves to illustrate our assumptions in Chap. 2. Resolution in thermionic emission is also temperature dependent in a manner illustrated in Eq. (2.26).

3.10 For field electron emission, it has been estimated that the tangential surface electron energy is roughly 0.1 eV for field strength $E_z > 10^7$ volts/cm. Assuming the resolution to be diffraction dependent, and considering emission from a tip-end form to occur at angles equivalent to $\alpha \simeq \pi/2$, show that the resolution is given approximately as $\delta_{opt} \simeq 20$ Å as stipulated in Chap. 2.

3.11 Draw a simple sketch of a specimen-area signal portion as detected by an image-intensification system, showing a prominent contrast feature associated with background noise, and show, using a simple alteration of the sketch, the elimination of unwanted noise by the modification of the voltage level of the signal.

3.12 Silver is evaporated from a heated tungsten filament in a surface emission source. If the average work function for the tungsten filament is 4.5 eV (compare with Table 2.1) and the ionization potential is 7.6 eV, to what temperature must the filament be heated to provide a positive ion/neutral ion ratio of 10^{-16}?

3.13 An ion probe employing U^+ ions having a mass of 235 amu (atomic mass units) utilizes a divergent beam directional focusing system with a magnetic field of 100 gauss, and a 60° geometry. For an accelerating potential of 8 kV, find the distance of the source and focal point from the magnet face. Sketch this system to some scale.

REFERENCES

1. E. Bruche and O. Scherzer, *Geometrische Elektronenoptik*, Springer-Verlag OH6, Berlin, 1934.

2. F. H. Nicoll, *Proc. Phys. Soc. (London)*, *50:* 888(1938).

3. A. Recknagel, *Z. Physik*, *104:* 381(1937).

4. V. K. Zworykin et al, *Electron Optics and the Electron Microscope*, 4 ed., John Wiley & Sons, Inc., New York, 1957.

5. H. Fernandez-Moran, *Proceedings of the Electron Microscopy Society of America*, p. 10, Claitor's Publishing Division, Baton Rouge, La., 1967.

6. B. M. Siegel et al, *Proceedings of the Sixth International Conference for Electron Microscopy*, p. 151, Maruzen Company, Tokyo, 1966.

7. A. Septier, R. Barer and V. E. Cosslett (eds.), *Advances in Optical and Electron Microscopy*, vol. 1, p. 204, Academic Press Inc., New York, 1966,

8. O. Scherzer, *J. Appl. Phys.*, *20:* 20(1949).

9. N. Niehrs, *Fifth International Congress for Electron Microscopy*, paper AA-2, vol. 1, Academic Press Inc., New York, 1962.

10. E. Ruska, *Fifth International Congress for Electron Microscopy*, paper A-1, vol. 1, Academic Press Inc., New York, 1962.

11. P. Williams, et al.; *Anal. Chem*, *49:* 1399 (1977).

12. C. W. Magee, *J. Electrochem. Soc.*, *126:* 660(1979).

13. I. Langmuir and K. H. Kingdom, *Proc. Roy. Soc.*, *A107:* 61(1925).

14. J. Orloff and L. W. Swanson, *J. Vac. Sci. Technol.* *12:* 1209(1975).

15. L. W. Swanson, G. A. Schwind, and A. E. Bell, *J. Appl. Phys.*, *51:* 3453(1980).

16. R. Herzog, *Z. Physik*, *89:* 786(1934).

17. W. H. Escovitz, T. R. Fox, and R. Levi-Setti, *Proc. Natl. Acad. Sci., USA*, *72:* 1826(1975).

18. J. Orloff and L. W. Swanson, *J. Vac. Sci. Technol.*, *15(3):* 845(1978).

SUGGESTED SUPPLEMENTARY READING

Barer, R., and Cosslett, V. E. (eds.): *Advances in Optical and Electron Microscopy*, Academic Press, London, 1987.

Born, M., and E. Wolf: *Principles of Optics*, Pergamon Press, New York, 1959.

Bozorth, R. M.: *Ferromagnetism*, D. Van Nostrand Company, Inc., New York, 1951.

Buseck, P. et al. (eds.): *High Resolution Transmission Electron Microscopy and Associated Techniques*, Oxford University Press, New York, 1989.

Cosslett, V. E.,: *Introduction to Electron Optics,* 2d ed., Clarendon Press, Oxford, 1950.

Grivet, P.: *Electron Optics,* Pergamon Press, New York, 1965.

Harting E. and Read, F. H.: *Electrostatic Censes,* Elsevier, New York, 1976.

Hawkes, P. W.: *Electron Optics and Electron Microscopy,* Taylor and Francis, Ltd., London, 1972.

Hertzberger, M.: *Modern Geometrical Optics,* Interscience Publishers, Inc., New York, 1958.

Hopkins, H. H.: *Wave Theory of Aberrations,* Oxford University Press, Fair Lawn, N.J., 1950.

Kerwin, L.: Ion Optics, in C. A. McDonald (ed.), *Mass Spectrometry,* McGraw-Hill Book Co., Inc., New York, 1963.

Pierce, J. R.: *Theory and Design of Electron Beams,* D. Van Nostrand Company, Inc., New York, 1959.

Reimer, L.: *Transmission Electron Microscopy, Physics of Image Formation and Microanalysis,* 2d ed., Springer, Berlin, 1989.

Ruska, E.: Past and Present Attempts to Attain the Resolution Limit of the Transmission Electron Microscope, in R. Barer and V. E. Cosslett (eds.), *Advances in Optical and Electron Microscopy,* vol. 1, Academic Press Inc., New York, 1966.

Sturrock, P. A.: *Static and Dynamic Electron Optics,* Cambridge University Press, New York, 1955.

Zworykin, V. K. et al.: *Electron Optics and the Electron Microscope,* 4th ed., John Wiley & Sons, Inc., New York, 1957.

4
ELECTRON AND ION
PROBE MICROANALYSIS

4.1   INTRODUCTION

In Chap. 2 we attempted to develop a rather logical trend in the analysis
of a material's surface features by treating primary electron emission and
related ionization phenomena as emission microscopy. In all emission microscopy modes of analysis, the specimen is an active element in the imaging system. In Chap. 3 we developed some of the basic features of the
electron and ion optical systems, specifically treating the imaging aspects
of a passive object (specimen) inserted into the path of a suitably controlled electron or ion beam, and the formation of electron and ion probes
as well as the detection of specific particles through mass or energy selection. We were concerned with the formation of an "image" of the "passive" object in the electron optical system, and as a consequence we were
really dealing with a transmission characteristic of the specimen.

If we momentarily digress to a second glance at Fig. 3.8, it will be
observed that aside from the imaging of the transmitted electron "signal,"
the bombardment of a material with a primary electron beam causes (aside
from normal transmission of a portion of the incident electrons) a small
portion to be reflected from the surface and also excites other (secondary)

electron emissions within the irradiated material. In addition, as we recall from Chap. 1, the energizing of electrons in a solid causes characteristic x-rays to be emitted as the energized electrons return to their original energy states after excitation. Thus, while the formation of an image in an electron optical system can give direct information bearing on the object geometry, the image itself contains no information concerning the chemical composition or atomic species of the imaged area or volume. We should also be aware that, with specific reference to Fig. 3.8, the image results as a consequence of object transmission or diffraction, and we therefore observe its volumetric geometry. It is also possible to form an image of electrons "reflected" or emitted from the surface of a specimen that, in view of the depth-of-field property of electron optical systems, can give a three-dimensional reconstruction of object-surface topography.

Figure 4.1 depicts the more prominent physical processes that occur when an electron beam strikes a solid specimen. It should of course be realized that the transmission characteristics are contingent on the thickness of the object, while the activity associated with the incident surface is primarily independent of the thickness. We might also observe that the backscatter, secondary electron emission, characteristic x-rays, etc., result within the area irradiated by the primary electron beam. Thus, where these radiations are associated with a specific object surface area, the emission resolution depends on the smallness of the area onto which the incident beam can be focused. Similar processes can occur when an ion beam replaces the electron beam, but in addition transmitted ions occur along with reflected and secondary ions. The secondary ions represent a sputtered mass loss from the specimen itself and this process alters the specimen surface in addition to providing a depth profile in comparison to the original, unaltered specimen surface.

4.2 ELECTRON PROBE MICROANALYSIS

Electron probe microanalysis really developed as a consequence of the successful design and applications of electron microscopy before 1950, and our treatment of this topic at this stage is therefore chronologically out of step. Nevertheless it is best dealt with at this point - before the treatment of the analytical modes of electron microscopy - because of the many fundamental concepts involved. We observe in Fig. 4.1 that because of the processes that occur when an electron beam strikes a solid, a number

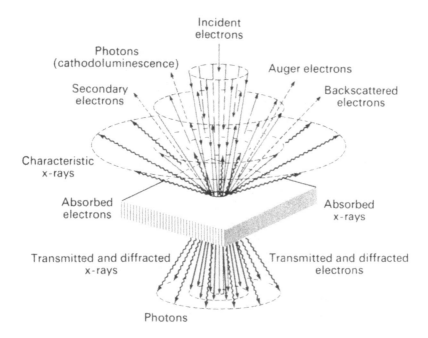

FIG. 4.1 *Reaction and interaction of electron beam encountering a solid.*

of different analytical modes are immediately presented. Namely, the characteristic x-rays emitted can be analyzed with regard to the area (or volume of origin); backscattered electrons, absorbed electrons, and secondary emitted electrons (that is, those electrons originally in the solid that are ejected as a result of primary electron collisions, etc.) can be monitored and used in the analysis of surface composition, charge, topography, etc.; specimen fluorescence as a result of photon production in the collision area can be observed; diffracted electrons can be used in the analysis of object crystallography and composition; and primary x-rays can be used to excite secondary x-rays (x-ray fluorescence) that can in turn be used to gain information relevant to the composition of the interaction zone.

The conventional electron probe microanalyzer consists mainly of that upper electron optical system of Fig. 3.8 consisting of a condenser-lens arrangement and an objective. In a conventional probe arrangement the objective serves the purpose of focusing the electron beam onto an area of the specimen, rather than the direct imaging of the specimen detail as in an image-forming system. Consequently the specimen is positioned at the focal length of the objective lens. With reference to Fig. 3.8, this would

mean that the specimen must be inserted within the lens, or the lens must
be inverted. These options are indicated schematically in Fig. 4.2.

It is perhaps already apparent that by restricting the zone of interaction of the electron beam with the specimen we increase the resolution
of the object composition, etc., within that zone. Thus the ability to focus the beam on the object determines the ultimate resolution of the probe
- susceptible, for the most part, to the same restrictions and lens aberrations as discussed generally in Chap. 3. In addition, the thickness of the
specimen will have an effect on the beam broadening or the effective size
of the interaction volume in the specimen as we will describe in Section
4.4.2.

In the most general analysis of secondary radiations resulting at an
electron focal zone of an object, the beam is fixed to a spot. While this
would of course give an analysis of a desired (attainable) resolution of a
fixed area, the analysis of this area with respect to its surroundings, or
the investigation of partitional compositions within a zone, cannot be easily obtained without moving the beam to these areas, or by a translation
or rotation of the object area with respect to the fixed-focus beam.

Electron microprobe analysis was first proposed in the early 1940s in
several patents filed by J. Hillier of RCA; it was developed in the late
1940s by R. Castaing [1,2] and A. Guinier [1] initially in the form of a
static electron microbeam x-ray probe which, as in all static beam instru-

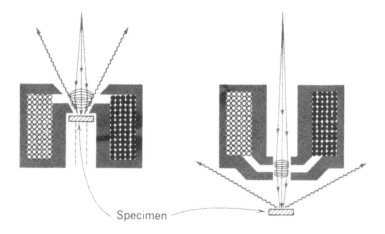

FIG. 4.2 *Objective-lens arrangements for electron probes
(general schematics).*

ments, relied on conventional light microscopy to position the specimen areas for analysis. In the more conventional microprobe designs, the incident electron beam is scanned over a specimen area by a set of deflection coils below the condenser lens, connected to modulate the detected signal with a cathode-ray tube; this enables a direct display of the x-ray or electron emission over the scanned area or the relative backscatter intensity profile. Many of the developmental features, instrumentation, and applications of electron microprobes have been reviewed by Castaing [2], K. F. J. Heinrich [3], D. B. Wittry [4], and others; for a general survey of the various microprobe modes, see the Suggested Supplementary Reading at the end of this chapter.

4.2.1 X-RAY EMISSION MICROPROBE ANALYSIS

The principle of x-ray analysis employed in the detection of characteristic x-ray emission from the electron focal zone on a specimen surface mainly derives from the classical x-ray spectroscopy features first exposed by H. Moseley [5]. In its simplest form, when a specimen is irradiated with energetic electrons, the constituent atoms are excited to energy levels dependent on the quantum state of the bound electrons energized. As we observed in Chap. 1, the return of a bound electron from state i to j is expressed as

$$E_i - E_j \equiv \Delta E = h\nu' \tag{4.1}$$

where ν' is the characteristic frequency of the emitted x-rays. Thus the characteristic x-ray wavelength becomes

$$\lambda' = \frac{hc}{\Delta E} \tag{4.2}$$

or

$$\Delta E(keV) = 12.4/\lambda'(\overset{\circ}{A}) \tag{4.2a}$$

where c is the velocity of light. If now we consider a transition from an excited state i to some initial state j, the energy transition wavelength emission can be expressed simply in terms of Eq. (1.40) considering multi-electron atoms in the form

$$\frac{1}{\lambda'} = \frac{\Delta E_n(\ell)}{2\pi hc} \tag{4.3}$$

or

$$\frac{1}{\lambda'} = \frac{m_o e^4 (Z - \sigma)^2}{\epsilon_o^2 c (4\pi h)^3}$$

$$\cdot \left\{ \left[\frac{1}{n_j^2} + \frac{\zeta}{n_j^4} \left(\frac{n_j}{\ell_j + \frac{1}{2}} - \frac{3}{4} \right) \right] - \left[\frac{1}{n_i^2} + \frac{\zeta}{n_i^4} \left(\frac{n_j}{\ell_i + \frac{1}{2}} - \frac{3}{4} \right) \right] \right\} \quad (4.4)$$

which is a more exact form of Moseley's law in the general solution

$$\frac{1}{\lambda'} = C(Z - \sigma)^2; \quad \Delta E = C'(Z - \sigma)^2 \tag{4.5}$$

where C and C' are constants for any discrete transition, and σ is a nuclear screening constant[†] having values varying from approximately 3 for atoms with Z equal to 30 or more to about 5 at Z equal to 90 for K-series emission spectra; it increases for higher-energy-state emission spectra, for example, L, M, N, etc. In the case of heavier elements we can simply assume the characteristic energies to be proportional to Z^2.

The characteristic emission spectra are designated with regard to the initial or j quantum level. Thus all transitions to the K level are K-series emissions, transitions to the L level are L-series emissions, etc. The characteristic line spectra are denoted by α, β, γ, and δ, etc., for transitions from higher levels i to a level j will be governed chiefly by the azimuthal quantum number ℓ, a feature readily observed in referring to Eq. (4.4). This feature is also readily apparent if we digress momentarily to review Fig. 1.7. It is also apparent that the energy levels are sufficiently spaced so that the characteristic emission lines and associated spectra have an easily recognizable displacement in wavelength as well as prominent line intensities.

The first emission series or K series has two prominent transitions from the n = 2 level to the ground state n = 1, denoted $\lambda_{K\alpha_1}$ and $\lambda_{K\alpha_2}$.

[†]*The screening constant included adjusts for the effective charge of the nucleus resulting from considering many electrons as opposed to a single negative charge; and in effect it characterizes the electric field perturbation on the nuclear energy by the multielectron cloud. Consequently the adjustment for effective Z becomes larger as we consider quantum levels at increasing distances from the nucleus.*

These occur from the L sub-shells. Similarly, prominent characteristic emissions for $\lambda_{L\alpha}$ and $\lambda_{L\beta}$ occur when the atoms in the zone of a focused microprobe electron beam are excited to quantum levels above n = 2. There are of course an enormous number of transitions possible but only those cited occur with any particular prominence. We shall reserve for Sec. 4.4 a more detailed treatment of the efficiency of x-ray excitations and the dependence of detected signal intensities on absorption mechanisms, etc., which must in actual practice be incorporated in the form of a correction factor in the final evaluation of element concentration.

We observe from Eq. (4.5) that the wavelength for a particular value of n becomes smaller with increasing atomic number Z while the energy increases. Thus by simply measuring the wavelength or energy of the emitted x-rays, we can identify the elements in the reaction zone of the specimen using graphical data as shown in Fig. 4.3. By comparing the relative intensities of emitted x-rays corresponding to the same quantum state of different elements, the concentrations of each individual atom species can be determined.

X-ray excitation in solid specimen targets Let us initially examine in more detail the focusing of electrons onto a specimen surface by the objective lens as illustrated in Fig. 4.2. If we consider the maxwellian distribution of the electrons leaving the heated filament of a thermionic emission electron gun (Fig. 3.9), the maximum current that can be concentrated into a focused spot on the specimen is given by D. B. Langmuir [6] in the general form

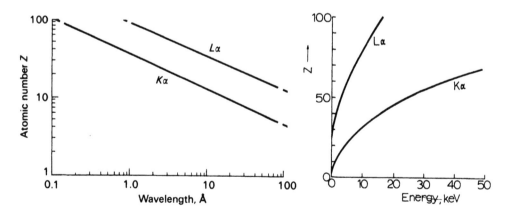

FIG. 4.3 *Atomic number Z versus wavelength and energy for Kα and Lα x-ray emission.*

$$J_f = J \left(\frac{eV_o}{kT} + 1 \right) \sin^2 \alpha \tag{4.6}$$

where J is the thermionic emission current essentially as expressed in Eq. (2.15), and α is for the present the objective aperture angle. Obviously since α is very small, we can assume $\sin \alpha \simeq \alpha$; and substituting for α, we can write Eq. (4.6) in the form

$$J_f = J \left(\frac{eV_o}{kT} + 1 \right) \alpha^2 \tag{4.7}$$

for the specimen surface at the defocused gaussian image plane of the objective lens. Since

$$I_f = \frac{4J_f}{\pi \delta_s^2} \tag{4.8}$$

where δ_s is the gaussian image of the electron source (shown in Fig. 3.12), we can express the total beam current at the focused spot on the specimen as

$$I_f = \frac{4J}{\pi \delta_s^2} \left(\frac{eV_o}{kT} + 1 \right) \alpha^2 \tag{4.9}$$

The depth of penetration (in microns) of electrons at the spot onto which they are focused is given approximately by the Thomson-Whiddington law in the form

$$d_p = 11 \times 10^{-9} \frac{W_A V_o^2}{Z\rho} \tag{4.10}$$

where W_A is the atomic weight of the specimen focal zone,* and ρ is the density of material in this zone. We now observe from Eqs. (4.9) and (4.10) that both spot intensity and depth of penetration are largely dependent on the accelerating potential experienced by the electron beam; and since the efficiency of characteristic x-ray emission mostly depends on incident-beam penetration and intensity, it depends generally also on the accelerating

---

*We shall use the expression specimen focal zone to denote that area onto which the electron beam is focused.

voltage V_o. This is also true for a field emission electron source as well, and this is implicit in Eqs. (2.20) or (2.23) for example, which would characterize J in Eq. (4.9).

Figure 4.1 has already depicted the fact that radiation from the focal zone of the specimen is uniformly distributed with respect to the incident electron beam, with the exception of backscattered electrons if the incident electron beam is not normal to the specimen surface. In this respect it should be realized that the characteristic x-ray emission also depends most directly on the relative excitation efficiency of the incident electron beam. If we consider that roughly only 2 percent of the incident-beam energy remains after absorption and conversion to heat, the efficiency of the process is obvious. We can express the energy loss dE characteristic of an incident electron in a solid specimen of density ρ as

$$dE = \rho f(E)dx \qquad (4.11)$$

where $f(E)$ will be assumed to represent an energy function that expresses the relative absorption coefficients of the constituents elements in a penetration distance (or path length) dx. If we now use dn to express the number of atomic ionizations in the focal zone of the solid creating Kα characteristic radiation, we can write

$$dn = C_A \rho_A \psi_A(E, E_{K\alpha})dx \qquad (4.12)$$

where C_A is the concentration of an element A of density ρ_A, and $\psi_A(E, E_{K\alpha})$ is the Kα ionization-potential function for A atoms. Now from Eq. (4.11) we can write

$$dn = C_A \frac{\psi_A(E, E_{K\alpha})}{f(E)} \qquad (4.13)$$

If we assume scattering of the focused incident electron beam to be negligible, we can relate the number of ionizations directly to the characteristic emission intensity; and the intensity ratio of the specimen containing element A, compared with the intensity from a pure standard of element A (denoted A') is then expressed by

$$\frac{I_A}{I_{A'}} = C_A \frac{\int_{E_{K\alpha}}^{E_o} \frac{\psi_A(E, E_{K\alpha})}{f(E)} dE}{\int_{E_{K\alpha}}^{E_o} \frac{\psi_A(E, E_{K\alpha})}{f(E)} dE} \qquad (4.14)$$

with the incident beam energy denoted by E_0. Simply writing the integral ratio as some function $g(E)^\dagger$ then results in the approximate expression

$$\frac{I_A}{I_{A'}} = C_A g(E) \tag{4.15}$$

which indicates quite clearly that the characteristic emission of the focal zone can be ideally related to the concentration of elements composing the zone provided a pure standard is used as a reference for each element A, B, C, etc., detected.

While Eq. (4.15) defines quantitatively the analytical capabilities of the electron probe operating in the characteristic x-ray emission mode, the analysis is mainly contingent on the actual resolution and detection of the x-radiation of wavelengths (λ_A, λ_B, etc.) or energies specific to a particular quantum state (K, L, etc.) of the elements of compositions C_A, C_B, C_C, etc., in the probe focal zone of the specimen. So we will now, before developing further the analytical applications of electron probe x-ray microanalysis, treat briefly the mechanics of characteristic x-ray resolution and detection.

Wavelength - Dispersive Spectrometry   The actual resolution of characteristic x-radiation dispersed due to electron probe ionization is performed by inserting an analyzing crystal (reference crystal) between the emission point on the specimen and the actual detector system. The "spectrometer" - analyzing-crystal relationship is basically described by Bragg's law

$$2d \sin \theta = n\lambda \tag{1.9}$$

Thus, for a crystal of known atomic spacing, the detection of a strong signal diffracted at some Bragg angle θ will uniquely identify the incident wavelength λ. Figure 4.4 illustrates the basic features of characteristic x-ray resolution using an analyzing crystal having a known crystallographic surface (hkℓ) that acts as the x-ray "reflection" plane according to Eq. (1.9). We observe from Fig. 4.4 that by collimating the x-rays incident on the analyzing crystal, or reflected from the crystal, or by collimating both incident and reflected signal modes, all but one specific wavelength can be excluded from the detector. We should of course be cog-

---

†*Here we must also regard g(E) as an energy-loss ratio for the corrections due to mass absorption, and will depend to an appreciable degree on the electron accelerating potential V_O.*

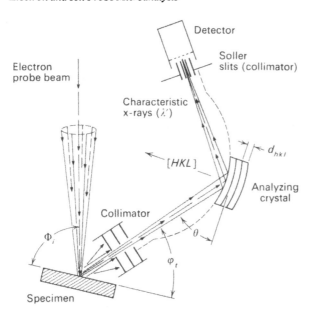

FIG. 4.4 *Simple crystal monochromator and col-*
limating system.

nizant of the fact that the intensity of a desired wavelength, if in fact
it exists in the emission profile, will depend to a large degree on the
angle of incidence Φ_i, and the corresponding intensity I_f, of the electron
beam focused onto a selected area of a specimen. Where I_f is defined as in
Eq. (4.9), we can appreciate the dependence on the accelerating voltage V_O.
It can also be seen that the intensity of the collimated wavelength profile
will vary according to the position of sampling the characteristic emission
as evidenced by the takeoff angle (ϕ_t of Fig. 4.4). Thus, where intensity
ratios are utilized in the analysis of element concentrations according to
Eq. (4.15), the angles Φ_i and ϕ_t as well as the accelerating potential must
be constant for calibrations based on pure standards of the elements detected. It should also be apparent from the simple schematic of Fig. 4.4
that the angular relationships of the various components can be mechanically positioned and the angles read directly as in conventional x-ray
diffractometer arrangements [7]. Indeed, electron probe x-ray wavelength
microanalysis is simply microscale x-ray wavelength spectrometry.

A number of more sophisticated analyzing crystal geometries are also
employed in the microprobe resolution of characteristic x-ray wavelength
emission. These designs, several of which are illustrated in Fig. 4.5,
allow the characteristic x-rays to be focused onto a detector, and in

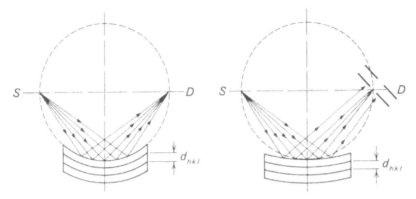

FIG. 4.5 *X-ray focusing with bent crystal monochromators.*
(oversimplified 2-dimensional schematic).

this sense these crystals constitute a simple x-ray optical device. As
Fig. 4.5 indicates, these "focusing" monochromators are simply curved (bent)
crystals that as a consequence of their geometry, and the continuous satisfaction of the Bragg condition [Eq. (1.9)], serve to concentrate the divergent (or dispersive) emission profile for all rays of a particular λ. Following Fig. 4.5, we observe that geometrically perfect focusing results
when the source S, analyzing crystal, and detector D lie on the arc of a
circle. Figure 4.5 also illustrates the ability selectively to focus and
collimate the characteristic x-rays by only partially fulfilling this condition for the analyzing crystal.

The ease of resolution, the ability to shape an analyzing crystal, and
other intrinsic features of the monochromator system depend to a large extent on the crystal properties; and the range of element analysis also depends on the spread of the necessary angular relationships, particularly
the Bragg angle, especially where these relationships affect the physical
shape and disposition of the analyzing crystal and detection systems. Table 4.1 indicates a number of the more common analyzing crystals along with
the relevant data.

Signal detection and scanning wavelength - dispersive microanalysis The
actual detection of x-rays and x-ray intensity profiles in the electron
microprobe from characteristic wavelength spectrometry is, for all practical purposes, identical to standard techniques of x-ray diffraction and
spectroscopy [7]. The basic categories of detectors are generally of the
proportional-counter types, including gas-type proportional counters such
as that shown schematically in Fig. 4.6a; Geiger counters (also character-

TABLE 4.1 *Common analyzing crystal characteristics*

Crystal designation	Reflection plane, hkℓ	$2\ d_{hk\ell}$, Å
Muscovite (mica)	002	19.840
Ammonium dihydrogen phosphate (ADP)	101	10.640
Low quartz (SiO$_2$)	101	6.687
Calcite (CaCO$_3$)	104	6.071
Sodium chloride (NaCℓ)	200	5.641
Lithium fluoride (LiF)	200	4.027
Topaz	303	2.712

ized in Fig. 4.6a); and scintillation counters or photomultiplier-type detectors shown schematically in Fig. 4.6b. In the proportional-type detectors of Fig. 4.6a, x-rays entering the cylindrical section are absorbed by a gas contained there, causing electron ejection and subsequent ionization of the gas atoms. The collection of the electrons under the influence of an applied field and the migration of the ions toward the cylindrical cathode shell causes a current to flow in the detector circuit. The intensity of current flow will depend on the incident x-ray intensity; and cur-

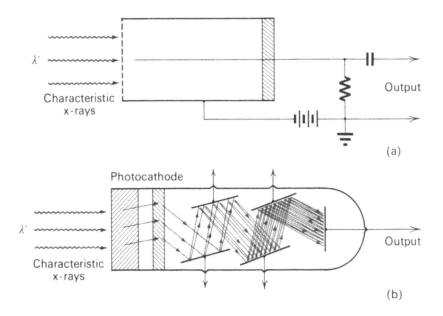

FIG. 4.6 *X-ray detector schematics. (a) Ionization-chamber detector; (b) scintillation-photomultiplier detector.*

rent flow can thus be related directly to x-ray intensity. The scintillation counter indicated in Fig. 4.6b relies on the visible fluorescence excitation of x-rays on striking certain substances. The technique simply
involves photomultiplication of the crystal emitted photon by the ejection
of a photoelectron from the photocathode (for example, a cesium-antimony
intermetallic compound). This is simply a modification of the secondary
electron multiplier shown previously in Fig. 3.24.

In the case of a static microprobe beam focused onto a specimen-sur-
face area, the detector signal will remain effectively constant for any
particular setting of the probe beam with the output signal intensity corresponding to the intensity spectrum of the characteristic x-radiation.
However, if the probe beam is scanned across an area of the specimen, the
detector output signal will show a corresponding change in magnitude as the
x-ray emission profile varies. If the detector output signal associated
with the excited x-rays of the scanning electron beam is amplified and used
to modulate a cathode-ray tube in synchronism with the probe scanning frequency of the specimen, the image on the screen will vary in brightness
according to the local specimen variation in characteristic x-ray emission
intensity. Figure 4.7 illustrates the design features of a scanning electron microprobe system. The scanning x-ray microprobe whose detector signal brightness modulates a cathode-ray display tube gives an enlarged image
of the specimen area, which is magnified by the ratio of display width to
the width of the scanning beam spot across the specimen. In addition, a
quantitative microanalysis of the elements composing the image is also directly observed with the detector "tuned" to an emission wavelength of a
desired element, as empirically expressed in Eq. (4.4). These features are
perhaps best understood by reference to photographs of cathode-ray-tube
images of specimen areas scanned for specific-element concentrations.
Figure 4.8 shows a typical sequence of such scans, which illustrate the
general specimen detail in addition to indicating the disposition and relative concentrations of the elements scanned in the specimen area.

X-ray energy-dispersive spectrometry (XEDS) You can observe in Eq. (4.5)
and from Fig. 4.3 that direct measurement of the x-ray photon energy (ΔE)
can allow a more direct relationship to be established between the x-ray
excitation spectrum and the elemental composition of an electron-irradiated
area in a probe beam. By considering the basic concepts of solid-state
energy bands as described briefly in Sec. 1.6, and the specific features

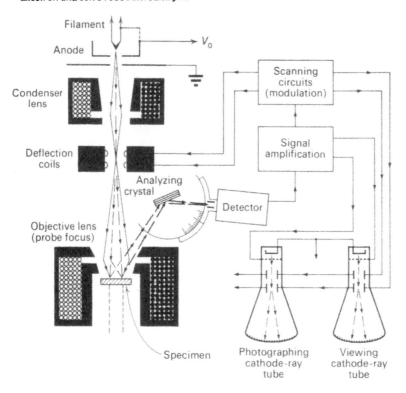

Filament

Anode → V_0

Condenser lens

Deflection coils

Analyzing crystal

Objective lens (probe focus)

Scanning circuits (modulation)

Signal amplification

Detector

Specimen

Photographing cathode-ray tube

Viewing cathode-ray tube

FIG. 4.7 *Scanning electron probe x-ray microanalyzer (schematic).*

of semiconductors illustrated schematically in Fig. 4.9, a solid-state semiconductor detection scheme was developed commercially and utilized successfully over the past decade. This technique is much simpler and far more compact than the wavelength spectrometry systems illustrated in Figs. 4.4 and 4.7. In the solid-state (semiconductor) detector (Fig. 4.9), x-ray photons dispersed from the ionization volume in the specimen enter a thin wafer of pure silicon doped with lithium (to induce extrinsic semiconduction). The doping establishes a favorable ionization which is not too small to allow intrinsic semiconduction by thermal excitation, yet not too large to prevent efficient x-ray ionization. The incident x-ray photon energy causes a proportionate number of ionizations which effectively create a conduction electron (and a hole) for each ionization event. The number of conduction electrons created is thus $\Delta E/E_i$, where E_i is the ionization energy for a single electron whole pair. Electrons created by a single x-ray photon are nearly instantaneously collected by an applied bias voltage, and integrated by a signal amplifier whose output is a series of

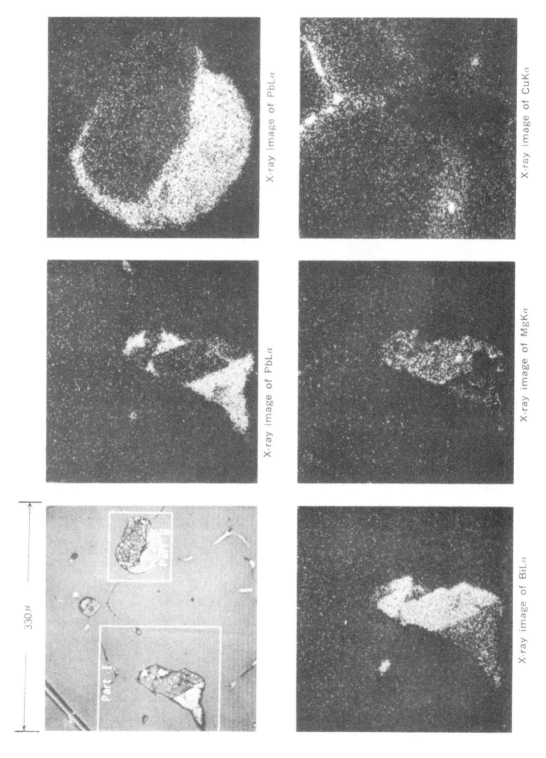

X-ray image of PbLα

X-ray image of CuKα

X-ray image of PbLα

X-ray image of MgKα

330μ

Part 1

X-ray image of BiLα

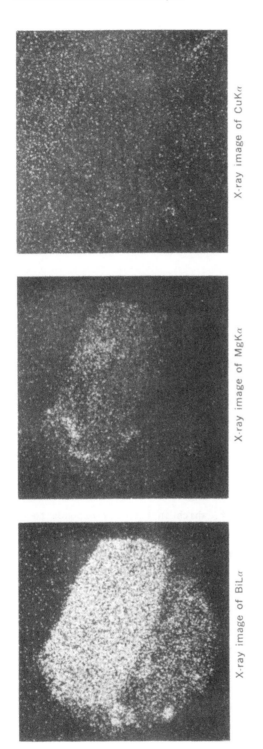

X-ray image of BiLα   X-ray image of MgKα   X-ray image of CuKα

FIG. 4.8 *Electron probe x-ray wavelength – dispersive microanalysis of the distribution of Pb, Bi, Mg, and Cu in Aℓ-Mg-Si sand-cast free-machining alloy.* [*From J. J. Theler, JOEL News, 4: no. 3 (1966)*].

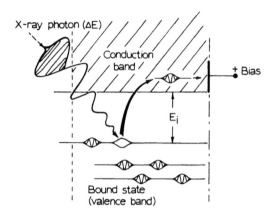

FIG. 4.9 *X-ray photon ionization of semi-conductor for lattice producing conduction electrons and holes in the valence band. This simple schematic ignores the details of the modified Si band structure due to Li doping. The effective gap energy is implicit in the ionization energy, E_i, shown. (Compare with Fig. 2.3)*

voltage pulses having a pulse height proportional to the energy of each corresponding x-ray photon.

These output pulses are measured and counted by a multichannel analyzer (MCA) which is a solid-state computer that can function by allowing specific pulses to charge up a small capacitor, and measuring the time required to discharge with a constant current. This time is used as an "address" to store a "count" in a specific channel in the computer memory. Each channel corresponds to a small energy range. Since this time is also proportional to the amplified pulse height, which in turn corresponds to a specific x-ray energy (ΔE), the accumulated signal in real time builds up a spectrum of characteristic x-ray counts versus x-ray energy which can be displayed or printed out from the memory signal profile. Figure 4.10 shows a schematic diagram of a typical energy-dispersive x-ray detector system along with a typical CRT display of an energy spectrum. The position of each point in specific memory channels corresponds to the total (relative) signal collected in real time. Some systems utilize continuous bars representing signal levels and other similar displays.

Since there is a probability of ionization creating conduction electrons due to thermal vibration of the lattice atoms in the silicon detector wafer (which will be proportional to $e^{-E_i/kT}$), it is necessary to keep the

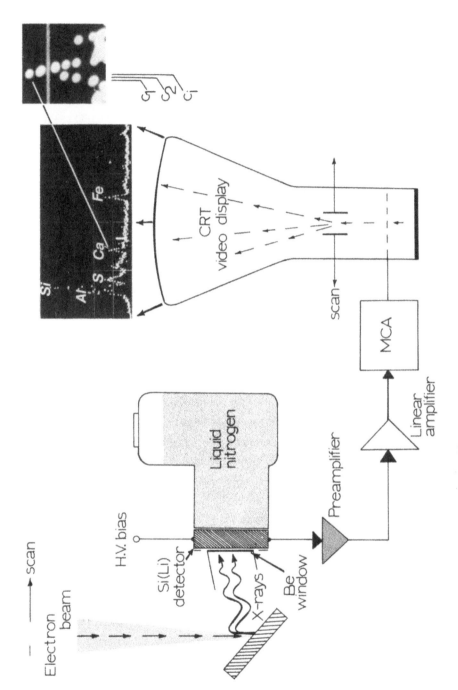

FIG. 4.10 *Schematic diagram of solid-state energy-dispersive x-ray (EDX) detection system with typical CRT multi-channel analyzer display. The enlargement of the signal spectrum portion shows the individual channel voltage level corresponding to the total counts stored in that channel (C_i).*

temperature as low as possible so that this thermal signal (E_T) will be small in comparison to the x-ray generated signal. The magnitude of E_T in the signal constitutes a noise level. Even at liquid nitrogen temperature (78°K), the x-ray signal is barely recognizable for elements below fluorine, and as a result elements below Z = 9 are normally not distinguished by the conventional Si(Li) detectors. Since liquid nitrogen is reasonably cheap and readily available, it is the most common coolant used in energy-dispersive x-ray (EDX) detector systems.

In principle, the detector system shown in Fig. 4.10 replaces the more cumbersome analyzing crystal system shown in Fig. 4.7. Otherwise the scanning electron probe x-ray microanalyzer utilizing energy-dispersive x-ray spectrometry is functionally unchanged, and characteristic x-ray energy maps of elemental distributions can be obtained which are identical to those shown in Fig. 4.8. This is done by limiting or blanking the scan of the specimen surface unless the energy region desired is triggered. This energy region of interest (ROI) is established electronically from the energy spectrum shown in Fig. 4.10, while the beam scan is displayed on an imaging CRT synchronized with the beam scan over the specimen surface area as shown schematically in Fig. 4.7. When the region of interest triggers the CRT display as the beam scans the specimen, the resulting illumination maps out the area from which the gated signal arises. This forms elemental x-ray maps identical to those illustrated in Fig. 4.8. These features of CRT display and the simplicity of solid-state detection make the energy-dispersive x-ray microanalysis (EDX) technique or energy-dispersive spectroscopy (EDS) a complimentary feature of scanning electron microscopy because observed surface features can be rapidly analyzed (at least in a qualitative way, if not in a quantitative way). We will describe these applications in our description of scanning electron microscopy in Chap. 5.

X-ray fluorescence and absorption microanalysis Up to this point we have focused our attention on the detection and analysis of characteristic x-rays, or those x-rays characteristic of a particular element emission after excitation to an appropriate energy state by an incident electron beam. Excitation of the compositional elements of a specimen to emit characteristic radiation can also occur by bombardment of an area with an x-ray beam composed of characteristic radiation from an appropriate reference element. Characteristic x-radiation generated in the specimen in this manner is termed x-ray fluorescence, and it can of course serve the purpose of microanalysis. The technique can, in its simplest form, be brought about simply by inserting a thin target situated above the specimen. Focusing the incident electron beam onto the target material will then produce a fine incident x-ray beam on the specimen or secondary target.

We have previously assumed the specimen to be an infinitely thick medium, with the excitation of characteristic x-rays occurring at a surface area; and we have also tacitly neglected absorption effects. Let us now suppose the specimen is a thin foil, penetrable by electrons and especially by x-rays. We can now treat both the characteristic x-rays generated as well as differentiating intensity profiles of the transmitted x-radiation as they are influenced by the characteristic absorptions of the elements composing the irradiated zone of the thin specimen. Figure 4.11 illustrates the essence of x-ray fluorescence and absorption microanalysis in a thin specimen.

The use of monochromatic x-rays for quantitative absorption microanalysis allows the incident and transmitted x-ray intensities to be related simply by

$$\frac{I_T}{I_o} = e^{-(\mu/\rho)_{eff}m} \tag{4.16}$$

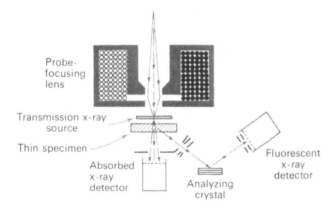

FIG. 4.11 *Microanalysis by x-ray fluorescence and absorption.*

where I_T = transmitted intensity

   I_0 = incident x-ray intensity detected with specimen removed (in Fig. 4.11)

$\left(\dfrac{\mu}{\rho}\right)_{eff}$ = effective or composite mass absorption coefficient

   m = mass per unit area of absorber

Since

$$\left(\frac{\mu}{\rho}\right)_{eff} = \quad c_i\left(\frac{\mu}{\rho}\right)_i \tag{4.17}$$

for an absorber or specimen area composed of n elements of concentration c_i, and from

$$\mu = K'z^{3.3}\lambda^3 \tag{4.18}$$

where K'is a constant, we can observe that substitution of these relation-ships into Eq. (4.16) allows the transmitted x-ray signal to be related to element identity and relative concentration.

It should be emphasized at this point that the determination of an element x in a specimen area containing a mixed composition (for example, a complex alloy system) is monochromatic although x-rays having two diffe-rent wavelengths are employed. That is, by selecting two different primary target elements as shown in Fig. 4.11 (which can be mechanically accom-plished by rotating a multitarget element selector in the incident electron beam path), we can simultaneously select λ_1 then λ_2 to irradiate a desired area. Then, writing Eq. (4.16) as

$$I_{T1} = I_{01}e^{-\left[(\mu_1/\rho)_x m_x + (\mu_1/\rho)'m'\right]} \tag{4.19}$$

and

$$I_{T2} = I_{02}e^{-\left[(\mu_2/\rho)_x m_x + (\mu_2/\rho)'m'\right]} \tag{4.20}$$

for the two incident wavelengths, we can eliminate the background element mass m', and solve for the desired element x:

$$m_x = \frac{\ln(I_{01}/I_{T1}) - (\mu_1/\mu_2)'\,\ln(I_{02}/I_{T2})}{(\mu_1/\rho)_x - (\mu_2/\rho)_x(\mu_1/\mu_2)'} \tag{4.21}$$

We observe from Eq. (4.21) that, for negligible variations of the background absorption coefficient with incident x-ray wavelength, we can substitute for the background absorption coefficient [as implied in Eq. (4.18)] and rewrite Eq. (4.21) as

$$m_x = \frac{\ln(I_{01}/I_{T1}) - (\lambda_1/\lambda_2)^3 \ln(I_{02}/I_{T2})}{(\mu_1/\rho)_x - (\mu_2/\rho)_x (\lambda_1/\lambda_2)^3} \tag{4.22}$$

In Eq. (4.22) the sensitivity of element determination increases with a difference $(\mu_1/\rho)_x - (\mu_2/\rho)_x$; the general trend for variations in this difference with radiation character and the atomic number can be readily observed in the curves of Fig. 4.12. It should also be apparent from Eq. (4.22) that the determination of an element, x, is also sensitive to the accuracies with which the x-ray intensities can be measured. It should also be pointed out that while we are considering, momentarily, only absorption, the detected signal will in reality contain some wavelength distortions due to fluorescence. Corrections in such measurements must therefore, of necessity, be made for such disturbances where a high degree of accuracy is required.

4.2.2 X-RAY MICROSCOPY

X-ray fluorescence and absorption microanalysis as discussed above are often referred to as analytical modes of x-ray microscopy. We shall now briefly discuss the concepts of basic x-ray microscopy, or the direct observation of structural detail of a specimen target by direct imaging of transmitted x-rays, or the recording of diffracted x-rays created by the direct bombardment of the specimen with a focused electron beam where the specimen acts as a self-consistent x-ray source. We can best visualize these modes of x-ray analysis by modifying Fig. 4.11 as shown in Fig. 4.13, where we simply replace the electronic detector by a photographic emulsion at a suitably extended projection distance from the point of excitation in the specimen. It is perhaps unnecessary to point out that the specimen, for either transmission radiography or transmission Kossel-line determination, must be sufficiently thin to allow for the x-ray penetration, while the reflection Kossel-line technique is equally effective for specimens of infinite thickness.

Figure 4.13 further shows that conventional specimen radiography is also accomplished by inserting a photographic emulsion in contact with the

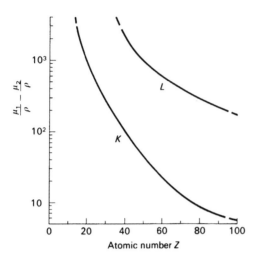

FIG. 4.12 *Difference in mass absorption coefficients for K- and L-state x-radiation as function of atomic number.* [*From A. Engstrom,* Acta. Radiol. *Suppl., 63 (1946).*]

specimen. In the case of projection x-ray microscopy, the effective x-ray source in the primary target can be made as small as 1 micron by using a fairly low electron accelerating potential. Consequently, for the specimen or secondary target situated very close to the emergent side of the primary target, the magnification attained in x-ray projection of specimen structure can be made quite high. The resolution associated with such projected x-ray images, is however, little better than that attainable with light microscopes.

4.2.3 VISIBLE FLUORESCENCE MICROANALYSIS: CATHODOLUMINESCENCE

Many materials exhibit fluorescence in the visible portion of the spectrum both on the incident and emergent side of a thin specimen irradiated by an electron beam. To be specific, however, the mechanism of this photoemission is similar to luminescence in the sense that it is stimulated by activator elements in trace amounts in the sepcimen area irradiated with the probe beam. Thus, just as normal luminescence can indicate, with regard to its intensity and color, the distribution of trace elements in a sample, excited luminescence by electron bombardment can give similar evidence. It must be cautioned at the outset, however, that cathodoluminescence cannot give direct compositional information of a specimen area,

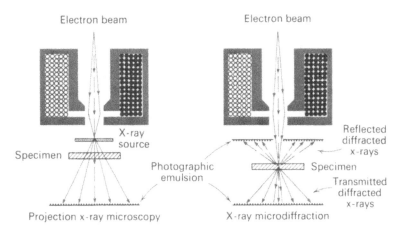

Projection x-ray microscopy

X-ray microdiffraction

FIG. 4.13 *Transmission x-ray microscopy and Kossel line x-ray diffraction microanalysis.*

but it can supplement the analysis with relative-distribution data for trace elements or phases whose luminescent behavior is known. It has also been shown that cathodoluminescence in semiconductor specimens is voltage dependent and, as a consequence, the surface states in certain semiconductors are amenable to study by cathodoluminescence. Numerous related semiconductor features have also recently been investigated including excess carrier lifetime, diffusion lengths, etc. [8,9,10]. The technique seems particularly well suited to the detection of trace elements in mineral samples and the like, and it can give some clues concerning the valence states in cases where the element distributions and relative concentrations are known. The spectral composition of the luminescence can also be analyzed in a conventional manner.

The actual design features to be incorporated in the electron optical system of a probe for cathodoluminescence observations are not extensive. A simple mirror system as indicated in Fig. 4.14 will suffice for observations by eye or by modulating the photodetected signal on the scanning beam, thereby developing a cathode-ray-tube image of the luminescence profile of a specimen area.

4.3 MICROANALYSIS BY ELECTRON INTERACTION SPECTROSCOPIES

4.3.1 *ELECTRON BACKSCATTER AND ABSORPTION*

In addition to the visible fluorescence created in specimens onto which an electron beam is focused, the electrons themselves can intrinsically inter-

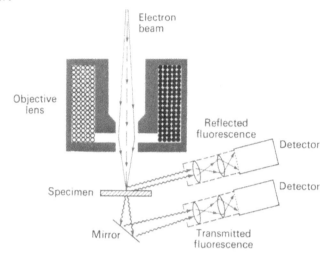

FIG. 4.14 *Cathodoluminescence detection.*

act with the irradiated area by reflecting from the area, by absorption in the area, and by the generation and emission of secondary electrons. In any of these alternative and simultaneous dispositions of the incident electrons, the degree of backscatter or absorption (which subsequently leads to secondary emission) is probabilistic in nature, with the composition and topography of the specimen area under study the main features determining the relative extent of either. Thus we can visualize phenomenologically that for a very rough surface morphology, scattering of electrons would be decidedly different from that for a perfectly smooth surface. In addition, if the area under investigation were composed of an element or elements of decidedly different atomic number, the degree of absorption would, as one might expect, become smaller with increasing element density (increasing Z). As a consequence, and neglecting surface topography, the number of reflected (backscattered) electrons will increase with increasing atomic number. This feature is ideally illustrated in the data reproduced from the early work of Palluel [11] in Fig. 4.15, where I_B is the backscattered electron intensity, and I_A is ideally the portion of the beam absorbed in the specimen (as detected by measuring local current).

The ideal situation depicted in Fig. 4.15, it must be cautioned, is subject to deviations caused by complex scattering where the surface topography in the area under investigation is irregular. There is really no reliable means of estimating the relationship of the degree or character of surface roughness, etc., with the backscatter probability or the effi-

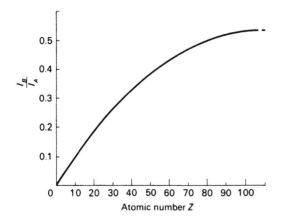

FIG. 4.15 *Backscatter-absorption ratio versus atomic number.* [*From P. Palluel, C. R.
Acad. Sci., 244: 1492 (1947).*]

ciency of absorption, as the case may be. Nevertheless we can appreciate
the fact that with reference to Fig. 4.15, backscatter and absorption can
be related to Z, which of course gives us an analysis of the area in question. Thus if we can eliminate, for the most part, the varying degree of
surface roughness by consistently polishing a specimen or specimens to a
smooth, flat finish, the detection of backscattered electrons in a static
or scanned electron beam can be related to element composition through Z.
Where the specimen area of interest consists of i elements, T. Danguy and
R. Quivy [12] relate the effective atomic number of the specific elements
by

$$Z_{eff} = \frac{\Sigma C_i Z_i^2}{\Sigma C_i Z_i} \qquad (4.23)$$

where C_i is the concentration (atomic fraction) of the ith element. Thus
in a simple two-element system (that is, a binary alloy), $C_1 = 1 - C_2$; and
if Z_{eff} is determined by measuring the overall reflection/absorption ratio
and then referring to Fig. 4.15, the concentration of element 2 can be determined from

$$C_2 = \frac{Z_1^2 - Z_{eff} Z_1}{Z_1^2 - Z_2^2 + Z_{eff} Z_2 - Z_{eff} Z_1} \qquad (4.24)$$

provided Z_1 and Z_2 are identified independently by x-ray emission scan as discussed in Sec. 4.2, or simply by knowing that Z_1 and Z_2 compose the sample. We observe from Eq. (4.24) that in the case of a simple binary system, the concentrations of the two composing elements can be accurately determined by backscattered electron analysis (see Prob. 4.3), and in this way backscatter analysis extends the resolution of the more conventional x-ray emission mode.

The absorbed electrons contribute to the effective specimen charge, and constitute a specimen current in a situation where the specimen is actively connected to an electric circuit. Thus one approach to the determination of the backscattered electron intensity is to simply measure the incident beam current in amperes, and to compare this with the specimen current in amperes. If the incident beam of a scanning-probe beam is then modulated with the specimen current, a visual display of signal attenuation with element absorption is obtained and gives contrast to those areas of the specimen containing various concentrations of elements, as well as to their locations. Where the specimen is biased positively with respect to a great reference, those low-energy electrons emitted from the surface (secondary electrons) will be unable to escape with the backscattered (high-energy) electrons and, as a consequence, the error in detecting secondary emission is curtailed. A biasing system of this type is illustrated in Fig. 4.16, which serves to illustrate the *modus operandi* of backscattered-adsorption analysis in general. The actual detection of backscattered electrons and a direct analysis of backscattered electron intensity profiles in an area scanned by a probe beam can, of course, give some indication of surface topography because of the variation in backscatter efficiency with varying degrees of surface roughness and specific morphological features.[†] By sampling the electron absorption, or sample current, the inverse (intensity considered) of the backscattered display image can be obtained, which gives in an optical sense a negative image of surface structure.

While we have treated surface-structure resolution in the previous discussion, it must be pointed out that charge disturbances, potential distributions, and related phenomena that contribute or detract from the sig-

---

[†]*In effect, the surface structure can be displayed visually as indicated in Fig. 4.38, and with sufficient resolution, can give an accurate interpretation of surface morphology. This feature is dealt with in detail in Chap. 5.*

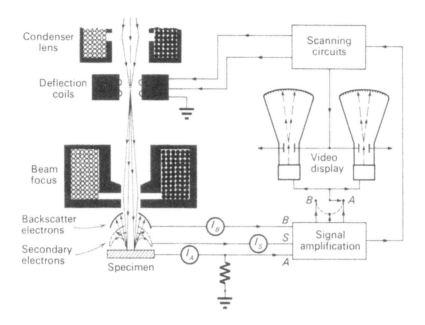

FIG. 4.16 *Specimen biasing for electron backscatter, absorption, and secondary emission current.*

nal detected can be studied with respect to image-contrast features, etc. Such analyses constitute a major analytical feature of scanning electron microscopy and are treated in more detail in Chap. 5.

4.3.2 SECONDARY ELECTRON EMISSION AND AUGER PROCESSES

Secondary electrons differ very markedly from backscattered electrons in energy. And, because of their relatively low energy, and wider range of values, secondary emission electrons can be very useful in providing information relevant to the surface topography and crystallography of a specimen area. Just as primary electron emission as discussed in Chap. 2 depends on the surface electron energy modes and crystallography, secondary emission is governed by very similar, and in some cases identical factors. Consequently, because electrons emitted in the various topographies of a specimen surface will normally possess varying velocities (energies), these areas will vary the intensity profiles of a modulated beam observed on a cathode-ray display tube. A depth of focus is thereby attained in which the surface is viewed in profile. Thus the detection and analysis of secondary electron emission constitutes a form of surface and scanning electron microscopy, which will be treated in Chap. 5.

We have previously discussed the fact that when a solid is irradiated with an electron beam, energy-state transitions as a result of element electron excitation produce characteristic x-rays. If we now look in retrospect at the accompanying absorption, backscatter, etc., processes (refer again to Fig. 4.1), we might logically question the efficiency of characteristic x-ray production. To be sure, the excited electron of an atom in a solid need not return to its lower energy state in order to lower the atomic energy level. The atom may, as an alternative or simultaneous process, return to the lower energy state by simply ejecting one of its electrons as an alternative to the quantized emission of x-radiation. Such a process can be termed internal conversion, and is commonly referred to as an Auger process, named for its discoverer [13].

Let us consider briefly the mechanics of the Auger process. We will consider an atom with a K electron excited to the L level. Now if an L electron is ejected as an Auger electron, the atom will be left doubly excited. If an L electron now drops into the K level, a characteristic x-ray photon is released of wavelength $\lambda_{K\alpha}$, as we noted previously. In addition the atom is still in the L excited state, and may return to its original energy state by an electron jumping down to the L level from a higher quantum level. The emitted radiation thus occurs as a satellite line (an emission line having a low amplitude and wavelength quite apart from the characteristic or main x-ray photon emitted). Obviously other combinations of event can occur, but regardless of the sequence of the atom's return to its orginal energy state, the ejected Auger electron possess a kinetic energy unique to the atom and the quantum level of its origin. In effect, Auger emission is a two-electron process involving three electron energy levels; not necessarily all different.

The detection of Auger electron signals can afford an analysis in many ways identical to the detection of characteristic or fluorescent x-rays. Auger electron spectroscopy is particularly useful for the analysis of low-Z elements. We should of course realize that Auger electrons differ markedly from secondary electrons emitted near the specimen surface because of the discreteness of their kinetic energies. Auger electrons also originate as a "reflected signal profile" in thick specimens, but they can occur in the transmitted signal (forward beam direction) where the specimens are sufficiently thin.

4.3.3 ELECTRON SPECTROSCOPIES

We have treated essentially three modes of electron beam interaction with solid materials: absorption, elastic scattering, and inelastic scattering. Elastic scattering processes can be regarded as giving information regarding spatial arrangements, while inelastic processes contain information concerning the electronic and chemical structure. Elastic scattering involves essentially no loss of the initial beam energy while inelastic scattering occurs with initial electron energy loss. Inelastic scattering of electrons from a solid can actually occur in a number of complex processes involving incident "particles" other than electrons, including x-ray photons, ultraviolet photons, ions, etc.* The inelastically scattered electrons arise by collisions of the incident probe "particle" with *atomic* or *molecular* electrons. These collisions cause the incident particle to loose some or all of its energy and to change its direction. As a consequence, both the energy loss or transfer and the excitation and detection geometries (angle of the incident beam and scattered signal relative to the specimen geometry) are important in optimizing the collection and analysis of characteristic emission spectra. In addition the inelastic scattering cross-section increases dramatically with decreasing atomic number. Consequently for low Z materials, the larger fraction of information resulting from the probe interacting with the specimen is contained in the inelastic scattering. The inelastically scattered electrons, because they arise by atomic or molecular ionization, contain information specific to the atomic or molecular structure. This can be elucidated by determining the specific energies of the inelastically scattered species.

There are basically three prominent electron emission processes associated with inelastic scattering from the atomic core region: photoelectron emission, Auger emission, and energy-loss electron emission. The energy-loss electrons are the initial electrons which have lost energy in the ionization process. Furthermore, although Auger electron emission is a "radiationless" transition, characteristic x-rays can be emitted along with the Auger electrons from the ionization volume below the focused probe on the specimen. These processes can all go on simultaneously, so that the efficiency of any one process is sometimes difficult to control except by

---

*In considering the "particle" nature of "electromagnetic photons" recall that from the deBroglie relationship $m = h/c\lambda$.

utilizing specific incident particles and particle energies as well as incident beam angles which may optimize the production of a specific emission species. The take-off angle for detection can also be important.

Although we have described the Auger process previously, it is appropriate to illustrate it in the context of the related inelastic electron scattering or emission processes, and the x-ray fluorescence process. These are shown in Fig. 4.17.

In Fig. 4.17(a) primary "particles" (which could include electrons, ions, x-rays, or high-energy photons) incident upon an atom or molecule produce photoelectrons, characteristic x-ray, and energy-loss particles if the incident energy is not absorbed in the ionization process. The atom or molecule becomes singly charged (M^+) as shown. In Fig. 4.17(b) an electron beam interacts with the atom or molecule causing photoelectron emission, Auger electron emission, and energy-loss electron emission if the primary electron energy is not absorbed in the ionization process. The atom or molecule becomes doubly charged (M^{2+}). In this process, the energy which is released as x-ray fluorescence in Fig. 4.17(a) is expended in the production of the Auger electron in Fig. 4.17(b). However at any moment after this event (the production of the Auger electron) electrons can occupy the E_2 level from higher energy states, and this "cascade", even to the E_1 level, can produce x-ray fluorescence as illustrated in Fig. 4.17(a). In Fig. 4.17(c), an incident x-ray photon of energy E produces a photoelectron of energy E_p. If this is the only process which occurs, then the ionization energy, E_1, can be determined as shown.

The production of Auger electrons as shown in Fig. 4.17(b) and of photoelectrons in Fig. 4.17(c) can occur, as implicit in Fig. 4.17(a), by the inelastic scattering of any energetic particle from an atom or molecule. However these simple sketches illustrate the "favorable" features of these specific processes. Although the energy-loss particles (if it does occur in the emission spectrum) can occur in the "reflected" portion of the emission signal, they normally occur in the "forward" (beam) direction. As a consequence, energy-loss spectra are normally very favorable in thin film microanalysis.

We might imagine that the three schematics in Fig. 4.17 are a portion of the contiguous ionization volume in a specimen material. Indeed, for a single-particle (monochromatic) beam focused upon a specimen area, all of these processes will occur simultaneously, but each with a specific efficiency determined by the incident beam parameters (energy, mass, geometry,

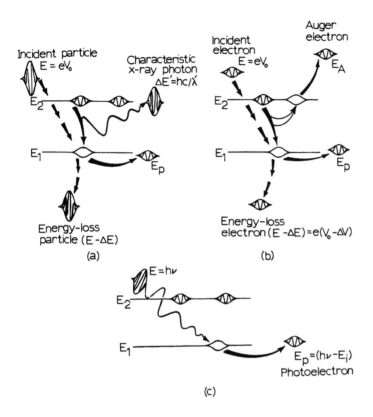

FIG. 4.17 *Atomic and molecular ionization processes*
involving single and multiparticle transitions asso-
ciated with inelastic scattering. (a) Primary parti-
cle of energy E producing photoelectron of energy E_p
and characteristic x-ray of energy E', and energy-loss
(exit) particle, (b) Primary electron of energy E pro-
ducing photoelectron of energy E_p, Auger electron of
energy E_A, and energy-loss (exit) electron, (c) Inci-
dent x-ray photon of energy E produces only photo elec-
tron of energy E_p.

etc.). Now imagine, in addition, that these atoms or molecules are organized in a "continuum" of "free" electrons. That is, electrons not bound to
the atoms or molecules; as depicted energetically in Fig. 2.3. The focused
primary beam will also produce secondary electrons from this regime, and
from the less energetically bound outer electrons associated with heavier
atoms ($Z > 8$), regardless of the type of energetic primary particles (electrons, ions, x-ray photons, etc.) For very energetic primary particles,
the atoms or molecules themselves can be desorbed (or sputtered) from the

surface. Elastic scattering such as primary particle back-scattering can also occur (e.g. electron backscatter for a primary electron beam).

We might mention that because there is a kind of connection between the electrons in this "free" electron continuum alluded to above, (and also certainly between them and the atoms, especially in metals, alloys, and semiconductors) these particle-beam interactions with the specimen give rise to plasma density oscillations in the electron gas. These oscillations are quantized, and are known as plasmons. The so-called bulk plasmon is a charge density fluctuation in the electron gas involving many electrons simultaneously in a coherent oscillation. The plasmon energy $(4\pi hne^2/m$; where n is an integer of charge), is of the order of 15 to 20 eV. Plasmons can appear in the electron energy-loss spectrum [14] (See Fig. 4.21).

Finally, since we are considering continuum effects we should also mention the phenomenon of "bremsstrahlung" or continuum radiation. This is electromagnetic radiation (photon emission) which is produced by charged particles deflected by the nuclear charge. This process can be described as radiative collision where the acceleration of the incident particle essentially causes this radiation. The bremsstrahlung is emitted with a range of energies from zero up to that of the primary particle beam. The bremsstrahlung intensity is proportional to $q^2 Z^2/M^2$, where q is the primary particle charge, and M is its mass (and of course Z is the atomic number). Since the bremstrahlung is superimposed on any detected spectrum, this signal "noise" will be reduced very rapidly as the incident particle mass is increased. This will be apparent in ion probe systems as compared to electron probe systems because of the enormous difference between the electron mass and the ion mass (recall that the mass of the proton is almost 1900 times greater than the electron mass!).

Auger electron spectroscopy (AES) When an electron beam of several keV (or greater or smaller) energy is focused onto a surface area at some acute incidence angle (<45°), the penetration is shallow and efficient Auger electron emission is detected in or near a few atomic layers from the surface. The emission spectra are characteristic of the ionized species and impurities on the surface, or in the surface region (a few atomic layers from the solid/vapor interface); a comparison of the relative intensities can give some indication of the impurity concentration. Using a field emission electron gun, electrons can be focused with enormous intensity onto relatively small areas where the fine structure in the Auger peaks can be related to the chemical environment on the specimen surface.

The detection system can vary, but an efficient signal collection and analysis scheme is illustrated in Fig. 4.18. In this simple system, the divergent Auger surface emission profile is focused into a 180° electric field sector analyzer utilizing the concept illustrated in Fig. 3.20. In this arrangement, Auger electrons having specific energies are separated into different paths having an effective radius of curvature R_j:

$$R_j = \left(\frac{2E_A}{e} \right)_j \left(\frac{1}{E} \right) \tag{4.25}$$

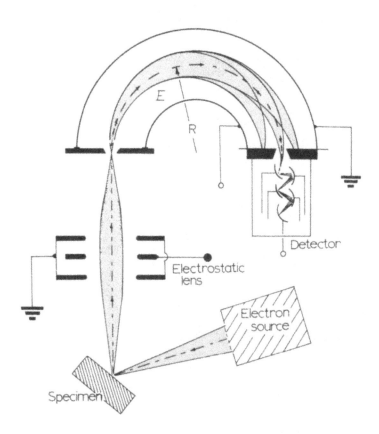

FIG. 4.18 *Simple electrostatic analyzing system for Auger electron spectroscopy (schematic).*

This assumes of course that the electric field is kept constant. By adjusting the electric field strength, E, the analytical sensitivity can be varied. The sensitivity is also increased by using the full 180° sector field because this provides for the maximum signal divergence (ΔE_A) at the detector. By continuously changing E it is possible to selectively and continuously vary the specific energy radii so that a continuous spectrum of counts can be detected through a single, fixed, exit aperture. In effect, the foci of Auger electron radii are adjusted to R_o shown in Fig. 4.18. Alternatively the field could be maintained constant and the detector slit moved to each point of focus. Figure 4.19 illustrates an example of a continuous Auger spectrum.

We might mention that the use of electrostatic lenses and electric field spectrometers is an efficient concept in Auger electron spectroscopy because the relatively low characteristic energies do not present any

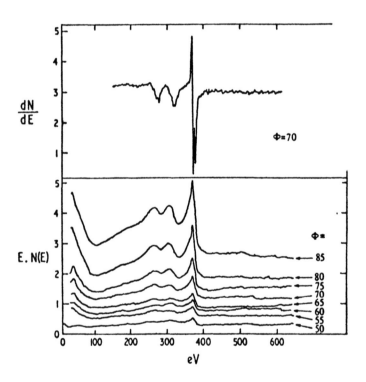

FIG. 4.19 *Auger electron spectra for Cd for various incidence angles, Φ the primary beam energy was 30 keV. dN/dE combines the implications of Eqs. 4.11 and 4.12 for Auger electron emission. From A. P. Janssen, C. J. Harland, and J. A. Venables,* Surface Sci., *62: 277 (1977)].*

serious design problems normally associated with very high energy electron optical systems such as the electron microscope. In addition, the resolving power of the 180° sector electric or magnetic field analyzer as conceptually illustrated in Fig. 4.18 is given generally by

$$\left(\frac{\Delta E}{E}\right) = \frac{W}{2R_0} \qquad\qquad (4.26)$$

where W is the total slit (aperture) width (entrance plus exit slit width), and R_0 is the axial radius of the spectrometer shown in Fig. 4.18.

You should recognize the possibility of incorporating an Auger electron detector (incorporating the essential features illustrated in Fig. 4.18) into a scanning electron microprobe system such as that shown schematically in Fig. 4.7. Here the crystal-analyzer detector arrangement would be replaced by the electron optical detector system shown in Fig. 4.18. The specimen could also be tilted into a favorable position of incident and take-off angle for maximum signal detection. This system can also be incorporated into a scanning electron microscope system, and we will describe these concepts in more detail in Chap. 5. Applications of scanning Auger microanalysis are also treated briefly in Section 4.5.3.

Auger electron spectroscopy is now a widely used technique for metal and inorganic surface analysis. Depth profile analysis can be coupled with the Auger electron spectrometer system to systematically desorb surface atoms by ion bombardment or high-energy electron bombardment (electron impact desorption) [15]. Furthermore, since the Auger electrons are detected in the first few atomic layers, the chemical information obtained applies to a depth of only about 10 Å. This is a very sensitive surface analysis method. Numerous examples of Auger electron spectroscopy applications have appeared regularly in the journal *Surface Science* over the past decade [16], and there is not a single issue of that journal which does not contain at least one article detailing an AES application. The technique has been a prominent metallurgical application for at least a decade [17], and Hawkins [18] has compiled a list of thousands of AES applications in the materials sciences up to 1975. Commercial AES units and modules have also been available for more than a decade from companies in the U.S.A. and Europe.

Electron spectroscopy for chemical analysis (ESCA), x-ray photoelectron spectroscopy (XPS), and ultraviolet photoelectron spectroscopy (UPS) ESCA is a general acronym which characterizes the analysis of elemental and

chemical phenomena in specimen surface layers. X-ray or ultraviolet photons in a primary beam of energy $h\nu$ interact with surface adsorbates or
surface layers in a specimen and eject inner shell photoelectrons with
characteristic energy E_p. By using U.V. or x-ray photons, the effective
mass will be so small that inner shell interactions with surface atoms is
favored, enhancing the photoelectron emission efficiency. If essentially
all of the primary photon energy (in the incident monoenergetic beam) is
expended in the creation of specific inner shell photoelectrons, then the
specific ionization energy, E_i, can be determined from the conservation of
energy principle:

$$E_i = h\nu - E_p; \quad E_p = h\nu - E_K - E_W$$

where E_K is the binding energy and E_W is the work function.

By measuring the spectrum of photoelectron emission from a sample (intensities and angular distribution of the photoelectron emission spectrum)
the electronic structure of atoms and molecules can be elucidated. This is
accomplished in a schematic arrangement as shown in Fig. 4.18, with the
electron source being either replaced or augmented by an appropriate x-ray
or U.V. source. In fact it is practical to utilize a common analyzer system as shown in Fig. 4.18 for all electron spectroscopies and even ion
spectroscopies by simply replacing the source by an appropriate module for
creating monoenergetic electrons, U.V. photons, x-rays or ions. You should
recognize that the inner-shell photoelectrons are not the same as the
photoemission electrons described in the emission procedures of Chap. 2.

Photoelectron emission spectroscopy has also become a very common
surface analysis technique over the past decade, and many applications
have been reviewed in the bibliography by Hawkins [18], and in recent work
of McRae [19]. Contemporary applications are also described in essentially
all issues of the *Journal of Electron Spectroscopy and Related Phenomena*
and *Surface Science*.

In some respects, ESCA, UPS, and XPS are not specifically microanalysis techniques because the probe beam size is generally in the micron
range. We include them here more for completeness, and the fact that they
can be combined with other microanalysis techniques.

Electron energy loss spectroscopy (EELS) Electron energy loss spectroscopy or generally energy loss spectroscopy (ELS) (which can also designate electron loss spectrometry) is generally concerned with a transmitted

signal (forward direction spectrometry) rather than a reflected signal profile. The significance of this feature is perhaps obvious upon perusing
Fig. 4.17 in retrospect. Consequently you can appreciate the fact that the
technique applies most effectively to thin film microanalysis. You can see
from Fig. 4.17 that information obtained with energy loss electrons is a
direct result of the fundamental excitation process, and does not suffer
the competitive complexities associated with characteristic x-ray or Auger
electron production. In effect, EELS has essentially the highest charac-
tersitic yield per inelastic scattering event of all the microanalysis
techniques discussed above (and implicit in Fig. 4.17). However, the technique can compliment x-ray energy dispersive microanalysis as a quantitative microanalysis tool. The most practical and significant aspects of
EELS involve the analysis and identification of light elements and high-
energy resolution measurements of local electronic structure or chemical
bonding.

Although a varerity of magnetic and electrostatic spectrometer systems
(Chap. 3) can be used to measure the energy distribution of energy-loss
electrons, the use of large uniform-field magnetic sector analyzers has
been successful, especially because large trajectory radii are able to
provide efficient resolution of beam divergence. In this scheme, shown
schematically in Fig. 4.20, monoenergetic electrons (E_0) generated in a
thin film volume are brought to a focus through a radius R. Electrons
having a spread in energy ($E_0 - \Delta E$) are bent through other radii as described in Eq. (3.29) and are focused at points displaced from the zero-
loss or monoenergetic focal reference. These displaced focal points can
be moved to the reference radius by adjusting the magnetic field of the
spectrometer [Eq. (3.29)]. If the separation of the two focused rays on
the focal plane in Fig. 4.20 is Δx, then the spectrometer dispersion is
defined by

$$D = \frac{\Delta E}{\Delta x} = \frac{2R}{E_0} \qquad (4.26a)$$

This expression is comparable to the expression shown previously in Eq.
(4.26). In fact, for a detector slit width, W_d, the spectrometer resolution in Fig. 4.20 would be W_d/D, although resolution cannot be improved
infinitely by reducing the slit size because signal aberrations are introduced beyond certain small slit sizes.

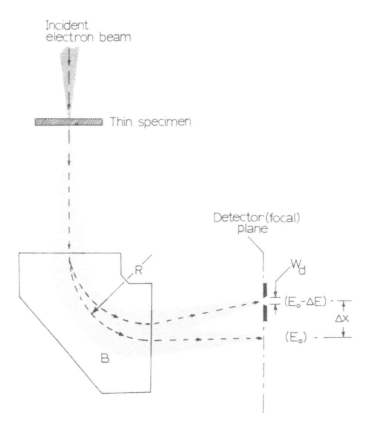

FIG. 4.20 *Schematic view of magnetic sector spectro-meter system for electron energy loss measurements.*

Figure 4.21 shows a segment of an energy loss spectrum recorded from a Be film (~ 0.05 μm thick) using an incident electron beam energy of 120 keV. The zero loss peak corresponds to the elastically scattered primary electrons, while the small signal features extending up to about 50 eV are the plasmon loss peaks which result from inelastic scattering of the primary electrons from conduction or valence band electrons as described previously. Above 50 eV, the spectrum is characterized by a generally decreasing background intensity with abrupt edges corresponding to energy loss phenomena associated with the ionization of inner-shell atomic electrons by the primary electrons. The threshold energy of these edges is in fact equal to the energy required to ionize a particular electronic level, and they are correspondingly labled K, L_{II-III}, M_{IV-V}, etc.

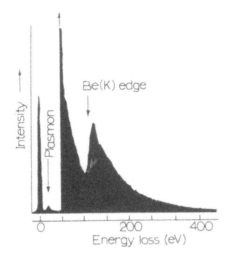

FIG. 4.21 *Experimentally measured electron energy loss spectrum for a 0.05 μm Be film. V_O = 120 kV* [*See Egerton* [20] *and Zaluzec* [21]].

For thin films composed of numerous elements, the EELS spectrum consists of the loss (or absorption) edges corresponding most prominently to the characteristic K-shell ionizations. Consequently the spectra, resembling Fig. 4.21, contain additional, abrupt edges corresponding to the elements present, while the intensity of the edge or the ratio of intensities are indicative of the elemental concentrations in the irradiated volume.

Electron energy loss spectrometry has become a standard analytical procedure coupled with transmission electron microscopy, especially scanning transmission electron microscopy to be discussed in detail in Chap. 7. We will describe some specific applications of the technique in the remainder of this chapter as well as in Chap. 7. You might also consult "Introduction to Analytical Electron Microscopy", edited by J. J. Hren, J. I. Goldstein and D. C. Joy (Plenum Press, New York, 1979) for more details of the EELS technique and some applications in both the physical and biological sciences.

4.4 RESOLUTION, CORRECTION FACTORS, AND ELEMENTAL ANALYSIS IN THICK SPECIMENS AND THIN FILMS

The design limitations and range of problem applications of microanalyzers can only be assessed with any real meaning if the various emission efficiencies are known, and if some assurance, through logical correction schemes, of the accuracies of signal detection can be obtained. If we limit our attention for the moment to the detection of characteristic x-rays, we might consider the simple mechanistic picture of x-ray excitation in a solid target as shown in Fig. 4.22. The point of excitation 0 might be considered to be the penetration depth as given previously in Eq. (4.10). The corresponding path lengths 0A and 0B, corresponding to the variations in takeoff angle ϕ_t, reflect a varying degree of signal detected as a result of absorption processes that increase with increasing path length and other processes contributing to the loss in signal emission efficiency. If we characterize the number of x-ray quanta generated per incident electron at point 0 of Fig. 4.22 as N_o, then the number generated per steradian per electron becomes $N_o/4\pi$ for isotropic emission from point 0. The actual number of x-ray photons available for detection at the solid surface will, however, depend on the absorption as related by

$$I_{eff} \equiv N_{eff} = \frac{N_o}{4\pi} \frac{\mu}{\rho} \csc \phi_t \qquad (4.27)$$

where it is observed that as ϕ_t, decreases, the intensity (efficiency) of characteristic radiation decreases. Quantitatively, the intensity of radiation at the point 0 of Fig. 4.22 can be expressed as

$$I_o = f(Z)(U-1)^{1.65} \qquad (4.28)$$

where f(Z) is a function dependent on the element atomic number, and also on other subtle interactions, such as backscatter, which are intimately

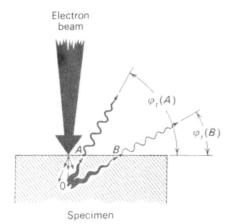

FIG. 4.22 *Characteristic x-ray excitation in solid target.*

related to Z for the most part,[†] and

$$U = \frac{V_o}{V_K}$$

where of course V_o is the electron beam accelerating potential, and V_K is the excitation potential for the characteristic x-ray emission lines as shown, for example, in Table 4.2 for $K\alpha$ and $L\alpha$ emission from the various elements.

Aside from the absorption of x-rays in the sample, the emission profile is also usually "contaminated" with fluorescent excitations that must be deleted from the signal intensity detected in order to obtain a primary intensity. In effect, if the characteristic $K\alpha$ radiation produced in a specimen has a shorter wavelength than that of the element under study, $K\alpha$ fluorescent emission lines result. The fluorescence correction for excitation of K lines by K lines follow several forms, among which the most successful seem to be those due to Castaing [22] and Wittry [23]. If we consider the intensity ratio measured in the form, say, of Eq. (4.15), then the true intensity is expressed by

$$\left[\frac{I_A}{I_{A'}} \right]_{true} = \left[\frac{I_A}{I_{A'}} \right]_{obs} \frac{(\mu/\rho_A \csc\phi_t}{(\mu/\rho)_{A'} \csc\phi_t + G} \tag{4.29}$$

where G is the correction factor for x-ray fluorescence, which in turn depends to some degree on atomic number, etc. The dependence of the relative correction factor G on the takeoff angle is illustrated in Fig. 4.23, which also serves to exemplify the relationship of the spectrometer-detector geometry as it affects the efficiency of signal analysis in the classical microprobe (wavelength dispersive spectrometer) and energy-dispersive x-ray spectrometers.

While we have considered specifically the fluorescence correction, which directly involves absorption lengths, ϕ_t, etc., it must be realized that basic data correction for very accurate quantitative analysis must also consider corrections involving observation time (exposure time), specimen contamination by the electron beam (a feature of concern in all electron optical systems employing oil-diffusion pumping or the introduc-

---

[†]*See empirical expressions derived for $K\alpha$ generation by M. Green and V. E. Cosslett,* Proc. Phys. Soc. (London) *78: 1206(1961).*

TABLE 4.2   *K- and L-excitation potentials (in kV)* [†]

Z	Element	V_K	V_L	Z	Element	V_K	V_L
6	C	0.28		54	Xe	34.6	5.50
7	N	0.40		55	Cs	35.9	5.71
8	O	0.53		56	Ba	37.4	5.99
9	F	0.69		57	La	38.7	6.26
10	Ne	0.87		58	Ce	40.3	6.54
11	Na	1.07		59	Pr	41.9	6.83
12	Mg	1.30		60	Nd	43.6	7.12
13	Al	1.55		61	Pm	45.3	7.40
14	Si	1.83		62	Sm	46.8	7.73
15	P	2.14		63	Eu	48.6	8.04
16	S	2.46		64	Gd	50.3	8.37
17	Cl	2.82		65	Tb	52.0	8.70
18	Ar	3.20	0.29	66	Dy	53.8	9.03
19	K	3.59	0.34	67	Ho	55.8	9.38
20	Ca	4.03	0.40	68	Er	57.5	9.73
21	Sc	4.49	0.46	69	Tm	59.5	10.10
22	Ti	4.95	0.53	70	Yb	61.4	10.5
23	V	5.45	0.60	71	Lu	63.4	10.9
24	Cr	5.98	0.68	72	Hf	65.4	11.3
25	Mn	6.54	0.76	73	Ta	67.4	11.7
26	Fe	7.10	0.85	74	W	69.3	12.1
27	Co	7.71	0.93	75	Re	71.5	12.5
28	Ni	8.29	0.99	76	Os	73.8	13.0
29	Cu	8.86	1.10	77	Ir	76.0	13.4
30	Zn	9.65	1.20	78	Pt		13.9
31	Ga	10.4	1.31	79	Au		14.4
32	Ge	11.1	1.41	80	Hg		14.8
33	As	11.9	1.52	81	Tl		15.3
34	Se	12.7	1.64	82	Pb		15.8
35	Br	13.5	1.77	83	Bi		16.4
36	Kr	14.3	1.90	84	Po		16.9
37	Rb	15.2	2.05	85	At		17.5
38	Sr	16.1	2.19	86	Rn		18.0
39	Y	17.0	2.36	87	Fr		18.7
40	Zr	18.0	2.51	88	Ra		19.2
41	Nb	19.0	2.68	89	Ac		19.8
42	Mo	20.0	2.87	90	Th		20.5
43	Tc	21.1	3.00	91	Pa		21.1
44	Ru	22.1	3.24	92	U		21.7
45	Rh	23.2	3.43	93	Np		22.4
46	Pd	24.4	3.64	94	Pu		23.1
47	Ag	25.5	3.79	95	Am		23.8
48	Cd	26.7	4.07	96	Cm		24.5
49	In	27.9	4.28	97	Bk		25.2
50	Sn	29.1	4.49	98	Cf		26.0
51	Sb	30.4	4.69	99	Es		26.8
52	Te	31.8	4.93	100	Fm		27.1
53	I	33.2	5.18				

tion into the vacuum system of carbon in any form), signal noise resulting from the associated electronics or from background radiation entering the detector (the signal-to-noise ratio is an important consideration). Quantitative element analysis in complex mineral or alloy systems becomes very difficult where corrections are concerned since there is absolutely no

---

[†]*Data from R. D. Hill, E. L. Church, and J. W. Mihelich,* Rev. Sci. Inst., *23: 523(1952).*

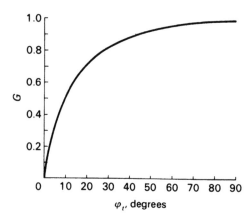

FIG. 4.23 *Variation of G with ϕ_t. (After D. B. Wittry, Univ. Southern Calif. School of Eng. Rept., 84-204, July, 1962).*

simple means for knowing the exact mechanisms of interaction and associated absorption functions. One is simply forced to experimentally determine in a systematic way the corrections necessary for each element detected (see Prob. 4.7).

The actual element resolutions in the electron microprobe utilizing wavelength dispersive x-ray spectrometry in terms of actual sensitivity are in the range of several hundred parts per million and employ analyzing spot sizes ranging from roughly one to several hundred microns. And, it must be emphasized that while we have observed that quantitative analysis is somewhat complicated, qualitative analysis by characteristic x-ray emission scans of small samples, as well as atomic-number contrast observed in scanning backscatter and absorbed electron images, provides a quick and very revealing display of the disposition and relative concentrations of elements in a specific (thick) sample area.

You should bear in mind that we are describing effects above which are related somewhat historically and practically to the conventional electron microprobe or wavelength dispersive x-ray spectrometer system which continues to be utilized in a great number of analytical laboratories throughout the world, and which continues to be manufactured commercially by numerous companies in the U.S.A., Europe, Japan, and elsewhere. Figure 4.24 illustrates the typical appearance of a commercial probe of this type.

Insofar as thick samples are concerned, the corrections and interactive processes indicated in Fig. 4.23 apply to wavelength and energy dispersive x-ray microprobe analysis. However energy-dispersive spectrometry is somewhat limited by the solid-state detection scheme described previously, although by comparison with wavelength spectrometry the facility is considerably simpler and more convenient to incorporate into an electron

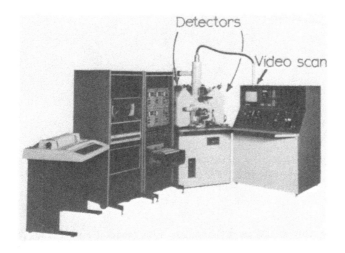

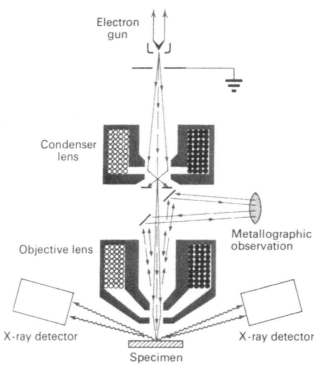

FIG. 4.24 *Commercial electron probe x-ray microanalyzer*
and schematic illustration of the prominent design and
operational features. (Photograph courtesy JEOLCO.,
U.S.A.

optical system.   Commercial energy dispersive x-ray analyzers have also
been available world-wide for more than a decade.   In addition computer
programs and tapes which provide for corrections due to interelement effects in an irradiated sample volume have also been available commercially.

4.4.1   *CALIBRATIONS FOR ENERGY DISPERSIVE X-RAY SPECTROMETRY (EDS or EDX)*

In order to perform quantitative analyses with reasonable facility utilizing a commercial EDS unit which is not equipped with correction routines
or other systems for performing rapid quantitative analysis, it is often
convenient to run calibrations on samples composed of elements which are
to be identified in an unknown specimen regime.   Unknown here refers to
either specific elements or specific compositions of known elements.   Such
calibrations can in effect simulate interelement effects, absorption, and
related phenomena at specific incidence (refer to Fig. 4.19) and take-off
angles, and under specific conditions of analysis: primary beam voltage,
counting rate and time for signal detection, etc.   Under these conditions,
the concentrations of elements present in the sample can be determined with
reasonable accuracy (at least 1%) by considering relative concentration
ratios determined either from the peak heights or included areas under
specific peaks corresponding to characteristic x-ray counts for elements
included in an energy spectrum display as shown in Fig. 4.10.

We will illustrate these features with the following example: polished copper-aluminum alloys are examined in a scanning electron microscope
fitted with an energy dispersive spectrometer.   The accelerating voltage
is maintained at 25 kV, and the angle of tilt between the incident beam
and the specimen normal is kept constant.   Furthermore, the counting time
for x-ray collection is kept constant.   Samples containing 5, 11, and 16
atomic percent aluminum are examined, and compared against pure copper.
This comparison consists in comparing the K-shell excitation peaks ($K\alpha$)
for aluminum and copper [$A\ell$ ($K\alpha$)/$Cu(K\alpha)$] as shown in Fig. 4.25 for both
peak height ratios and peak area ratios.   The areas under $K\alpha$ peaks were
determined by adding together the counts in all channels included within
the peak (as shown in Fig. 4.10).   An alternative method could involve
ratios of the total signal peak heights or areas ($K\alpha$ + $K\beta$) or ($K\alpha$ + $L\alpha$),
etc.   However there is usually little difference, and Fig. 4.25 is indi-
cative that in many cases simply comparing specific excitation peak areas
can provide for a fairly accurate representation of the elemental composition ratio if the appropriate calibration constant is known.

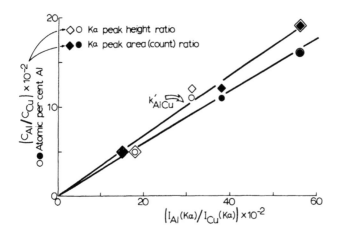

FIG. 4.25 *Concentration calibration from characteristic x-ray emission spectra ratios. The concentrations of copper is* $C_{Cu} = (100 - C_{Al})$ *in atomic
percent. So the ratio* $C_{Al}/C_{Cu} = C_{Al}/(100-C_{Al})$

From Fig. 4.24 we can see that

$$\frac{I_{Al}(K\alpha)}{I_{Cu}(K\alpha)} = K_{AlCu} \frac{C_{Al}}{C_{Cu}}$$

where I represents the (Kα or other excitation) peak area or peak height
measured from the x-ray spectrum, C represents the concentration, and K is
the proportionality constant obtained from the slope of an intensity ratio
plot shown typically in Fig. 4.25. You might recall from Eq. (4.15) that
the concentration of an element, A, can in general be related to the characteristic emission intensity by

$$I_A = C_A \, g(E) \, I_{A'}$$

or more generally we could consider

$$I_A(K\alpha) = C_A \, g[E(K\alpha)]_A \tag{4.15a}$$

where $g[E(K\alpha)]_A$ depends upon the ionization cross section, x-ray fluorescence yield, atomic weight in the ionization cross section, the relative
emission intensity or peak intensity (Kα) for the element A, and the detection efficiency for that particular x-ray. If we now write

$$\frac{I_A(K\alpha)}{I_B(K\alpha)} = k_{AB} \, \frac{C_A}{C_B} \quad ; \quad \frac{C_A}{C_B} = k'_{AB} \, \frac{I_A(K\alpha)}{I_B(K\alpha)} \tag{4.30}$$

we have a simple, general relationship for determining the specific elemental compositions when experimental plots like Fig. 4.25 can be obtained and $k'_{AB} = g[E(K\alpha)]_B / g[E(K\alpha)]_A$ can be evaluated in a simple (linear) way. When the sample contains more than two elements, the relative intensity ratios of successive elements is determined and by assuming $\Sigma C_i = 1$, we obtain a set of $N = (i - 1)$ equations with N unknowns which can be individually solved for specific compositions. For situations where $k_{AB} \simeq k_{BC} \simeq k_{AC} \simeq k_{Ax}$ etc., we can measure the concentration of unknown trace elements (x) by comparing ratios of larger, major peak intensities in samples of "known" compositions from

$$C_x \simeq C_A \left[\frac{I_{Ax}(K\alpha)}{I_A(K\alpha)} \right] , \qquad (4.31)$$

where $I_A(K\alpha)$ is the largest intensity peak corresponding to the largest element concentration. You must realize that concentrations can be determined in any convenient units and can be measured in consistent units (atomic percent, weight percent, etc.), and the example shown in Fig. 4.25 does not represent any particular standard. In most cases concentrations are treated in weight percents rather than atomic percent, although a simple conversion can be made from one to the other (atomic to weight percent or vice versa). You must also realize that the assumptions made above demand that the samples examined must have the same relative surface polish, must be of the same size (and certainly one cannot compare concentrations in a very thin film with those in a very thick specimen even if the surface topographies are similar or identical, etc.), and must be examined under the same conditions (primary beam potential), incident and take-off angles, etc.).

Signal efficiency and primary electron accelerating voltage The importance of the incident (primary) beam potential in any analysis system from which characteristic x-rays or characteristic electrons are to be emitted can be recognized in connection with the required excitation energy threshold and the depth of penetration. As a "rule of thumb", for a recognizable peak to occur in an energy dispersive x-ray spectrum, the primary electron accelerating voltage should be twice the excitation potential for the particular peak. This is illustrated in Fig. 4.26, with reference to Table 4.2. You should note in Table 4.2 that the excitation potential for $Cu(K\alpha)$ x-ray emission (V_K) is roughly 8.9 keV. You should also note that if a sample contains elements such as gold (for example a Cu-Au alloy), the $Au(K\alpha)$

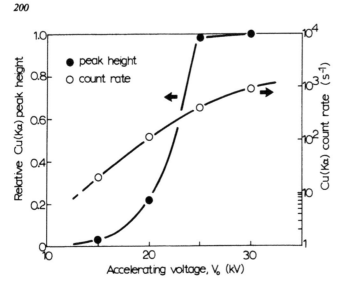

FIG. 4.26 *Effect of accelerating potential (V_o) on
the characteristic Cu(Kα) x-ray emission for a Cu -
16 atomic percent Aℓ alloy (Fig. 4.24). The Cu(Kα)
relative peak heights plotted are based on 1.0 at 30 kV,
and a constant counting time of 100 s. Specimen tilted
50°.*

emission peak would not occur in the spectrum. Only the Au(Lα) peak would
appear at voltages of 25 kV or larger (because V_L in Table 4.2 for Au is
14.4 keV). For many heavy elements, only the M lines or peaks occur in
the characteristic x-ray energy emission spectrum (uranium for example).

4.4.2 THIN FILM MICROANALYSIS AND THIN FILM APPROXIMATIONS

When an electron probe beam enters a specimen, interactive processes described above induce beam spreading. As the beam propagates into a material, the beam spreading gives rise to lateral dispersion and as a result
there is an enlargement of the volume in which excitation occurs, and which
contributes to the detected signal. This effect, as shown schematically
in Fig. 4.27, increases the signal volume and decreases the analytical resolution. What we mean by resolution here is the ability to confine the
signal detected to a small volume of the sample. Consequently, for very
thick specimens, the characteristic x-ray source size can be many times the
focused beam diameter. In thick specimens we normally consider the secondary electrons to originate near the entrance surface, with backscatter
electrons originating at greater depths in the specimen. The primary x-
ray excitation is assumed to occur below the backscatter region. Cosslett
and Thomas [22] were among the first to document the scattering of elec-

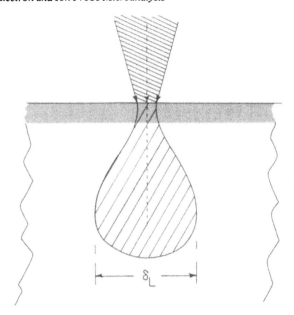

FIG. 4.27 *Activated volume schematic for electron probe on a solid specimen. The shaded region illustrates the thin-film approximation. The depth of penetration is largely determined by the Thomson-Widdington law [Eq. (4.10)]. The lateral resolution (due to beam spreading) in the interaction zone for thick samples is roughly* δ_L.

trons through different thicknesses of material. In thicker specimens, the backscattered electrons originating below the entrance surface can give rise to primary x-ray excitation as well, and this signal can originate close to the surface.

In general, bulk specimen microprobe analyses are based on a comparison of characteristic emission from an unknown and a standard, but we have demonstrated in Equations (4.30) and (4.31) that fairly accurate analysis can be accomplished by energy dispersive x-ray spectrometry using simple signal comparisons if the correction factors for generation efficiency, absorption, and secondary fluorescence can be lumped together and determined. Furthermore, since there is a dependence of secondary processes and absorption, etc. on the penetration depth determined by the Thomson-Widdington law [Eq. (4.10)], there are certain unique features of thin film analysis as originally discussed by Philibert and Tixier [23]. These include the ability to neglect absorption and fluorescence corrections, and the assumption that, for the case of characteristic x-ray generation, the x-ray intensity leaving the film is equal to the generated x-ray

intensity. This latter assumption is sometimes termed the "thin film criterion" or the "thin film approximation". Using this assumption, one can
show that for a homogeneous specimen consisting of elements A, B, C, etc.,
the characteristic x-ray intensity of say Kα emission is given by

$$I_A(K\alpha) = \varepsilon_A \cdot \kappa_A \cdot C_A \cdot t \ldots. \text{ etc.} \tag{4.32}$$

where ε_A is the x-ray detection efficiency, κ_A is a constant related to the
ionization cross-section, x-ray fluorescence, atomic weight, and relative
Kα emission intensity for element A, C_A is the composition (in weight percent) of element A, and t is the film thickness. Taking the ratios of the
x-ray intensities we can obtain

$$\frac{I_A(K\alpha)}{I_B(K\alpha)} = \left(\frac{\varepsilon_A \kappa_A}{\varepsilon_B \kappa_B} \right) \cdot \left(\frac{C_A}{C_B} \right) = k_{AB} \left(\frac{C_A}{C_B} \right) , \text{ etc.} \tag{4.33}$$

or

$$\frac{C_A}{C_B} = k'_{AB} \left(\frac{I_A}{I_B} \right) \tag{4.33a}$$

$$\frac{C_B}{C_C} = k'_{BC} \left(\frac{I_B}{I_C} \right) \tag{4.33b}$$

and we can recognize then that the concentrations must sum to 100 percent:

$$C_A + C_B + C_C = 100 \text{ (weight percent)} \tag{4.33c}$$

These equations are independent of the film thickness and are fundamentally
identical to Eq. (4.30). As we demonstrated previously k_{AB} or k'_{AB}
values can be measured from known thin films of known compositions (as in
Fig. 4.24). Cliff and Lorimer [24] have shown that if the composition of
a thin film is known, the specimen can be used as its own standard and
k'_{AB}, etc. determined as shown for bulk specimens in Fig. 4.25.

   If k_{AB} or k'_{AB} scaling factors are available for various combinations
of elements then weight fraction ratios of various elements in thin films
can be obtained simply by multiplying the measured intensity ratio (I_A/I_B)
by the appropriate k'_{AB} factor. Consequently, if k'_{AB} values are available
for various combinations of elements, thin film standards are unnecessary.

Cliff and Lorimer [25] have in fact determined experimental values of k'_{AB} (at 100 kV) for a series of elements A relative to Si (as element B), and the constant k'_{ASi} is commonly called simply k:

$$k = k'_{ASi} = \left(\frac{C_A}{C_{Si}} \right) \cdot \frac{I_{Si}}{I_A} \tag{4.34}$$

You must of course recognize that these values apply to specific emission lines, e.g., Kα, Lα, etc., i.e. k(Kα), k(Lα), etc. Obviously in many anallyses Si will not be present in a thin specimen. Thus, to obtain the correct k'_{AB} factor for any two elements the measured or calculated k value for element A relative to Si is divided by k for element B relative to Si, etc., i.e.

$$k'_{AB} = \left(\frac{k'_{ASi}}{k'_{BSi}} \right) = \left(\frac{C_A}{C_{Si}} \right) \cdot \left(\frac{I_{Si}}{I_A} \right) \cdot \left(\frac{C_B}{C_{Si}} \right) \cdot \left(\frac{I_{Si}}{I_B} \right) =$$

$$\left(\frac{I_B}{I_A} \right) \cdot \left(\frac{C_A}{C_B} \right) \tag{4.35}$$

Values of k'_{AB} can also be calculated as described by Goldstein, et al [26]. However these calculations must consider the particular detector efficiency and especially differences in the Be window absorption (refer to Fig. 4.10) for the various commercial energy dispersive x-ray detectors. In otherwords the results of microanalysis, and particularly calibration factors, will differ from one detector to another. Absorption is also a factor in thicker specimens as well, and there are obviously limitations to the thin film approximation. As a consequence, the ratios measured from Eqs. (4.33) and (4.35) must be corrected for the effect of absorption if any of the characteristic x-ray intensities vary appreciably with thickness in the range of specimen thickness being considered. These features have been discussed recently by Goldstein [27].

Beam spreading and spatial (lateral) resolution in thin films The actual spatial resolution in thin-film microanalysis is equal to the sum of the electron probe diameter at the specimen surface and the beam broadening, b. The beam broadening is a function of atomic number, specimen thickness, and the accelerating voltage, and Goldstein, et al [27] have estimated the beam spreading or broadening by assuming that elastic scattering of elec-

trons occurs at the center of the thin film, and that the electron beam is
a point source from

$$b = 625 \left(\frac{Z}{W_A V_o} \right) \cdot \left(\rho^{1/2} \, t^{3/2} \right) \tag{4.36}$$

where Z is the atomic number, W_A is the atomic weight, ρ is the density (in
g cm^{-3}), V_o is the accelerating voltage in kV, and t is the film or speci-
ment thickness in cm. Of course Z and W_A are the effective values for mul-
tielement regimes, and as shown in Fig. 4.28, the broadened beam (b in cm)
characterizes a cylindrical interaction volume which is effectively the
neck of the interaction volume shown in Fig. 4.27 (within the shaded region
characterizing the thin-film regime). You can readily observe from Eq.
(4.36) that the beam broadening varies inversely with the accelerating voltage and increases with film thickness. So it is a logical consequence to
increase the accelerating voltage in thicker films in order to compensate
for the associated beam spreading. Furthermore, since the efficiency of
x-ray production is so small, films must normally be at least a few hundred
Angströms thick, otherwise the signal becomes difficult to detect. In addition, in the limit of a zero-diameter probe, the maximum attainable resolution is essentially the spreading given by Eq. (4.36) and illustrated in

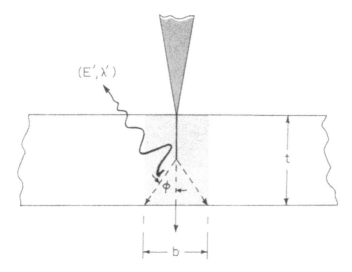

FIG. 4.28 *Interaction volume (shaded) for character-*
istic x-ray production in thin films as a result of
beam spreading to a diameter b shown. The scattering
angle is denoted by ϕ.

Fig. 4.28. For a 1000 Å thick film and an incident electron beam of 100 keV energy, for example, the calculated values of b for C, Aℓ, Cu, and Au single-component systems are approximately 40, 80, 180, and 520 Å, respectively.

As we have already mentioned, x-ray excitation competes with characteristic Auger electron production (See Fig. 4.17). In addition, these processes stimulate energy loss spectra which in thin films can complement x-ray or Auger electron spectroscopy. In fact, electron energy loss spectroscopy has the highest yield of characteristic information per inelastic scattering event because it is essentially the non-competetive sum of other excitation events, and the energy loss spectrum is strongly peaked in the forward direction (the direction of propagation of the incident or primary electron beam). Furthermore, as we noted previously, electron energy loss spectroscopy is sensitive to light element analysis while energy-dispersive x-ray spectrometry is not normally sensitive to light elements ($Z < 10$). Egerton [28] has shown that the number of atoms of element A (N_A) detected by energy loss spectroscopy for a thin specimen (~ 1000 Å thick) is given by

$$N_A = I_K^A(\alpha,\Delta)/[I_0(\alpha,\Delta) \cdot \sigma_K^A(\alpha,\Delta)] \qquad (4.37)$$

where $I_K^A(\alpha,\Delta)$ is the integrated intensity above background of the kth inner-shell edge of element A measured over an energy window Δ (or W_d in Fig. 4.20) with a spectrometer accepting electrons scattered over an angle 0 to α (the angular spread shown entering the magnetic sector spectrometer in Fig. 4.20 for example); $I_0(\alpha,\Delta)$ is the corresponding intensity measured over the zero-loss peak; and $\sigma_K^A(\alpha,\Delta)$ is the appropriate partial ionization cross section for the particular experimental conditions used. One can also consider the relative concentration ratio of two elements in the excited volume in the same manner that we demonstrated for energy-dispersive x-ray spectrometry above [Eq. (4.33) for example], i.e.

$$\frac{N_A}{N_B} = \left[\frac{\sigma^B(\sigma,\Delta)}{\sigma^A(\sigma,\Delta)}\right] \cdot \left[\frac{I_K^A(\alpha,\Delta)}{I_K^B(\alpha\ \Delta)}\right] = k''_{AB} \cdot \left[\frac{I_K^A(\alpha,\Delta)}{I_K^B(\alpha,\Delta)}\right] \qquad (4.38)$$

In Eq. (4.38) the signals must be compared for the same energy window (Δ). It is also interesting to note that like Eqs. (4.33), Eq. (4.38) is independent of the film thickness in the range of specimen thicknesses where the thin-film criterion holds. This has been discussed by Goldstein, et al [26, 27].

Zaluzec [21] has described two different detection limits for characterizing the sensitivity limit of thin film spectroscopy: the minimum detectable mass (MDM) and the minimum mass fraction (MMF). MDM refers to the ability to detect an element above the background (noise) signal while MMF refers to the smallest concentration of an element that is detectable in a matrix. Joy and Maher [28] have calculated values ranging from 5×10^{-20} to 5×10^{-19}g for a transmission electron microscope operating at 100 kV accelerating potential, and employing a "conventional" thermionic emission electron gun for elements in a specimen matrix ranging from Z = 10 to 40. For these conditions, MMF values range from 0.5 to about 2 weight percent. These values apply to energy-dispersive x-ray spectrometry for thin films, and you should compare these values with typical MMF values of 0.1 to 0.01 weight percent for the conventional thick specimen electron probe microanalysis techniques discussed at the beginning of this chapter. However we must recognize that in the thin film spectroscopies the probe size is considerably smaller, and as a consequence the spatial or lateral resolution is considerably better than in the electron microprobe. Similar sensitivities have been described for electron energy loss spectroscopy, however the sensitivity of EELS relative to EDX(EDS) under equivalent conditions can ideally be greater by factors of nearly 500 and 25 for MDM and MMF, respectively [21]. However the differences in the mode of data acquisition and the limitations to sensitivity imposed by even moderately thick (~ 300 Å) specimens in energy loss spectroscopy have not allowed this improvement to be realized in contemporary analytical instruments. This situation will undoubtedly be improved as instrumental sensitivities, etc. are improved.

You must certainly realize that the thin film techniques outlined in this section apply specifically to conventional and scanning transmission electron microscopy. These techniques will be described in detail in Chapter 7, and we will expand upon these concepts there, as well as emphasizing analytical applications particularly in the context of analytical electron microscopy (AEM). Duncumb [29] was one of the first to mount an energy-dispersive x-ray sepctrometer (Fig. 4.10) on the column of a transmission electron microscope. More recently, electron energy loss spectrometers (Fig. 4.20) have also been incorporated in the transmission electron microscope, affording the analytical comparison and complement we alluded to above [See for example "Introduction to Analytical Electron Microscopy", edited by J. J. Hren, J. I. Goldstein, and D. C. Joy, Plenum Press, New York, 1979].

4.5 APPLICATIONS OF ELECTRON PROBE MICROANALYSIS

Although we have demonstrated some of the simpler applications of many of the electron spectroscopies and microprobe analysis techniques in our development of and description of the basic principles, we have not always developed a sense of the wide range of applications which have been developed for a particular technique. We have also not forcefully illustrated some of the very powerful applications of certain techniques in the solution of many contemporary materials problems involved in the development and understanding of particular properties of materials. Some of these will be dealt with in more detail in this section and many will be repeatedly illustrated as we deal with the analytical aspects of scanning electron microscopy and conventional and scanning transmission electron microscopy.

We must also mention that techniques involving surface analysis or very small interaction volumes usually require exceptional vacuum conditions in order to reduce or avoid surface contamination. It is now a routine matter to achieve pressures below 10^{-10} Torr, but for bulk material analyses, pressures as high as 10^{-4} Torr are often satisfactory. This is true for conventional electron microprobes (wavelength-dispersive spectrometers) and energy-dispersive x-ray spectrometers.

4.5.1 SPECIMEN PREPARATION

The preparation of specimen materials for examination in an electron microprobe is generally identical to standard practice in metallography and mineralogy in that specimens are ground flat and polished uniformly. For metals and fairly good conductors, no additional treatment is required. However, poor conductors must be securely connected to the probe circuits by a conductive paint (for example, Aquadag or Silver Print, etc.). In the case of poorly conducting minerals, ceramics, or conductive materials containing nonconductive inclusions, vapor-deposited layers of C, Aℓ, Pd, or Au-Pd are cast onto the sample to enhance the analysis and signal detection. In many cases, evaporation of conductive layers is unnecessary, and it is undesirable where intrinsic electrical features are under investigation.

A major problem of investigating any sample materials in a microprobe, and in any evacuated electron or ion optical system, is the presence of contaminating agents that, as a result of interaction with the electron or ion beam, deposit on the area of study where beam intensities are greatest on the specimen. This causes a loss in resolution, and subsequently complicates both emission and absorption processes. The contamination problem has been reduced in recent years with advances in high-vacuum and particularly ultrahigh-vacuum technology.

Some of the more specific details of specimen preparation for electron probe microanalysis are included in Appendix B. It might be mentioned in passing, however, that specimen sizes are quite flexible, particularly where commercial probe units are concerned. Normally, specimens 1 in.^{2} and up to 0.5 in. in thickness are accommodated, while specimens as small as 1 mm^2 are also equally amenable to positioning and analysis in commercial units such as the one shown in Fig. 4.24.

Flat specimens are useful in many types of electron spectroscopies, but in many cases they are untenable. Surface irregularities contribute to beam scattering and introduce other complexities in terms of ionization efficiencies, etc. This results in certain irregularities in the emission spectrum, which can vary in intensity from region to region. This gives rise to losses in sensitivity in some cases. For many applications, no alteration of the specimen surface can be tolerated, and in some cases no coating can be allowed. However it is sometimes useful to have the coating signal appear in the spectrum to act as a calibration. Au-Pd coatings in fact provide for a wide calibration range in energy-dispersive x-ray spectrometry. For energy-dispersive x-ray spectrometry, carbon can be used as a conductive coating without contributing to the emission spectrum because, as we noted previously, elements below about $Z = 10$ are not detectable above the thermal noise of the solid-state detector. Consequently, carbon will be effectively invisible to the spectrometer.

For thin film specimens, the methods of preparation usually render the surfaces rather smooth. These specimens are usually prepared for transmission electron microscopy, and the methods are described in some detail in Appendix B.

4.5.2 ELECTRON-INDUCED DESORPTION (EID)

In the event of surface contamination involving the adsorption of impurities or reactions involving oxidation or other chemical changes, it may

become necessary to "clean" the surface through the removal of the adsorbed
species or reaction products. This can sometimes be accomplished by scanning a fine-focussed electron beam over the surface, causing thermal scattering (and spallation) and the desorption of the surface products. Elec-
tron-induced or electron impact desorption can also be applied to the removal of indigenous surface layers, that is the material itself in order to
provide for a kind of depth profile. This is sometimes helpful in the bulk
analysis of surface regions involving surface-sensitive techniques such as
Auger electron spectroscopy, and there are indeed many examples of this
technique [See for example M. Nishijima and T. Murotani, *Surface Science*,
32: 459(1972)].

We will also describe similar impact desorption schemes in connection
with ion bombardment and ion beam analysis techniques in later sections of
this chapter. Ion bombardment can also provide for depth profiling by
systematically removing material from a surface, and these features will
also be described in later sections.

4.5.3 SCANNING AUGER MICROANALYSIS (SAM)

By combining a beam scan system similar to Fig. 4.7 with the primary electron source shown in the Auger electron spectroscopy system shown in Fig.
4.18, scanning Auger elemental maps or elemental images can be obtained
similar to the characteristic x-ray maps shown in Fig. 4.8. This same
technique of primary beam scan coupled with synchronous CRT display of
the analyzed (detected) signal arising from the scanned specimen area can
be applied to any characteristic emission regime in order to produce elemental or characteristic emission displays over the scanned area. Characteristic CRT images displayed by synchronous signal detection using Auger
electron emission are sometimes called Auger images and since they are
essentially elemental surface displays, the technique is commonly referred
to as scanning Auger microscopy (SAM) and sometimes scanning auger electron
spectroscopy (SAES). It is ideally scanning Auger microanalysis, and the
device is also often referred to as a scanning Auger microprobe (SAM) since
the primary beam can be focused to a relatively small spot size, commensurate with and smaller than the electron microprobe.

Perkin-Elmer Physical Electronics Division (USA) has developed a commercial scanning Auger microprobe which utilizes a cylindrical-mirror
electron energy analyzer (CMA) with an integral, coaxially mounted electron

gun.   In fact Physical Electronics developed the first commercial SAM in
1973.   In this arrangement, the electron gun is mounted inside the electrostatic analyzer tube which provides for a direct focus of specific-
energy Auger electrons through field-gating of the electrostatic analyzer,
utilizing a computer interface scheme.   Electrons with divergent energies
focus outside the selective aperture at the detector.   Auger electrons
ejected over a wide range of angles which encompass the primary electron
beam can be collected, and the spectrometer collection efficiency is increased by allowing use of the entire 2π entrance cone of the analyzer.
This system, shown schematically in Fig. 4.29, utilizes a high intensity,
small focus LaB_6 filament.   Figure 4.30 illustrates Auger images of specific elemental distributions compared with energy-dispersive x-ray maps of
the same areas.   The difference in image intensity is indicative of the
more efficient Auger electron production in the competitive processes described previously, and illustrated schematically in Fig. 4.17.   You might
examine the recent paper by J. F. Moulder and A. Joshi, *Met. Trans., 12A:*
1140(1981) for some related examples of SAM applications.   It is also
recommended that you consult the *Auger Electron Spectroscopy Reference
Manual,* by G. E. McGuire, published by Plenum Press, New York in 1979.

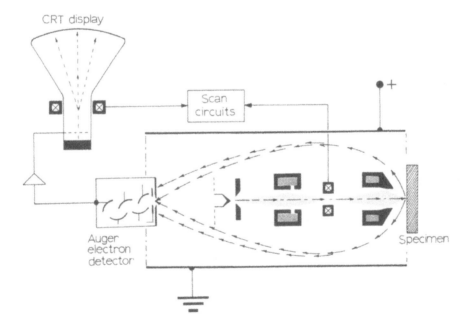

FIG. 4.29 *Scanning Auger microprobe utilizing an electrostatic, cylindrical-mirror electron analyzer with an internal (coaxial) electron gun
arrangement.*

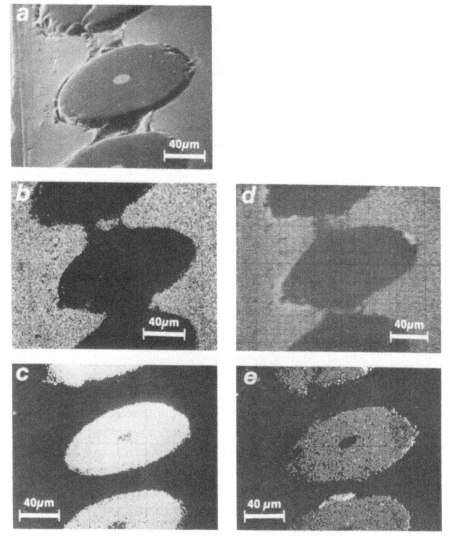

FIG. 4.30 *Comparison of Auger and EDX elemental maps for boron fiber-reinforced titanium. (A) SEM image, (b) and (c) show Auger electron images of Ti and B respectively, (d) and (e) show corresponding EDX images (maps). (Courtesy Dr. A. Joshi, Perkin-Elmer, Physical Electron Division.)*

4.5.4 QUALITATIVE AND QUANTITATIVE ELECTRON PROBE MICROANALYSIS (EPMA) EXAMPLES

Qualitative studies of inclusions Perhaps the most spectacular success of microprobe analysis, particularly the more conventional scanning modes, occurs in the qualitative identification of the relative composition and disposition of elements in inclusions. Figure 4.8 has already given us a fair idea of the power of the technique, and we have stressed continuously the rapidity and ease of such analyses. The microprobe approach has become routine in almost all investigations, and this approach has involved the conventional wavelength - dispersive analysis of x-rays, energy-dispersive x-ray microanalysis, Auger spectrocopy and scanning Auger microanalysis, and electron energy-loss spectrometry. In many applications two or more of these techniques have been employed simultaneously in order to confirm the analytical results. This is particularly true in analytical electron microscopy to be discussed in detail in Chapter 7. Figure 4.30 is an excellent example of this technique, while Fig. 4.31 and 4.32 identify pyrite crystal inclusions in a coal matrix and a mineral matrix respectively. These examples illustrate the use of an energy-dispersive x-ray spectrometer mounted on a commercial scanning electron microscope. We will treat these applications in more detail in Chapter 5.

We should point out in Fig. 4.32 that the analysis in Fig. 4.32(c) is made using a characteristic iron line scan across the two pyrite particles. This is achieved by limiting the detector energy window to the width of the iron peak illustrated for example in Fig. 4.31. Signal intensity admitted into the MCA (see Fig. 4.10) then drives the CRT line scan amplitude, with the x-ray emission synchronized with the electron beam position on the specimen. There are three basic modes of spatial analysis available in a scanning beam device: point-mode analysis where the beam dwells on a single point on the specimen, line-mode analysis where the beam traverses a line across the specimen surface as illustrated in Fig. 4.32(c), and an area-mode analysis where the electron beam is scanned across as shown in Figs. 4.30 and 4.31.

While Figs. 4.30 to 4.32 illustrate some analytical examples for inclusions in thick (or bulk) specimens, Fig. 4.33 illustrates this type of analysis in thin films utilizing electron energy-loss spectrometry. As we mentioned above, the energy-dispersive x-ray mapping features can also be combined with the energy-loss spectrometry features in the transmission electron microscope (either conventional or scanning transmission electron microscopes), and the combined approach, constituting the analytical electron microscope (AEM) is especially facilitated in the scanning transmission electron microscope (STEM). In the example shown in Fig. 4.33, both inclusions have the same cubic (NaCl-type) structure and the lattice parameters are very similar [4.32 Å for TiC in Fig. 4.33(a) and 4.24 Å for TiN

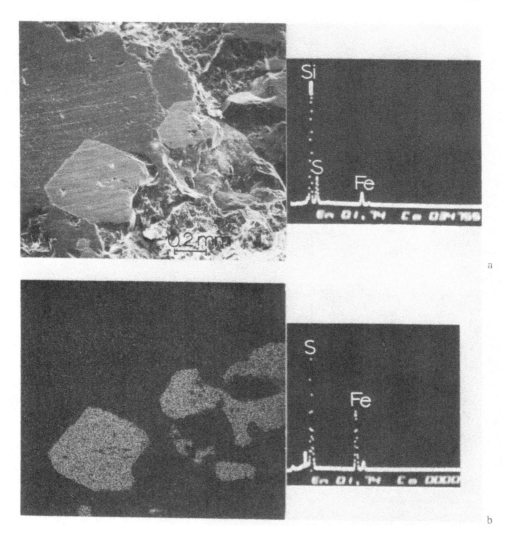

FIG. 4.31 *Pyrite inclusions in a quartz mineral matrix. (a) Scanning electron microscope (seondary electron) image, (b) Elemental x-ray map using or admitting only the iron (FeKα) and sulfur (SKα) signal data. The x-ray characteristic energy spectra for the complete area is shown in (a) while in (b) the spectra are for the pyrite inclusions only.*

in Fig. 4.33(b)]. Consequently electron diffraction analysis can be some-what inconclusive or questionable.

There are now many examples of electron spectroscopic analyses and indeed they are routine in most examinations involving phase identifica-tion or other similar examples of microchemical analysis. It has already been pointed out that extensive bibliographics of electron spectroscopies

(a)

(b)

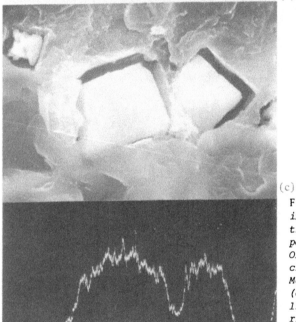

(c)

FIG. 4.32 *Pyrite inclusions
in coal. (a) Scanning electron micrograph of pyrite
polyhedral crystals in an
Ohio coal (b) Tetrahedral
crystals of pyrite in New
Mexico (Carthage) coal, (c)
(characteristic) FeKα x-ray
line scan across the two pyrite crystals in (b).*

exist (see for example ref. 18). Indeed nearly all comtemporary scientific and technical journals contain a continuous array of papers examining
microchemical phenomena in thin films, bulk films, and on surfaces which
involve a full range of the electron probe microanalysis techniques outlined in this chapter. It might be of interest to recognize how extensively

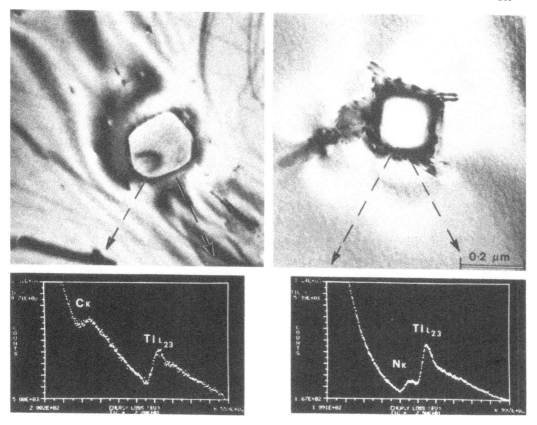

FIG. 4.33 *Example of phase (inclusion) identification in a thin-electron-transparent specimen utilizing electron energy-loss spectrometry in a scanning transmission electron microscope. TiC particle is on the left, TiN particle is on the right. (Courtesy Drs. N. J. Zaluzec and P. J. Maziasz; After Ref. 30).*

some of these techniques have been used for decades [31-33]. It is especially interesting to note that microanalysis employing electron energy loss concepts, and as illustrated in Fig. 4.33 was suggested as early as 1943 by Hillier [34], and Hillier and Baker [35], and indeed some of their results of nearly four decades ago are as significant as those shown in Fig. 4.33. Delays in implementing these techniques on a routine basis occurred mainly because of technological (instrumental) limitations. This is in fact the cause for delays in implementation of a wide number of analytical and diagnostic concepts. However it is significant to recognize the fact that many very fundamental and very significant concepts arise long before commercialization. This is one of the reasons we emphasize the historical developments in many areas of this book.

Diffusion, segregation and reaction phenomena Segregation, like inclusion studies, is readily amenable to study by the various microprobe analytical modes, and considerable advance has been made in the study of phase equilibria, segregation in metals, and related properties in ceramics. Some of the pioneering work in segregation studies is due to J. Philibert and co-workers [24], M. J. Fleetwood [25], and D. A. Melford [26], to name only a few. And, as indicated previously, the current journals generally contain numerous examples of work of this nature. Figure 4.34 presents an adequate illustration of the revelation of grain-boundary segregation, and related studies in ceramic materials have become invaluable in the evaluation of additives, insofar as their relative morphology and disposition in the matrix are concerned. The resolution of interfacial phenomena of this type is of course improved by utilizing thin films where the spatial (or lateral) resolution for interfaces oriented perpendicular to the specimen surfaces can be approximated from Eq. (4.36). Some of these features will be illustrated more fully in later chapters.

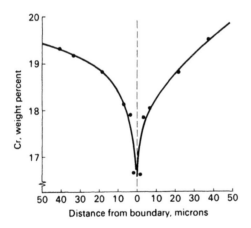

FIG. 4.34 *Variation of Cr concentration in grain boundaries of Nimonic 80A (80:20 Ni:Cr alloy precipitation hardened by Ni_3Al except at grain boundaries where carbides precipitate in association with Cr depletion shown).* [*After M. J. Fleetwood, J. Inst. Metals, 90: 429(1962).*]

Diffusion analysis using the many forms of electron probe microanalysis is usually quite straightforward, especially in the conventional electron probe (wavelength-dispersive spectrometer) utilizing the line-scan mode. Figure 4.35 shows rather typically the diffusion analysis capability using the characteristic x-ray mode of detection.

There are instances where it is necessary to examine and characterize reaction products on surfaces. These can include corrosion and other oxide layers. If the layers can be removed by stripping them off, they can be examined by scanning or transmission electron microscopy or analytical electron microscopy. However for layers which cannot be separated from the substrate, electron spectroscopies can be employed along with electron - induced desorption or ion sputtering to systematically expose successive surface layers in a kind of depth profile. We will describe some of these techniques in combination with ion spectroscopies to be discussed in later sections of this chapter. However for thicker surface layers and reaction products which can result by the reaction of small particles to form new

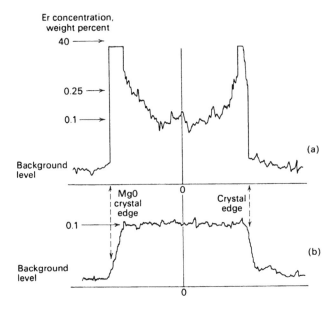

FIG. 4.35 *Er diffusion microanalysis. (a) Er Lα scan of 0.6-mm thick MgO wafer onto which 3 microns of Er was vapor deposited on each side and then heated in an open furnace for 100 hr at 1400°C. (b) Er Lα scan of similar crystal heat 130 hr at 1495°C.*

reaction products which coat the particles, techniques involving selective
primary beam incidence can be employed as illustrated in Fig. 4.36.   In
this illustration, chalcopyrite particles leached in acid-potassium dichromate solution have formed an elemental sulfur layer around the particles.
This is confirmed by scanning the center of the particle at 10,000 x magnification and comparing the corresponding energy-dispersive x-ray spectrum
from this area with that for an equivalently magnified area at the edge of
the particle.   Beam penetration into the unreacted core produces $CuFeS_2$
emission spectra along with the spectrum of elemental sulfur.   A 25 keV
incident electron beam energy was employed in the scanning electron microscope used to undertake the analysis shown in Fig. 4.36.

General qualitative analysis, topographical, and atomic-number contrast
observations   It should be emphasized that considerable information regarding the character of a material or its composition can be gleaned from microprobe observations and various signal detections, even without the elucidation of the actual elements present.   In many cases the concentration
of impurities at grain boundaries, or the size of precipitates, etc., will
be beyond the resolution of the x-ray detector.   Nonetheless sufficient
contrast in atomic number may result so that backscattered electrons may
indicate at least an abnormality, or cathodoluminescence from the suspect
areas may provide some valuable information, at least to the point of presenting some logical possibilities.   In many cases materials are doped with
suitable activators that, as a consequence of diffusion, segregate on precipitates existent in the matrix, or defects contained herein, and excite
local cathodoluminescence or allow for sufficient contrast resolution in
the backscattered electron image.

Figures 4.37 and 4.38 demonstrate rather convincingly the contrast
features that could arise for the segregation of vacancies and/or interstitials near dislocation arrays in a material.   Information provided in
the manner indicated in Fig. 4.37 thus serves a useful purpose when attempting to relate structural or compositional features to the residual
physical or electrical properties of a material.   Figure 4.37 essentially
expands a study, to be treated in Fig. 5.39, of electrical resistivity inhomogeneities in CdS single crystals.[†]   Figure 4.37a shows the CdS surface
light optically, while Fig. 4.37b and c indicate the extent of homogeneity

---

[†]*It may be informative to read Sec. 5.6 somewhat out of context at this*
point.

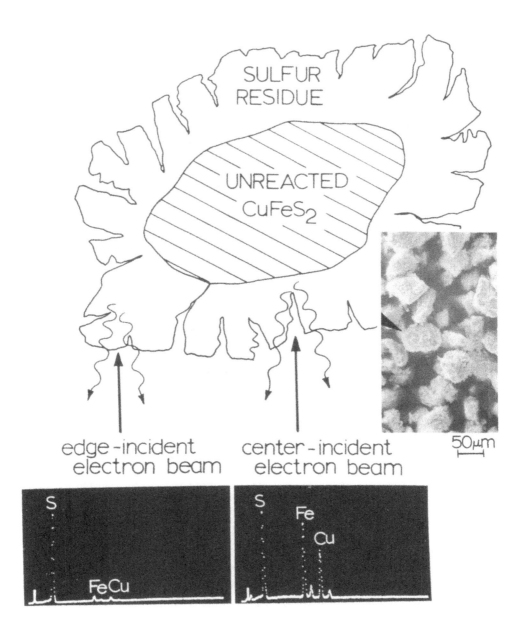

FIG. 4.36 *Identification of a sulfur reaction product forming as a coating on chalcopyrite (CuFeS₂) particles after leaching in acid-potassium
dichromate solution for 30 min. The unreacted particle acts like a
shrinking core. The positions of beam scans in the scanning electron
microscope at 10,000 x magnification are shown along with the characteristic elemental x-ray spectra. The SEM insert shows the actual particle
examined (After Murr and Hiskey [39]).*

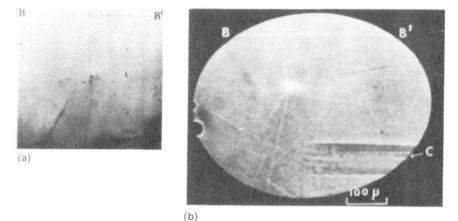

(a)

(b)

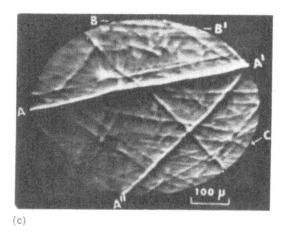

(c)

FIG. 4.37 *Surface potential inhomogeneities on (0001) CdS single-crystal surface. (a) Optical micrograph. (b) Electron mirror micrograph at zero specimen potential. (c) Electron mirror micrograph at 150 V applied to specimen. (From F. L. English and M. K. Parsons in D. G. Thomas (ed.)* Proc. Int. Conf. on II-VI Semiconducting Compounds, *Benjamin, Inc., N.Y., 1967.)*

in the surface potential distributions by electron mirror microscopy. Figure 4.38 illustrates both the cathodoluminescent (a) and backscattered electron microprobe images (b) of the identical area shown in Fig. 4.37, and presumably is indicative of some segregation phenomenon occurring along the regions of potential discontinuity.

We have attempted to create an awareness of the capability and applicability of the many modes of electron microbeam analysis. A great variety of information concerning a material's composition, element disposition, and related features is available in one mode of microprobe operation or

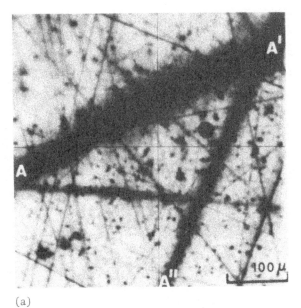

(a)

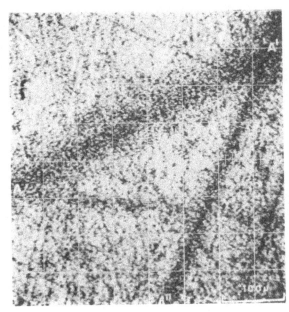

FIG. 4.38 *Continuation of investigations of Fig. 4.37. (a) Fluorescence (cathodoluminescence) mode scan photograph of same area as Fig. 4.37 examined in electron probe microanalyzer. Note again references A-A' and A'-A" indicating surface discontinuities. (b) Same area as a observed by scanning electron backscatter. (From F. L. English and M. K. Parsons in D. G. Thomas (ed.),* Proceedings of the International Conference on II-VI Semiconducting Compounds, *Benjamin, Inc., New York, 1967.)*

(b)

another; and there remain numerous "untapped" capabilities in microprobe analyzers, perhaps most notably the energy-analysis potential. Furthermore, the topographical information that can be readily obtained by an electron microprobe is in many cases inaccessible by any other means. We shall discuss this unique capability in some detail in Chap. 5.

4.6   ION MICROPROBES AND MICROANALYSIS

You should recall from Chapters 1 and 3 that although there are some phenomenological similarities between electrons and ions, they are fundamentally
very different, particularly with regard to particle mass. Furthermore,
the deBroglie wavelength of ions is such that in both ion microscopy and
ion probe microanalysis the effect of diffraction on spatial (or lateral)
resolution is less than the lattice spacing. As a consequence, diffraction
effects are insignificant. There is an intimate intrinsic connection between electrons and ions in the collision cascade caused by an energetic
positive ion penetrating a solid because ions created in the sample (and
composing the sample) are associated with electron ejection described in
previous sections of this chapter. However this process differs from electron probe phenomena in that rather than being excited to emit some characteristic (secondary) signal, the atoms themselves are dislodged (desorbed)
by the violent interaction associated with the collision cascade.

Figure 4.39 illustrates the principal features of primary ion beam
interaction with a solid specimen. In comparison to electron beam interaction shown in Fig. 4.1, the ion beam impact involves violent collisions.
In principle the phenomenological aspects of the emission characteristics
shown in Fig. 4.1 for electrons occur for ions: secondary and backscatter
ion signals occur, ion transmission takes place, and atom excitation by
inelastic collisions can create outer-shell characteristic x-rays, luminescence, and secondary electrons. In principle, energy-loss ions can also
be detected, but the loss is sometimes beyond detection when compared to
the incident ion energy.

There is a general loss (or erosion) of material in the impact zone
of the ion probe by particle sputtering, and although only a small fraction
of the sputtered particles are in an ionized state, they can be effectively
collected and focussed electrostatically and analyzed by mass spectroscopy
as described in Chapter 3. Although ion probe microanalysis is one of the
most destructive surface analytical techniques, it is one of the most sensitive because there is little or no noise background, and ideally there
are no particles other than the primary species (as backscattered ions) and
those composing the sample surface (as secondary ions). In addition, all
elements are detectable including hydrogen. Isotopes are also detectable.

The main elements of an ion probe microanalysis system involve the
coupling of a source as shown in Fig. 3.23 with an analyzer as shown in
Fig. 3.22. In effect the source arrangement of Fig. 3.23 can be coupled

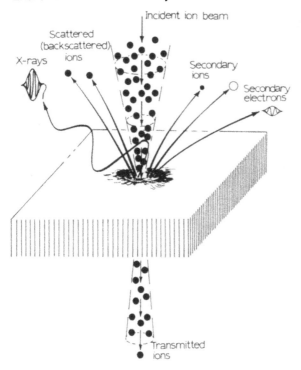

FIG. 4.39 *Principal reaction and interaction features of ion beam encountering a solid.*

in reverse with one unit providing primary ions to interact with the sample, and a second unit to focus the ion emission spectrum into a magnetic sector analyzer for elemental (ion) discrimination and quantification [40].

4.6.1 SCANNING ION MICROANALYSIS

While, like electron probe microanalysis an ion beam can effectively monitor the composition of a sample area by dwelling on that area (as a concentrated point), it is more useful and more representative to scan the primary beam across an area of the specimen, and to record the emission characteristics in such a way as to allow synchronous display of elemental composition on a CRT or in some other specific display mode. This two-dimensional characterization of a surface can be expanded to involve three-dimensional solid characterization by systematic erosion of the rastered sample area, giving rise to composition depth profiling.

Figure 4.40 illustrates the essential features of a scanning ion microprobe, or scanning ion microscope (SIM) as it is sometimes called [41]. In this particular design, a field ionization (FI) source [42] is employed to produce a brightness of roughly 10^8 amperes/Sr cm^2 [see Eq. (3.33)]

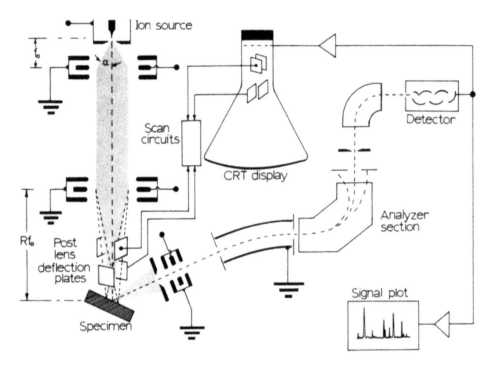

FIG. 4.40 *Schematic representation of a scanning ion microprobe (or micro-scope) design utilizing a field ionization (high brightness) source.*

using gases such as H, He, Ar or Xe. Almost any gaseous material can be ionized as shown conceptually in Fig. 2.22. With light gases such as H (where a proton beam is generated) or He, micro-sputtering is reduced, and the scattered protons can be detected and images of the surface formed by synchronous display of the scanned area on a CRT as depicted in Fig. 4.40. However by selecting specific elements detected, selective composition maps can be obtained similar to those obtained for characteristic electron (Auger) and x-ray microanalysis (as shown for example in Fig. 4.30). In addition, the SIM can be operated in the secondary electron mode, with constrast produced by the angular dependence of the secondary electron yield (which includes or involves the angle of incidence of the ion beam on the specimen surface), and secondary electron current detection by a fixed channeltron detector[*] coupled to a CRT to produce synchronous-scan

---

[*] *An image intensification system employing a channel-plate arrangement for electron detection and signal amplification.*

images as in the conventional scanning electron microscope (discussed in detail in Chap. 5).

Figure 4.41 illustrates the secondary emission image features which can be obtained in a scanning ion microprobe system as shown in Fig. 4.40, and shows the appearance of a typical field emitter. The FI source is easily fabricated from polycrystalline Ir or W (as shown in Fig. 4.41).

In the ion probe lens arrangement shown in Fig. 4.40, the objective lens collimates the beam to a diameter given by [43].

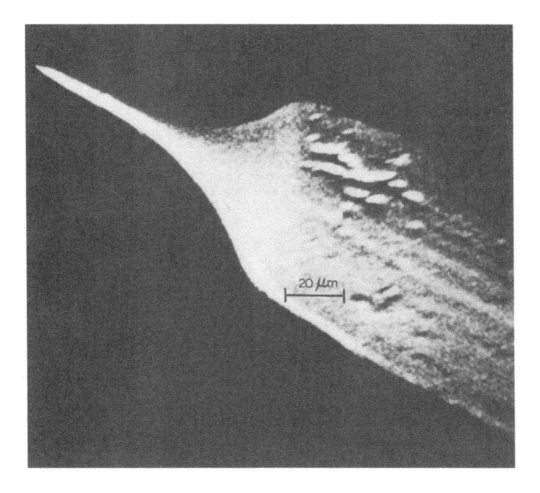

FIG. 4.41 *Secondary electron image of tungsten field emitter observed in a scanning ion microprobe utilizing a hydrogen ion beam accelerated through a potential difference of 15 kV. The ion beam was created using an iridium field emitter in the ion microprobe. (Courtesy of Jon Orloff and Lyn Swanson; after Ref. 43).*

$$d_1 = 2M_1 \left[\rho^2 + \left(C_{S1}\alpha^3/4 \right)^2 + \left(C_{C1}\frac{\Delta E}{E_o}\alpha \right)^2 \right]^{1/2} \tag{4.39}$$

where ρ is the virtual source radius (this is about 10 $\overset{\circ}{A}$ for field emitters such as that shown in Fig. 4.41), C_S and C_C are the spherical and chromatic oberration coefficients of the objective lens (1), α is the aperture angle (defined in Fig. 4.40), and ΔE is the ion beam energy spread, E_o is the initial beam energy (eV_o), and M_1 is the objective magnification. The beam diameter on the specimen (beam focus or spot diameter) is

$$d_2 \equiv d_{beam} = \left\{ (M_2\, d_1/2)^2 + \left[\left(\frac{C_{S2}}{4} \right)\left(\frac{\alpha}{R} \right)^3 \right]^2 + \left(C_{C2}\frac{\Delta E\alpha}{E_o R} \right)^2 \right\}^{1/2} \tag{4.40}$$

where M_2 (projector lens magnification)$=R/M_1$ ($R = M_1 M_2$, the product of the objective and projector lens magnifications), and C_{S2} and C_{C2} are the projector spherical and chromatic aberration coefficients respectively. You should refer to Fig. 4.40 and note that f_o shown is the objective lens focal length (in mm). You should refer to the original work of Orloff and Swanson [43] for additional details.

Field emission sources not only provide for high brightness, small diameter (focused spot) primary beams in ion microprobes, but in electron microprobes as well. We will treat field emission electron sources in more detail in Chapters 5 and 7. However before we leave the discussion of high brightness field-ion sources for ion probes we should briefly mention the liquid metal ion (LMI) source which emits remarkably high intensity ion beams of a wide variety of elements [44]. The mechanism of operation of LMI sources involves the formation of a field-stabilized cone of liquid metal (known as the Taylor cone) from which field evaporation and/or field ionization occur. In the case of liquid Ga sources, Swanson, et al [45] observed primarily field evaporation as the mechanism of ion formation. The supply of ionizable species to the cone apex occurs by viscous flow motivated by capillary forces which occur along the emitter shank as well as electrostatic gradients near the emitter tip [45]. This source concept has also been referred to as an electrohydrodynamic source [44].

High brightness or high intensity field-ionization sources can provide beam diameters [d_2 in Eq. (4.40)] of less than 0.2 µm. However there are some low energy ion spectroscopy (LEIS) applications where the simple ion source features described in Section 3.4.1 find continued use, especially in forming very small primary beam diameters (~ 100 Å). There is some special significance with regard to depth and spatial (or lateral) resolution in stationary and scanning ion probe microanalysis because of the erosion capability of the primary beam in producing depth profiles. This can be controlled both by the source function (including polarity) and the ion species (i.e. the choice of heavy ions as opposed to lighter ions - high Z versus low Z). We will discuss minimum beam diameter and spatial resolution in more detail below.

4.6.2 *ION SCATTERING SPECTROSCOPY (ISS)*

Because of the ability to control the primary ion beam, it is possible to vary the sensitivity of scattered ion microanalysis by varying the primary beam energy. In low energy ion scattering spectroscopy (LEIS) the incident ions (usually low mass species such as H^+, He^+, Ar^+) possess energies in the range from 25 to 25,000 eV. At the lower end of this range, the resolution depth is on the order of a monolayer or two of the sample surface, and adsorbed impurity atoms can be detected along with the elemental composition of the surface layers. The signal analyzed is that of the primary ion beam or inert ions which have been scattered (or backscattered) from the sample, and by selectively controlling the erosion of the surface a semiquantitative depth profile can be obtained. As we intimated above, the primary species can be chosen from any of the 90 natural elements, and this can have an important consequence on the analytical capabilities. For more specific details you are encouraged to consult "Inelastic Ion Surface Collisions", edited by W. Heiland and C. W. White, and published by Academic Press (New York) in 1977. Modifications of these concepts are also implicit in the surface analysis of neutral and ion impact radiation (SCANIR) [46], and in most ion scattering spectroscopies the primary beam is defocused to reduce sample (surface) damage.

In high energy ion scattering spectroscopy (HEIS), the analytical regimes described above are modified through the application of more energetic primary ion beams (> 25 keV) which can penetrate up to a micron into the sample. The same (primary) ion species is emitted and detected as in LEIS, but average bulk composition can be determined along with crystallographic orientation. The limit of HEIS involves very high energy primary beams in excess of 1 MeV. This is known as Rutherford backscattering (RBS) and is, as a result of the very high beam energies used (~ 2 MeV) a very specialized technique. It is also a very destructive method of analysis like many other ion probe spectroscopies. Bulk detection limits with HEIS or RBS rarely exceed 100 ppm, and lateral (or spatial) resolutions are not better than about 1 μm [47]. However the techniques can detect surface impurities.

Rutherford backscattering utilizes H^+ or He^+ ions exclusively, and the take-off angle is usually around 20° from the specimen surface. By comparison, the take-off angle for ISS at low primary beam energies is usually at right angles to the primary beam direction. The primary beam is not normal to the specimen surface of course. The backscattered ion energy can be simply related to the mass from which it scattered through the relationship

$$E_{B(i)} = \left(\frac{M_{S(i)} - M_i^o}{M_{S(i)} + M_i^o} \right)^2 E_p$$

where E_p is the primary ion energy, M_i^o is the primary ion (and backscattered ion) mass, and $M_{S(i)}$ is the mass of the scatterer (specimen or adsorbed material, etc.). For additional details you are directed to the recent book of W-K. Chu, et al. "Backscattering Spectroscopy", published by Academic Press, (New York) in 1978.

4.6.3 SECONDARY ION MASS SPECTROMETRY (SIMS)

Secondary ion mass spectrometry, illustrated conceptually in Fig. 4.40 for a field-ionization, scanning ion-probe-type system, is one of the more sensitive and more prominent contemporary mass spectrometric techniques for solid surface analysis. As implicit in Fig. 4.40, the SIMS approach complements electron probe microanalysis, and indeed can be combined with electron spectroscopic techniques as discussed in a later section.

Secondary ion mass spectrometry is like many of the electron spectrometry schemes described in previous sections, a relatively old technique [48], and the first commercial SIMS instrument evolved from developments of the early 1960's [49], and became available in the late sixties. SIMS is one of the most straightforward chemical analysis techniques which can be applied to solid surfaces because rather than being excited to emit some characteristic (secondary) signal, the surface atoms are themselves analyzed after being desorbed by the primary ion collision cascade. These emitted atoms are in an ionized state after being ejected from the surface, and can be collected (by an electrostatic accelerating field)) and focussed into a mass spectrometer for elemental discrimination and quantitative comparison as shown schematically in Fig. 4.40. Although a combined electric and magnetic sector (or double focusing) mass analyzer is usually employed in signal (secondary ion) descrimination and a combination magnetic sector system as shown in Fig. 4.40, very low energy ion mass discrimination is often facilitated by using a quadrupole mass filter. This is a non-magnetic quadrupole arrangement of electrodes which produces a hyperbolic field configuration combining a d.c. potential with a radio-frequency (r.f.) potential, and the signal "beam" is transmitted through the quadrupole arrangement to produce an ion beam divergence according to the ion mass. This arrangement does not require a geometrical sector, and can be combined with time of flight effects in a linear beam transmission profile to achieve additional ion energy discrimination [50].

SIMS differs from electron spectroscopies in that in SIMS, outer-shell ionizations predominate, and this interaction is highly dependent upon the chemical environment of the atom. Interelement effects also confuse quantification. Like ion probe microanalysis in general, SIMS can be performed using low energy and high energy primary ions. You might recognize that if low energy primary ions are employed, the interaction mechanisms are restricted to the surface. This is termed *static* SIMS, and is also especially useful in the analysis of organic (polymer) solids where fragmentation or molecular re-arrangement can occur. When higher primary ion energies are used, surface erosion occurs, and this is termed *dynamic* SIMS. This technique facilitates in-depth chemical profiling of particular elements. Noble gas ions (particularly Ar^+) are preferable for static SIMS applications because they produce little sample disturbance at moderate energies (10-20 keV). In the dynamic SIMS mode, the high-energy primary ion species become inplanted in the surface layers to high concentrations. This phenomena can be used to influence the chemical environment of the sample atoms in a desired manner, e.g. to stabilize the secondary ions and to increase their yield from the sample.

The secondary-ion detector (Fig. 4.40) is usually an open electron multiplier as shown in Fig. 3.24. Atomic concentrations C_A and C_B of two elements A and B can be related to their respective ion count rates (sensed by the detector) n_A and n_B by the expression [51]:

$$\frac{C_A}{C_B} = \left(\frac{n_A}{n_B}\right)\left(\frac{a_A \varepsilon_A}{a_B \varepsilon_B}\right) \qquad\qquad (4.41)$$

where a_A and a_B are the isotopic abundances of the elements A and B, and ε_A and ε_B are their respective ion yields. The ion yields are defined as the fraction of sputtered atoms that are in the appropriate ionized state (A^+, B^+ for example). Isotope abundances are usually constant and accurately known, and in order to obtain a quantitative analysis for element A when the concentration of element B is known (as a reference element) it is only necessary to determine the ion yield ratio $\varepsilon_A/\varepsilon_B$. This can be determined from a measurement on a standard sample, and is conceptually identical to the standardization procedures and determination of calibration constants discussed in previous sections for electron spectroscopies. You should compare Eq. (4.41) with Eqs. 4.30 and 4.34 for example. In this context, n_A/n_B for the secondary ion signal represents the corresponding signal intensity ratio detected.

The first ion microprobe mass analyzer (IMMA) was built by H. Liebl around 1966 [52] and employed a duoplasmatron ion source together with a double-focus mass spectrometer for secondary ion discrimination. The duoplasmatron primary source, employing an active gas supply for the plasma-ionization process, is still in use in many contemporary, commercial SIMS units. However as the spot size is reduced and higher beam intensities are required, the field-ion sources illustrated schematically in Fig. 4.40 are becoming increasingly popular.

Spatial resolution and minimum detectable concentration We discussed spatial resolution in the context of thin films as lateral resolution in Section 4.4.2 above. Spatial resolution really involves both depth and lateral features. For electron probes, the depth features are described approximately by the Thompson-Widdington law [Eq. (4.10)]. This can apply in principle to ions with appropriate adjustment for the variation of ion energy with mass as well as accelerating potential. In general, ion penetration increases for heavier ions and for higher accelerating potentials (V_o). Just as in electron probe systems, the intrinsic lateral resolution in ion probe systems is determined by the extent of the collision cascade as illustrated schematically in Fig. 4.42 which shows the ion interaction zone in a solid specimen similar to that shown for electron interaction in Fig. 4.27.

In general, the ion beam diameter is one of the more important factors determining the lateral resolution. In focusing ion beams to fine probes, the same basic laws applicable to electron microprobes apply: the current which can be focused into a small spot depends on the source brightness and optical aberrations (Chapter 3, Section 3.4.2). For a duoplasmatron ion source a beam diameter of 0.1 μm can be obtained at a total current of 10^{-10} amperes for a beam brightness (B) of about 200 amperes/cm^2 Sr. This can be achieved in a focal length as long as 3 cm [53]. This brightness is much less for the electron impact or gas discharge sources (c f. Fig. 3.18). However, as we noted previously, field-ion (FI) sources can provide a brightness as high as 10^6 A/cm^2 Sr for LMI sources and up to 10^8 A/cm^2 Sr for Ir or W solid field emitters. FI sources can generate a total current of 10^{-8} to 10^{-7} amperes. Consequently FI sources produce beam currents

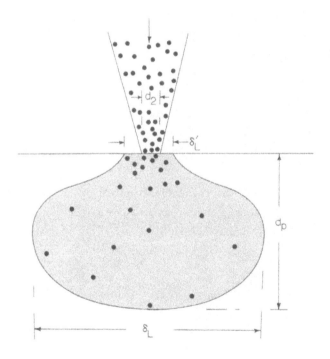

FIG. 4.42 *Interaction zone for ions focused onto a solid specimen. The beam diameter is designated d_2 while the lateral resolution is δ_L. The intrinsic lateral resolution is determined by the extent of the collision cascade at the surface, and is denoted δ_L. The maximum cascade depth is denoted d_p.*

which are 10^2 - 10^3 times higher than the more conventional sources, and which provide brightnesses which are as high as a million times the conventional ion sources (Fig. 3.8).

For low energy primary beams (3 - 10 keV ions) used in many SIMS applications, the lateral resolution is essentially the intrinsic lateral resolution ($\delta_L \simeq \delta_L'$ in Fig. 4.41) because the effected zone is within the first few atomic layers of the surface. However for deep penetration of the primary beam, the interaction zone buldges outward, decreasing the lateral resolution (because $\delta_L > \delta_L'$). Since the minimum detectable concentration C_{min} is proportional to the ratio of the minimum detectable ion current divided by the product of the total primary ion beam currents and the secondary ion yield

$$C_{min} = \frac{K \, I_{min}}{I_p \, S^+ + \eta} \tag{4.42}$$

very low concentration detection requires high beam currents, and SIMS units must utilize FI sources for high sensitivity. Ignoring the constant K in Eq. (4.42), if the minimum detectable current (I_{min}) is 10^{-18}A, the primary beam current is 10^{-12}A, and the secondary ion yield, S^+, is 3×10^{-2} (number of secondary ions per primary ion; with η, the overall spectrometer transmission,[*] = C = 1), C_{min} is roughly 3×10^3 ppm. Obviously increasing I_p by a factor 10^3 will reduce the detection limit by a similar factor to near 1 ppm.

You should recall from Eq. (3.32) the connection between probe resolution, beam current, and beam brightness. This also determines the range of lateral resolution in the SIMS [δ_{opt} in Eq. (3.32)].

4.6.4 SIMS APPLICATIONS

SIMS has become a relatively routine surface analysis technique. Liebl [54] estimated that in 1980 there were approximately 200 dedicated SIMS apparatus in use worldwide, and there were probably an equal number of SIMS modules combined with other microanalytical techniques in 1982. Examples of SIMS applications abound in almost all scientific and technical journals, especially *Scanning*, a journal originated in 1978.

---

[*]*This is the transmission or efficiency of transmission of the secondary ion species through the spectrometer.*

Although there is generally a need for high vacuum in SIMS operation, there are no specific specimen preparation procedures. Indeed, specimen surface preparation is not desirable and unnecessary. Typical, automated spectral displays are obtained as is typical in other similar mass spectra. By sampling for fixed conditions at different locations on a surface, the mass concentrations relative to some fixed profile such as a grain or interface can be determined. This is illustrated in Fig. 4.43 for phosphorus boundary segregation to a grain boundary in a large-grain copper sample.

Concentration depth profiles are a useful and somewhat unique application of SIMS in bulk specimens where chemical gradients are suspected or where concentration profiles have been created by such processes as ion bombardment. With duoplasmatron primary ion sources the sputter yield (or rate of sample erosion) is typically several microns per hour. This compares with $0.1~\mu m~h^{-1}$ for electron impact desorption for example.

Figure 4.44 shows a composition (concentration) depth profile of ion-implanted boron in silicon. The boron ions were implanted at a beam energy of 70 keV and a fluence of 10^{16} ions/cm^2. Similar applications are also useful in combination with other surface analytical techniques discussed previously, and these applications will be described in Sec. 4.8.

4.7 ATOM-PROBE FIELD-ION MASS SPECTROSCOPY

4.7.1 ATOM-PROBE FIELD-ION MICROSCOPY

Around 1966, E. W. Müller and co-workers conceived a novel application and modification of the field-ion microscope. In this particular modification the FIM was combined with a mass spectrometer making possible the analysis

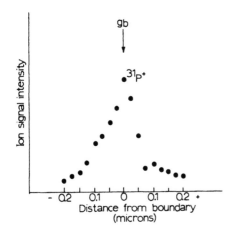

FIG. 4.43 *Secondary ion mass analysis of $^{31}P^+$ along a line approximately normal to a grain boundary trace intersecting a bulk, large-grain copper sample [average grain size of 200μm. utilizing a field-ionization source (H^+) of primary ions].*

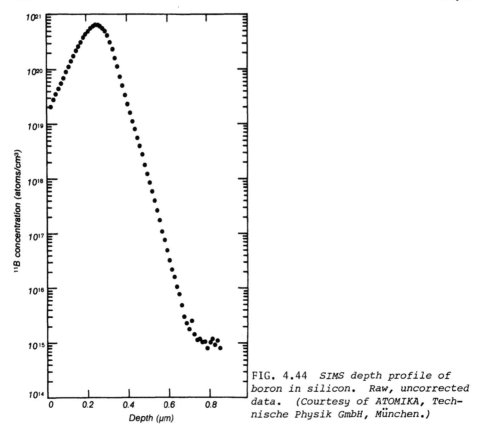

FIG. 4.44 *SIMS depth profile of boron in silicon. Raw, uncorrected data. (Courtesy of ATOMIKA, Technische Physik GmbH, München.)*

of the chemical nature of individual surface atoms on an emission end form [55].

The technique basically involves attaching at time-of-flight mass-spectrometer system to the screen section of a field-ion microscope (see Fig. 2.21) having a small opening in the central screen section; the vacuum is maintained between these continuous sections. The needle specimen holding assembly of the FIM is then modified so that the tip may be tilted about the axis of the microscope. With the FIM operating normally with an imaging gas (as described in Chap. 2), a particular atom of interest can be selected for analysis. The specimen is then tilted so that the image of the atom of interest falls in the probe hole in the FIM fluorescent screen. The imaging gas is then withdrawn from the system, and the tip pulsed with a high voltage (pulsed field evaporation, see Sec. 2.6.3) in order to desorb (field evaporate) the atom of interest. The ionized atom then enters the probe hole and encounters the spectrometer incorporating a detector with single-particle sensitivity. Figure 4.45 illustrates in a somewhat exaggerated schematic view the operation of the atom-probe FIM.

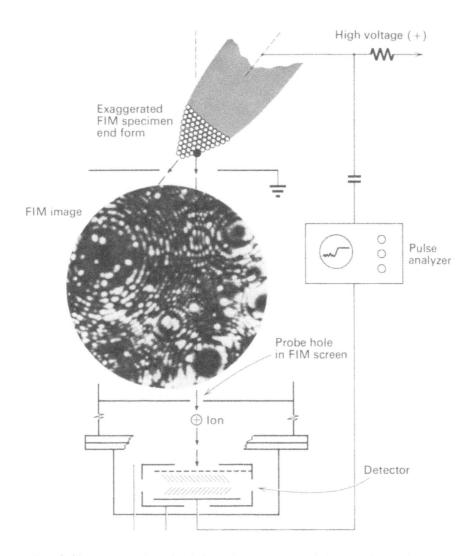

FIG. 4.45 *Atom-probe field-ion microscope. Field-ion image is from carbon-steel end form, and you are to imagine that circled impurity atom is chosen for analysis. Shown is axial propagation of this impurity species into detector section of mass spectrometer. (Field-ion image courtesy O. Nishikawa and E. W. Müller.)*

The charge-to-mass ratio of the detected ion in Fig. 4.45 can be calculated from

$$\frac{1}{2} M_i v^2 = ZeV_e \tag{4.43}$$

and

$$v = \frac{d}{t} \tag{4.44}$$

where M_i = ion mass

v = velocity

Ze = ion charge

V_e = desorption (evaporation) voltage in kilovolts (see Table 2.4)

d = distance between emitter tip and detector

t = observed time of flight

Once M_i has been calculated, the atomic species can then be identified.

The atom-probe FIM is the ultimate in analytical sensitivity since it
can identify a single atom. It should be apparent, considering the treatment of field-ion microscopy in Sec. 2.6.3, that detailed studies of short-
range order in alloys, the chemical nature of precipitates and inclusions,
and the identification of surface adsorbates can be pursued with an unprecedented accuracy using this technique.

4.7.2 FIELD DESORPTION AND IMAGING ATOM - PROBE MASS SPECTROSCOPY

In 1972, R. J. Walko and E. W. Müller again modified the basic imaging
features of the field-ion microscope by applying a high field to desorb
surface species as positive ions using a single, specific field pulse [56].
The first desorption images obtained by Walko and Müller [56] for tungsten
displayed the dominant W^{3+} field evaporating species but also included the
less abundant W^{4+}, as well as impurity atoms in the bulk and adsorbed surface impurities. As a consequence, it was not possible to identify specific species and specific atom image features associated with particular
surface regimes (crystallographic orientations). Walko and Müller later
recognized the need for time-gated desorption as a means to determine the
crystallographic variation in abundance associated with each desorbing
species. This concept was later incorporated into an imaging atom-probe
technique described by J. A. Panitz [57] where, in principle, the desorption microscope technique was combined with the technique of the atom-
probe FIM.

In the imaging atom-probe (IAP) technique shown in a simplified schematic in Fig. 4.46, a detector capable of amplifying and displaying the impacts of individual ion species desorbed from the emission tip surface is placed a distance R from the ion acceleration region (electrode) shown. The detector is curved to maintain an essentially constant distance between the end-form surface and the detector surface, and R defines a so-called driff region, which is field free. A positive d.c. bias voltage, V_{dc} and a positive, high voltage pulse having a voltage amplitude V_p are applied to the emission tip to produce the field required for desorption of the ion species, which desorb as positive ions and drift over the distance R with a kinetic energy given by [58]:

$$\frac{1}{2} M_i v^2 = q(V_{dc} + V_p) = ne(V_{dc} + V_p) \qquad (4.45)$$

where M_i is the ion mass, n is its charge state, and v is the ion velocity

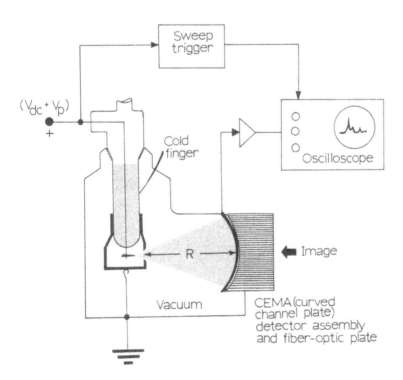

FIG. 4.46 *Schematic view of an imaging atom-probe mass spectrometer system. This is fundamentally a field desorption microscope similar in design to the FIM shown in Fig. 3.25.*

at the entrance to the drift region (the electrode opening shown in Fig. 4.46).

The time required for the ion to traverse the field-free drift region (the distance R) is from Eq. (4.44)

$$t = \frac{R}{v} = R \left[2e(V_{dc} + V_p)(\frac{n}{m}) \right]^{-\frac{1}{2}} \tag{4.46}$$

If desorption pulse (V_p) triggers the sweep of the oscilloscope shown in Fig. 4.46, t (the arrival time for the specific ion) can be measured directly. Consequently, the mass-to-charge ratio or the specific ion mass can be determined from

$$M_i = \frac{0.193\ n}{R^2} \ (V_{dc} + V_p) t^2 \tag{4.47}$$

if V_{dc} and V_p are in kV, t is in µs, and R is in meters. Equation (4.47) is indicative of the fact that different ion species (M_i) will not arrive at the detector simultaneously but sequentially in time, with the smallest M_i (or M_i/n in atomic mass units, AMU) arriving first. This feature allows the image of a preselected species to be separated from the composite desorption image by a process called time-gating where the detector is activated coincidentally with the arrival of the species of interest [58].

There are many very unique applications of time-gated desorption imaging in spite of the very severe restrictions in specimen composition and geometry, not to mention the lack of readily available (or accessible) instrumentation. One particularly exciting application involves the analysis of grain boundaries. There are numerous incidents where grain boundary segregation is so minute (a Gibbs adsorption layer is almost impossible to readily ascertain by other microanalytical techniques) that it would go undetected. In field-ion microscopy of grain or phase boundaries there are many cases of apparent elemental (atomic) decoration at or near grain boundaries. Figure 4.47 shows some excellent examples. You can appreciate that although it would be possible to position the emitters shown in Fig. 4.47 so that each "decorated" species observed could be individually positioned over the probe hole in the atom-probe of Fig. 4.45, the analysis of a representative number of them might be tedious. However with the time-gated field desorption features implicit in the design shown in Fig. 4.46, specific ions segregated or trapped along the boundary can be recorded in a single desorption image. This capability is demonstrated in the remark-

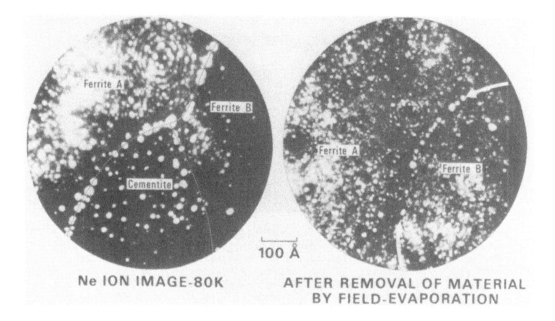

Ne ION IMAGE-80K   AFTER REMOVAL OF MATERIAL
BY FIELD-EVAPORATION

FIG. 4.47 *Field-ion micrographs of decorated (by impurity atom segregation to) interphase boundaries in neutron irradiated A302B pressure vessel
steel (Courtesy of M. K. Miller and S. S. Brenner; presented at the 28th
International Field Emission Symposium, Portland, Oregon, 1981).*

able image sequence shown in Fig. 4.48. Similar results of oxygen segregation to a grain boundary in molybdenum have also been obtained by Waugh
[59].* While Fig. 4.48 illustrates an essentially linear time-gated desorption image, the adsorption or inclusion of a particular atom species
over a surface (crystallographic, etc.) area will appear as an area map or
in principle an elemental (atomic) map conceptually similar to those elemental maps obtained in the Auger spectrometer, the energy-dispersive x-ray
spectrometer, and similar microanalysis regimes as shown for example in
Fig. 4.30. When time-gated desorption imaging is combined with field evaporation, a systematic depth profile can be obtained as well. Some of
these examples are shown in the excellent review by Panitz [58] along with

*The details of grain boundary and interfacial structure, etc., can be
found in L. E. Murr, Interfacial Phenomena in Metals and Alloys, *Addison-
Wesley Publishing Co., Reading, Massachusetts, 1975; reprinted by TechBooks,
Herndon, Virginia, 1990.*

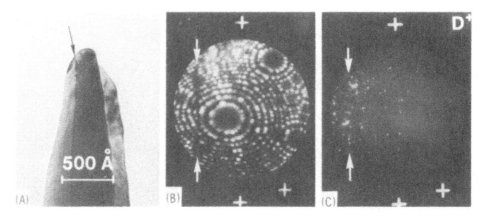

FIG. 4.48 *Image sequence showing deuterium trapping at a grain boundary in tungsten (A) Transmission electron micrograph of emission end form. The arrow shows the location of the grain boundary as well as the direction for surface imaging in the IAP, (B) FIM image showing the grain boundary (between the arrows), (C) IAP desorption image gated for D⁺* showing defect trapping along the boundary. [Courtesy G. L. Kellogg and J. K. P. Panitz in Appl. Phys. Letters, 37: 625(1980)].

the details of the instrumentation involved in the imaging atom-probe technique and related desorption phenomena.

Some interesting image features can also be obtained by non-gated, multilayer desorption images. These contain crystallographic and related contrast features but no specific ion abundance information [60]. They are also different from multilayer field-ion images obtained by recording a continuously evaporating tip. This latter image is also lacking in atomic detail.

Pulsed-laser atom-probe field-ion microscopy The time-gated desorption pulsing of emission end-forms in the IAP or atom-probe field-ion microscope has certain limitations in applications involving non-metallic specimens because of the inability to transmit the nanosecond, high-voltage pulses through such high resistance end-forms. In an investigation of photon-stimulated field ionization, Tsong, et al [61] observed controlled field evaporation of the surface layers of tungsten and aluminum oxide emitters using pulsed lasers. These observations led to the recognition of the advantages of using laser-assisted field desorption, including an increase in the mass resolution, by eliminating the energy spread which can occur in the high-voltage pulse desorption process due to evaporation of surface ions during the pulse rise time [62].

Kellogg and Tsong [63] have recently described the development of a time-of-flight atom-probe FIM utilizing pulsed-laser assisted field desorption. In this arrangement, a laser beam is focused onto the field emitter as shown conceptually in the arrangement of Fig. 2.28. The laser pulse replaces the pulsing voltage (V_p) in Eqs. (4.45) - (4.47), and the detector is triggered by a photodiode detector connected to the CEMA detector shown in the schematic of Fig. 4.45. Since the rate of field-evaporation is temperature dependent, the brief laser-induced temperature rise initiates field evaporation. This technique has allowed pulsed-laser atom-probe mass spectra for semiconductors such as silicon to be obtained as shown in Fig. 4.49, which also illustrates the general nature and appearance of atom probe spectra. The upper curve in Fig. 4.49 is an entire spectrum of high-purity Si taken in the presence of $\sim 4 \times 10^{-7}$ Torr H_2. The lower curve shows the double peak in an expanded time scale. The voltage shown is the d.c. bias, V_{dc} in Eq. (4.45).

The PLAP also has certain advantages in studying molecular adsorption because in high-voltage pulse desorption, adsorbed molecules often dissociate, making it impossible to study the field-free adsorption state. Kellogg [64] has recently reported on the systematic investigation of pulsed-laser stimulated field desorption of molecular hydrogen from molybdenum. These results indicated that at temperatures of 60 K and fields of 2.0 V/Å and higher, hydrogen is field adsorbed on molybdenum surfaces in molecular form.

4.8   INTEGRATED MODULAR MICROANALYSIS SYSTEMS

Modern (contemporary) analysis and microanalysis, or more specifically the approach to microanalysis, is frequently interdisciplinary with respect to the approach, and synergistic with respect to the actual undertaking. This is necessary because of the limitations in specific techniques outlined previously, and because the overlap and duplication provide additional confidence in the solutions obtained. The microanalysis techniques outlined in this chapter require either an electron or ion primary beam or some related excitation source. Many detectors can be utilized in discriminating mass-to-charge ratios, specific ion mass, and variations in charged-particle energy, particularly specific electron energy. As a consequence, it is sometimes an effective approach to combine several techniques in order to expand the range of information obtainable. This can be accomplished by adding specific modules to a basic ultra-high vacuum system. For example,

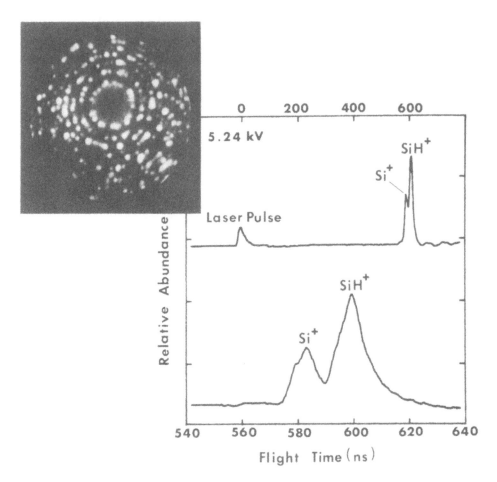

FIG. 4.49 *Pulsed-laser atom-probe spectrum of a high-purity silicon end-form with ~ 4 X 10^{-7} Torr H_2 in the system. The spectra correspond to Si^+ SiH^+ as indicated. No H^+ signal is detected indicating that field dissociation does not occur. The FIM image insert shows the silicon end-form imaged with hydrogen ions. (FIM insert courtesy of Dr. G. L. Kellogg; the Si spectrum is from G. L. Kellogg and T. T. Tsong* [63]*).*

if one begins with a basic ion-probe system as shown in Fig. 4.40, a secondary electron detector can be added to allow secondary electron imaging as shown in Fig. 4.41. By adding an electron gun to the system and probing either intermittently or simultaneously the ion-irradiated area, Auger electrons can be detected using either the magnetic-sector analyzer or a separate electrostatic analyzer (Fig. 4.18 for example). In addition, a solid-state detector system for discriminating characteristic x-ray emission (Fig. 4.10) could also be added.

Figure 4.50 illustrates this concept. In the system shown schematically in Fig. 4.50, modular units (sources and detectors) are inserted into a symmetrical (spherical) ultra-high vacuum chamber in a convenient arrangement to optimize, as far as possible, the necessary incident and take-off angles. In the arrangement shown, the modules are not necessarily on a specific circumferential line, but are placed along some optimum incident or take-off direction (diameter). This concept permits the use of rapidly exchangeable or interchangeable sources and detectors which can allow for rapid switching from one microanalysis mode to another.

Although completely efficient integrated modular systems are sometimes difficult to achieve, several commercial units are currently available around the World. New developments in lens designs and efficiency, along with concomitant improvements in focal lengths, signal collection and analysis efficiency, detector sensitivity, etc., are also aiding in achieving remarkably efficient and sensitive systems. Figure 4.51 illustrates somewhat

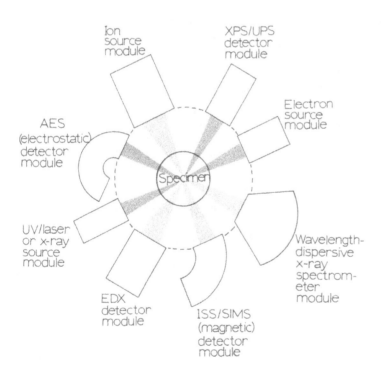

FIG. 4.50 *Schematic (conceptual) arrangement for integrated modular microanalysis system utilizing a spherically symmetric module array.*

FIG. 4.51 *Commercial surface analysis system for XPS, UPS, AES, SIMS, ISS, and SAM. (Courtesy INFICON Leybold-Heraeus, Inc.)*

typically the appearance of a contemporary commercial system for surface analysis which you can readily recognize to be conceptually similar to the schematic arrangement shown in Fig. 4.50.

The concept of integrated modular microanalysis discussed here can also be combined with imaging and image-display features to obtain a complete description of specific shapes, surface structures, compositions, crystal structures, and other specific characteristics. These features will be described in more detail in Chapter 5 and 7. In Chap. 7, this concept will be treated as analytical electron microscopy where integrated modular systems are utilized in an electron optical arrangement to permit imaging by transmission and scanning electron microscopy and microanalysis by selected-area and convergent-beam electron diffraction, EDX, and EELS. This approach applies specifically to thin film specimens while the approach outlined in this section and illustrated in Fig. 4.50 applies most practically to thick films or bulk specimens.

4.9 SUMMARY AND COMPARISON OF ELECTRON AND ION PROBE MICROANALYSIS TECHNIQUES

Having arrived at this point in this chapter by a complete reading of the sections, or even by a casual perusal of select topics, you must surely recognize that nearly all of the analytical techniques involve surface analysis. Many are not limited to the actual surface layers, but can penetrate deep into the bulk, while others continuously create new surface to produce a systematic depth profile of chemical composition, and still others combine these features with specific or near surface microanalysis of continuously or intermittantly exposed new surface.

We mentioned earlier that in most cases an ultra-high vacuum (> 10^{-9} Torr) is preferred, and in some cases where impurity adsorption is to be detected, even greater vacuum. In most cases no specimen treatment is permitted in order to avoid the creation of artifacts or compositional or structural alterations. In most cases minimum detectable mass (MDM) and minimum mass fraction (MMF) define the sensitivity. The MDM is limited by the probe current density and the noise level of the signal to be detected. The MMF is limited by the availability of a large number of counts in order for small peaks in the spectrum to be statistically significant above the background. This is not usually consistent with the highest spatial resolution. Resolution is also important in regard to both depth and the spreading of the beam or lateral resolution. The ultimate course involves single-atom detection which is achieved in the atom-probe field-ion microscope or a modification thereof utilizing time-of-flight mass spectrometry.

It would seem instructive and convenient at this point to compare many of the electron and ion microprobe techniques described in this chapter. This is done in Table 4.3 which contains data based upon the recent work of Rudd [65] and Werner [66]. The volume edited by Czanderna [67] is also recommended as a very informative reference which can give additional perspective to the surface analysis techniques, and provide additional and still contemporary application examples. The recent work of Castaing [68] is also recommended because this describes imaging and probing techniques for surface studies. We will treat some of these imaging techniques in Chap. 5. Finally, you might peruse the recent article by Holloway and McGuire [69] which describes the characterization of electron devices.

PROBLEMS

4.1 Calculate the Kα x-ray wavelength for Er. What is the spread in the wavelength, Δλ, between the Kα and Kβ lines? Compare this wavelength difference with that which would occur for Kα radiation from a sample containing Er and Tm.

4.2 Calculate the Bragg angle θ for the detection of Ag Kα radiation in an electron microprobe arrangement using a sodium chloride analyzing

TABLE 4.3 Summary of the principal features of different electron and ion microprobe techniques

Acronym	Technique	Incident (exciting) radiation	Particle emitted	Sample Penetration	Lateral resolution	Detection Limit	Detectable element range	Materials applications and information obtained
AES	Auger electron spectroscopy	electron at 0.1-5 keV	Auger electron	0-10 Å	0.1-03 µm	0.1%	≥Li	any solid matter, adsorbate analysis; elemental analysis of surface
EDS/EDX or EDXA	Energy dispersive x-ray spectroscopy	(10-30 keV)	Characteristic x-ray photon	>1µm	>1µm	0.1%	>Na	any solid matter elemental analysis of bulk or near surface regime
EID	Electron-induced desorption	electrons at 20-1300 eV	neutral particle or ion absorbed	none	——	<0.1%	——	solid surfaces
ELS/ EELS	Electron energy loss spectroscopy	electron* at 750 keV	inelastically scattered electron	0.01 -0.1µm	<0.04µm	<0.1%	≥Li	any solid thin film; elemental analysis
EPMA/ WDS	Electron probe microanalysis/ wavelength dispersive (x-ray) spectrometry	electron at 0.5- 40 keV	characteristic x-ray photon	0.02- 2µm	>1µm	<0.1%	≥Li	any solid matter elemental analysis (near-surface composition)

*Field-emission source. Note that in most cases listed the conventional electron or ion emission source is implicit. Use of field-ionization or field-emission sources will improve the performance.

ESCA	Electron spectroscopy for chemical analysis	U.V. photons	photo-electron	~5-20 Å	0.2-1μm	0.1%	>Na	any solid, gases, frozen liquids; elemental composition, chemical binding, electronic states
FDS	Field-desorption spectroscopy	(see imaging atom-probe mass spectroscopy next)						
IAPMS/APFIM	Imaging atom-probe mass spectroscopy/atom-probe field-ion microscopy	———	ion or ions desorbed from surface	individual layers can be pulse desorbed	single atom	single atom	all elements and isotopes	many metal and semiconductor whiskers, adsorption phenomena, depth profile, crystallographic data, etc.
HEIS	High-energy ion scattering	(see RBS)						
ISS	Ion scattering spectroscopy	ion at 25-10^5 eV	same as incident	first exposed layer at any time	1-100μm	0.1% to 100 ppm (bulk)	$\geq$Li	all solid matter, elemental composition, location of adsorbed species, semiquantitative depth profile
LEIS	Low-energy ion scattering	ion 20-2000 eV	same ion as incident	5-10 Å	<100μm	0.1%	$\geq$Li	any solid material, elemental composition, depth composition profile, adsorbed species.

TABLE 4.3 (continued)[†]

Acronym	Technique	Incident (exciting) radiation or particle	Particle emission	Sample penetration	Lateral resolution	Detection	Detectable element range	Materials applications and information obtained
PIXE	Proton-induced x-ray emission	ion (proton) 2–200 keV	characteristic x-ray photon	<1μm	<500μm	0.1–10	>Na	any solid material, bulk elemental composition; depth composition profile
RBS (HEIS)	Rutherford backscattering	ion 10^5–10^7 eV	same ion as incident	<10μm	1–3μm	0.1%	>C	Thick solids, quantitative elemental composition, crystal orientation, surface stoichiometry
SAM	Scanning Auger microscopy/microanalysis	(see AES)						
SCANIIR	Surface compositional analysis by neutral and ion impact radiation	ion <4 keV	sputtered excited neutral particle or ion	exposed layer	<100μm	>0.1%	>Li	solid materials, approximate elemental composition of surface layers
SIMS	Secondary ion mass spectroscopy	ion 10^2–10^5 eV	secondary (surface) ion	exposed layer (<10 Å)	0.1–1μm	0.1 ppm	>H	solid materials; elemental composition of the surface; depth composition profile (sample destruction)

STM/STS	scanning tunneling microscopy/ scanning tunneling spectroscopy	tunneling electrons	tunneling electrons	~1 Å	~1 Å	----	----	analysis and micro-analysis in UHS or air, adsorbed molecules on surface structures of solids even in aqueous media can be examined.
UPS/XPS	ultra-violet photoelectron spectroscopy/ x-ray photoelectron spectroscopy	U.V. or x-ray photon	photo-electron	10-20 Å	<10^3 µm	1%	>He	solid materials; work function analysis, adsorbed molecules, electronic states of surface, approximate chemical composition.

---

†*You should note that the sensitivities, etc. given refer ideally to a dedicated technique on analytical system. There is sometimes a loss of sensitivity, etc. when the technique is part of an integrated-modular system. (See Ref. 69 for additional details and applications of SIMS, ISS, AES, XPS, etc.).*

crystal (assume a screening constant for silver nuclei of 3.7). If the incident probe beam is normal to the specimen surface, and the detector axis is inclined 60° with respect to the incident electron beam, calculate the takeoff angle ϕ_t for the characteristic x-rays assuming a simple, bent analyzing crystal (see Fig. 4.4).

4.3   The investigation of a polycrystalline copper-gold alloy by electron backscatter indicates a nonhomogeneous alloying of the specimen since differing intensity profiles are observed for a scan of several grains having an average diameter of 10 microns. If the grains are identified as A, B, C, and D in terms of categories of backscatter intensity, estimate the composition (concentration) of Au and Cu of the respective grains, considering the data recorded in the table below.

Grain Designation	Relative Electron Backscatter (I_B/I_A)	C_{Au}	C_{Cu}
A	0.30	——	——
B	0.35	——	——
C	0.38	——	——
D	0.41	——	——

4.4   The interpretation of intensity profiles for characteristic x-ray emission sometimes hinges on corrections for absorption within the solid, and on the actual displacement of the excitation zone from the incident surface. If we were to examine foils, say, of a number of different elements (varying Z), methodically altering the thickness, it might be possible, by utilizing several approaches, to elucidate the extent and character of x-ray absorption. Following these guidelines, outline a procedure for systematically checking this point.

4.5   A sample undergoing investigation is known to contain large precipitates containing Ti and Aℓ. If the Aℓ in some precipitates segregates to certain crystallographic zones, how can this be detected by using backscattered electrons? If the microprobe is equipped with a mica-crystal monochromator, indicate the necessary excitation potentials in kilovolts for producing Kα x-ray lines for these elements. Calculate the angle the detector must make with the analyzing-crystal surface in order to detect these elements.

4.6 Calculate the electron energy required for $K\alpha$ and $L\alpha$ characteristic x-ray emission for Cu and Mo.

4.7 An annealed 70:30 brass (Cu-Zn) alloy sample is used as a test specimen for studying corrections for takeoff absorption in concentration analyses by x-ray microprobe techniques. Using a pure Cu standard, the detector counts per second for Cu $K\alpha$ averaged 400, 550, 620, 810, 980, and 1060 for takeoff angles of 5, 10, 20, 25, and 30°, respectively. For the brass sample, the detector counts per second for Cu $K\alpha$ averaged 28, 63, 100, 186, 322, and 490 for the same respective takeoff angles and a constant accelerating potential. Referring to Eq. (4.15), plot g(E) versus ϕ_t.

4.8 Calculate the $M\alpha$ x-ray line excitation potentials (in kilovolts) for Mo, Te, Pr, and Dy. Calculate the efficiency ratio for producing Mo $K\alpha$ radiation and Mo $M\alpha$ x-radiation for a constant takeoff angle of 45° and an electron accelerating potential of 30 kV. What can you say about this ratio in terms of microprobe x-ray analysis?

4.9 If depth of penetration of incident electrons decreases for decreasing accelerating potential, and excitation of characteristic x-rays becomes increasingly efficient with a reduction in absorption effects, discuss the advantages of employing fairly high accelerating potentials (in excess of 20 kV). How about protons?

4.10 Refer to Figs. 4.37 and 4.38. If we assume that no characteristic x-ray emission was detected for a scan of the areas showing potential discontinuities, indicate in a two-point argument the evidence displayed for possible vacancy segregation in these areas.

4.11 Let us suppose we wish to study phosphorus embrittlement by grain-boundary segregation in Cu. Describe how to proceed on the basis of electron microprobe analysis and ion microprobe analysis. Compare the sensitivities.

4.12 A scanning ion microprobe utilizing a primary proton (H^+) beam is characterized by the following parameters: f_o (objective lens focal length) = 5.5 mm, C_{S1} = 88 mm, C_{S2} = 4000 mm, C_{Cr1} = 9 mm, C_{Cr2} = 75 mm, ρ = 10 Å, ΔE = 4eV (for H^+), E_o = 12 kV, α = 0.012, and R = 6. Find the probe diameter.

4.13 Refer to Fig. 4.49. If the travel distance in the field-free region of the laser-induced desorption atom-probe is 15 cm, find the velocity of the Si^+ ion within this region. What total voltage must be applied to impart this velocity?

4.14 A 300 Å carbon film builds up a contamination layer of carbon or hydrocarbon polymer in an electron probe of 50 Å. What additional beam broadening does this induce if the accelerating voltage is 125 kV. Express the broadening as a percent increase.

REFERENCES

1. R. Castaing and A. Guinier, *Proc. Delft Conf. Electron Microscopy*, 60(1949).

2. R. Castaing, *Advan. Electron. Electron Phys.*, 13: 317(1961).

3. K. F. J. Heinrich, *Bibliography on Electron Probe X-ray Analysis and Related Subjects*, John Wiley & Sons, Inc., New York, 1965.

4. D. B. Wittry, *ASTM Spec. Tech. Publ.* 349: 128(1964).

5. H. Moseley, *Phil. Mag.*, 26: 1024(1913).

6. D. B. Langmuir, *Proc. IRE*, 25: 977(1937).

7. B. D. Cullity, *Elements of X-ray Diffraction*, Addison-Wesley Publishing Company, Inc., Reading, Mass, 1956.

8. D. B. Wittry and D. F. Kyser, *J. Appl. Phys.*, 38: 375(1967).

9. D. F. Kyser, doctoral dissertation, University of Southern California, 1967.

10. W. J. Campbell, J. D. Brown, and J. W. Thatcher, *Anal. Chem.*, 38(5): 425R (1966).

11. P. Palluel, *Compt. Rend.*, 244: 1492(1947).

12. T. Danguy and R. Quivy, *J. Phys. Rad.*, 17: 370(1956).

13. P. Auger, *Compt. Rend.*, 180: 65(1925).

14. C. H. Chen, J. Silcox, and R. Vincent, *Phys. Rev.* B12: 64(1975).

15. M. Nishijima and T. Morotani, *Surface Sci.*, 32: 459(1972).

16. K. Veda and R. Shimizu, *Surface Sci.*, 36: 789(1973).

17. G. J. Dooley and T. W. Haas, *J. Metals*, 22(11): 17(1970).

18. D. T. Hawkins, *Auger Electron Spectroscopy: A Bibliography*, 1925-1975, Plenum Press, New York, 1977.

19. E. G. McRae, *Rev. Mod. Phys.*, 51: 541(1979).

20. R. F. Egerton, *SEM/1978*, Vol. I, O. Johari (ed.), SEM, Inc., AMF O'Hare, Illinois, 60666, p. 13.

21. N. J. Zaluzec, *Thin Solid Films*, 72: 177(1980).

22. V. E. Cosslett and R. N. Thomas, *Brief J. Appl. Phys.*, 15: 883(1964).

23. J. Philibert and R. Tixier, *J. Phys. D*, 1: 685(1968).

24. G. Cliff and G. W. Lorimer, *Proc. 5th European Congress on Electron Microscopy,* Institute of Physics, Bristol, p. 140 (1972).

25. G. Cliff and G. W. Lorimer, *J. Microscopy, 103:* 203(1975).

26. J. I. Goldstein, J. L. Costley, G. W. Lorimer, and S. J. B. Reed SEM/ 1977, O. Johari (ed.), IITRI, Chicago, p. 315(1977).

27. J. I. Goldstein, Chap. 3 in *Introduction to Analytical Electron Microscopy,* J. J. Hren, J. I. Goldstein, and D. C. Joy (eds.), Plenum Press, New York, p. 83(1979).

28. D. C. Joy and D. M. Maher, SEM/1977, O. Johari (ed.), SEM, Inc., AMF O'Hare, Illinois, p. 325 (1977).

29. P. Duncumb, J. *de Microscopic, 7:* 581(1968).

30. N. J. Zaluzec, *Proc. Electron Microscopy Soc. of America,* G. W. Bailey (ed.), Claitor's Publishing Div., Baton Rouge, La., p. 98 (1980).

31. J. Philibert, *J. Inst. Metals, 90:* 241(1962).

32. K. F. J. Heinrich, *Symposium on X-ray and Electron Probe Analysis,* ASTM Special Tech. Publ. 349(1964).

33. E. W. White, P. J. Denny, and S. M. Irving, Quantitative Microprobe Analysis of Microcrystalline Powders in McKinley, Heinrich, and Wittry (eds.), *Electron Microprobe,* John Wiley & Sons, Inc., New York, 1966.

34. J. Hillier, *Phys. Rev., 64:* 318(1943).

35. J. Hillier and R. F. Baker, *J. Appl. Phys., 15:* 663(1944).

36. J. Philibert et al., *Compt. Rend., 251:* 1289(1960).

37. M. J. Fleetwood, *J. Inst. Metals, 90:* 429(1962).

38. D. A. Melford, *J. Inst. Metals, 90:* 217(1962).

39. L. E. Murr and J. B. Hiskey, *Met. Trans., 12 B:* 255 *(1981).*

40. H. Liebl, *Anal. Chem., 46:* 23A(1974).

41. R. Levi-Setti, Proc. 7th Annual SEM Symposium, 11T Research Institute, O. Johari (ed.), p. 125 (1974).

42. J. Orloff and L. W. Swanson, *J. Vac. Sci. Technol., 12:* 1209(1975).

43. J. Orloff and L. W. Swanson, *Scanning Electron Microscopy/1977,* Vol. 1, O. Johari (ed.) IIT Research Institute, Chicago, p. 57(1977).

44. V. E. Krohn and G. R. Ringo, *Appl. Phys. Letters, 27:* 479(1975).

45. L. W. Swanson, G. A. Schwind, and A. E. Bell, *J. Appl. Phys., 51*(7): 3453(1980).

46. N. H. Tolke, D. L. Simms, and E. B. Foley, *Radiation Effects, 18:* 221(1973).

47. J. W. Mayer and J. F. Ziegler, *Ion Beam Surface Layer Analysis,* Elsevier Sequoia, Lausanne, 1974.

48. F. L. Arnot and J. C. Milligan, *Proc. Roy. Soc. (London), A156:* 538 (1936).

49. H. Liebl and R. F. K. Herzog, *J. Appl. Phys., 34:* 2893(1964).

50. K. Wittmaack, *Scanning, 3:* 133(1980).

51. S. J. B. Reed, *Scanning, 3:* 119(1980).

52. H. Liebl, *J. Appl. Phys.*, *38:* 5277 (1967).

53. H. Liebl, *Vacuum, 22:* 619(1972).

54. H. Liebl, *Scanning, 3:* 79(1980).

55. E. W. Müller, J. A. Panitz, and S. B. McLane, *Rev. Sci. Instr., 39:* 83(1968).

56. R. J. Walko and E. W. Müller, *Phys. Stat. Sol. (a), 9:* K9(1972).

57. J. A. Panitz, *J. Vac. Sci. Technol., 11:* 206(1974).

58. J. A. Panitz, *Progress in Surface Sci., 8:* 219(1978).

59. A. R. Waugh, 24th Intl. Field Emission Symposium, Oxford, England, 1977 [see also the sequence by A. R. Waugh and M.J. Southon (Fig. 45) in the article by Panitz above (Ref. 58)].

60. A. R. Waugh, E. D. Boyes, and M. J. Southon, *Nature, 253:* 342(1975).

61. T. T. Tsong, J. H. Block, M. Nagasaka, and B. Viswanathan, *J. Chem. Phys., 65:* 2465(1976).

62. T. T. Tsong, *Surface Sci., 70:* 211(1978).

63. G. L. Kellogg and T. T. Tsong, *J. Appl. Phys., 51* (2): 1184(1980).

64. G. L. Kellogg, *J. Chem. Phys., 74* (2): 1479(1981).

65. E. J. Rudd, *Thin Solid Films, 43:* 1(1977).

66. H. W. Werner, *Mater. Sci. Engr., 42:* 1(1980).

67. A. W. Czanderna, (ed.), *Methods of Surface Analysis*, Elsevier, Amsterdam, 1975.

68. R. Castaing, *Mater. Sci. Engr., 42:* 13(1980).

69. P. H. Holloway and G. E. McGuire, *Applications of Surface Sci., 4:* 410(1980).

SUGGESTED SUPPLEMENTARY READING

Barer, R., and Cosslett, V. E. (eds.): *Advances in Optical and Electron Microscopy*, vol. 10, Academic Press Inc., New York, 1987.
Benninghoren, et al. (eds.), *Secondary Ion Mass Spectrometry - SIMS II*, Springer-Verlag, Berlin, 1979.
Carlson, T. A., *Photoelectron and Auger Spectroscopy*, Plenum, New York, 1975.
Castaing, R.: *Advances in Electronics and Electron Physics*, vol. 13, Academic Press Inc., New York, 1961, p. 317.
Castaing, R.: Probing and Imaging Techniques for Surface Studies, in *Mater. Sci. Engr., 42:* 13(1980).
Caudano, R., and Verbist, J. (eds.), *Electron Spectroscopy: Progress in Research and Applications*, Elsevier, New York, 1974.
Cosslett, V. E., A. Engstrom, and H. H. Pattee (eds.), *X-ray Microscopy and X-ray Microanalysis*, Elsevier Publishing Company, Amsterdam, 1960.
Czanderna, A. W. (ed.): *Methods of Surface Analysis*, Elsevier, Amsterdam, 1975.
Egerton, R. F.: *Electron Energy Loss Spectrometry in the Electron Microscope*, Plenum Press, New York, 1986.
Elion, H. A.: *Instrument and Chemical Analysis Aspects of Electron Microanalysis and Macroanalysis*, Pergamon Press, Ltd., London, 1966.

Hren, J. J., Goldstein, J. I., and Joy, D. C. (eds), *Introduction to Analytical Electron Microscopy,* Plenum Press, New York, 1979.

McGuire, G. E., *Auger Electron Spectroscopy Reference Manual,* Plenum Press, New York, 1979.

McKinley, T. D. (ed.): *The Electron Microprobe,* John Wiley & Sons, Inc., New York, 1966.

Meyer, O., and Kappeler, F. (eds.), *Ion Beam Surface Layer Analysis,* Plenum, New York (2 vols.), 1976.

Miller, M. K., and Smith, G. D. W., *Atom Probe Microanalysis: Principles and Applications to Materials Problems,* Materials Research Society, Pittsburgh, Pennsylvania, 1991.

Rabalais, J. W., *Principles of Ultraviolet Photoelectron Spectroscopy,* Wiley-Interscience, New York, 1977.

Reimer, L.: *Transmission Electron Microscopy, Physics of Image Formation and Microanalysis* (2nd ed.), Springer-Verlag, Berlin, 1989.

Thomas, J. P., and Cachard, A., *Material Characterization Using Ion Beams,* Plenum Press, New York, 1978.

Tolk, N. H., Tully, J. C., Heiland, W., and White, C. W. (eds.), *Inelastic Ion Surface Collisions,* Academic Press, New York, 1977.

(In addition to the references cited above you might consult recent issues of the Journals *Analytical Chemistry* and *Thin Solid Films* for microanalysis applications and also the *Journal of Mass Spectroscopy and Ion Physics.*)

5
ELECTRON AND ION
MICROSCOPY OF SURFACES

5.1 INTRODUCTION

Most of our previous observations of material properties, chemical composition, and structure have been of a surface nature. A limiting factor, except in the special case of field-ion microscopy and the atom-probe FIM has
been resolution. And while resolution on a fine scale is attainable in the
field emission microscope, the form of the specimen as well as the conditions involved in its imaging are not applicable to materials as they exist
in the technological environment; nor are the responses noted always unambiguously interpretable on the basis of bulk solid properties of engineering significance. In the surface microscopy modes of analysis outlined in
Chap. 2, the analysis was also limited by the methods of specimen preparation, which in effect represented varying degrees of destructive testing;
and this was also the case where selected-area composition was desired by
electron and ion probe microanalysis. The exception is of course the scanning tunneling microscope (STM) where the probe (emission end form) is a
consideration.

The technological and engineering importance of electron and ion optics, and ion and electron microscopy in particular, are manifest primarily
in the ability of these approaches to detect impending structural failure
at its inception, and to provide a mechanistic basis for the description of

fracture processes. We therefore require that surface structure be observed in situ, and that the material remain chemically and morphologically intact. In other instances it might be desirable to follow the course of failure of either a mechanical or electrical nature as it relates to surface morphology, crystallography, chemical composition, and defect concentrations and character.

In the present chapter we shall treat the examination of materials basically as electron and ion metallography; and where fracture analysis is of prime concern, we will effectively treat the subject of electron fractography. There are basically three prominent modes of materials surface analysis utilizing electron optical techniques. These involve replica electron microscopy where electron transmission through a thin "impression cast" of the surface forms a density-contrast image of surface topography; low-angle-reflection electron microscopy performed directly within an electron microscope; scanning electron microscopy where surface topography is imaged by analyzing secondary emission electrons; backscattered electrons, specimen current due to absorbed electrons; and characteristic x-ray and electron (Auger) mapping and electron mirror microscopy and electron holography. Similarly, there are several similar modes of surface microscopy and analysis involving ion optical techniques. These involve secondary ion microscopy (in particular scanning ion microscopy) utilizing both secondary ion and secondary electron emission for surface image formation, and elemental concentration maps; and ion mirror microscopy (or related techniques employing electrostatic mirror concepts).

5.2 THE ELECTRON MICROSCOPE

The electron microscope had its beginnings in the original electron diffraction experiments of C. J. Davisson and L. H. Germer [1] and the later electron optics developments of Davisson and C. J. Calbick [2], E. Brüche and H. Johannson [3], and M. Knoll and E. Ruska [4]. The original prototype electron microscope, claiming resolutions in excess of ordinary light microscopes, and employing the magnetic-lens design, was constructed by E. Ruska about 1934; and 4 years later B. von Borries and Ruska [5] described the first practical (as a scientific instrument) electron microscope. Almost simultaneously in 1939, A. Prebus and J. Hillier [6] constructed a similar electron microscope. The microscope originally designed by von Borries and Ruska was actually the prototype Siemens instrument, while that of Prebus and Hillier, later modified to include stable power supplies designed by A. W. Vance [7], became the RCA prototype. In 1940 an electron microscope designed by Hillier and Vance was announced as a commercially available unit by RCA; within 5 years, commercial instruments were being manufactured by Siemens, RCA, Hitachi, Ltd., JEM, and others. Thus by 1945, electron microscopes capable of object resolution on the order of 10 Å were readily available.

Figure 5.1a is typical of conventional electron microscope designs that, from the standpoint of the materials scientist, operate nominally at 100 kV to 200 kV; and Fig. 5.1b shows schematically the electron optical system and lens-design characteristics. Since we have treated the important aspects of electron optics in Chap. 3, we will presume a working knowledge of the principles of operation of the electron microscope, and consequently dispense with any elaborate treatment of specific design features. Instrumentation and design of power supplies will also be excluded from our treatment since they are not germane to the utilization of commercial instruments in materials science research. However, we will resume a detailed treatment of high-voltage power-supply design in Chap. 8 where this feature contributes rather intrinsically to the ultimate application of high-voltage electron microscopy in the analysis of thick materials.

We should realize at the outset that in electron microscopy, as in other forms of surface microscopy treated in Chap. 2, the ultimate goal insofar as the researcher is concerned is the interpretation of images and the elucidation of object detail, structure, and morphology. In most studies involving the electron microscope, the electron beam either traverses the material of interest or a representative surface "cast" of the material. Thus, in either case, the image formation occurs as a result of electron transmission. The two modes of image formation of immediate concern are described schematically in the sketches of Fig. 5.2, which shows the image formation by the normal direct beam transmission or bright-field mode, and the dark-field image formation utilizing diffracted electrons. In each of the situations, we observe that the final image is formed by electrons scattered, or generally diffracted, from the object. Because of this scattering process, the image contrast and clarity are very much dependent on the elimination of extraneous signal information at the aperture plane (back focal plane). We must make a point of the fact that the distribution of scattered electrons at the aperture plane entirely determines the intensity distribution at the gaussian image plane, and also the contrast features in the final image on the fluorescent viewing screen or photographic

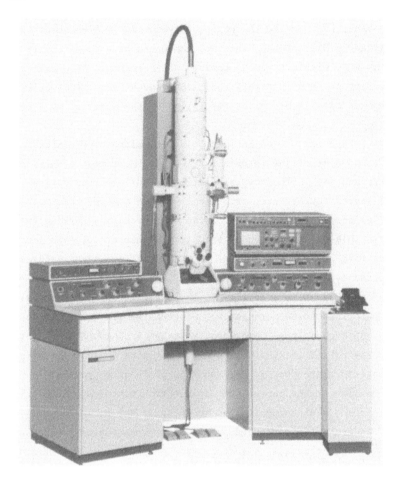

FIG. 5.1 *(a) Commercial electron microscope (JEM-100 CX).*

plate. It must also be realized that the primary signal information for
the formation of the image occurs in the objective-lens section of the
electron microscope. Thus, where resolution and other electron optical
features are concerned, Chap. 3 should provide an adequate basis for understanding the important functions and design variables contributing to the
overall lens performance.

5.2.1 *IDEAL IMAGE FORMATION IN THE ELECTRON MICROSCOPE*

We have observed in Chap. 4 (see especially Fig. 4.1) that a multitude of
interactions and residual processes occur when electrons encounter such
solid matter; and we have examined some of the important consequences of

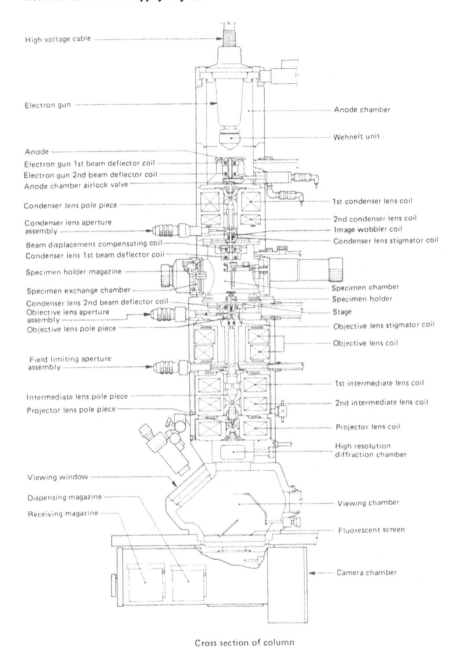

High voltage cable

Electron gun

Anode chamber

Wehnelt unit

Anode
Electron gun 1st beam deflector coil
Electron gun 2nd beam deflector coil
Anode chamber airlock valve

Condenser lens pole piece

1st condenser lens coil

Condenser lens aperture
assembly

2nd condenser lens coil
Image wobbler coil
Condenser lens stigmator coil

Beam displacement compensating coil
Condenser lens 1st beam deflector coil

Specimen holder magazine

Specimen exchange chamber
Condenser lens 2nd beam deflector coil
Objective lens aperture
assembly
Objective lens pole piece

Specimen chamber
Specimen holder
Stage
Objective lens stigmator coil

Field limiting aperture
assembly

Objective lens coil

Intermediate lens pole piece

1st intermediate lens coil

Projector lens pole piece

2nd intermediate lens coil

Projector lens coil

High resolution
diffraction chamber

Viewing window

Dispensing magazine

Receiving magazine

Viewing chamber

Fluorescent screen

Camera chamber

Cross section of column

FIG. 5.1 *(b) Commercial electron microscope (JEM-100 CX). Schematic view of electron microscope showing lens design. (Courtesy Japan Electron Optics Laboratory Co., Ltd.)*

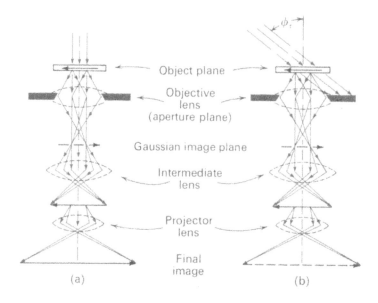

FIG. 5.2 *Bright- (a) and dark-field (b) electron ray diagrams for image formation in electron microscope.*

processes. Now we wish to examine only the total response of an electron encountering a solid, where the incident electrons are altered by scattering processes or absorption. In this ideal case we shall consider the solid object encountered by an energetic electron beam to be totally noncrystalline. Consequently the image formed of such an object in the electron microscope will arise by contrast resulting from differential scattering from the various parts of the specimen because of the associated mass-density variations, that is, local variations in thickness or atomic number Z.

Scattering in the object can occur at a single point of collision of an electron with an atom center in the object (single scattering), or a single electron can be scattered by two or more atoms in traversing the object (plural scattering). Where the object is very thick, a single electron originally scattered from an atom near the incident side of the object may undergo numerous scattering processes (multiple scattering) before finally emerging on the exit side. These scattering processes can of course occur in two modes. If no energy is lost, the scattering is said to be completely elastic; however, where the incident electron looses a portion of its kinetic energy between scattering incidences, the processes are described as inelastic scattering. Within the irradiated zone of a specimen, electrons can be scattered by collision with a nucleus, or by the direct encounter of another scattered electron.

We will consider that our ideal solid, composed of a completely random collection of atoms as shown in Fig. 5.3, scatters electrons from a point 0 over a range of solid angles Ω, with the scattered intensity decreasing with an increase in the scattering angle. Since the solid angle Ω is related to the electron-scattering angle α_s by

$$\Omega = 2\pi(1 - \cos\alpha_s)$$

this relationship defines Ω with respect to Fig. 5.3. Let us then describe the total elastic-scattering cross section per atom (at a point of reference 0 in Fig. 5.3) in terms of the differential cross section D(S) by

$$\sigma_a = \int D(S)\,d\Omega$$

or for that portion elastically scattered outside the objective aperture

$$\sigma_a = 2\pi \int_\alpha^\pi |f(\mu)|^2 \sin\alpha_s \, d\alpha_s \tag{5.1}$$

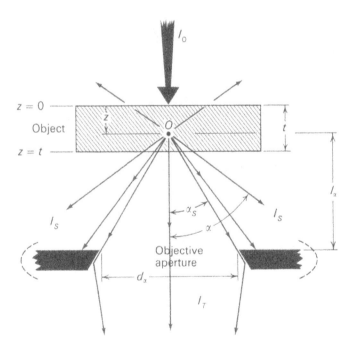

FIG. 5.3 *Electron scattering in solid object.*

where $f(\mu)$ is the electron scattering amplitude, and where α in the lower limit of the integral is ideally the objective aperture angle (Fig. 5.3). Consequently, for an agglomerate solid of i atom species each having a corresponding concentration (volume fraction) C_i, and a corresponding scattering cross section σ_i, we can express the total scattering cross section as

$$Q = N_A \sum_i \frac{\rho_i C_i \sigma_i}{w_i} \qquad (5.2)$$

where N_A is Avogadro's number, and w_i is the atomic or molecular weight associated with that scattering portion of atoms or molecules of the ith species of density ρ_i.

For the solid object of Fig. 5.3, we might imagine that electrons that undergo inelastic scattering are thrown outside the objective aperture, that is, that $\alpha_s > \alpha$[Eq. (5.1)], and it is therefore these electrons that cannot be adequately focused, and that would ultimately disclarify the image. We can represent the bright-field-image intensity probability by the region between z and z + dz of the object of Fig. 5.3 as

$$\frac{dI}{I} = - Q \, dz \qquad (5.3)$$

where at z = 0, $I = I_o$; at the emergent side, I = I for z = t. Integrating Eq. (5.3) between these limits, we obtain

$$I = I_o e^{-Qt} \qquad (5.4)$$

which represents that portion of the incident beam actually reaching the fluorescent screen in the bright-field mode. The intensity of Eq. (5.4) is actually the transmitted intensity, and the intensity component of electrons elastically scattered into the objective aperture is obtained from

$$I = I_T + I_S$$

or

$$I_S = I_o (e^{-Qt} - e^{-Q_T t}) \qquad (5.4a)$$

Equation (5.4a) describes the intensity in the dark-field image (Fig. 5.2b) with I_T eliminated by the objective aperture either by tilting the incident beam as shown in Fig. 5.2b, or by simply shifting the aperture to stop the transmitted beam while allowing I_S to pass. We can observe on comparing

Eqs. (5.4) and (5.4a) that the image intensity effectively reverses. The extreme case is of course where a specimen area approaches zero thickness, or where a hole exists in the object. The bright-field intensity is then, from Eq. (5.4), $I = I_o$; and for the dark-field image, $I = 0$.

You should observe that Q has dimensions of reciprocal distance. Thus Q is simply the number of scattering events per unit travel distance for the electrons. If we now define the mean free path between scattering incidences as

$$\Lambda = \frac{1}{Q}$$

we can observe from Eq. (5.4) that for $t = \Lambda$, the incident beam intensity I_o is reduced by $1/e$. That is, the intensity of the beam transmitted through the objective aperture from a specimen of thickness Λ and striking the final image screen has an intensity $0.37\ I_o$, while $0.63\ I_o$ represents that portion scattered outside the objective aperture. If the scattering acts are assumed to have a Poisson distribution, the probability of n_S, scattering events for the object shown in Fig. 5.3 can be expressed by

$$P_S = \frac{1}{n_S!}\left(\frac{t}{\Lambda}\right)^{n_S} e^{-t/\Lambda} \tag{5.5}$$

and for single scattering $n_S = 1$, implying that the maximum probability for such collisions is zero. Thus

$$\frac{dP_S}{d(t/\Lambda)} = 0 = e^{-t/\Lambda}[1 - (t/\Lambda)]$$

from which we observe again that $t = \Lambda \equiv t_c$. We can consider this situation to represent a critical object mass-thickness where, for a single atomic species, $i = 1$, and

$$(\rho t)_c = \frac{w}{N_A \sigma_a} = \rho\Lambda \tag{5.6}$$

which is the critical mass-thickness as originally defined by von Borries [8]. The significance of Eq. (5.6) is simply that it defines an upper bound for considering single scattering in a noncrystalline object. Table 5.1 illustrates several numerical values for critical mass-thickness based on experimental measurements of numerous materials by C. E. Hall [9] at 65 kV operation in an electron microscope without an objective aperture, while Fig. 5.4 illustrates the variation of the critical mass-thickness for

TABLE 5.1 *Critical mass-thickness for numerous materials at 65 kV[†]*

Material	Z	ρ, g/cm^3	$\Lambda = t_c$, Å	$(\rho t)_c$, g/cm^2
Be	4	1.73	1900	3.3 x 10^{-5}
C	6	2.00	1390	2.8
Al$_2$O$_3$	—	3.70	600	2.2
Si	—	2.20	1220	2.6
Cr	24	7.10	195	1.4
Ge	32	5.36	260	1.4
Pd	46	12.00	185	2.2
Pt	78	21.45	155	3.3
U	92	18.50	164	3.0

[†]*After C.E. Hall,* Introduction to Electron Microscopy, *pp. 210-213, McGraw-Hill Book Company, New York, 1966.*

polystyrene films employing aperture angles α and accelerating potentials as shown.

We can relate the number of electrons scattered for any corresponding scattering angle α_s with the total number of electrons, using the Rutherford scattering equation in the form

$$\frac{dN(\alpha_s)}{N} = \frac{\rho N_A Z^2 e^2}{wV_0^2 \alpha s^2}\left(1 + \frac{1}{Z}\right) dz \tag{5.7}$$

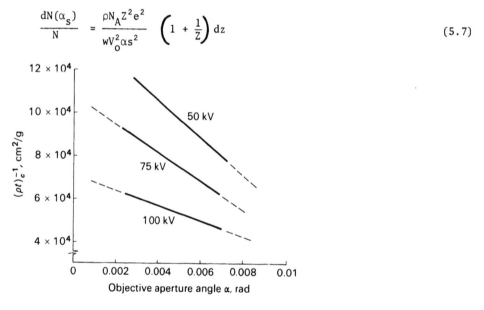

FIG. 5.4 *Variation in critical mass-thickness versus α for polystyrene observed at several accelerating potentials. (After C.E. Hall,* Introduction to Electron Microscopy, *pp. 210-213, McGraw-Hill, New York, 1966).*

where $N(\alpha_s)$ is the number of electrons scattered out of the objective aperture as shown in Fig. 5.3 (that is, $\alpha_s > \alpha$) and N is the total number of electrons in the incident beam. We observe of course from Eq. (5.7) that the proportion of scattered electrons decreases for increasing accelerating potential V_o. In fact, we can equate Eq. (5.7) directly to image intensity and observe that the bright-field intensity increases for increasing potential V_o or low atomic number Z.

Our treatment of scattering intensity so far has assumed only simple scattering processes, and that the object is structureless. What this implies is that the image produced by normal bright-field illumination (Fig. 5.2a) of a noncrystalline thin film in an electron microscope will be formed by the local satisfaction of Eq. (5.6) as the composition (density) and film thickness vary. The image produced by such an object is thus simply an intensity-contrast image, which for a material composed of a single atomic species arises where local object thickness changes.

It must be cautioned that most materials do in fact have some order, and even so-called amorphous solids usually possess some degree of crystallinity. In a real solid, scattering is not limited to single scattering, nor is it entirely elastic. Consequently the image for ordered solids will result primarily by diffraction contrast, which in turn will be influenced by the degree of inelastic scattering. We shall reserve until Chap. 7 the details of diffraction contrast. For the present we are simply interested in dealing with image contrast as it is affected by Eq. (5.6) when applied to image intensity as described by Eq. (5.4). However, it might be useful to look ahead at Fig. 7.8 in order to compare the prominent distinctions between mass-thickness and diffraction contrast phenomena.

5.2.2 *INTENSITY CONTRAST IN NONCRYSTALLINE THIN-FILM IMAGES*

If we consider a noncrystalline object of uniform density ρ and thickness t as depicted in Fig. 5.3, the bright-field image (Fig. 5.2a) appearing on the fluorescent screen of the electron microscope will possess a uniform intensity; such an image is simply a brightness governed entirely by Eqs. (5.4) and (5.7). Obviously, without a brightness standard of known Q and thickness t, no information is contained in this brightness pattern, and of course there is no image contrast. Contrast arises in the object for variations in thickness, or where the local densities vary appreciably as a result of different atomic species to cause an appreciable variation in Q.

Figure 5.5 illustrates ideally a thin film containing variations in Q and t. With reference to Eq. (5.4), the bright-field-image intensity in regions X_{1C} is given by

$$I_1 = I_o e^{-Q_1 t_1}$$

In region X_2, the atomic species is characterized by a density ρ_2 where we will assume $\rho_2 > \rho_1$. Thus the difference in scattering in this region as compared with that in regions X_{1C} is governed by $Q_2 - Q_1 = \Delta Q$. The corresponding image intensity in region X_2 will be

$$I_2 = I_o e^{-(Q_1 + \Delta Q)t_1}$$

The intensity change across the region X_2 with reference to regions X_{1C} will be

$$\Delta I_2 = I_1 - I_2 = I_o e^{-Q_1 t_1}(1 - e^{-\Delta Q t_1}) \tag{5.8}$$

Contrast arising in region X_2 can then be defined as the intensity changes relative to the background intensity I_1, that is,

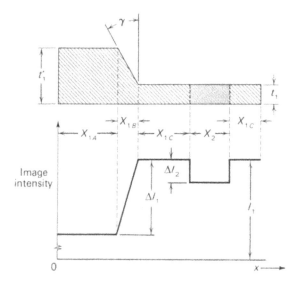

FIG. 5.5 *Variation of Q and t in object showing corresponding bright-field-image intensity profile.*

$$\text{Contrast} \equiv \frac{\Delta I_2}{I_1} = 1 - e^{-\Delta Q t_1} \tag{5.9}$$

For $\Delta Q t_1 < 1$, we can approximate

$$\frac{\Delta I_2}{I_1} \simeq \Delta Q t_1 \tag{5.9a}$$

If we now consider the image intensity corresponding to that portion of the object designated X_{1A} in Fig. 5.5 to be given by

$$I_1' = I_0 e^{-Q_1 t_1'}$$

the intensity difference between regions X_{1A} and X_{1C} will be

$$\Delta I_1 = I_1 - I_1' = I_0 (e^{-Q_1 t_1} - e^{-Q_1 t_1'})$$

If we set $\Delta t = t_1' - t_1$ as the thickness difference causing an appreciable intensity variation, we can then rewrite

$$\Delta I_1 = I_0 e^{-Q_1 t_1} (1 - e^{-Q_1 \Delta t}) \tag{5.10}$$

The contrast arising between regions X_{1A} and X_{1C} can now be determined as

$$\frac{\Delta I_1}{I_1} = 1 - e^{-Q_1 \Delta t} \tag{5.11}$$

where again, if $Q_1 \Delta t < 1$, we can approximate

$$\frac{\Delta I_1}{I_1} \simeq Q_1 \Delta t \tag{5.11a}$$

Where the total scattering cross section varies with the object thickness, we can also write for the general case

$$\frac{\Delta I}{I} = 1 - e^{\Delta Q \Delta t} \simeq \Delta Q \Delta t \tag{5.12}$$

We can conclude from Eq. (5.12) that the contrast will be enhanced by increasing the density and the atom cross-section since Q is directly proportional to the product of ρ and σ as shown previously in Eq. (5.2). Because of the direct dependence of contrast on Q, we can in fact vary σ for any particular area simply by changing the aperture angle α, which in the electron microscope involves making the aperture size smaller, thereby changing the limit of the integral in Eq. (5.1). We can observe from Fig. 5.3 that the objective-aperture angle α is given simply by

$$\alpha \simeq \frac{d_\alpha}{2\ell_\alpha} \tag{5.13}$$

where d_α is the objective-aperture size or hole diameter, and ℓ_α is the mean distance from the object to the objective aperture. Consequently decreasing the objective-aperture size for a particular object area will increase the contrast up to a point where the intensity admitted to the final image screen is not seriously depreciated. Figure 5.6 illustrates this feature for several common aperture sizes used in a conventional electron microscope.

Since the practical limit of observable contrast in an image is about 10 percent,[†] we can presume $\Delta I/I$ of Eqs. (5.9a), (5.11a), and (5.12) to be a minimum and write

$$\left(\frac{\Delta I}{I}\right)_{\min} \simeq 0.1 \simeq \Delta Qt \simeq Q\Delta t \simeq \Delta Q\Delta t \tag{5.14}$$

from which we can estimate the limitations on thickness or scattering cross-section detection. We can infer from Eq. (5.14) that the minimum detectable thickness variation is given by

$$(\Delta t)_{\min} \simeq \frac{0.1}{Q} \simeq \frac{0.1w}{N_A \rho \sigma_\alpha} \simeq 0.1\Lambda \tag{5.14a}$$

for a single atomic species composing the object. Equation (5.14a) attests to the feature previously elaborated on and demonstrated in Fig. 5.6, namely, that the detectability increases for decreasing aperture angle α, since this results in a corresponding increase in σ. You should of course recognize that Eq. (5.14a) corresponds to the conditions in bright-field illumination.

5.2.3 *DARK-FIELD-IMAGE CONTRAST FOR NONCRYSTALLINE OBJECTS*

In our treatment of contrast thus far we have been concerned with the bright-field image. It is instructive at this point to treat briefly the image contrast arising in the dark-field mode (Fig. 5.2b), especially

---

[†]*Ten percent is essentially the practical limitation although it is possible to detect 5 percent.*

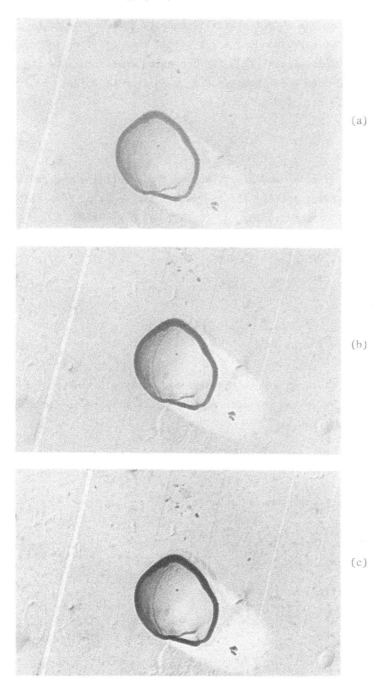

FIG. 5.6 *Contrast enhancement by reducing objective aperture d_α. (a) No aperture; (b) d_α = 70 microns; (c) d_α = 30 microns. (See Probs. 5.10 and 5.11 for details of image density.)*

since the limit of thickness detectability is usually an order of magnitude
smaller than that of the corresponding bright-field image.   If we follow
the notation of R. D. Heidenreich [10], the intensity scattered into the
aperture in the tilted-beam mode shown in Fig. 5.2b from an object of
thickness t as shown in Fig. 5.3 is given by

$$I = I_0 \pi N D(\phi_t) \alpha^2 \int_0^t e^{Qz} \, dz \tag{5.15}$$

where N is the number of atoms and $D(\phi_t)$ is a constant defining the radial
intensity in the scattering pattern for any particular angle of beam tilt
as depicted in Fig. 5.3.   Integrating Eq. (5.15), we then obtain

$$I = I_0 \pi N \frac{D(\phi_t)}{\sigma_\alpha} \alpha^2 (e^{Qt} - 1) \tag{5.15a}$$

which can be written

$$I = I_0 \pi \alpha^2 \frac{D(\phi_t)}{\sigma_\alpha} (e^{Qt} - 1) \tag{5.15b}$$

Equation (5.15b) thus represents the dark-field-image intensity.   If we
again refer to Fig. 5.5, we can consider the intensity difference from regions X_{1A} to X_{1C} to be given by

$$\Delta I_1 = I_1 - I_1' = I_0 \pi \alpha^2 \frac{D(\phi_t)}{\sigma\alpha} (e^{Q_1 t_1} - e^{Q_1 t_1'}) \tag{5.16}$$

Again, where $\Delta t = t_1' - t_1$, we can rewrite Eq. (5.16) as

$$\Delta I_1 = I_0 \pi \alpha^2 \frac{D(\phi_t)}{\sigma_{\alpha 1}} e^{Q_1 t_1} (e^{Q_1 \Delta t} - 1) \tag{5.16a}$$

Consequently the dark-field-image contrast becomes

$$\frac{\Delta I_1}{I_1} = \frac{e^{Q_1 t_1} (e^{Q_1 \Delta t} - 1)}{e^{Q_1 t_1} - 1} \tag{5.17}$$

If we assume in Eq. (5.17) that $Q_1 t_1$ and $Q_1 \Delta t$ are appreciably less than
unity, the contrast becomes

$$\frac{\Delta I_1}{I_1} \approx \frac{\Delta t}{t_1} \tag{5.17a}$$

Again assuming the minimum perceptible contrast to be 10 percent, we can
set

$$\frac{\Delta I_1}{I_1}\bigg|_{min} = 0.10 \simeq \frac{\Delta t}{t_1}$$

and observe that the minimum perceptible thickness change is given by

$$(\Delta t)_{min} \simeq 0.10t \tag{5.17b}$$

We can observe on comparing Eq. (5.14a) for the bright-field image and
Eq. (5.17b) for the dark-field image, that the dark-field image is not
limited by the critical thickness Λ. Consequently, where $t_1 < \Lambda$, the perception of thickness contrast in an object is enhanced in the dark-field
image. (See Prob. 5.2, which develops similar equations for the addition
of a vapor deposit to a substrate).

5.3 REPLICATION ELECTRON MICROSCOPY

We were exposed in Chap. 2 to a mode of object surface analysis (thermionic
emission microscopy) in which the object forms an active component in the
electron optical system. We will take up a similar mode of active object
surface imaging at a later portion of the present chapter when we shall
discuss the imaging of object surface features using secondary and backscattered electrons as an extension of the analytical modes discussed in
Chap. 4. We will now discuss the observation of object surface topography
by the formation of an intensity contrast image in the electron microscope
by transmission through a replica film. This method, originally initiated
by H. Mahl [11] at the time the first commercial electron microscopes were
being made available, mainly involves making a thin cast of a specimen surface. Such surface replication can be accomplished in a variety of ways,
and we shall treat only the most successful of these in any detail. We
must state at the outset that replication, as a technique, forms an intrinsic function in replica electron microscopy; consequently we shall deal with
the technique in some detail in the text rather than relegating it to summarized form in Appendix B.

In a very broad sense, replicas are of two types: single or two stage.
These can be produced with a variety of materials, usually by a contact
flow process where a liquid plastic forms a solid cast over the object

surface, or by vapor deposition of a desired thickness of a material onto the object surface, or by a combination of these processes.

5.3.1 SINGLE-STAGE REPLICATION

Figure 5.7 illustrates the modes of single-stage surface replication. In Fig. 5.7a, a plastic replica is made by forming a liquid such as Formvar, Parlodion, Collodion, or other suitable material. Image formation in the electron microscope results as a consequence of thickness contrast as described by Eqs. (5.11) and (5.17). In Fig. 5.7b, a metal or conductive element such as carbon is vapor deposited onto the object surface in a vacuum evaporator with the object surface effectively normal to the evaporation source. The object then intercepts the evaporating vapor stream along a portion of uniform sphere of emission from the source. The thickness of material vapor deposited normal to any surface section is given ideally by

$$t_n \simeq \frac{W}{4\pi\rho d_s^2}$$ (5.18)

where W = total weight of material evaporated

ρ = density of material evaporated

d_s = normal distance of object surface from evaporation source

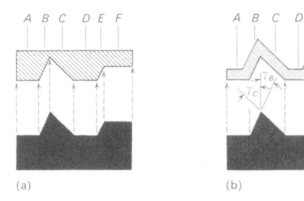

(a)   (b)

FIG. 5.7 *Idealized single-stage replicas. (a) Plastic (nonconductive) flow replica. (b) Metal (or conductive) vapor-deposited replica.*

You are cautioned that Eq. (5.18) as used in practice is only a good estimate. Furthermore, it assumes a point source. In otherwords a specific
mass of material is assumed to evaporate over a spherical surface to a uniform thickness t_n. If metal is evaporated from a confined geometry such
as a tungsten boat, the thickness must be corrected for the specific surface area evaporated upon (see Prob. 5.6).

In Fig. 5.7b, t_n is the background thickness for the replica; background intensities arise from sections A, D, and F as shown. The contrast
arising in sections B, C, and E of Fig. 5.7b occurs as a result of the effective thickness in these areas. The magnitude of the effective thickness,
which occurs for surface topography and relief variations, can be calculated with reference to Fig. 5.7b by considering the ramp angle of the relief.
Thus, in section B of Fig. 5.7b, the effective thickness is observed to be

$$t_{eff}^{(B)} = t_n \csc \gamma_B \tag{5.19}$$

and in section C, the effective thickness is similarly

$$t_{eff}^{(C)} = t_n \csc \gamma_C \tag{5.19a}$$

Thickness contrast arising in these areas is therefore governed by

$$\Delta t = t_{eff} - t_n = t_n (\csc \gamma - 1) \tag{5.19b}$$

as substituted into Eqs. (5.11) and (5.17). A similar interpretation of
contrast arising from the plastic replica also obtains; in this case the
background reference is the thinnest section, or section F in Fig. 5.7a.

The plastic replica suffers many disadvantages, the most severe being
the image distortion arising from charges accumulated on the nonconducting
surfaces. However, the evaporated replicas are more difficult to make and
in most cases cannot be stripped from the specimen surface to be observed.
Carbon replicas condensed by vaporization of an arc source in vacuum, as
originated by D. E. Bradley [12,13], have become more or less standard in
electron microscopy. These replicas are stable (noncharging), mechanically strong, and result in fairly recognizable contrast features. They are
normally very fine grained, nearly completely free of crystalline structure. As a consequence, the optimum in topographical resolution is usually on the order of 20 Å. Generally, replica image contrast is readily interpreted according to Eqs. (5.11) and (5.17), provided the crystal size
of an ordered domain is less than the expected resolution. Grain sizes of

evaporated metals, provided the object surface is not crystalline, are less than 100 Å.

If the object surface to be replicated is crystalline, the evaporation of a metal normally crystallizing in the same system, say an fcc metal evaporated onto a fcc object surface, may tend to condense epitaxially. The replica feature is then lost since contrast arising from such foils is predominantly diffraction contrast (as we shall see in Chap. 7). Where the grain size borders on the limiting conditions for mass-thickness contrast, a granularity may appear in the image and contribute artifacts to the image interpretation. It is advisable in studying crystalline surfaces using single-stage metal replication to evaporate metals having a crystal system different from that of the object, or to alter the evaporation parameters in order to suppress epitaxial tendencies and to reduce the size of replica film crystallites to less than 100 Å.[†] Figure 5.8 illustrates the use of Ti(hcp) as a single-stage replica for observing growth steps on the (100) surface of NaCl. The contrast at the individual ledges arises by the variations in Ti thickness, and the three-dimensional nature of the image results from a difference in the displacement of the "pits" from the normal to the evaporation source, giving rise to "shadows" (discussed in Sec. 5.33). The Ti film vapor deposited onto NaCl as shown in Fig. 5.8 was readily removed from the surface or "stripped" simply by immersing the crystal and surface film in water, allowing the Ti replica to float free. Carbon or other metals are similarly removed from object surfaces, even where the surface to be replicated is not soluble in water or another solvent.

Oxide films formed on the surface of numerous metals and alloys can also be stripped for direct observation as a single-stage replica in the electron microscope [11]. Specific applications of stripped oxide replicas are found in the work of Heidenreich [14], and a more complete discussion of replicating techniques is given in an article by Heidenreich and C. J. Calbick [15].

[†]*See, for example, the variation of metal grain size in evaporated thin films with deposition parameters as presented in L. E. Murr and M. C. Inman, Phil. Mag., 14: 135(1966).*

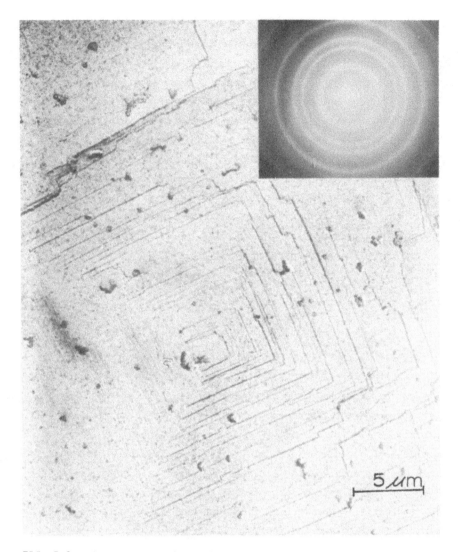

FIG. 5.8 *Single-stage Ti replica showing contrast image of "growth" steps on (100) surface of NaCl single crystal resulting from hygroscopic surface dissolution. Average film thickness as estimated from Eq. (5.18) was $t_n \simeq 400$ Å. Film itself is noncrystalline (grain size ~ 100 Å) as shown by selected-area electron diffraction pattern inserted (see Chap. 6). Accelerating potential was 100 kV.*

5.3.2 IMAGE-MAGNIFICATION CALIBRATION

In order to make any quantitative measurements of surface or related structural morphology in electron microscope images, whether of a replica nature or not, it is important to have an accurate measure of the image magnification. Normally, magnification can be calibrated from any standard of size

or dimensions, and such calibration is usually performed simply by relating
a reference image size or dimension to the corresponding projector- or intermediate-lens current. It is common practice in electron microscopy involving magnifications above several thousand times to maintain a constant
projector-lens current and to vary image magnification by simply varying
the intermediate-lens current, while concurrently maintaining the image
focused at a relatively constant objective-lens current.

In recent years it has become standard practice to employ a single-
stage replica of a very fine ruled grating to calibrate magnification.
Such gratings, in the form of vapor-deposited carbon or metal replicas of
etched gratings on a glass surface, are available commercially.[†] Figure
5.9 illustrates typical examples of magnification calibration using a single-stage metal-replica grating ruled to 54,800 lines per inch. Magnification is calculated as

$$M = \frac{N_1}{N_1'} \tag{5.20}$$

where N_1 is the number of lines per unit length of the grating and N_1' is
the actual number of lines per unit-length measured from the replica image.
The magnifications measured from the images of Fig. 5.9 are then plotted
against the intermediate-lens current in order to obtain an accurate calibration curve.

Since we have already observed in Chap. 3 that the effective lens focal lengths are dependent on the accelerating potential, separate calibration curves for magnification must be made for each operating voltage.
Similarly, where the objective-lens focus current varies appreciably as a
result of incorporating various specimen stage designs in the electron microscope, magnification must also be calibrated for each of these situations, particularly where the distance from the object plane to the objective-lens plane changes.

5.3.3 TWO-STAGE REPLICATION AND ELECTRON METALLOGRAPHY

We can observe from Figs. 5.8 and 5.9 that the single-stage evaporated replica suffers two severe limitations in general practice. The first is

---

[†]*Diffraction-grating replicas are available from E. F. Fullam, Inc.,
Schenectady, New York.*

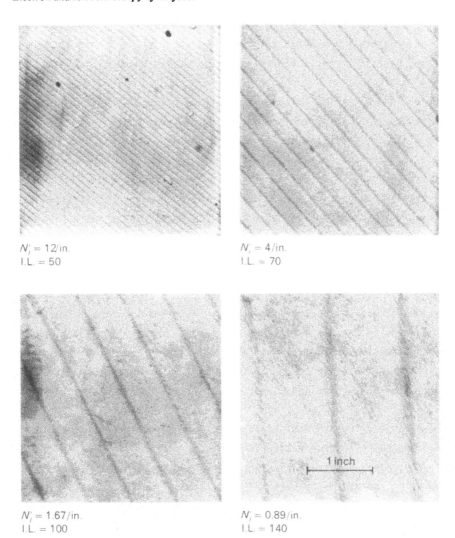

$N_i' = 12/\text{in.}$
I.L. = 50

$N_i' = 4/\text{in.}$
I.L. = 70

$N_i' = 1.67/\text{in.}$
I.L. = 100

$N_i' = 0.89/\text{in.}$
I.L. = 140

1 inch

FIG. 5.9 *Magnification calibration using diffraction-grating replica of N_1 = 54,800 lines per inch in Hitachi H.U. 125 electron microscope operated at 100 kV with fixed projector-lens current of 130 mA. Measured value N_1' and corresponding intermediate-lens current (I. L.) in milliamperes is given for each image. Note image rotation with increasing intermediate-lens current.*

the inability to remove or strip evaporated replicas from many surfaces, particularly those which are not smooth, and those to which the film strongly bonds. The second is the very limited contrast and image prominence afforded by the single-stage replica where contrast is limited solely by thickness changes. The ability to remove the replica from the sur-

face is actually the greatest single factor in the observation of surface
detail, and certainly a general method is desirable wherein any desired
surface can be readily replicated, regardless of its chemical or physical
nature.

The two-stage replica is in effect a description of the final surface
cast, although in actual practice this is formed on a third (actually first)
stage or plastic. The technique consists simply in first making a plastic
replica of the surface, stripping the plastic replica from the surface,
evaporating carbon or some other suitable material onto the plastic replica, and finally evaporating a more dense metal onto the carbon-plastic replica at a suitable angle so as to enhance the particular topographical
features such as spikes or depressions by creating contrast shadows. The
plastic base is then dissolved from the carbon-metal shadowed portion,
leaving this composite as the final two-stage surface replica. This technique of course has the advantages of mechanical stripping combined with
contrast enhancement of surface topography. Figure 5.10 illustrates the
essential steps in the preparation of a two-stage replica for electron microscopy observations of surface topography. It also shows the placement
of the specimen grid directly on the initial plastic, which then serves as
a supporting stage for the subsequent evaporations, treatment, and observations.

The removal of the initial plastic (Parlodion, Collodion, etc.) and
embedded specimen grid from the surface to be replicated is usually done
mechanically, using a soluble sticky tape to simply strip the plastic free.
The removal of the stripping tape and plastic-base replica following the
evaporation of the final replica stages is then accomplished by using a
suitable solvent. In actual practice the plastic base is removed by a vapor wash using a suitable solvent as indicated in Appendix B. The vapor
wash can be accomplished by using a simple petri dish and holder for the
replica on the support grid, or by using various reflux systems available
commercially.[†]

The technique of shadowing a base replica such as carbon to enhance
image contrast of surface topography is detailed in the sketch of Fig.
5.11. Observe that if the angle between the surface replica plane and the
evaporation source is acute, the deposition thickness of metal in the vapor

---

[†]*Replica-washing apparatus is available from E. F. Fullam, Ted Pella
(Pelco), and other suppliers of microscopy accessories.*

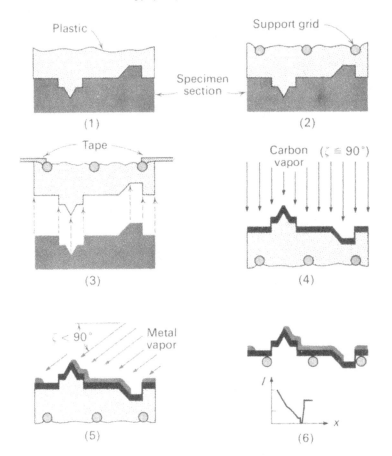

FIG. 5.10 *Basic steps in preparation of a carbon-metal shadowed two-stage replica of a surface for electron microscopy. (1) Liquid plastic placed on surface. (2) Locating-support grid placed on plastic drop and allowed to set. (3) Plastic impression stripped from surface using suitable tape subsequently dissolved away. (4) Plastic impression shadowed with carbon or similar cast by vacuum vapor deposition. (5) Low-angle metal shadow by vacuum vapor deposition followed by solvent vapor wash to dissolve plastic. (6) Final observation of two-stage replica on support grid showing portion of bright-field-image intensity profile.*

stream will occur effectively in the form of Eq. (5.18) considering the surface not normal to the evaporation source, or

$$t_n' = \frac{W}{4\pi\rho d_S^2} \sin \zeta \qquad (5.18a)$$

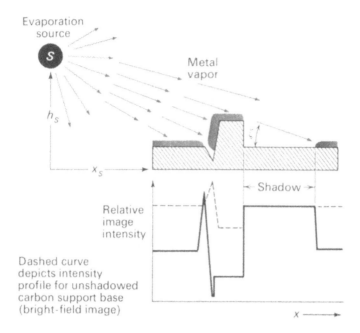

FIG. 5.11 *Contrast enhancement by shadow casting.*

where ζ is the angle of shadow as depicted in Fig. 5.11. The angle of shadow actually changes over the replica surface, but where the dimensions of
concern are extremely small, the angular variation is unimportant insofar
as image contrast is concerned. We can effectively calculate

$$\zeta = \tan^{-1} \frac{h_S}{x_S} \tag{5.21}$$

where h_S is the height of the evaporation source above (or below) the replica plane, and x_S is the horizontal distance of the source to the replica. Thus, where h_S and x_S are very large in comparison with the actual
replica area, ζ can be considered effectively constant over the surface.
It should also be observed that d_S in Eq. (5.18a) is given by

$$d_S = h_S \csc \zeta$$

From the simple geometry of Fig. 5.11 it is also possible to estimate
the height of surface protrusions or the depths of surface cavities, as the
case may be, simply by considering

$$h_p \simeq \tan \zeta \times \text{shadow length} \tag{5.22}$$

where h_p is the optimum vertical dimension of a protrusion or cavity. This feature is illustrated in Fig. 5.12, which shows a tiny crystal on a carbon base shadowed with Pt at a 15° angle, that is, $\zeta = 15°$. We can measure the height of the crystal directly from Fig. 5.12 considering Eq. (5.22) to be approximately 2000 Å; and from the direct measurement of the dimensions in the plane of the image, we can effectively reconstruct the three-dimensional form of the crystal rather precisely, provided of course an accurate magnification calibration has been performed as outline previously in Sec. 5.3.2.

With regard to the geometrical recognition of surface topography and similar features in the electron microscope as shown in Fig. 5.12, it may be of interest to illustrate the limitation of shape acuity in electron microscopy. Von Borries and G. A. Kausche [16] actually derived an expression for minimum size of a regular polygon of n sides recognizable as such in the image plane:

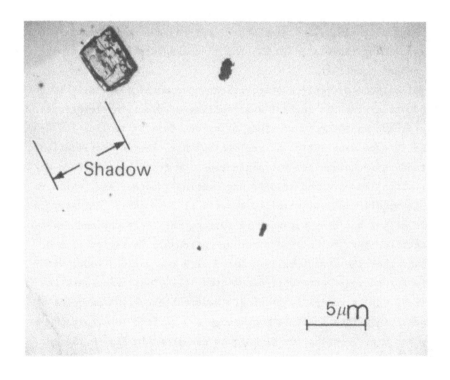

FIG. 5.12 *Platinum-shadowed crystal on carbon support film* $(\zeta_{Pt} \approx 15°)$.

$$\frac{D_{min}}{\delta_{opt}} = 2\left(1 + \cos\frac{\pi}{n}\right)\sqrt{\frac{n}{\pi}\cot\frac{\pi}{n}} \tag{5.23}$$

For the crystal of Fig. 5.12 (where n = 4), we observe that $D_{min}/\delta opt$ = 3.7. Thus, assuming an optimum resolution of 10 Å, the minimum dimension for such a crystal to be recognizable would be 37 Å.

It should be mentioned that even in the preparation of a single-stage replica, the vapor deposition of a metal at some angle ζ with respect to the surface will enhance contrast, as previously shown in Figs. 5.8 and 5.9. In the two-stage replica, any combination of normally deposited base and shadowed contrast metal may be employed; Appendix B lists a number of the more successful. It is also advisable in some instances to shadow-cast the contrast metal first, and then the carbon or similar base. In this way the contrast at surface protrusions is somewhat more sharp at fine detail; provided of course that the depth of shadow is not too large. In this respect we should recall from Eq. (5.7) that the scattering diminishes with increasing accelerating potential V_0. As a consequence, contrast is reduced at very high potentials; in cases where replicas are to be imaged with higher voltages, for example $\geq$ 100 kV, the shadowing depth should be increased to enhance contrast.

The applications of replica observations have mounted steadily since their introduction with the initial availability of electron microscopes. Nearly any surface is amenable to study using one form of replication or another, and surface examination by replication has, in many applications, become a standard technique for the nondestructive testing of surface damage in structural members and related engineering facets. And, while replicas can be readily made of metallic as well as nonmetallic surfaces, a considerable effort has been expended in surface metallography and fractography. Fractures have in fact been rather thoroughly investigated over the years, and the fine structure associated with the various modes of fracture are for the most part well documented [17][+], while mechanistic interpretations remain generally lacking. Numerous excellent examples of the uniqueness of surface fracture morphology and surface metallography as observed by two-stage replication techniques are shown in Fig. 5.13.

[+]*See* Fractography and Atlas of Fractographs, *vol. 9 of the Metals Handbook (Eigth Edition), American Society for Metals, Metals Park, Ohio, 1974.*

Figure 5.13a and b illustrates typical microstructures and surface-rupture morphology for a Mg casting, respectively, while c and d show a similar comparison for a Ti alloy. Figure 5.13e and f illustrates the cleavage-type surface of an Al alloy. All of the replicas of Fig. 5.13 were plastic (cellulose acetate) stripped, and carbon shadowed ($\zeta \neq 90°$) with a final Cr shadow at $\zeta \approx 45°$. Electron metallography and related techniques are essentially identical to conventional light metallography, especially where surface preparation is concerned. Thus we must realize that replica contrast features arise at phases and similar structural phenomena as indicated in Fig. 5.13 as a result of preferential surface etching prior to replication.

Figure 5.14 illustrates the investigation of cleavage and surface stress-corrosion phenomena in steels employing a plastic (cellulose acetate) and carbon - 45° Cr-shadowed replica. It should be emphasized at this point that the shadowing angle, ζ, is somewhat critical in producing contrast at various surface features, particularly where fracture and related severe distortions occur. In effect, where the surface features are rather steep, or where intricate and deep cavities appear, a portion of such features may be devoid of contrast, and therefore go unnoticed if the vapor stream of shadowing metal cannot impinge on these features. Thus, as a general rule, for very heavily distorted and extensive surface features, a high shadow angle is best, while for fine scale features, low shadowing angles might be more revealing insofar as replica contrast is concerned. In many instances the replica can also be rotated during the shadowing operation to effect a heavy cast on all sides of a protrusion or cavity on a surface, thereby enhancing the overall contrast at such features. In many applications, particularly metallurgical, the surface features can be enhanced by chemical etching or similar polishing treatments. This is particularly helpful when investigating grain-boundary phenomena, phase morphology, etc. In the case of stress corrosion as illustrated in Fig. 5.14b, the surface of the 9-4-30 steel was etched with Nital to enhance the grain-boundary contrast and the fine corrosion surface damage.

It should be cautioned that, in certain specific applications, the surface replica may be considerably unreliable insofar as the image it produces, particularly where certain surface facets or other complex topography prevents a representative stripping, causing the plastic to tear away from the surface and thus be unrepresentative of the fine structure

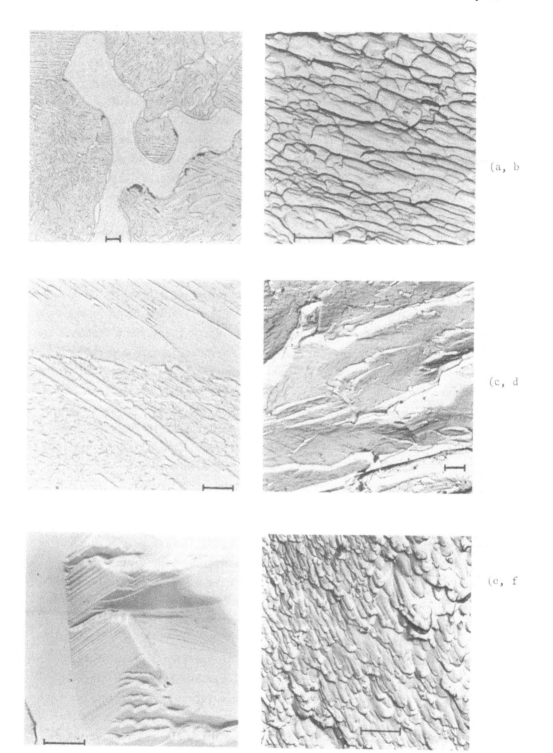

(a, b

(c, d

(e, f

present. Porous surfaces, particularly fractured ceramics and other related materials, are most apt to produce this effect. Considerable care
must therefore be exercised in the interpretation attached to certain surface replica images. In many instances it is best to resort to more reliable modes of analysis, such as scanning electron microscopy discussed
in Sec. 5.5.

5.3.4 SELECTED-AREA REPLICATION

The investigation of slip and deformation in metals and alloys is particularly amenable to study using shadow-cast replication. Heidenreich first
observed slip steps at high resolution using anodically produced oxide
replication on aluminum, and these studies were later extended using shadow-cast replication by H. G. F. Wilsdorf [18,19], D. E. Bradley, and
numerous others. Of particular interest in deformation studies has been
the direct observation of surface alterations, for example, change in the
density and character of slip lines, surface-cavity formations, surface
fractures, etc., with a deformation variable. In this respect it is desirable to continuously reexamine the same surface area in order to relate
specific topographical variations to the variable under study. Techniques
that accomplish the repeated examination of a specific area can be thought
of as selected-area replication [20-22].

Selected-area replication mainly consists in devising a scheme for
locating rather precisely the same area to be replicated, usually by some
mechanical means, and then devising the replica itself to facilitate the
final location in the electron microscope. Standard locating replica grids
have been available for some time; they consist of a geometrical array of
grid spacings, one of which, usually the center spacing, is used as the
locating field. The area to be replicated on the surface of a test specimen is located by a consistent positioning of the specimen, say in a mi-
crometer location stage of an optical metallograph or metallurgical micro-

---

FIG. 5.13 *Cellulose acetate and carbon-shadowed* ($\zeta \neq 90°$) *replicas showing surface microstructure and fracture characteristics for numerous materials at 75 kV. (a) AZ92A-Mg casting microstructure; (b) fracture topography of same Mg casting as a following simple tensile rupture;
(c) Ti-8Aℓ-1 Mo-IV microstructure; (d) cleavage-type fracture morphology
for Ti-8Aℓ-1 Mo-IV alloy; (e) Aℓ$_2$O$_3$ cleavage fracture; (f) tensile surface
of a 7079-T-6 Aℓ alloy. Magnification marker equals 2 microns. (Courtesy
R. F. McElwee and W. Fritze).*

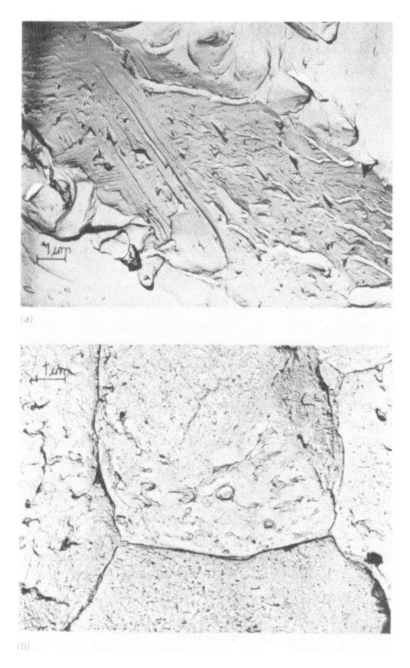

FIG. 5.14 *Stress-corrosion fracture surfaces. (a) Carbon 45°
Cr-shadowed replica of stress corrosion cleavage-type fracture
in 431 stainless steel. (b) carbon 45° Cr-shadowed replica of
the stress corrosion fracture surface of 9-4-30 steel.
(Courtesy C. H. Avery).*

scope. A liquid plastic such as a 1 percent Collodion solution is then dropped on the specimen surface within an indicated area of the size of the replica grid. The replica grid is placed on the plastic drop and the plastic is dried. The hardened plastic and the replica grid are then stripped from the surface area, and the two-stage final replica is prepared as outlined in Fig. 5.10.

The dynamic changes in surface topography, particularly where dynamic stress or strain modes are involved, are especially amenable to study using a selected-area replication technique. The fatigue of metals and alloys is typical of this application, and several fairly successful attempts have been made to study cumulative surface damage as a result of fatigue [23,24]. Figure 5.15 illustrates the successive observation of the accumulation of primary and cross-slip lines on the surface of an Al specimen fatigued at room temperature in push-pull (reversed strain of $\pm$ 0.002) at a cycling rate of 15 Hz (cycles per second) [24]. The annealed surface is observed to be free of slip lines or any other evidence of deformation as revealed by Collodion-carbon replication with a 15° Pt shadow ($\zeta \approx 15°$). Following the accumulation of a number of fatigue cycles as indicated, slip lines are observed to form; at 26,000 cycles, surface microcracks begin to appear in the more densely slipped areas.

Considerable effort has been expended over the past several decades in attempts to elucidate the mechanism of fatigue. The atomistic processes continue to remain vague. Early experiments by R. D. McCammon and H. M. Rosenberg [25], W. A. Wood [26], N. Thompson and coworkers [27], and others led to the general conclusion that fatigue was a surface phenomenon, with microcracks presumably originating from surface cavities. Dislocation mechanisms were proposed to ultimately create surface intrusions or extrusions, but these have since proven unsatisfactory. It is generally demonstrated that fatigue can occur at low temperature [23,25] in the same basic manner as at room temperature, and the source for surface cavitation thus appears to be the creation of vacancies or interstitials transported by the movement of dislocations, since the process now appears to be independent of thermal variables, but is stress dependent even at low temperatures [23]. Recent work by transmission electron microscopy has had as its goal the elucidation of the detailed structure of slip lines, and similar fatigued surface topography [28-31].

Figure 5.16 illustrates the formation of surface cavities on an Al surface fatigued in liquid nitrogen using the fatigue device described by

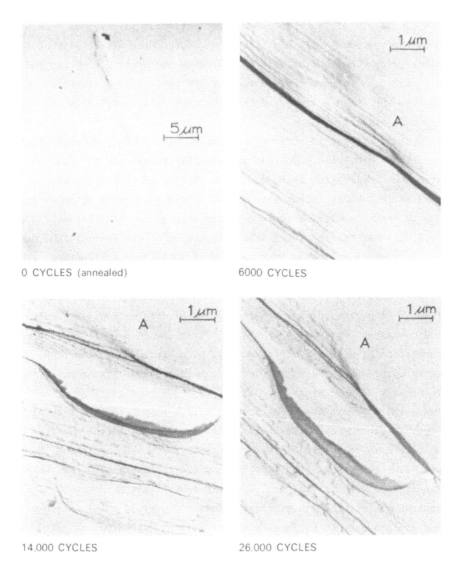

0 CYCLES (annealed)

6000 CYCLES

14,000 CYCLES

26,000 CYCLES

FIG. 5.15 *Selected-area replication of fatigued Al. Room-tempera-*
ture fatigue with cumulative number of cycles as indicated. Loca-
tion zone is indicated as A. Carbon-Pt-shadowed ($\zeta = 15°$) replicas.

Wilkov and Murr [24] and a zero-mean-strain (push-pull) mode of cyclic
stress application. The images of Fig. 5.16a and b are from appreciably
disposed areas of the fatigue specimen. The linking of the surface cavities to form microcracks is apparent from the insert in Fig. 5.16b. The
images result from a Collodion-carbon-Pt-shadowed ($\zeta \simeq 15°$) replica observed at 80 kV in an electron microscope operating with single condenser

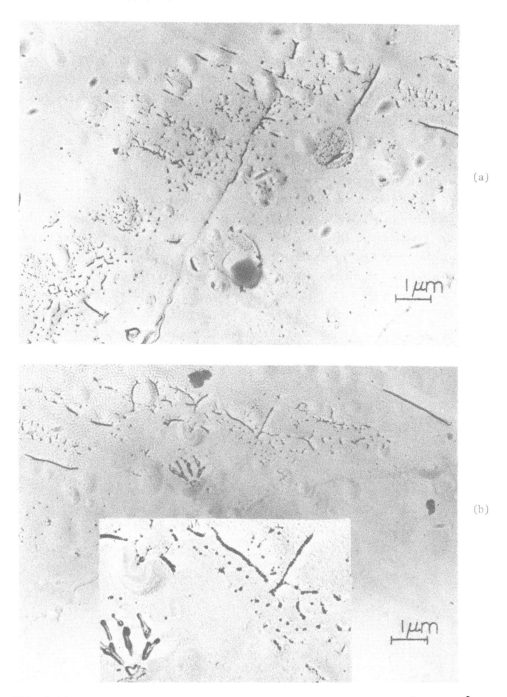

FIG. 5.16 *Surface cavities and microcracks on Al fatigued at 78°K for 10⁵ cycles (ε = ± 0.002). (a) Cavities associated with slip lines and randomly disposed areas within slip systems. (b) Apparent joining of cavities to form surface microcracks; insert shows details of enlarged cavitation features.*

illumination. The white shadows appearing at the cavities occur because of
the Pt being shadowed on the stripped-surface side, following the procedure
of Fig. 5.10. It is interesting also to observe from Fig. 5.16 that the
cell-like arrangement of cavities in several instances matches the dimensional disposition of dislocation subcells observed by transmission electron microscopy [29,30].

5.3.5 *EXTRACTION REPLICATION*

If a surface is etched prior to the application of a plastic base for replication, or prior to shadowing, inclusions present in the matrix and in-
soluble in the etchant will protrude from the surface and become "embedded"
in the replica. If the solid section onto which a replica has been cast in
this manner is then immersed in the etchant a second time, the replica and
the "embedded" inclusions can be floated free, with the inclusions contained in a supporting film (the replica) as they are disposed in the actual
matrix. We say the inclusions have been "extracted" from the surface matrix. Figure 5.17 illustrates the steps in the extraction of inclusions
from the surface matrix of a solid.

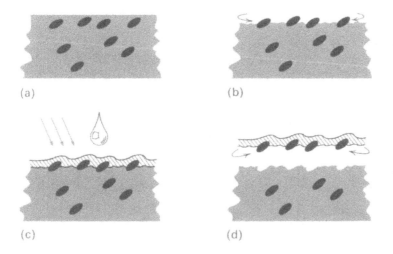

(a)   (b)

(c)   (d)

FIG. 5.17 *Extraction replication of matrix inclusions
showing section views of basic steps involved. (a) In-
clusion-bearing matrix section; (b) initial surface etch
to expose inclusions or precipitates; (c) shadow-cast
(carbon) or liquid plastic replica formed over exposed
inclusions or precipitates; (d) reetching treatment to
free surface replica containing embedded inclusions
(replica sections float free from matrix section submerged in etchant).*

This technique, originated by R. M. Fisher [32] as a means for identifying precipitates in steels, serves the purpose of a replica only in the sense that the contrast is of course enhanced at the thick precipitates or their differing density as compared with the intensity of the support film as background. Figure 5.18 illustrates a typical example of the extraction of $M_{23}C_6$ precipitates from the grain-boundary regions of a high-carbon steel using a direct carbon-replication technique and an H_2SO_4 etch.

The unique advantage of extraction replication does not lie in the enhanced image contrast but rather in the ability to identify the chemical nature and crystal structure of precipitates removed from the matrix by using either selected-area electron diffraction techniques [33] (discussed in more detail in Chap. 6) or the various suitable analytical modes of the electron or ion microprobe as outlined in Chaps. 4 and 7. The usefulness of extraction replication therefore really lies in the microscopy of the intrinsic particle structure, and we shall therefore defer additional comments until we have familiarized ourselves with electron diffraction, diffraction contrast and analytical electron microscopy in the succeeding chapters. We might also mention that where stripping of solid surface sections can be accomplished, it is also a form of extraction. For example, the stripping of very thin layers from mica by using a sticky tape or plastic set.

Extraction replicas can also be made from vapor-deposited aluminum, or SiO_x films deposited onto a surface prepared as in Fig. 5.17(b), and these films are particularly useful in analyzing carbides in the analytical electron microscope to be described in Chap. 7.

5.4 REFLECTION ELECTRON MICROSCOPY

In most surface studies, replica techniques of one form or another enable the materials scientist to achieve sufficient resolution of topographical detail on virtually any reasonably smooth surface. However, in certain specific cases, where small fragments or related very delicate object surfaces require detailed study, replication is impossible or unreliable. In such situations the surface must be observed intrinsically, if it is to be observed at all, and this feature is accomplished by imaging electrons reflected from such surfaces.

Surface examination using reflected electrons was introduced at the time commercial electron microscopes were being made available. Two modes of analysis were introduced to study surfaces in the electron microscope by

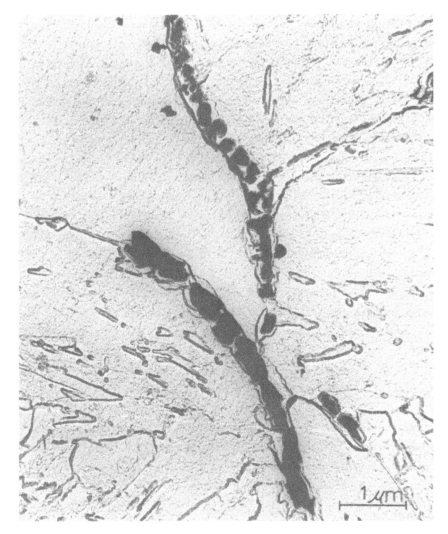

FIG. 5.18 *Direct carbon-replica extraction of $M_{23}C_6$ in Am 350 steel.
Initial etch (Fig. 5.17) was in 20% H_2SO_4 distilled water followed
by vaporization of carbon at incident angle of 75° while rotating
specimen. Final (release) stage employed 5 to 10% H_2SO_4 solution.
(Courtesy C. H. Avery).*

the reflection method. These included the normal or high-angle reflection
mode of E. Ruska and H. O. Müller [34], and the low-angle or glancing mode
of von Borries [35]. As it turned out, the method of Ruska and Müller en-
tailed rather severe limitations on resolution and image intensity, while
that of von Borries became an accessory operation on commercial instruments
and was modified to its present stage, which allows resolutions on the or-

der of several hundred angstroms or less to be obtained rather routinely without the necessity for altering significantly the electron optics or mechanical design features of conventional electron microscopes. Figure 5.19 illustrates the basic features of electron reflection. In this schematic, electrons emerging from the condenser lens of the electron microscope are deflected off axis by an external deflection system (A). These deflected electrons are then bent back onto the specimen surface inclined nearly parallel with the electron optic axis by a second set of deflection coils (B). Reflected electrons then pass into the objective lens and are focused and imaged in the conventional manner in the electron microscope.

The reflection-image interpretations for objects having-extremely exaggerated topographical features suffer considerable distortions as a result of diffraction and other inelastic scattering as depicted in Fig. 5.20. Consequently, reflection electron microscopy must consider such features, especially where a considerable portion of the image contrast results from diffraction and the low-energy electrons resulting from inelastic scattering.

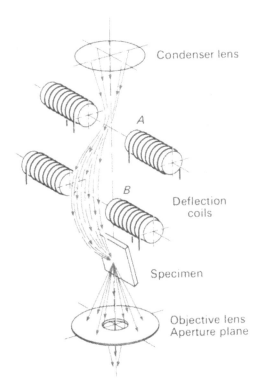

Condenser lens

A

B

Deflection coils

Specimen

Objective lens
Aperture plane

FIG. 5.19 *Surface examination by reflected electrons in electron microscope.*

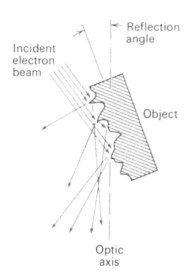

FIG. 5.20 *Electron reflection from very rough surface.*

As we mentioned at the outset, reflection electron microscopy of surfaces is only advantageous when more conventional replication techniques
are not feasible, since the resolution and image contrast of surface features resulting from the reflected electrons is not generally better than
replication, or scanning electron microscopy (discussed in Sec. 5.5). This
feature is observed on comparing these various modes of imaging as shown in
Fig. 5.21. Figure 5.21a shows the reflected electron image of slip steps
in Fe corresponding to a reflection angle of 15° (with reference to Fig.
5.20), while Fig. 5.21b and c illustrate typical pearlite structures as
imaged by replication electron microscopy and scanning electron microscopy,
respectively. Figure 5.21d shows the corresponding crystal structural details of pearlite as observed by thin-film high-voltage electron transmission microscopy (discussed in detail in Chap. 8).

The particular advantages of reflection electron microscopy are perhaps best illustrated in cases where surface topography is to be associated
with internal or related structural features of a physical or chemical nature. Typical applications can be found in the characterization of deformation mechanisms on the basis of internal dislocation or related defect
structures, and the corresponding specimen surface topography. L. E. Murr
[36] has described the design of a special specimen-stage arrangement for
an early electron microscopy intended for the observation of surface areas
in thin-foil specimens viewed directly by transmission electron microscopy.
The stage was constructed so that an area observed by electron transmission

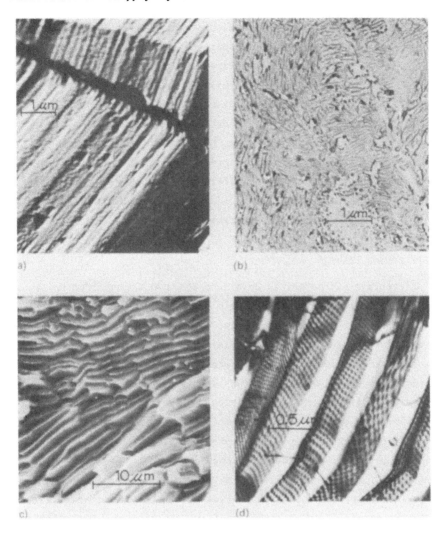

FIG. 5.21 *Comparison of electron optical imaging modes. (a) Surface
steps in Fe imaged by reflection electron microscopy at approximately 15° incidence; (b) replication electron microscope image of pearlite; (c) scanning (secondary) electron microscope image of pearlite;
(d) high-voltage (850 kV) transmission electron microscopy image of
pearlite. (a,c,d courtesy of Japan Electron Optics Laboratory Co.,
Ltd., Tokyo; b courtesy of R. F. McElwee).*

could be rotated nearly parallel to the optic axis for electron reflection
analysis. The specimens of particular interest were those foils electrolytically polished from bulk fatigue specimens (see Appendix B) or thin
films of annealed metals intended to be fatigued directly within the electron microscope [30,31,37]. This technique, it was hoped, would allow the

defect structure and associated surface topography at various stages of
deformation to be observed in situ. Because of the etching irregularity
of electropolished thin films, it is extremely difficult to prepare reliable replicas of thin-film surfaces. Similarly, very fragile objects which
would have their surface destroyed or distorted by the stripping of replicating plastics or similar casts might be amenable to study by electron
reflection. For many such observations, however, scanning electron microscopy offers equivalent resolutions, increased depth of field, and considerably greater ease in handling versatility; thus the merits of the various surface techniques available should be carefully evaluated as they
relate to a specific problem under investigation.

5.4.1 REFLECTION ELECTRON HOLOGRAPHY

Reflection electron microscopy as described above has been effective in observing surface phase transitions, atomic steps, and dislocation emergence
on a surface [38-40]. These studies have revealed that many surface structures are in fact imaged by phase contrast where electron wave phase differences over a surface area are used to create contrast.

In electron holography, two regions in a reflection electron image of
a surface area at glancing angle incidence are overlapped by means of an
electron biprism to form an off-axis electron hologram. The optically reconstructed interferogram displays the phase distribution of the diffracted
electron wave which reflects surface topography.

In light holography, a coherent light wave scattered from an object is
brought to interfere with a reference wave, and the resulting interference
pattern can be magnified and recorded on film. By shining monochromatic
light (of any wavelength) on this film wavefronts are reconstructed which
are "copies" of the original.

The formation of electron holograms requires a coherent electron
source and an electron beam splitter to divide the coherent electron wave
into two beams to create an interference condition. A coherent electron
source can be created with a field emission (FE) electron gun illustrated
schematically in Fig. 5.22(a)[41]. This coherent electron beam collimated
within about 2×10^{-6} rad is then deflected onto a specimen surface as shown
schematically in Fig. 5.19. A Bragg-reflected beam is then selected by the
objective aperture to form the REM image which is split into a two-beam
interference pattern to form the reflection hologram as illustrated sche-

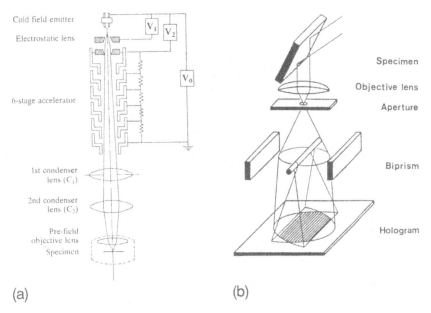

(a) (b)

FIG. 5.22 *Schematic features for the formation of reflection electron holograms. (a) Coherent electron probe-forming system in a field-emission (FE) electron microscope. (b) Electron optical system for reflection electron hologram formation (After ref. 43).*

matically in Fig. 5.22(b) [42,43]. The electron biprism shown in Fig. 5.22 (b) (and invented by Möllenstedt and Dücker [44]), consists of two grounded plates with a fine wire filament between them having a positive potential of approximately 100 volts. The electrostatic potential applied to the wire filament acts on the electron beam like an optical biprism. A reflected wave from a "flat" portion of the specimen surface is used as a reference which can be considered a plane wave at the image plane. The two waves overlap in the lower plane to form an interference pattern between the two waves as shown in Fig. 5.22(b). The electron holography process actually involves two steps as shown in Fig. 5.23(a). First a hologram is formed as illustrated in Fig. 5.22(b). Then an image is reconstructed by shining light on the hologram. This reconstruction is performed with an optical system shown schematically in Fig. 5.23(b); the camera recording the resultant hologram is set obliquely to compensate foreshortening in the reflected-beam direction.

A reflection-electron interferogram of a screw dislocation emerging on a GaAs(110) surface is reproduced in Fig. 5.24[43]. The vertical dis-

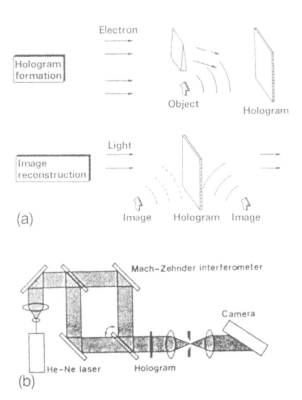

FIG. 5.23 *Electron holography process. (a) Schematic of two-step holography process: hologram formation and reconstruction. (b) Details of optical reconstruction system (After ref. 44).*

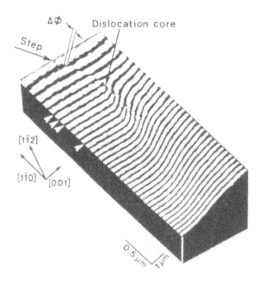

FIG. 5.24 *Reflection-electron interferogram of a screw dislocation emerging on a gallium-arsenide (110) surface. (Courtesy A. Tonomura; from ref. 43).*

placement around the dislocation can be observed as a shift of the interference fringes from the parallel fringes. The interferogram shows the
vertical displacement viewed from the direction perpendicular to the surface. A monatomic step can be observed as a small displacement of fringes
along the direction indicated by the arrow "step" in the figure, although
only a fraction ($\Delta\phi$) of the total phase shift is detected.

We shall treat electron holography in more detail in Chapter 7 in the
context of transmission electron holography.

5.5 SCANNING ELECTRON MICROSCOPY

Historically, the scanning electron microscope is as old as the conventional transmission electron microscope, having been suggested on a theoretical
basis by M. Knoll [45] only a few years after the development of the first
transmission electron microscope [4]. Working models of scanning electron
microscopes [46,47] were in existence at the time transmission electron
microscopes were being made available commercially, but their early com-
plication, poor vacuum facilities that contributed to image loss and specimen deterioration, and their generally poor resolution prevented them from
flourishing. It was not until the postwar developments of D. McMullan [48],
K. C. A. Smith and C. W. Oatley [49], and Oatley and coworkers [50] that
the scanning electron microscope became available as a commercial instrument with a capability of imaging surface topography and morphology with an
unprecedented advantage of depth of field and a capability for studying any
surface in its original, unaltered state.

You are cautioned that while a surface is viewed directly in the scanning electron microscope, the specimen size is limited by the viewing section, or specimen chamber. This means that very large specimens must be
partially destroyed or physically altered. This particular feature represents a limiting condition for certain scanning electron microscope applications. Consequently, for nondestructive viewing of surface features of
large objects, replication techniques must be utilized with the SEM.

In effect, the development of modern scanning electron microscopes
closely paralleled that of the electron microprobe (see Chap. 4). It is,
for all practical purposes, nearly identical in basic electronic design
features to the scanning electron probe microanalyzer as discussed in Chap.
4 and outlined schematically in Fig. 4.14. Commercial scanning electron
microscopes have been available from a number of manufacturers since the

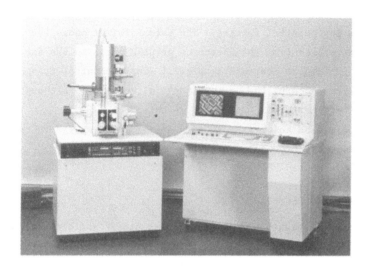

FIG. 5.25 *Commercial (Hitachi S-4000 field emission) scanning electron microscope. (Courtesy Hitachi, Ltd.).*

1960's. Figure 5.25 illustrates the physical appearance of a typical instrument, while Fig. 5.26 shows the corresponding system block schematic. We are here reminded again that the instrument is basically a simple electron optical-beam focusing system with a capability to detect secondary emissions from a bombarded specimen, and to present a synchronous visual display of the signal on a cathode-ray tube screen. In most cases, the addition of an x-ray energy-dispersive spectrometer to the specimen chamber makes the conventional scanning electron microscope an analytical tool (see Fig. 4.10). The use of a field-emission electron source, improves brightness and electron beam coherence which increases the image resolution. The scanning electron microscope shown in Fig. 5.25 for example can attain a spatial resolution of 15 Å.

From the standpoint of actual surface microscopy, the scanning electron microscope normally images topographical details with maximum contrast and depth of field by the detection, amplification, and display of secondary electrons. Inasmuch as the origin and nature of secondary electron emission is germane to an understanding of surface image contrast in the SEM (scanning electron microscope), we will briefly treat the physical features of secondary emission processes.

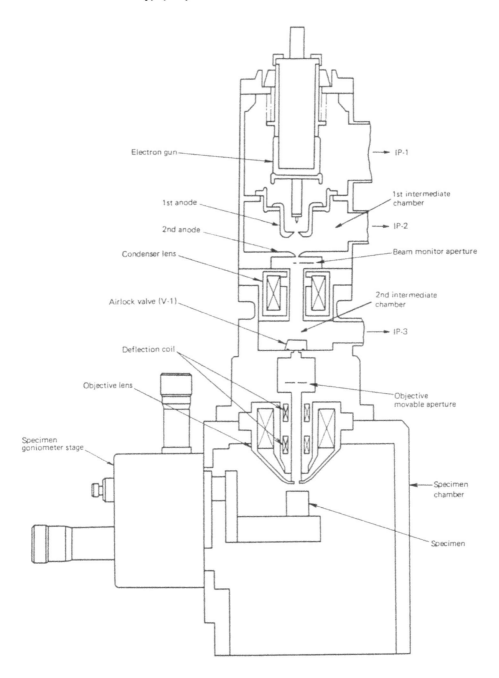

FIG. 5.26(a) *Schematic view of Hitachi S-4000 scanning electron microscope shown in Fig. 5.25. (Courtesy Hitachi, Ltd.).*

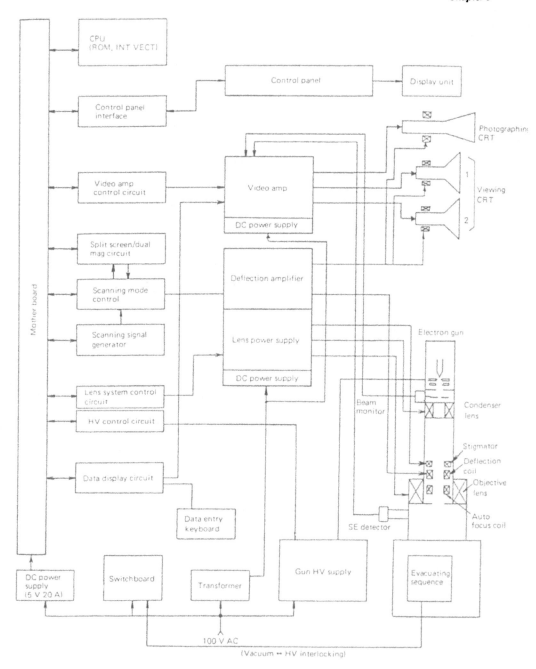

FIG. 5.26(b) *Block diagram of display unit showing operating principles*
of Hitachi S-4000 scanning electron microscope in Figs. 5.25 and 5.26(a).
(Courtesy Hitachi, Ltd.).

5.5.1 SECONDARY ELECTRON EMISSION FROM SOLIDS

When a solid is bombarded with an incident (primary) electron beam characterized by an accelerating potential of V_0 as shown in Fig. 4.1, a portion
of the emitted radiation consists of low-energy electrons originating in
the solid as a result of atomic excitation by the high-energy primary beam.
The electrons emitted by the solid as a consequence of primary excitation
are termed secondary electrons, and are characterized by their relatively
low energy as compared with the incident electron energy, or the energy of
elastically scattered (backscattered) electrons or those electrons that
emerge from the solid as a result of inelastic collisions with a kinetic
energy appreciably less than the incident electrons (Auger electrons).
Figure 5.27 illustrates the distribution of electron emission from Ag as
measured experimentally by E. Rudberg [51] who used a magnetic-deflection
method and a primary electron energy of 155 eV. The secondary electron
emission is characterized by the maxima E_S, while the elastically scattered
(backscattered) energies are observed at $E_e(B)$, and the inelastically scattered emission profile (Auger electrons) observed as a small maxima at E_i
(B). As we noted in Chap. 4, it is possible to unambiguously distinguish
those inelastically scattered electrons from the secondary electrons. As
a general rule, electrons are classified as secondary electrons if their
energy is less than about 50 eV.

The secondary emission yield for a solid is characterized as the ra-

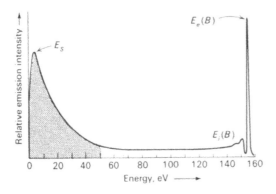

FIG. 5.24 *Energy distribution of electron
emission from Ag.* [*After E. Rudberg*, Phys.
Rev., *45: 764(1934).*]

tio of secondary electron current to primary electron current as defined
by

$$\delta = \frac{I_S - I_B}{I_o - I_B} \tag{5.24}$$

where I_S is the secondary emission current, and I_B is the backscattered
electron current. Although we define the secondary emission yield δ in
Eq. (5.24), it must be pointed out that δ is independent of primary electron current density, as opposed to the thermionic emission processes discussed in Chap. 2. The emission yield depends almost solely on the primary
electron energy (accelerating potential), the density, topography, and to a
small extent, the crystallography of the emission surface as the work function is altered. In this sense we might express the secondary emission
yield by

$$\delta = \eta e^{-E_W/E_S} \tag{5.25}$$

where η is a constant. The dependence of δ on the work function as generally depicted in Eq. (5.25) is observed experimentally as shown in Fig. 5.28,
which also serves to illustrate the dependence of δ on the incident electron energy (accelerating potential).

Figure 5.29 illustrates the typical response of secondary emission
yield for solids (particularly metals)   [See also the same response for
Ge at accelerating potentials as high as 5 kV, in J. B. Johnson and K. G.
McKay, *Phys. Rev.*, *93:* 688(1954)] as a function of the incident electron
beam potential V_0. We can discuss this response in a little more detail
by examining this δ and V_0 data [52] for Ag as a complement to Fig. 4.25.
Here again, as in Fig. 5.28, a maximum in δ appears at some incident electron potential, which we can describe as $V_0(max)$. This feature arises
primarily because, as the primary electron energy is increased, more secondary electrons are emitted. However, if we look back momentarily to Eq.
(4.10), the depth of penetration is also observed to increase with increasing V_0. Consequently the depth of release of secondary electrons increases
to a point where absorption processes begin to deplete the total secondary
emission current, and therefore a balance of optimum secondary emission
current is observed to occur for some $V_0(max)$. This is of course, as we
can surmise on a mechanistic basis, a universal effect for all solids, metals, or nonmetals. And, since the emission is governed to this extent by

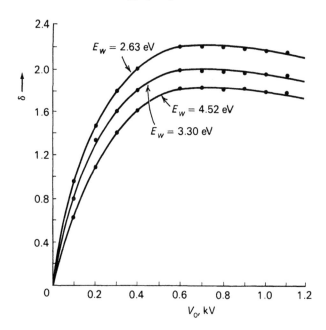

FIG. 5.28 *Variation of secondary electron emission yield from tungsten activated with Th of varying thicknesses to produce residual changes in effective work function.* [*From K. Sixtus,* Ann. Physik, *3: 1017(1929).*]

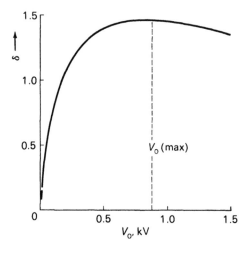

FIG. 5.29 *Secondary electron emission yield of Ag.* [*After R. Warnecke,* J. Phys. Radium, *7: 270(1936).*]

the Thomson-Whiddington law,

$$z \equiv d_p = 11 \times 10^{-9} \frac{W_A V_o^2}{Z\rho} \tag{4.10}$$

we can express the primary electron energy at some depth z in the solid in the more general form

$$E_o(z) = \sqrt{e^2 V_o^2 - Kz} \tag{5.26}$$

where K is a constant proportional to $Z\rho/W_A$ as shown in Eq. (4.10). We observe of course that for $z = d_p, E_o(z) = 0$, giving Eq. (4.10). We would expect therefore, in the basis of Eq. (5.26), that V_o(max) and the corresponding δ(max) would differ for various solids; and this feature is in fact observed experimentally as outlined in the data reproduced in Table 5.2. It should also be apparent that while the primary beam is the principal beam is the principal excitation mechanism for secondary electron production, backscattered electrons which occur deeper in the specimen also create secondary electrons. In the z-direction along the primary beam path, Auger electrons are first excited, then secondary and finally backscattered electrons. Characteristic x-ray production occurs at the deeper levels of penetration, although some x-ray signal originates throughout the excitation time.

Let us now look in retrospect at Eqs. (2.1) and (2.12) which characterize the necessary quantum-mechanical considerations for electron emission from a solid surface. We observe generally that the true secondary electron emission current occurs by the transfer of primary electron energy [as determined by Eq. (5.26)] to the lattice electrons; and where the conditions of Eqs. (2.1) and (2.12) are met, a secondary emission current occurs. However, the primary electron energy decays on penetration, and the transfer process thus decreases in efficiency with penetration. In addition, the secondary electrons created well below the surface must overcome absorption and scattering processes in addition to the surface work-function barrier. The resulting secondary electronic emission yield responds as depicted in Figs. 5.28 and 5.29.

We have to this point established, hopefully, a simple mechanistic description of secondary electron emission, as it is influenced by the surface work function (and crystallography), and the incident electron energy, as it is influenced in the energy-transfer process by the density

TABLE 5.2 *Secondary electron emission properties of solids*[†]

Material	δ_o(max)[‡]	V_o(max), [§]kV
Ag	1.50	0.80
Al	1.00	0.30
Au	1.40	0.80
B	1.20	0.15
Ba	0.80	0.40
BaO	2.30 - 4.80	0.40
BaF_2	4.50	—
Be	0.50	0.20
BeO	3.40	2.00
C diamond	2.80	0.75
graphite	1.00	0.30
soot	0.45	0.50
Cu	1.30	0.60
Cu_2O	1.20	0.40
Fe	1.30	0.40
Glasses	2.00 - 3.00	0.30 - 0.45
Mica	2.40	0.35
Mg	0.95	0.30
MgO crystal	20.00 - 25.00	1.50
NaCl crystal	14.00	1.20
Si	1.10	0.25
SiO_2	2.10 - 4.00	0.40
ZnS	1.80	0.35

[†]*Data from N. R. Whetten in* Methods of Experimental Physics, *Vol. IV. Academic Press, Inc., New York, 1967.*

[‡]δ_o(max) *is the maximum secondary electron emission yield at normal electron beam incidence for* $\phi = 90°$.

[§]V_o(max) *is the incident electron accelerating potential at maximum emission yield for* $\phi = 90°$.

and chemical identity of the irradiated solid. Completely descriptive theoretical treatments continue to be lacking despite the fact that numerous attempts have been partially successful [53, 54]

5.5.2 THE INFLUENCE OF OBJECT INCLINATION AND SURFACE MORPHOLOGY ON THE SECONDARY ELECTRON EMISSION YIELD

We should finally discuss the effect of inclination of the object surface with respect to the incident electron beam. We can observe that where the depth of penetration as given by Eq. (4.10) is depicted as an attenuated path length in the solid, the secondary electron emission yield will be influenced by the actual depth from the surface. For example, let us con-

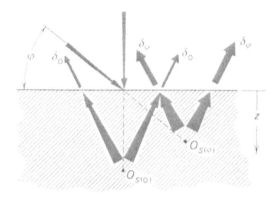

FIG. 5.30 *Secondary electron emission
yield with primary beam incidence.*

sider the situation in Fig. 5.30. Here the depth of penetration at normal
beam incidence characterizes the mean depth of origin of secondary electron
liberation $O_{S(0)}$ while for an inclination of the primary electron beam, the
mean depth of origin of secondary electron liberation $O_{S(0)}$ is a factor 1
- sin ϕ less than the normal incidence situation. Thus, where we restrict
our attention to some mean depth Z_m below the surface where N_S secondary
electrons are generated, the mean secondary electron current yield as it is
influenced by absorption processes can be expressed by

$$\delta_o = N_S e^{-\eta Z_m} \tag{5.27}$$

for normal incidence, and

$$\delta_\phi = N_S e^{-\eta Z_m \sin\phi} \tag{5.28}$$

for an incident angle of ϕ as depicted in Fig. 5.30. Consequently we can
express the effective secondary electron emission yield for a primary beam
incidence of ϕ as

$$\delta_{eff} \equiv \delta_\phi = \delta_o e^{\eta Z_m(1 - \sin\phi)} \tag{5.29}$$

where η is a constant for any particular solid. As a limiting condition,
we can of course let $Z_m = d_p$ [Eq. (4.10)]. The implications of Fig. 5.30
and the general form of Eq. (5.29) are in fact supported experimentally as

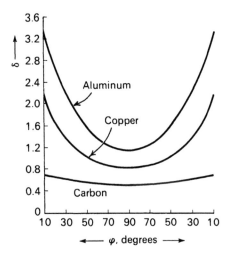

FIG. 5.31 *Secondary electron emission yield as function of primary electron beam incidence* ϕ. [*After H. O. Müller, Z. Physik, 104: 475 (1937).*]

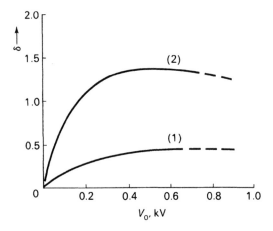

FIG. 5.32 *Secondary electron emission yield for granulated ("black") Ni surface (1) and a smooth Ni surface (2).* [*After H. Bruining,* Philips Tech. Rev., *3: 80 (1938).*]

shown in the effective secondary electron emission yield data of Müller
[55] reproduced in Fig. 5.31.

A similar difference in secondary electron emission will also occur
for very rough surface topographies and similar morphological features where
these features tend to absorb the portion of the secondary electrons emitted
or where they are scattered on striking complicated features at the various
emission surface sites. The implication of this particular phenomenon are
clearly evident from the secondary electron emission yield data for a granulated Ni surface and a smooth Ni surface obtained by H. Bruining [56] and
reproduced as Fig. 5.32.

5.5.3 *SURFACE IMAGE FORMATION IN THE SCANNING ELECTRON MICROSCOPE USING SECONDARY ELECTRONS*

As we have observed in the foregoing discussions, secondary electron emission is dependent on the following features:

1 The accelerating potential of the incident electrons
2 The surface morphology and especially the angle the incident electrons
   make with any particular surface site
3 The density of any surface site, which influences penetration of the
   incident beam and absorption of the secondary electrons
4 The surface chemistry and crystallography, which influence the potential
   barrier (work function) to secondary emission
5 The local surface-charge accumulation.

In the scanning electron microscope, as the incident beam is swept across a
surface, the secondary electron emission yield, δ, will vary according to
the conditions outlined above; consequently the secondary electron detector
current will change concurrently. This change will appear as brightness
variations over the surface as the amplified signal modulates the video
scan of the cathode-ray display tube (see Fig. 5.26). Where surface morphology and dimensions change with respect to the incident beam, the secondary electron emission current will provide image dimension to the video
display (or depth of field) as discussed in Chap. 3. The resulting secondary electron image thus takes on a three-dimensional nature by virtue of
the current contrast and depth of focus of the emission site.

Figure 5.33 illustrates very dramatically the image character and contrast for a fractured 85/15 TiMo wire observed in a Cambridge Stereoscan
scanning electron microscope using secondary electrons. Figure 5.33a shows

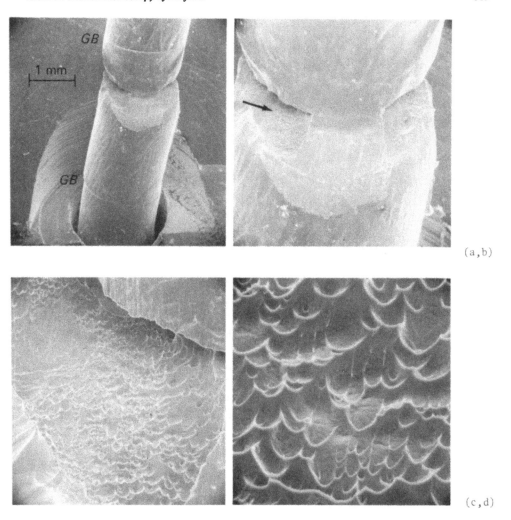

(a,b)

(c,d)

FIG. 5.33 *Shear-fracture characteristics at single-crystal portion of 85/
15 TiMo wire. (a) Low magnification view of mounted wire showing sheared
crystal portions. Crystal (grain) boundaries are indicated intersecting
wire along GB. (b) Magnified view (2.5 X a) showing central flow characteristics and peculiar fracture surface indicated by arrow. (c) Details
(4 X b) of fracture surface marked by arrow in b, (d) High-magnification
observation (5 X c) showing fracture-surface characteristics in c. Accelerating potential was 20 kV. (In collaboration with N. Hodgkin, NMH Co.,
Santa Ana, Calif.)*

the wire (after a vacuum-strain anneal to produce large, axially oriented
crystals) sheared across a single crystal portion. The peculiar ductile
appearing shear fracture associated with the surface portion marked with
an arrow in Fig. 5.33b is elaborated in the details of higher-magnification

observations in c and d. The base of the wire shown inserted into a mounting hole in Fig. 5.33a was painted with a Ag conductive paint to make contact with the observation pedestal. Similar observation of nonconductive
samples would also require uniform preshadowing with thin vapor deposits
of Au, Pd, or both, or some other suitable conductive vapor or sputter
coating (see Appendix B).

The most efficient and consequently the most common electron collection device for the SEM is the scintillator, light-pipe, photo-multiplier
system originated by Everhart and Thornley [61]. This system, illustrated
conceptually in Fig. 5.34, consists of a metal mesh Faraday cage which establishes a potential difference to accelerate secondary or reflected electrons so as to strike a scintillator screen attached to the end of a light
pipe. The scintillator screen is simply a thin (~ 0.05 μm thick) aluminum
film vapor-deposited or sputtered onto the hemispherical end of the light
pipe. The light pipe is usually about 1 cm in diameter, and photons produced at the scintillator screen travel along the light-pipe to the photomultiplier located outside the specimen chamber.

5.5.4 SECONDARY ELECTRON IMAGE RESOLUTION, CONTRAST, AND DEPTH OF FOCUS

Resolution in the scanning electron microscope is determined by the same
conditions of objective-lens design and concomitant aberrations as outlined
previously for a general electron optical system in Chap. 3. Because the

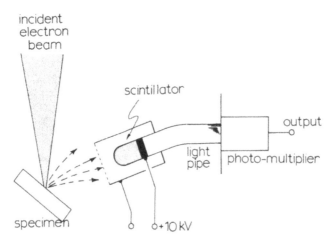

FIG. 5.34 *Schematic diagram of common electron
emission collection system in the SEM.*

nature of surface observation depends to a large degree on the focus of the incident beam to a fine spot, beam current also becomes an important factor in the space-charge limiting of resolution; and we can effectively express the theoretical limit of focus (surface resolution) by combining the space-charge limitations with Eq. (3.21a) to obtain

Spot Diameter $\equiv$ d(min) =

$$\sqrt{\frac{6I_o kT}{\pi J_o eV_o \alpha^2} + \left(\frac{C_S \alpha^3}{2}\right)^2 + \left(C_C \frac{\Delta V}{V_o}\right)^2 + 4 \delta_{pp}^2} \qquad (5.30)$$

where I_o = incident beam current

   k = Boltzmann's constant

   J_o = current density at cathode corresponding to a temperature T

As it turns out, the objective- or focusing-lens design becomes a limiting factor in the scanning electron microscope since in order to collect secondary electrons emitted from a scanned surface, the magnetic-field experienced at the surface cannot exceed about 240 amp/m [57]. This feature of course places restrictions on the objective-lens design, with the result that the corresponding aberrations are larger than those normally encountered in conventional electron microscopy. This of course applies only to thermionic emission, and Eq. (5.30) should be recognized to apply to thermionic emission. More generally, we could write an equation for the minimum beam diameter at the specimen surface as

$$d(min) = \sqrt{\frac{P}{\alpha^2} + Q\alpha^2 + R\alpha^6} \qquad (5.31)$$

where

$$P = \frac{4 I_o}{\pi^2 \beta} + (0.61 \lambda)^2$$

$$Q = \left(\frac{C_C \Delta V}{V_o}\right)^2$$

$$R = (0.3 C_S)^2$$

Here I_0 is the beam current and β is the brightness as described in Eq. (3.33). Optimum values for α (Fig. 5.35) occur near 10^{-3} radians and by controlling I_0 and β it is possible to achieve near atomic resolutions as demonstrated by Crewe and colleagues [58-60] using field emission guns.

Lafferty [62] also demonstrated more than three decades ago that materials such as LaB_6 possessed superior thermionic-emitting properties, and Broers [63] some twenty years later described an electron source using a pointed LaB_6 rod in a scanning microscope. The advantages of LaB_6 include its much lower operating temperature and significantly higher emission current than tungsten. Similar features have also been observed for other metal hexaborides, namely Ce, Ba and Sm [64]. Lanthanum hexaboride has been developed into one of the more useful high-brightness cathodes commercially available in a variety of electron optical systems, including scanning and transmission electron microscopes.

The diffraction limitation on resolution included in Eq. (3.20) is also greater in the scanning microscope than in conventional electron microscopes because of the fact that in order to minimize the incident electron penetration depths so as to enhance the secondary electron emission, lower accelerating potentials must be employed. The result is therefore longer wavelengths, and increased diffraction effects - and a measurable increase in the chromatic aberration.

Because the image of a surface in the scanning electron microscope results from a recombination of an amplified signal, rather than a direct imaging process as in conventional electron microscopy, the nature of the

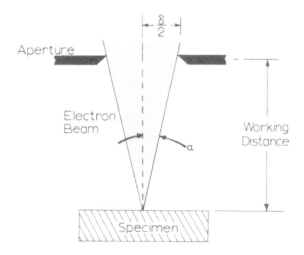

FIG. 5.35 *Simple diagram illustrating relationship between working distance, final aperture size, and beam divergence (aperture angle α) in the scanning electron microscope or any scanning electron beam system. This is conceptually and geometrically similar to Fig. 5.3.*

signal, and the introduction of electronic variables, will have an important influence on the image quality and resolution. We must therefore consider the signal-to-noise ratio of the image signal, as well as factors contributing to electronic noise between the surface emission sites and the cathode-ray display tube.

The signal, as it reaches the display tube, is recognizable if the signal-to-noise ratio is appreciably greater than unity. Thus, where we denote the signal-to-noise ratio as S, we can write

$$S = \frac{V_{sig}}{(V_{noi})^2} \qquad (5.32)$$

where V_{sig} is the signal voltage, and $(V_{noi})^2$ is the mean-square-noise voltage. The noise component of the signal arises in the detection process and is amplified with additional contributions arising in the detector-amplification coupling and at the output of the amplification networks. Noise originates in the detection stage as a result of statistical fluctuations in the arrival of electrons at the object surface, the inclusion of backscattered electrons in the secondary electron emission profile, and the unvoidable inclusion of secondary electrons excited by the emergence of backscattered electrons. Conventional scanning microscopes employing accelerating electrode-scintillator photomultiplier tubes as electron detectors collect and accelerate secondary electrons by a positive potential of approximately 10 kV on the accelerating electrodes; in this way the scintillator is excited to a sufficient level to activate the photomultiplier tube (Fig. 5.34). Backscattered electrons, on the other hand, are energetically capable of exciting the scintillator without acceleration, and the accelerating electrode in this mode of detection is maintained at ground potential. This provides an effective selection scheme for excluding either secondary or backscattered electrons emitted from the scanned object, depending on the desired mode of operation.

At higher magnifications, secondary electrons emerging from the object surface as a result of backscattered electron excitation do so at points widely disposed with respect to the incident probe beam, and to those secondary electrons excited by the primary electron beam. The resulting signal contains considerably more noise and the final image display lacks resolution. Normally the secondary emission current required to form an image is of the same magnitude as the primary beam current ($\simeq 10^{-12}$ amp) The final picture produced is normally composed of 250,000 points requiring as long as several minutes of scan time to develop on the cathode-ray tube. In this sense the number of picture lines and the scanning speed serve to define the final image resolution.

As a result of the mechanical and electronic design limitations imposed on the scanning electron microscope, practical resolution in commercial instruments on the order of 50 to 100 Å has become feasible. Any additional reduction in the probe size must involve a field emission scheme for increasing the incident beam intensity. Commercially available scanning instruments now claim an ultimate resolution ranging from 50 to 250 Å, and field-emission guns are used in those instruments operating in the very high resolution range.

W. P. Dyke and J. K. Trolan [65] were among the first to point out that currents of 10^6 A/cm^2 could be obtained with field emission filaments as opposed to 10 A/cm^2 with the more conventional, thermionic emission filaments. Utilizing this concept, Crewe and co-workers [58,59] developed a field emission source for electron microscopy, and such source concepts are now common practice in many scanning electron microscope designs, as well as transmission electron microscopes, especially scanning transmission electron microscopy (see Chap. 7). As shown in Chap. 3 [Eq.(3.20)], if all other design and noise features are eliminated, resolution in terms of minimum spot size is dependent only upon spherical aberration and diffraction effects, and will occur when the beam divergence (α in Fig. 5.35) is $0.8\ C_S^{-1/4}\lambda^{1/4}$. Thus, for an instrument with $C_S \cong 1$ mm and $\lambda = 0.078$ Å (at $V_o = 25$ kV), the minimum spot size will approach 10 Å. Broers and Brandis [66] originally obtained 50 Å resolution with LaB$_6$ filaments while Crewe [67], using a cold field-emission tungsten filament, obtained 30 Å, or less in the scanning transmission mode. More recently, better resolutions have become somewhat routine in scanning electron microscope designs.

As illustrated in Fig. 5.25 and 5.26 field-emission (FE) sources are now routinely utilized in commercial scanning electron microscopes because cold field emission achieves much higher spatial resolutions (10-20 Å), has a brightness 100 times greater than LaB$_6$, and often requires replacement only after hundreds of hours of operation. Since the operating voltage for scanning electron microscopy is usually below 30 kV, the necessary high vacuum conditions and discharge protection in the electron gun have been routinely achieved. It is only recently that higher voltage operation of FE sources has become commercially feasible in transmission electron microscopy where the vacuum conditions are 10^{-8} Pa (10^{-10} Torr) in the electron gun and the source brightness of 10^9 A (cm^2 str) and an energy spread of about 0.3 eV.

Contrast reversal and the no-contrast angle M. D. Coutts and E. R. Levin [68] have pointed out that on tilting samples in the SEM, image contrast between two elements A and B in a flat surface plane often decreases to zero at some critical angle ϕ_0. Above this angle, the contrast is reversed. If the contrast between these two elements is given by

$$\frac{\Delta I}{I} = \frac{I_A - I_B}{I_A} = 1 - \left(\frac{I_A}{I_B} \right) \tag{5.33}$$

where I_A and I_B are the total number of reflected and secondary electrons from these areas, then for element A we can have

$$I_A \propto \frac{I_0 - I_{aA}}{I_0} = \xi_A + \delta_{oA} \, e^{\eta_A (1 - \sin \phi)} \tag{5.34}$$

In Eq.(5.34) I_0 and I_{aA} are the incident beam and specimen absorbed current respectively in area A, ξ_A is the reflected electron coefficient (backscatter coefficient), and δ_0, η, and ϕ are as described in Eq. (5.29). Neglecting the corresponding backscattered electron contribution when $\xi_A \simeq \xi_B$ we can show that at the no-contrast angle, ϕ_0

$$\delta_{oA} \, e^{\eta_A (1 - \sin \phi_0)} = \delta_{oB} e^{\eta_B (1 - \sin \phi_0)} \tag{5.35}$$

and

$$\phi_0 = \cos^{-1}[1 - (\ln \delta_{oA}/\delta_{oB})/(\eta_A - \eta_B)] \tag{5.36}$$

Equations (5.35) and (5.36) hold true only when $\xi_A \simeq \xi_B$, and this occurs only rarely. Coutts and Levin [68] demonstrated these conditions for Au and Pt and found that in all cases (over a range of accelerating voltages), contrast reversal occurred around 10°. While this phenomenon is complex and Eq. (5.36) is an approximation which could be expected to apply only in special circumstances, the implications are that the specimen angle for ideal conditions (flat, polished specimens) in the SEM can have an important influence on contrast. It is therefore not good practice to simply set the specimen angle at 45°. It is in fact advisable to seek the optimum contrast conditions by varying the specimen tilt angle. You should realize that this condition may not apply if a specimen is heavily coated. This means in fact that for most normal SEM viewing of coated specimens, this feature may not be important because the emission of electrons will occur

primarily in the coating. It might also be instructive to point out that
the change of coating structure (integrity, uniformity, lack of agglomera-
tion, etc.) with average thickness is also of considerable importance as
described originally by Hodgkin and Murr [69].

Voltage contrast Contrast formation as a result of the variation in secondary electron emission has been previously outlined, and thus where the
detection of intrinsic secondary electron yield constitutes the basis for
image formation, the final cathode-ray display results from basically to-
pograhical and compositional contrast. The secondary electron image can
also be influenced to a large degree by the potential distribution on a
specimen surface. This feature,when utilized specifically in imaging surface characteristics, is called voltage contrast.

   A potential difference generated at the specimen surface can thus aid
in the analysis of specific features, notably those associated with semiconductor and related devices. Typical of this mode of surface analysis
and observation is the voltage contrast arising at a transistor as shown
in Fig. 5.36. In Fig. 5.36a, the collector-base voltage is zero, and no
voltage contrast occurs. The image is formed solely by topographical and
compositional contrast. In Fig. 5.36b, a potential difference of 2 volts
is established across the collector and base of the p-n-p transistor, and
the collector now appears darker as a result of the additional voltage contrast. The video signal in such situations can also be derived solely from
the voltage-dependent emission. In such cases the voltage applied to the
junction of transistors and related devices possesses a known frequency.
If the video amplifier is then tuned to this particular frequency, only
that voltage-dependent component is amplified and subsequently displayed.
It is possible in this way to analyze voltage differences as small as 0.005
volt [70].

Absorbed current mode contrast In related semiconductor applications, the
video signal can also be derived from electromotive forces generated in the
device specimen by the creation of free carriers in the junction regions as
a result of the incident electron bombardment. The resulting images show
an enhanced voltage contrast for the various components serving as signal
generators. In this respect, absorbed current in general (defined by incident beam current less the emissive modes current) can provide for an alternative or simultaneous contrast scheme because the absorbed current at
the point of beam raster can be detected and synchronously scanned. Thus,

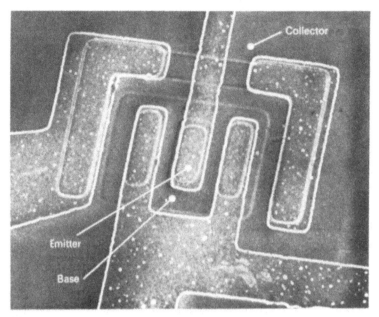

(a)

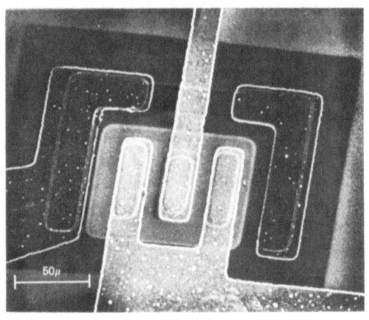

b)

FIG. 5.36 *Voltage contrast arising in secondary emission image of a transistor. (a) Transistor with zero bias; (b) reverse bias of 2 volts. (Courtesy Japan Electron Optics Laboratory Co., Ltd., Tokyo.)*

variations in the absorbed current or in the absorption-promoted specimen
signal can be synchronously scanned on the CRT display tube or on a paral-
lal display. (See Figs. 5.25 and 5.26).

Figure 5.37 shows for comparison the corresponding images of a PROM
device portion observed in the secondary electron emission and absorbed
electron current modes, and illustrates the general features of the absorbed electron current mode of image contrast in the SEM. It should be pointed out that the images in Fig. 5.37 were obtained on a non-dedicated SEM
portion of an integrated modular microanalysis system as illustrated schematically in Fig. 4.52.

Depth of focus in the SEM One of the most striking features associated
with the secondary electron emission images observed in the scanning electron microscope such as that of Fig. 5.33, is the vertical distance on the
object surface through which the focus appears sharp. This depth of focus
originates generally as discussed in Sec. 3.3.2, and can be expressed simply as

$$\text{Depth of focus} \simeq D_f \frac{1 + M}{M^2} \tag{5.37}$$

where D_f is the depth of field, as given in Eq. (3.23), with the aperture
angle determined by the objective (final focusing lens) bore and the working distance to the specimen (Fig. 5.35), and M is the image magnification.
Image magnification in the scanning electron microscope is simply the ratio
of the scanning width of the beam in the cathode-ray display tube to that
of the electron beam incident on the specimen. Since in the scanning electron microscope the aperture angle is simply the ratio of the objective-
lens pole-piece radius (or aperture radius) to the focal length (or distance from the specimen surface to the objective base), depths of focus of
nearly 100 microns are attainable at a magnification of 1000 times.

The depth of field [Eq. (3.23)] can also be expressed in terms of the
beam current and spot brightness by substituting for α in Eq. (3.33) in
terms of the brightness to obtain

$$D_f \simeq \left(\frac{2.5 \ B}{I_o} \right)^{1/2} d(\min) \tag{5.38}$$

Figure 5.38 illustrates the relationship between magnification, depth
of focus, and resolution attainable in commercially available scanning
electron microscopes utilizing thermionic emission guns.

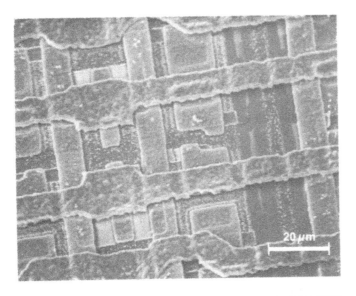

FIG. 5.37 *SEM images of programable read-only memory (PROM) device portion. Secondary electron image (top); absorbed electron current image (bottom). (Courtesy of Perkin-Elmer Physical Electronics Division).*

5.5.5 *ELECTRON BACKSCATTER IMAGES: ATOMIC NUMBER CONTRAST*

Backscattered electrons are influenced by surface conditions similar to those outlined for secondary electrons, with the additional feature that the intensity of the backscattered electron yield is proportional to the

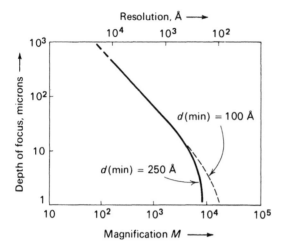

FIG. 5.38 *Relation between magnification, depth of focus, and resolution for a commercial scanning electron microscope. (Courtesy Japan Electron Optics Laboratory Co. Ltd., Tokyo.)*

atomic number Z of the emission surface as discussed previously in Sec. 4.3.1. Images formed in the backscatter mode in the scanning electron microscope are thus essentially identical to those formed in the conventional electron probe microanalyzer. The scanning electron microscope images utilizing electron backscatter are, however, considerably more revealing because of electron optical refinements that result in resolutions on the order of 10 times better than those attainable in the scanning electron probe microanalyzer discussed in Chap. 4.

The resolution of object detail is somewhat reduced in the electron backscatter mode of analysis as compared with the secondary electron mode since resolution in the backscattered signal depends on the accelerating potential and the specimen chemistry. Figure 5.39 illustrates this difference for the coincidence of materials of different Z. While the topographical image clarity is reduced in the backscattered mode, the atomic number contrast arising in zones of different Z gives enhanced acuity to the recognition of composition differences in the surface.

In terms of the characteristic energy loss for the elastically scattered (backscattered) electrons, we can write

$$\frac{\Delta E}{E} = \frac{2m}{M} (1 - \cos \phi')$$ (5.39)

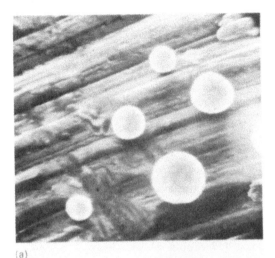

(a)

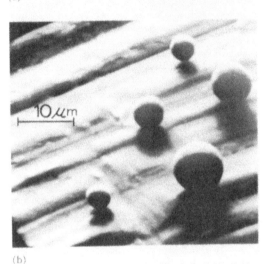

(b)

FIG. 5.39 *Scanning electron microscope images of copper particles on an aluminum substrate. (a) Secondary electron image; (b) backscattered electron image. (Courtesy Robert Jernigan).*

where m is the electron mass, M is the nuclear mass of the scattering zone (or of any particular scattering zone on the specimen surface), and ϕ' is the deflection angle. Values of ΔE are indeed small. For example for $\phi' = 30°$ in aluminum, $\Delta E = 0.107$ eV.

5.5.6 *SPECIMEN ANALYSIS USING ALTERNATIVE DETECTION MODES*

Just as in the electron probe microanalyzer, the fluorescent properties of certain compounds and related materials, or of a matrix containing impurities, can be visually observed using a light microscope focused on the specimen surface can also be employed to provide the signal energy, with the result that absorption-current variations over the specimen are used to form the cathode-ray display as described above and illustrated in Fig. 5.37(b). The mechanics of both these operations is identical to that of the electron microprobe discussed in Chap. 4, and we will assume that no further clarification is necessary.

The scanning electron microscope can also incorporate a detection mode for transmitted electrons when the specimen material is thin enough to accommodate this feature. Several designs for incorporating the basic features of the conventional transmission electron microscope in a scanning electron microscope have been reported [71]; and commercial scanning instruments now incorporate this feature as an alternative analysis mode with a maximum accelerating potential of 50 kV. This mode of analysis combined with the conventional secondary electron mode could allow the surface features of defects and related phenomena to be related directly with the internal crystal structure as observed by electron transmission. Thus the limitations expressed in Sec. 5.4 could be overcome. These features are also incorporated into the analytical electron microscope to be described in Chap. 7.

The energy-loss properties of transmitted electrons can also be investigated in the scanning electron microscope as discussed in Chaps. 4 and 7. The detector for transmitted electrons can be identical to that employed in the case of secondary and backscattered electron detection, and a simple aperture arrangement permits dark-field image formation and video scan.

5.5.7 DIFFRACTION EFFECTS IN THE SEM

When crystalline materials are examined in the scanning electron microscope, it is possible to produce reflective-mode images which arise as a direct consequence of the effects produced by the interaction of the incident electron beam with the crystal lattice, or more specifically with the atomic planes and their orientation relative to the specimen surface and the angular features of the incident electron beam. As shown in Fig. 5.40, there are two prominent effects connected with the interaction of the electron beam with the crystal lattice planes. One involves the channeling of the electrons between the atomic planes, and the other involves systematic diffraction from the crystal planes when the Bragg conditions implicit in Eq. (1.9) are satisfied. These effects are sometimes regarded as the consequence of the interaction of two Bloch waves [72], and when the incident

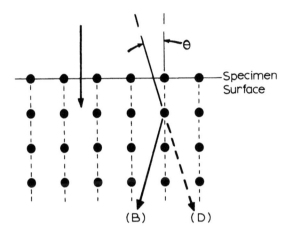

FIG. 5.40 *Illustration of electron diffraction and channeling effects in the SEM.*

wave satisfies the Bragg angle, both waves have the same amplitude. For angles off the Bragg angle [θ in Eq. (1.9)] the channeling wave is preferentially excited for angles greater than θ, while the scattering wave is excited for angles less than θ (the Bragg angle). As a consequence of these systematic diffraction effects, electron backscatter (reflection) or secondary emission intensity is reduced on one side of any specific lattice plane or set of planes (Fig. 5.40) and increased on the other side. This gives rise to reflected signal intensities which are spatially, geometrically, and crystallographically unique, and possessing intensity variations alluded to above. A simpler concept can be gained in two ways: one can simply expect different intensities for reflected electrons which channel and those which do not, or one can visualize differences in the two internal wave components designated as bright (B) and dark (D) in Fig. 5.40. The bright and dark portions are, as shown conceptually in Fig. 5.40, related to 2θ. The effects observed in the resulting images can be reversed in intensity, if the absorbed signal is utilized in forming the image. This is shown conceptually in Fig. 5.37.

Coates [73], in SEM examinations of semiconductor crystals, was the first to observe weak bands of contrast similar to Kikuchi bands observed in electron diffraction patterns from thick crystal specimens in the transmission electron microscope. The origin of these bands was qualitatively explained by Booker, et al. [74]. These patterns in the SEM arise by conditions of operation which are not the same as in conventional SEM imaging.

These patterns are referred to as electron channeling patterns (ECP's) and occur when the incident parallel electron beam is rocked through a sufficiently large angle which promotes channeling contrast from a number of different families of crystal planes as a result of momentary satisfaction of the Bragg diffraction conditions (Fig. 5.40). The beam rocking can be achieved independent of beam scan but the incident beam can be fixed at a point on the specimen surface. This produces a selected-area electron channeling pattern (SAECP) or selected area channeling pattern (SACP).

Figure 5.41 shows a signal differentiated ECP from a (111) - oriented silicon single crystal surface and a simple schematic illustration of some of the more apparent crystallographic features. It should be apparent that the ECP's are phenomenlogically identical to Kikuchi electron diffraction patterns to be treated in detail in Chapter 6. As a consequence, it is possible to obtain crystallographic information for crystal specimens examined in the SEM. This information includes surface orientation, crystallographic directions, and crystal geometry as well as crystal lattice perfection. Either emitted or absorbed electrons can form the ECP.

In addition, there are other features which, as a result of their influence on the crystal lattice, can provide observable deviations and variations in the ECP. These effects include local elastic and plastic strains, stoichiometric or crystallographic variations, etc. Consequently, from SACP analysis in the SEM it is possible to determine lattice parameter and crystallographic orientation information on a local basis (for example in individual grains or inclusions) along with other features such as strain intensity. These applications will be treated in more detail in Chapter 6 (see Sec. 6.6.1), where we will also compare ECP's with Kikuchi patterns and other types of electron diffraction patterns normally associated with the transmission electron microscope.

5.5.8 APPLICATIONS OF SCANNING ELECTRON MICROSCOPY

The applications of the scanning electron microscope in the investigation of surfaces and related structural features of interest to the materials scientist are essentially limited only by his or her imagination and inventiveness, or more accurately, by their lack of these characteristics. A number of the particular modes of analysis have already been demonstrated during the course of discussion and exposition, and we shall simply amplify these impressions with some examples and references of interest to the materials and mineral scientist.

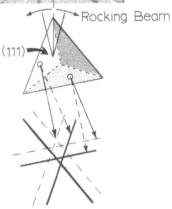

FIG. 5.41 *Signal differentiated ECP from (111) silicon single crystal (Courtesy of Hitachi, Ltd., Tokyo, Japan). The schematic diagram illustrates the origin of the main band from low index crystal planes.*

Specimens possessing considerable surface relief are of course the most prevailing examples of application, which is unique since replication techniques, as discussed previously, are unable to provide the necessary reliability in exposing the intricacies of such surface features. The observation of surface replicas also limits the observable area because of the design limitations in the specimen chamber of conventional electron microscopes as well as the necessity to support the replica on a grid mesh that "hides" up to 40 percent of the replicated surface area. These limitations are overcome in the scanning electron microscope since entire specimen sections can be directly observed and tilted for virtually any incident beam coincidence.

Figure 5.42 illustrates quite strikingly the study of intricate fracture surface features that, for the most part, would not be amenable to detailed observation using any form of surface replication. The figure shows the ductile-brittle fracture transition in HY-80 steel at various low temperatures, with the transition occurring very close to - 136°C. Figure 5.42 also provdes a good example of SEM applications in fractography, an area where it has become a routine and indispensable tool.

We are led, through an examination of applications such as that illustrated in Fig. 5.42, to speculate on the possibility for using the scanning

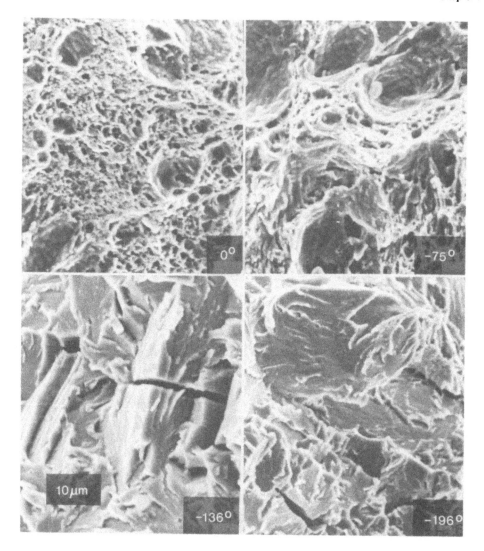

FIG. 5.42 *Fracture topography of HY-80 steel showing ductile-brittle
transition at low temperature. Accelerating potential was 25 kV.
(Courtesy Japan Electron Optics Laboratory Co., Ltd., Tokyo).*

electron microscope for the dynamic analysis of fracture and related phenomenological activity. While such experiments could be facilitated in commercially available instruments, a limitation exists on the rapidity one
can record simultaneous events. A visual video scan of an image on the
cathode-ray display tube usually requires from 0.1 to 10 sec. while corresponding photographic exposures require on the order of 100 sec. It will
therefore require the use of fast pulse sampling techniques for such work

in order to photographically record events occurring within intervals of less than 1 sec. These limitations appear to be capable of adjustment, and it may utlimately become possible to observe the development of microcracks and crack propagation directly within a scanning electron microscope with a resolution approaching that nominally attained in replication microscopy, and with the unprecedented depth of focus afforded by the scanning microscope. Sequential observations of microdeformation can of course be performed as shown for example in N. Gane and F.P. Bowden, *J. Appl. Phys., 39:* 1432(1968).

The scanning electron microscope in some instances lacks the clarity of replication, but compensates as indicated above. Figure 5.43 shows a surface section of a fracture in a single crystal of Zn accomplished in liquid nitrogen. The resolution and contrast at this magnification is of course equal or superior to that attainable in replication of such a surface topography. Figure 5.44 provides some additional examples of these features. It is especially evident in Fig. 5.44(a) that complex facets would be difficult to replicate without tearing and creating artifacts. Figure 5.44(b) provides additional evidence for the remarkable depth of focus associated with SEM images, and illustrates some readily apparent features of intergranular, brittle fracture phenomena.

Fracture topography - microstructure correlations It is frequently necessary to simultaneously observe the underlying microstructure associated with any particular surface topography or specific fracture features such as fatigue zones, fracture differences in transformed phases or chemically and structurally different regions and the like. The simultaneous observation of fracture features and microstructure revealed by optical metallography can be accomplished by the isolation of polished and etched regions within or adjacent to fracture surfaces. Sasaki and Yokota [75] devised a technique for comparing these two regimes which consisted in masking the fracture surface and the specimen and sectioning, polishing, and etching on a plane perpendicular to the fracture surface. More recently, Chesnutt and Spurling [76] developed a modification of this technique which is illustrated in Fig. 5.45. In this arrangement, the fracture surface to be examined is cut into a convenient size for use in the SEM. A portion of the surface is coated with a suitable stop-off lacquer by simply dropping the lacquer in specific areas to be examined. The exposed surface regions are then electropolished and finally chemically etched to bring out specific microstructural features if these are not already apparent from electro-

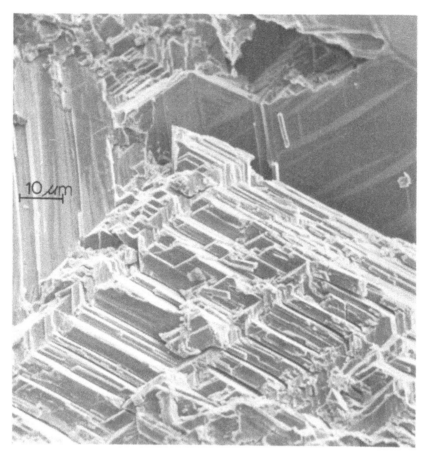

FIG. 5.43 *Cleavage fracture topography of a single crystal of Zn failed at - 196°C. V_O = 25 kV. (Courtesy Japan Electron Optics Laboratory Co., Ltd., Tokyo.)*

polishing. The lacquer is then removed by immersion in a suitable solvent. As shown in Fig. 5.45, this technique produces regions of microstructure which taper away from the fracture areas of interest, and provide for very direct correlations between the fracture surface topography and the microstructure as it is revealed by metallographic etching which produces contrast distinctions between different phases and at grain and phase boundaries.

General applications Studies involving dynamic features other than fracture studies are of course also tractable. These include oxidation, chemical reactions, crystal growth, etc. Typical examples of these applications can be found in the early observations of chemical reactions,

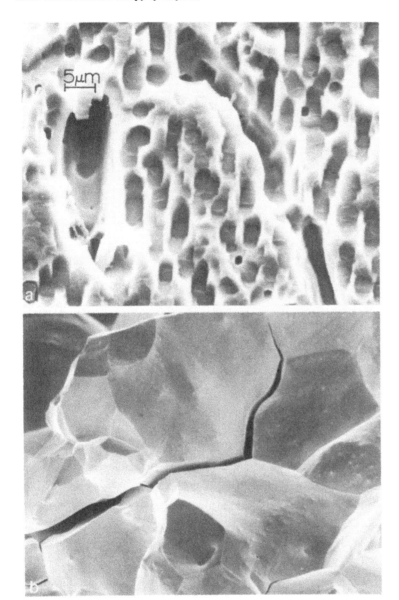

FIG. 5.44 *Secondary electron emission images in the SEM.
(a) Crystallographic etch pits in pyrite surface following
leaching in acid-ferric sulfate solution. (b) Intergranular fracture in polycrystalline iridium wire (room temperature fracture). Magnification for (a) and (b) is the same
as shown in (a).*

crystal growth by I. Minkoff [77] oxidation by R. F. W. Pease and coworkers
[78] and surface-topography changes during ion bombardment by A. D. G.
Stewart [79].

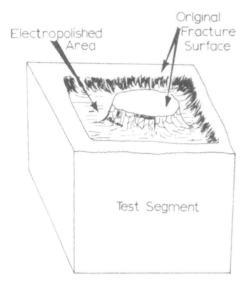

FIG. 5.45 *Fracture topography - micro-*
structure correlation. Schematic dia-
gram illustrating method of stopping
off fracture surface areas and electro-
polishing adjacent areas on a test
segment. The original fracture surface
is preserved by a stop-off lacquer.
Unlacquered areas are electropolished.

Ceramic materials, either conductive or nonconductive, are equally
amenable to surface studies in the scanning electron microscope. The examination of ceramic surfaces in general has been outlined two decades ago
by R. F. M. Thornley and L. Cartz [80] while phase separations and related
phenomena have been studied by A. J. Majumdar et al [81]. At this writing,
examples of the applicability of the scanning electron microscope appear
frequently, and routinely in the ceramic journals.

Because of the contrast facility available in semiconductor investigations as illustrated generally in Figs. 5.36 and 5.37, one of the most
important applications of the scanning electron microscope has been in the
semiconductor and related microelectronic-device fields. Contact phenomena, interfacial defects, other interfacial phenomena, and related features of microelectronic-circuit components are also amenable to study by
scanning electron microscopy; and numerous investigations of this nature
have already been performed routinely in laboratories throughout the world.
Some of these features are illustrated in Fig. 5.46.

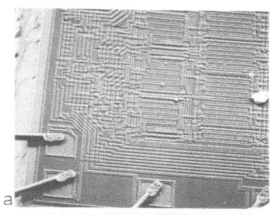

FIG. 5.46 *Secondary electron SEM images of*
PROM circuit portions. (a) low magnifica-
tion view of circuit components and contacts.
(b) Contact electrode region. Note granular
appearance of aluminum interconnects. For
magnification reference note that the wire
diameter is 0.033 mm. (c) Magnified view of
aluminum agglomerates in the conducting
interconnects [see Ref. 82].

As we demonstrated in Chapter 4, the scanning electron microscope is a powerful analytical and microanalytical tool when energy-dispersive x-ray microanalysis and Auger spectrometry are utilized in conjunction with secondary electron imaging of specific features. This is especially true when specific-energy mapping techniques are utilized as shown in the examples in Figs. 5.47 and 5.48. In Fig. 5.47 sulfur crystals are formed by the acid-bacterial (β) reaction of a pyrite inclusion in a quartz (monzonite) host rock as shown. These are readily identifiable in the characteristic sulfur x-ray maps shown. In Fig. 5.48, characteristic elemental Auger maps (or AES images) illustrate the nature of inclusions and selective segregation on or near the fracture surface of a modular cast iron embrittled at 480°C. As we noted earlier, the utility of Auger electron images lies not only in the clear micro chemical (elemental) distinctions possible but also in the ability to clearly display carbon (indicative of carbides), nitrogen (indicative of nitrides), boron (indicative of borides) and other low-Z constituents which are not detectable by energy-dispersive x-ray spectrometry (refer to Fig. 4.10).

The scanning electron microscope is a very common and even necessary tool in almost all contemporary technical areas, including production and quality control aspects of microcircuit and related microelectronics fabrication, and a host of other technical areas including failure analysis and fractrography (see for example ASM Handbook vol. 9 *Fractrography and Atlas of Fractographs,* published by the American Society for Metals in 1974). An enormous collection of results can be found in the annual SEM Symposia edited by O. Johari since about 1969. These volumes are now published annually by SEM, Inc., AMF O'Hare, Illinois, and the details are outlined in the Suggested Supplementary Reading list which follows at the end of this chapter. You might also peruse any current issue of the journal *Scanning.*

The applications of the scanning electron microscope as outlined briefly above are certainly not exhaustive. At this writing, several thousand papers have been published in areas dealing specifically with materials science applications such as surface metallurgical studies, ceramic studies, electrical properties, surface physics and chemistry, and related areas. Hopefully this treatment has allowed the reader to form sufficient associations so that he or she may readily apply the basic techniques outlined to the solution of problems unique in his or her own field or fields of interest.

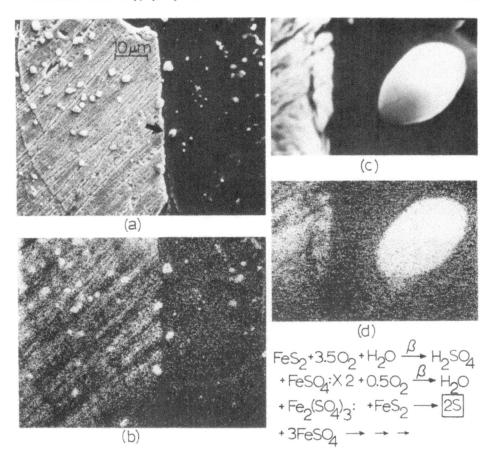

FIG. 5.47 *Sulfur crystals formed on the surface of a leached quartz-mineral containing a reacted pyrite inclusion. (a) Secondary electron (SEM) image of leached pyrite inclusion in quartz matrix showing sulfur crystal precipitates formed on the surface as a reaction product according to the reactions shown, (b) sulfur characteristic energy-dispersive x-ray map of the area in (a). Note the sulfur image features in the pyrite region as well along etch pits and preferential etching along polishing lines, (c) magnified view of individual sulfur particle on the quartz surface to the right of the pyrite inclusion [(arrow) in (a)]. (d) sulfur characteristic x-ray map of (c).*

It must be pointed out in concluding our treatment of scanning electron microscopy, and its applications that considerable thought and manipulation of the sample may be required in attaining optimum performance for any particular application. For example, with the sample tilted for a very acceptable secondary emission image at some particular operating voltage, the x-ray emission signal may be weak or, as a result of insufficient excitation voltage, particular x-ray spectra may not appear. Corres-

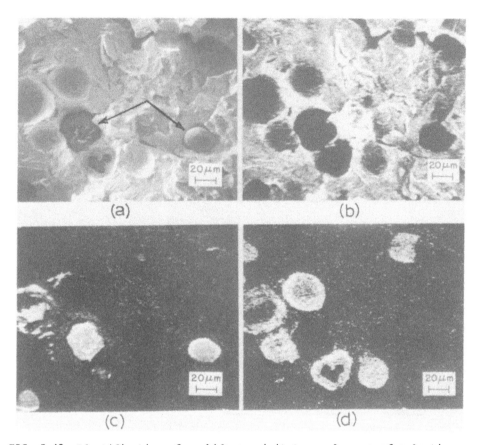

FIG. 5.48 *Identification of carbide precipitates and zones of selective segregation of elements on the fracture surface of a modular cast iron in the SEM. (a) Secondary electron image of the fracture surface, (b) Characteristic Fe-Auger image (46 eV Auger electron map), (c) characteristic C-Auger image showing the precipitates in (a) (arrows) to be carbides (272 eV Auger electron map), (d) Characteristic Sb-Auger image (460 eV Auger electron map). (Courtesy of Dr. A. Joshi [76].)*

pondingly, a good electron channelling pattern may not be obtained from some area without either altering the incident beam angle or the accelerating voltage. It is important to carefully consider the nature of the specimen and the operating conditions to insure that information contained in the sample is being optimally recorded. Very often each specific signal profile will require a different incident beam or detector take-off angle or a different accelerating potential, or both. You should not rely on standard specimen settings or operating parameters. Refer to Table 4.2 when you are concerned about specific element x-ray excitation. Remember that V_o must equal V_K or V_L for K_α or L_α peaks to appear.

5.6 ELECTRON MIRROR MICROSCOPY

As discussed in Sec. 3.2.3, electrons encountering a retarding field in a lens system of the electrostatic type are reflected. In effect these electrons are mirrored from the retarding-potential zone and are influenced to a marked degree by the magnitude of the retarding field as well as its homogeneity. As a consequence, pictorial representation of the electrical-potential distributions and/or magnetic-field distributions associated with ferromagnetic domains, grain boundaries, surface inclusions, or related solid-state phenomena that influence the surface characteristics of the retarding electrode can be obtained by photography of electrons mirrored from the retarding field adjacent to such surfaces. In effect, if a specimen material of interest is made the active biasing electrode in the electron mirror system, surface features can be observed. This electron optical system therefore assumes the features of an electron mirror microscope.

Figure 5.49 illustrates schematically the prominent features of an electron mirror microscope system consisting primarily of a conventional electron optical system terminated by a specimen-mirror arrangement. In the arrangement shown, a low-energy electron beam is either focused or spread near the specimen surface, which is a few tenths of a volt negative with respect to the mirror-field cathode. Thus negative charges or topographic protrusions that alter this intrinsic retarding potential create a stronger divergence in the locally mirrored zones and cause a concomitant loss in electron density of equivalent points on the photographic emulsion or fluorescent screen. These areas therefore appear darkened, while topographic valleys or positively charged zones cause a concomitant convergence in locally mirrored electrons, reinforcing the electron density and causing these representative areas on the fluorescent viewing screen to appear bright. Surface contrast therefore results, with a resolution dependent, for the most part, on the distance from the surface at which the electrons are actually mirrored. Resolution is also dependent on the ability to focus the electron beam onto the mirror area, a function that, because of the conventional electron optical lens design, is hampered by spherical and chromatic aberrations. Chromatic aberrations in the mirror field itself also contribute to loss in resolution. Object magnifcation is a function of the mirror distance D_m, and the specimen negative bias potential V_r:

$$M = \frac{k_m D_m}{V_r} \qquad (5.40)$$

where k_m is a mirror-field constant. The surface detail resolved in the electron mirror microscope therefore depends on the distance from the surface at which the electrons are reflected, the optimum of course being reflection from the surface directly. Electron mirror microscopy is by its nature a microprobe mode of analysis since the object surface is not directly imaged by contact reflection.

There has been some success in the application of the electron mirror microscope to the study of semiconductor-junction properties, ferromagnetic domains, and grain boundaries [84-87] and L. Mayer [88, 89] has been a pioneer in the development of the instrument as a research tool. The usefulness of electron mirror microscopy in the study of ferroelectric surfaces such as barium titanate and lead zirconate titanate was demonstrated by F.L. English [90] who observed electrical potential distributions presumably resulting from ferroelectric polarization of these materials. Figure 5.50 illustrates the surface domain structure in barium titanate made visible by electron mirror microscopy. It should of course be readily apparent that resolution as such is a limitation in electron mirror microscopy where, as evidenced by Fig. 5.47, the resolution is not appreciably better than that attainable in a light-optical system. Nonetheless, considerable information concerning domain structures, interface potential distributions, and related surface potential and polarity phenomena could ultimately be attained using electron mirror microscopy. Figure 5.51 attests to this by clearly indicating the existence of surface potential discontinuities in CdS [91].

Several other interesting features of electron mirror microscopy are to be found in association with such operations as electron beam machining, etc. In these operations an electron beam is focused onto an area of a few microns in diameter with sufficient intensity to form a hole. The formation of a hole or other micromachining feature can be followed by the direct observation of the local image formed by mirroring of the low-energy electrons penetrating the work area. In this way microbeam electron machining operations can be directly observed. Numerous related operations involving the imaging of mirrored low-energy electrons might also have important applications in many other areas of materials analysis, especially those areas involved in the interpretation of object composition, geometry,

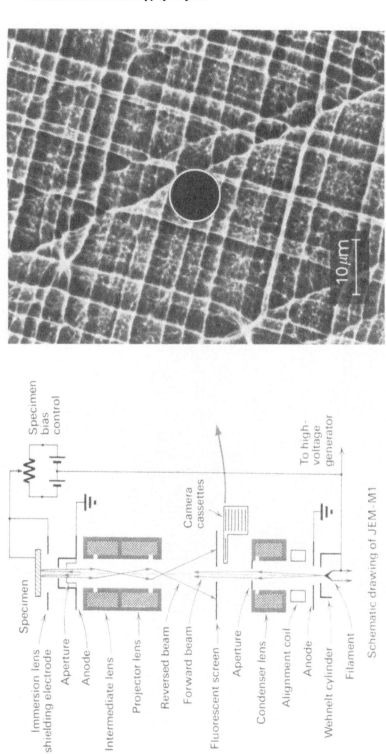

FIG. 5.49 *Electron mirror microscope schematic and representative surface image for Au evaporated onto NaCl cleavage surface; 35 kV operation. Microscope is commercial JEM-M1 electron mirror microscope. (Courtesy Japan Electron Optics Laboratory, USA.)*

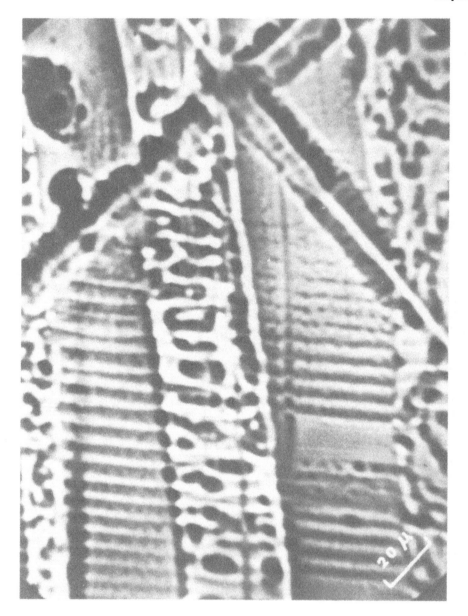

FIG. 5.50 *Electron mirror micrograph of ferroelectric-domain surface structure of BaTiO₃ heated above Curie temperature and cooled to room temperature. Electrical potentials present on surface as result of ferroelectric-induced variations in the bound surface charge produce dark regions (corresponding to negative hills) and bright regions (corresponding to positive holes or valleys). Dark bands are thought to be surface a domains, with the bulk c domains characterized by bright areas.* (Courtesy *F.L. English,* and J. Appl. Phys. *39: 128, 3231(1968).)*

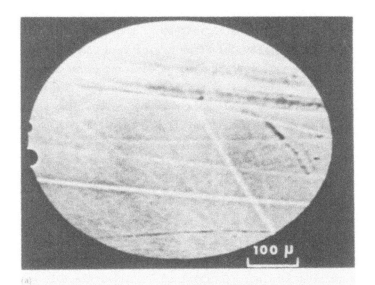

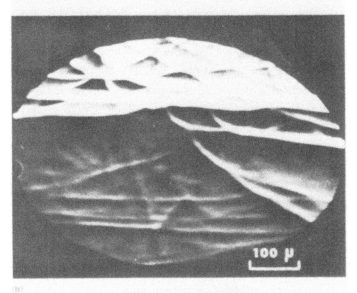

FIG. 5.51 *Electron mirror micrographs of (0001) surface of single-crystal CdS showing regions of inhomogeneous resistivity when electric field is applied normal to c axis. (a) No voltage applied; (b) 300 volts applied. (From F. L. English and M. K. Parsons in D. G. Thomas (ed.),* Proceedings of the International Conference on II-VI Semiconducting Compounds, *Benjamin, Inc., New York, 1967.)*

or related features giving rise to selected electron energy losses. Insofar as the general topic of energy-loss microscopy is concerned, the applications of this feature should be obvious after having read Chap. 4 (see,
especially, Prob. 4.12). Electrical properties of semiconductors and other
device components, especially potential distributions, are also amenable to
study by electron mirror microscopy. Adsorption phenomena can also be observed or at least detected as a result of work function variations which
can be imaged as contrast differences in the electron mirror microscope
[92].

If we digress momentarily to inspect the details of Figs. 5.50 and
5.51, it should be immediately obvious that the field emission microscopy
mode of surface analysis, much of the detail replies on the interpretation
of the image. This is particularly true when we are forced to assign to
the electron mirror micrograph intensity profiles some reliable topographical feature, or some related intrinsic function of the local electric or
magnetic field distributions associated with the specimen surface. And,
where the resolution is not high enough to ensure an unambiguous interpretation of surface morphologies and the like, the inherent difficulties in
image interpretation certainly, at this stage, outweigh the surface topography analytical attributes of electron mirror microscopy. We shall see
that in a great deal of the electron optical applications, image or object
signal interpretation will pose a formidable problem; that interpretation
becomes in fact an intrinsic feature of, for example, electron diffraction
and transmission microscopy.

5.7 ION MICROSCOPY OF SURFACES

The scanning ion microscope was treated conceptually in Chap. 4 (Sec.
4.6.1) in connection with scanning ion microanalysis. You might recall
that in our treatment of scanning ion microanalysis it was pointed out that
mass analysis of solid surfaces could be performed in a fashion similar
to the elemental analysis of areas in the scanning electron microscope by
gating or filtering specific ion signals as opposed to the selection of
characteristic x-rays. In many respects, the scanning ion microscope is
based on the concept of ion sputtering mass spectrometry, and is capable
of performing both surface and interior (bulk) microanalysis. As noted in
Sec. 4.6.1, both lateral images of the surface and concentration profiles
can be obtained. Here we will be concerned primarily with the ability of

the ion microscope to provide spatially resolved mass images of solid surfaces.

We should point out that while scanning ion microscopy of surfaces can be achieved as illustrated schematically in Fig. 4.40, ion microscopy of solid surfaces can be achieved through an arrangement conceptually similar to the photoelectron emission microscope (Fig. 2.6) where an ion source replaces the optical source, and secondary ions emitted from the sample surface are accelerated and separated according to their M_i/e ratio to produce an image corresponding to a given element. Such images are actually secondary ion images.

Figure 5.52 shows a schematic representation of an ion microscope. Normally the specimen is bombarded over an area of roughly 200 μm or more using any convenient ion species or specific ions where special secondary ion emission characteristics may be desired. Argon ion bombardment leaves the chemistry of the sample unchanged and is therefore well suited to revealing chemical (compositional) features in the sample. In general, positive ion images are more representative of a sample but for certain applications negative secondary ion images are useful.

The secondary ions originating from the ion-bombarded specimen area in Fig. 5.52 are accelerated by the uniform electrostatic field of the immersion lens to form a global ion image. This image of the surface area actually contains many superimposed ion images; one for each isotope present. A mass spectrometer is then used to separate the ions according to their M_i/e ratio to produce a specific ion image corresponding to a given element. This is accomplished by the use of a convex electrostatic mirror consisting of a first electrode at ground potential, followed by a control electrode and repelling electrode. The mass-resolved beam enters the mirror and ions entering along the mirror axis are decelerated, stopped, and returned in the reverse direction. The unwanted ions having higher (or lower) kinetic energies hit the repelling electrode and are neutralized and eliminated from the beam. Consequently, the electrostatic mirror acts as a low-energy filter. The energy-filtered ions "reflected" from the mirror enter the magnetic prism and describe a circular trajectory which produces a filtered beam in the same direction as the incident, unfiltered beam from the specimen surface.

The filtered ion image, which is a virtual image as it leaves the magnetic prism shown in Fig. 5.52, is projected by a projector lens onto the cathode of an electron image converter which in effect constitutes a

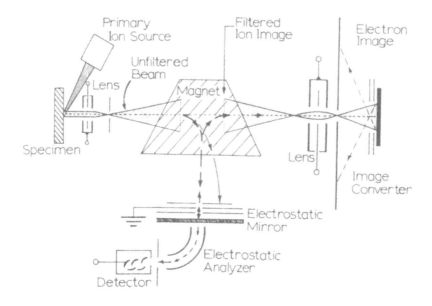

FIG. 5.52 *Schematic diagram of an ion microscope.*

second immersion lens. The secondary electrons emitted from the cathode
are focused by this lens into an equivalent electron image. This image
can either be displayed on a fluorescent screen or directly recorded photographically. It is important to realize that this final image results
by specific ion excitation of the photocathode in the photomultiplier
(similar in design to Figs. 3.24 or 4.6 for example). Consequently, the
image constitutes an elemental map or an elemental concentration map of
the specimen surface.

The illumination of the image is defined as the number of specific
M_i ions arriving per unit area in unit time at the detected image, and in
principle, the total number of M_i ions collected from a sample area A_o
during time t can be computed. If we define the practical ion yield, τ,
as the ratio of the number of specific ions, M_i, collected to the number
of atoms removed from the target, M, the image illumination can be related
to concentration by

$$i = \tau \, C_m \, Si_p G^{-2} \tag{5.41}$$

where C_m is the atomic concentration of element M_i, S is the total sputtering yield (number of atoms of any kind removed per incident particle)
i_p is the number of primary particles arriving per unit time and unit area

on the specimen surface, and G is the magnification. The product Si_p is the sputtering rate. Image quantification is ion microscopy involves an accurate relationship between illumination and concentration, and this is accomplished by measuring the optical density of the image features of interest. Rapid computer scan systems which perform this function have been devised as described by Steiger and Rüdenauer [93] and Drummer and Morrison [94].

Frequently an aperture is used in front of the photomultiplier counting system is localized probe analysis to limit an area A of the magnified image (where $A = A_0 G^2$). In this case the measured secondary ion current, I, is related to the concentration, C_m, by the equation [95]

$$I = \tau C_m S i_p A_0 \qquad\qquad\qquad (5.42)$$

where i_p is now defined as the primary beam current density, and A_0, as described above, is the sample area. The aperture area A can be adjusted for any specific sampling requirements for any given analytical problem.

5.7.1 APPLICATIONS OF ION MICROSCOPY

The ion microscope was first proposed by Castaing and Slodzian around 1962 [96], and the technique has found numerous applications in geology, metallurgy, materials science, biology and analytical chemistry [97-100]. Improvements in ion optics have made it possible to combine an elemental sensitivity in the 10^3 ppm range with a surface resolution of the order of 0.3 μm point-to-point.

Normally, samples for examination in the ion microscope must be relatively flat as in photoelectron or thermionic electron emission microscopy discussed in Chap. 2. In studies involving analytical microscopy where more precise concentration data is desired, samples are usually polished flat. Insulating materials can be examined but usually require a conducting metal grid to be deposited on the area of interest with a mesh size around 150 x 150 μm. This conducting grid serves to eliminate charging of the surface area. Alternatively the entire area can be covered by a very thin conducting film, and then the primary ion beam can be used to eliminate a portion of this film to form a window exposing the specimen surface area of interest.

A commercial ion microscope has been available from CAMECA (CAMECA IMS 300) since the early 1970's. The specific design features of this

instrument are illustrated schematically in Fig. 5.52 In addition, it is
possible to utilize the various scanning ion microscope designs such as
that shown in Fig. 4.40 in imaging modes similar to that outlined in Fig.
5.52 by adding the necessary filtering and detection systems. Many ion
imaging schemes utilize the secondary emission signal from the bombarded
area to form an image much like the scanning electron microscope, as shown
in Fig. 4.41 [101]. However the special feature of specific ion images is
the ability to provide specific ion maps or concentration maps of the specimen surface.

PROBLEMS

5.1 Calculate the critical thickness for a SiC section consisting of equal
weights of each element for an intensity-contrast-image interpretation
in electron microscopy. If it is stipulated that 60 percent of the
incident electron beam must penetrate this thickness of SiC, calculate
the necessary accelerating potential by considering the quantum mechanical considerations of Prob. 1.7. Compare this solution with an
approximation using the Thomson-Whiddington law, stating any other
assumptions made.

5.2 Consider a crystal cube of side t_C , supported on a substrate film of
thickness t. If the total scattering cross sections for the film and
crystal are Q_S, and Q_C, respectively, show that the dark field contrast
for this crystal will be given by

$$\frac{\Delta I}{I} = \frac{D_C(\phi_T)\sigma_S}{D_S(\phi_T)\sigma_C} \frac{e^{Q_C t_C}-1}{e^{Q_S t}-1} e^{-Q_S t} + e^{-Q_C t_C} - 1$$

Show also that by making the proper assumptions, the contrast is proportional to the ratio of the object thicknesses.

5.3 Referring to the data of Table 5.1, construct a curve relating density
and critical object thickness for various materials. Derive a general
expression relating density and critical film thickness for the admission of noncrystalline image-contrast interpretation. Calculate
the minimum thickness variation detectable in a noncrystalline (or
microcrystalline) Ti foil (such as that of Fig. 5.8) assuming a 10
percent optimum contrast perception.

5.4 Estimate the theoretical difference in contrast resulting for a 1000 Å carbon film observed at 50 kV in an electron microscope using first a 100-micron objective aperture and then a 20-micron aperture if the object plane is 20 mm above the aperture plane. Calculate the minimum thickness variations detectable in this film when imaged with the corresponding aperture sizes. (Assume a scattering amplitude of 0.5.)

5.5 A platinum film is vapor deposited onto a 1 in.^{2} glass substrate in a vacuum evaporator. Estimate the expected thickness change from the center of the substrate to the edges if a 3-mg strip of Pt is vaporized on a tungsten wire located directly beneath the substrate at a distance of 5 cm. Estimate the image contrast arising at a scratch in the glass 0.5 mm from the edge of the substrate and having a ramp angle of 45° and a depth of 1 micron when a section of the platinum film is removed from this area and observed in the electron microscope. (Q for Pt is 65 x 10^4 cm^{-1}.)

5.6 The following powder charges of Au and Ag were evaporated onto NaCl substrates at low vacuum in order to produce a very fine grain (essentially noncrystalline) film: 2, 4, 6, 8, 10 mg. The substrate was located 5 cm from the source consisting of a tungsten boat fashioned in a 30° V. Make a graph of approximate film thickness versus charge weight for these metals. Compare these curves with those obtained experimentally by L. E. Murr and W. R. Bitler, *Mater. Res. Bull, 2:* 787(1967), and discuss any discrepancies. Estimate the minimum Au and Ag sample weight that can be evaporated under these conditions to ensure a mass-thickness image-contrast interpretations provided the films are essentially structureless (refer to Prob. 5.3).

5.7 Thin films of Al$_2$O$_3$ are observed in the electron microscope at 65 kV. Approximate thickness of 100, 200, 325, 550, and 700 Å are observed with a constant incident beam current. Plot the relative bright-field-image intensity change versus thickness, assuming Q = 13.7 x 10^4 cm^{-1} (reference 9). Estimate the bright-field-image contrast of carbon granules on the Al$_2$O$_3$ films having an assumed constant radius of 0.01 micron, and plot the values against the corresponding film thicknesses (assume Q for carbon to be 7.2 x 10^4 cm^{-1}). Many of the carbon granules have irregular shapes. Calculate the maximum number of irregular sides observable in an image if the instrument resolution is 20 Å.

5.8   In the stress-corrosion replica images of D6AC steel shown in Fig.
       P5.8a, and 7075 T-6 aluminum shown in Fig. P5.8b, assume the obser-
       vation were both made under the conditions stipulated for Fig. 5.9,
       but an intermediate lens-current setting of 55 ma. The replicas were
       plastic and carbon with a 45° chromium shadow, and the technique was
       essentially identical to that depicted in Fig. 5.10. Make a histro-
       gram showing the distribution of corrosion depths along the grain
       boundaries for these two alloys. Discuss a comparison of the data
       for each alloy assuming the condition or orginal surface treatment
       were identical. Estimate the mean depth of the respective corrosion
       sites in each alloy.

5.9   Aircraft landing gear are often susceptible to impact damage. Des-
       cribe a nondestructive method for checking such damage. In particu-
       lar, suppose a particular landing gear is to be tested in situ after
       numerous landing operations. What criteria could be used to estimate
       a safety factor or similar precaution for economical parts replace-
       ment?

5.10   Refer to Fig. 5.6. The operating conditions involved in the photo-
       graphing of Fig. 5.6 were the same as those depicted for the series
       of calibration micrographs of Fig. 5.9, but in Fig. 5.6 the interme-
       diate lens current was 58 mA and the image was enlarged by 2.3 in the
       print. Assuming the intrusion shown to be hemispherical, calculate
       the angle of Pt shadow ζ. What evidence is contained in the image
       of Fig. 5.6 to support the assumption that the particle is approxi-
       mately hemispherical? Estimate the time of exposure of the area of
       Fig. 5.6 to the electron beam, assuming a rate of contamination
       build-up of 100 Å/min. Outline your approach and any assumptions
       made.

5.11   Refer again to Fig. 5.6. The base and shadow composite of this
       figure are actually a plastic, carbon, Pt-shadowed replica of a
       stainless steel (type 304) surface fatigued at ambient temperature
       for 75,000 cycles. The regular nonconcurrent striations observed
       in the background of the shadowed particle are therefore slip mark-
       ings. What are the possible crystallographic surface orientations
       of the bulk surface-area replicated? Explain the fact that the fre-
       quency of nonconcurrent slip markings as observed in Fig. 5.6 is
       considerably less than that observed in Fig. 5.15 for considerably
       fewer stress cycles (assume the strain the same for both cases).

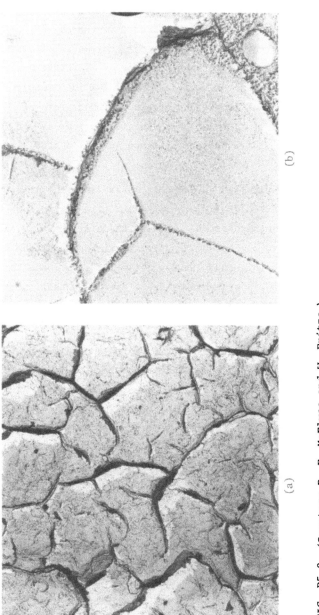

FIG. P5.8 *(Courtesy R. F. McElwee and W. Fritze.)*

5.12 Calculate the theoretical size (resolution) of the spot onto which a
scanning electron microscope beam can be focused if the objective
lens has a spherical aberration coefficient of 1 cm, and a chromatic
aberration coefficient of the same magnitude for an effective aperture angle of 0.003. The desired beam current is 2×10^{-12} amp for
an accelerating potential of 20 kV, and a cathode temperature of
2300°C. Discuss the result in relation to the resolutions of 200 to
300 Å attainable in commercial designs. (Assume a voltage stability
of 10^{-5} volts.)

5.13 Based on the background provided in Chap. 5, devise an experiment to
investigate the relationship of dislocations with slip markings observed on deformed specimen surfaces. Outline in detail the limitations of specimen design, dimensions, and related instrumentation and
techniques to accomplish this end. Devise a similar experiment to
investigate the relationship of surface impact damage to the initiation of internal defects by direct area correlation. Discuss any
limitations on the mode of analysis you choose.

5.14 Contrast for a fracture surface of MgO observed in the secondary
emission mode of the scanning electron microscope can be assumed to
occur primarily by the variation in the incident beam angle with respect to the particular emission surface site. Thus the contrast in
the video display intensity profile might be assumed to be of the
form

$$\frac{\Delta I}{I} \; \alpha \; \frac{\Delta \delta}{\delta_o}$$

where δ_o is the secondary electron emission yield for $\phi = 90°$. Consider the cross section in Fig. P5.14 to represent a surface section
of an MgO fracture, and calculate the contrast arising at A, B, and
C, assuming N to represent the background emission yield. The accelerating potential is 20 kV, and the normal coincidence secondary
electron emission yield at N is $\delta_o \simeq 10$. Assume the mean depth of
penetration to be given by the Thomson-Whiddington law; and the electron beam to be directed normal to the surface at N. What other assumptions must be made in order to describe contrast solely on the
basis required below?

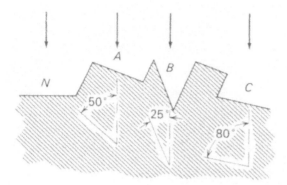

FIG. P5.14

5.15 The scanning electron micrograph below illustrates a rotated (side)
      view of the crystalline fracture shown in Fig. 5.30. For an accele-
      rating potential of 20 kV, and a magnification of 43 times, estimate
      the depth of field and prove this with reference to views shown in
      Fig. 5.30. Considering the fracture to be primarily ductile, esti-
      mate the residual strain in the portion shown. Discuss the mode of
      fracture mechanistically, utilizing dislocation theory, and speculate

FIG. P5.15

on the characteristic features observed in Fig. 5.30. Compare the appearance of the fracture surface shown in Fig. 5.30c with that observed in Fig. 5.13f, considering the direction of fracture indicated in Fig. 5.30 and the view shown below.

5.16 In an electron mirror microscopy arrangement, a Ge surface is imaged with a retarding potential of -0.20 volt. If the mirror field constant is 20 kV/m, find the effective image-plane distance above the specimen for an image magnification of 400 times.

5.17 The analysis of some of the alterations in a brass (70 Cu - 30 Zn) high-energy switch electrode is shown in Fig. P5.17. Discuss the alterations which are apparent and the advantages and necessity of utilizing AES and ESCA in addition to energy-dispersive x-ray spectrometry (you might refer back to Chap. 4). Discuss the apparent processes taking place during service. Also discuss the use of materials such as brass, and similar alloy systems, in the design of spark-gap electrodes in pulsed-power applications such as fusion reactors and the like. What features of electrode damage might be amenable to study by ion microscopy? Discuss your answer.

REFERENCES

1.   C. J. Davisson and L. H. Germer, *Phys. Rev., 30:* 705(1927).

2.   C. J. Davisson and C. J. Calbick, *Phys. Rev. 38:* 585(1931).

3.   E. Bruche and H. Johannson, *Naturwiss., 20:* 353(1932).

4.   M. Knoll and E. Ruska, *Ann. Physik, 12:* 607(1932).

5.   B. von Borries and E. Ruska, *Z. Wiss. Mikroskopie, 56:* 317(1939).

6.   A. Prebus and J. Hillier, *Can. J. Res., A17:* 49(1939).

7.   A. W. Vance, *RCA Rev., 5:* 293(1941).

8.   B. von Borries, *Z. Naturforsch. 4a:* 51(1949).

9.   C. E. Hall, *Introduction to Electron Microscopy,* pp. 210-213, McGraw-Hill Book Company, New York, 1966.

10. R. D. Heidenreich, *Fundamentals of Transmission Electron Microscopy,* John Wiley & Sons, Inc., New York, 1964, p. 35.

11. H. Mahl, *Metall., 19:* 488(1940); and *Z. Tech. Physik., 21:*

12. D. E. Bradley, *Brit. J. Appl. 5:* 65(1954).

13. D. E. Bradley, *J. Inst. Metals, 83:* 35(1954-1955).

14. R. D. Heidenreich, in R. D. Heidenreich and W. Shockley (eds.), *Report of a Conference on Strength of Solids,* Physical Society, London, p. 57, 1948.

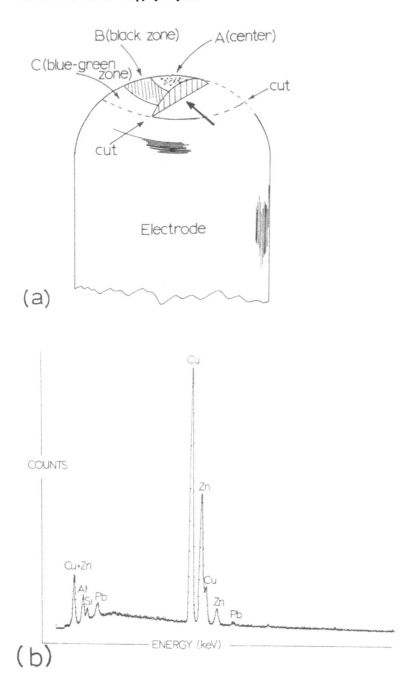

FIG. P5.17 *Analysis of spark-damaged pulsed-power brass electrode (a) Schematic view showing specimen preparation from electrode end, (b) Energy-dispersive (EDX) analysis showing bulk electrode (undamaged) composition at base of cut shown in (a).*

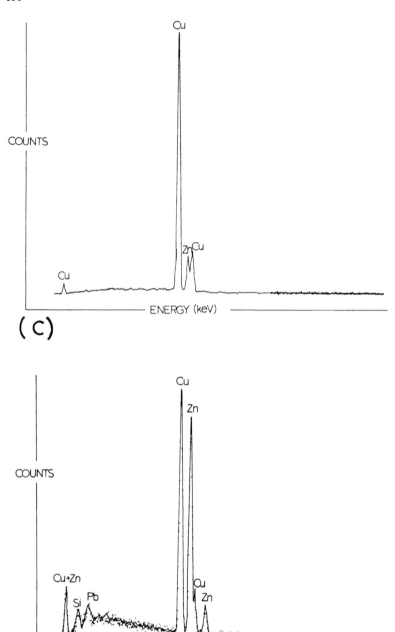

FIG. P5.17 (continued) *(c) EDX analysis of A (center), (d) EDX analysis in zone B.*

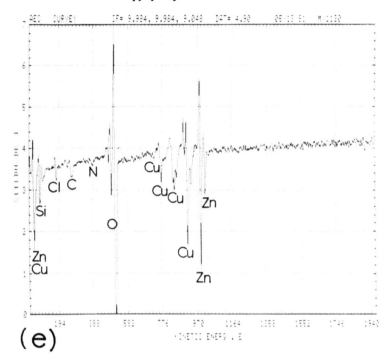

(e)

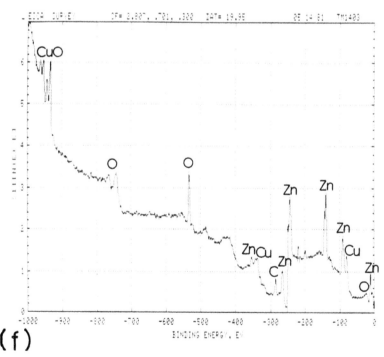

(f)

FIG. P5.17 (concluded) *(e)* *AES analysis of zone B surface,*
(f) ESCA analysis of zone B of electrode surface in (a) (Cour-
tesy F. L. Williams)

15. R. D. Heidenreich and C. J. Calbick, in W. G. Berl (ed), *Physical Methods of Chemical Analysis,* vol. I, Academic Press Inc., New York, 1960.

16. B. von Borries and G. A. Kausche, *Kolloid-Z., 90:* 132(1940).

17. A. Phillips, V. Kerlins, and B. Whiteson, *Electron Fractrography Handbook,* Technical Report, MLTDR-64-716, Air Force Materials Laboratory, Wright Patterson Air Force Base, Dayton, Ohio, January 31, 1965.

18. H. G. F. Wilsdorf and D. Kuhlman-Wilsdorf, *Zeit. Angewandte Physik, 4:* 361(1952).

19. *Ibid.* 418(1952).

20. J. A. Nankivell, *Brit. J. Appl. Phys. 4:* 141(1953).

21. E. D. Hyam and J. Nutting, *Brit. J. Appl. Phys., 3:* 173(1952).

22. A. L. Stuijts and H. E. Haanstra, *Third Inter. Conf. Reactions of Solids,* Madrid, 671(1957).

23. M. A. Wilkov, *Proc. ASTM, 60:* 540(1960).

24. M. A. Wilkov, and L. E. Murr, *Materials Research and Standards, ASTM, 4:* 285(1964).

25. R. D. McCammon and H. M. Rosenberg, *Proc. Roy. Soc. (London), A242:* 203(1957).

26. W. A. Wood, *Proceedings of the International Conference on Fatigue of Metals,* p. 531, Institute of Mechanical Engineers, London, 1956.

27. N. Thompson, N. J. Wadsworth, and N. Louat, *Phil.Mag., 1:* 113(1956).

28. R. L. Segall, P. G. Partridge, and P. B. Hirsch, *Phil. Mag. 6;* 1493 (1961).

29. C. E. Feltner, *Acta Met., 11:* 817(1963).

30. J. C. Grosskreutz, *J. Appl. Phys., 34:* 372(1963).

31. L. E. Murr and P. J. Smith, *Trans. AIME, 233:* 755(1965).

32. R. M. Fisher, *Symposium Techniques in Electron Metallography, ASTM Spec. Tech. Publ. 155:* 49,(1953).

33. W. C. Bigelow , J. A. Amy, and L. O. Brockway, *Proc. ASTM, 56:* 945 (1956).

34. E. Ruska and H. O. Müller, *Z. Physik, 116:* 366(1940).

35. B. von Borries, Z. *Physik, 116:* 370(1940).

36. L. E. Murr, master's thesis, Pennsylvania State University, University Park, Pa., 1964.

37 L. E. Murr and M.A. Wilkov, *J. Sci. Instr., 40:* 594(1963).

38. N. Osakabe, Y. Tanishiro, K. Yagi, and G. Honjo, *Surface Sci., 97:* 393(1980).

39. N. Osakabe, Y. Tanishiro, K. Yagi, and G. Honjo, *Surface Sci., 102:* 424(1981).

40. N. Osakabe, Y. Tanishiro, K. Yagi, and G. Honjo, *Surface Sci., 109:* 353(1981).

41. A. Tonomura, T. Matsuda, J. Endo, H. Todokoro, and T. Komoda, *J. Electron Microsc., 28:* 1(1979).

42. N. Osakabe, T. Matsuda, J. Endo, and A. Tonomura, *Japanese J. Appl. Phys., 27(9):* L1772(1988).

43. N. Osakabe, J. Endo, T. Matsuda, A. Tonomura, and A. Fukuhara, *Phys. Rev. Lett., 62(25):* 2969(1989).

44. G. Möllenstedt and H. Dücker, *Z. Phys., 145:* 375(1956).

45. M. Knoll, *Z. Tech. Physik, 16:* 467(1935).

46. M. von Ardenne, *Z. Physik, 109:* 553(1938).

47. V. K. Zworykin, J. Hillier, and R. L. Synder, *Bull. ASTM, 117:* 15 (1942).

48. D. McMullan, *Proc. Inst. Elec. Engr., 100:* 245(1953).

49. K. C. A. Smith and C. W. Oatley, *Brit. J. Appl. Phys., 6:* 391(1955).

50. C. W. Oatley, W. C. Nixon, and R. F. W. Pease, *Advan. Electron. Electron Phys., 21:* 181(1965).

51. E. Rudberg, *Phys. Rev., 45:* 764(1934).

52. R. Warnecke, *J. Phys. Radium, 7:* 270(1936).

53. J. J. Brophy, *Phys. Rev., 82:* 757(1951).

54. A. J. Dekker and A. van der Ziel, *Phys. Rev., 86:* 755(1952).

55. H. O. Müller, *Z. Physik, 104:* 475(1937).

56. H. Bruining, *Phillips Tech. Rev., 3:* 80(1938).

57. R. F. W. Pease and W. C. Nixon, *J. Sci. Instr., 42:* 81(1965).

58. A. V. Crewe, *Quarterly Review of Biophysics, 3:* 137(1970).

59. A. V. Crewe, J. Langmore, J. Wall, and M. Beer, *Proc. Electron Microscopy Soc. Ameri.,* C. J. Arceneaux (ed.), Claitor's Publishing Div., Baton Rouge, La., 1970, p. 250.

60. A. W. Crewe, *Science, 168:* 1338(1970).

61. T. E. Everhart and R. F. M. Thornley, *J. Sci. Instr., 37:* 246(1960).

62. J. M. Lafferty, *J. Appl. Phys., 22:* 299(1951).

63. A. N. Broers, *J. Sci. Instr. (J. Phys. E.), 2:* 273(1969).

64. L. W. Swanson and D. R. McNeely, *Surface Sci., 83:* 11(1979).

65. W. P. Dyke and J. K. Trolan, *Phys. Rev., 89:* 799(1953).

66. A. N. Broers and E. K. Brandis, *Scanning Electron Microscopy Symposium,* IIT Research Institute, Chicago, 1969, p. 15.

67. A. V. Crewe, *Scanning Electron Microscopy Symposium,* IIT Research Institute, Chicago, 1969, p. 11.

68. M. D. Coutts and E. R. Levin, *Proc. Electron Microscopy Society of America,* C. J. Arceneaux (ed.), Claitor's Publishing Division, Baton Rouge, La, 1972, p. 398.

69. N. M. Hodgkin and L. E. Murr, in *Microstructural Science, vol. 2,* G. P. Fritzke, J. H. Richardson, and J. L. McCall (eds.), Elsevier, New York, 1974, p. 129.

70. G. A. Separin, G. V. Spivak, and S. J. Stepanov, *Proc. Intern. Congr. Electron Microscopy.*

71. K. C. A. Smith and C. W. Oatley, *Brit. J. Appl. Phys., 6:* 177(1955).

72. P. B. Hirsch, A. Howie, and M. J. Whelan, *Phil. Mag., 7:* 2095(1962).

73. D. G. Coates, *Phil. Mag., 16:* 1179(1967).

74. G. R. Booker, A. M. B. Shaw, M. J. Whelan, and P. B. Hirsch, *Phil. Mag., 16:* 1185(1967).

75. G. Sasaki and M. J. Yokota, *Metallography, 8:* 265(1975).

76. J. C. Chesnutt and R. A. Spurling, *Met. Trans., 8A:* 216(1977).

77. I. Minkoff, *Acta Met., 14:* 551(1966).

78. R. F. W. Pease, A. N. Boers, and R. A. Ploc. *Proc. Fourth Regional Conf. Electron Microscopy (Prague),* 389(1964).

79. A. D. G. Stewart, in S. S. Breese (ed.), *Proc. Fifth Intern. Conf. Electron Microscopy (Philadelphia),* New York, D12(1962).

80. R. F. M. Thornley and L. Cartz, *J. Am. Ceram. Sco., 45:* 425(1962).

81. A. J. Majumdar, R. W. Nurse, S. Chatterji, and J. W. Jeffry, *Nature, 211:* 622(1966).

82. L. E. Murr, *Microelectronics Journal, 10:* 12(1979).

83. J. F. Moulder and A. Joshi, *Met. Trans., 12A:* 1140(1981).

84. L. Mayer, *J. Appl. Phys., 28:* 259(1957).

85. L. Mayer, *J. Appl. Phys., 31:* 346(1960).

86. L. Mayer, *J. Phys. Soc. Japan, 17 (Suppl. B-1):* 547(1960).

87. G. V. Spivak et al. *Soviet. Phys. Cryst., 4:* 115(1959).

88. L. Mayer, *J. Appl. Phys., 24:* 105(1953).

89. L. Mayer, R. Rickett, and H. Steneman, *Fifth Intern. Conf. Electron Microscopy, vol. I,* Academic Press, Inc., New York, 1962.

90. F. L. English, *J. Appl. Phys., 39:* 128, 3231(1968).

91. F. L. English and M. K. Parsons, in D. G. Thomas (ed.), *Proc. Intern. Conf. II-VI Semiconducting Compounds,* W. Benjamin, Inc., New York, 1967.

92. A. B. Bok, in S. Amelinckx, et al. (eds.), *Modern Diffraction and Imaging Techniques in Materials Science,* American Elsevier, New York, 1970, p. 655.

93. W. Steiger and F. G. Rüdenauer, *Anal. Chem., 51 (13):* 2107(1979).

94. D. M. Drummer and G. H. Morrison, *Anal. Chem., 52:* 2147(1980).

95. G. H. Morrison, in *Characterization of Metal and Polymer Surfaces, vol. 1,* Academic Press, N.Y., 1977, p. 351.

96. R. Castaing and G. Slodzian, *J. Microsc. 1:* 395(1962).

97. G. Slodzian, *Surface Sci., 48:* 161(1975).

98. G. H. Morrison and G. Slodzian, *Anal. Chem., 47(11):* 943(1975).

99. R. Castaing, *Mater. Sci. Engr., 42:* 13(1980).

100. B. K. Furman and G. H. Morrison, *Anal. Chem., 52(14):* 2305(1980).

101. K. Lam, T. R. Fox, and R. Levi-Setti in *Proc. 28th Intl. Field Emis. sion symposium,* L. Swanson and A. Bell (eds.), Oregon Graduate Center, Beaverton, Or, 1981, p. 59.

SUGGESTED SUPPLEMENTARY READING

Bethge, H. and J. Heydenreich (eds.): *Electron Microscopy in Solid State Physics,* Elsevier, New York, 1987.

Brown, L. M. (ed.): *Electron Microscopy and Analysis,* Taylor & Francis, Ltd. London, 1988.

Bruining, H., *Physics and Application of Secondary Electron Emission,* Pergamon Press, New York, 1954.

Crewe, A. V.: *Scanning Electron Microscopes: Is High Resolution Possible? Science,* vol. 154, pp. 729-738, 1966 (this article provides an interesting perspective).

Goldstein, J. I., et al.: *Scanning Electron Microscopy and X-ray Microanalysis,* Plenum Publishing Corp., New York, 1981.

Haine, M. E. and V. E. Cosslett: *The Electron Microscope,* Interscience Publishers, Inc., New York, 1961.

Hall, C. E.: *Introduction to Electron Microscopy,* McGraw-Hill Book Company, New York, 1966.

Heidenreich, R. D.: *Fundamentals of Transmission Electron Microscopy,* Interscience Publishers, Inc., New York, 1964.

IIT Research Institute: *Proceedings of the Annual Scanning Electron Microscope Symposia,* Chicago (1969 - 1982), O. Johari (ed.).

Joy, D. C., and D. L. Davidson: *Physical Scanning Electron Microscopy,* Academic Press, Inc., New York, 1982.

Kay, D. (ed.): *Techniques for Electron Microscopy,* Blackwell Scientific Publications, Ltd., Oxford, 1961.

Larsen, P. K. and P. J. Dobson (eds.): *Reflection High-Energy Electron Diffraction and Reflection Electron Imaging of Surfaces,* Plenum Publishing Corp., New York, 1988.

Lawes Scanning Electron Microscopy and X-ray Microanalysis (Analytical Chemistry and Open Learning Series), Wiley, New York, 1987.

McKay, K. G.: Secondary Electron Emission, *Advances in Electronics,* I, 66 (1948).

Oatley, C. W.: The Scanning Electron Microscope, *New Scientist,* p. 153, June 12, 1958; and *The Scanning Electron Microscope,* Cambridge University Press, England, 1972.

Thornton, P. R.: *The Scanning Electron Microscope, Sci. J.,* vol. 1, pp. 66-71, 1965. *See also: Scanning Electron Microscopy,* Chapman & Hall, Ltd., London, 1968.

Watt, I. M.: *Principles of Electron Microscopy,* Cambridge University Press, New York, 1989.

Zworykin, V. K. et al.: *Electron Optics and the Electron Microscope,* John Wiley & Sons, Inc., New York, 1954.

6
ELECTRON DIFFRACTION

6.1 INTRODUCTION

In Chap. 4 we considered the analysis of materials with regard to the interaction of an incident beam with the atomic constituents. Essentially
we considered the effect of the nuclear charge and energy as they influenced electron scattering, absorption, or the production of characteristic x-
rays unique to both these features. In Chap. 5 we developed the basis for
replica image contrast based on the density and thickness of the material,
which also forms the basis for image contrast for any noncrystalline,
structureless material observed by transmission electron microscopy.

   For the most part, materials of technical, industrial, or engineering
importance are crystalline to some degree, and the interaction of an incident or analyzing electron beam with such materials will be influenced measurably by this crystal structure, or some other regularity in the arrangement of the constituent atoms. We shall in fact consider a material to be
structureless only if it does not possess a crystal unit cell, since we
will observe in the treatment to follow that the unit cell defines the
characteristics of electron diffraction. In this chapter we will consider
the intrinsic properties of electron diffraction from solid matter as dis-

cussed initially in Chap. 1. Our treatment will consist in first establishing the basis for the structural mechanisms that characterize electron diffraction, so we can define the notation for describing the diffraction profile. We will then discuss the various applications of electron diffraction, and the interpretation of the various diffraction phenomena.

6.2 ELECTRON SCATTERING FROM ATOMS IN A CRYSTAL LATTICE

In the simplest approximation, we can consider diffraction from a crystalline lattice as a kinematic scattering process ideally described by wave reflection from {hkℓ} planes as expressed in the Bragg equation [Eq. (1.9)]. However, whereas x-ray diffraction can be described as photon reflection from a spherical nucleus, the scattering of an electron from a lattice atom must consider its potential distribution. We can also recall from Sec. 1.3.2 that the total electron wave front will be characterized by the phase relations between components. Consequently the characteristics of the diffraction wave front for electrons scattered by a crystal lattice will depend on the individual phase relationships, and the corresponding magnitudes of the components. We can illustrate this concept as follows: Let us initially consider the character of individual electron waves as they originate from various scattering centers (atoms) in a lattice to be depicted as rotating vectors of phase ϕ as depicted in Fig. 6.1. If their amplitudes are described by $f_j(\mu)$, the atomic scattering factor, then the diffracted wave amplitude A [the same A as in Eq. (1.20)], in the general case, is

$$A = \sqrt{X_j^2 + Y_j^2}$$

or

$$A = \sqrt{\left[\sum_j f_j(\mu)\cos\phi_j\right]^2 + \left[\sum_j f_j(\mu)\sin\phi_j\right]^2} \tag{6.1}$$

The diffraction intensity in the general case is then

$$I = |A|^2 = \left[\sum_j f_j(\mu)\cos\phi_j\right]^2 + \left[\sum_j f_j(\mu)\sin\phi_j\right]^2 \tag{6.2}$$

and since $|A|^2 = A^*A \equiv |\psi|^2 = \psi^*\psi$, we can write equivalently

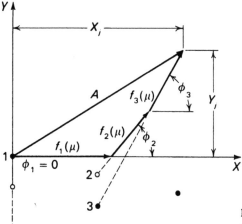

FIG. 6.1 *Fresnel construction for calculating wave amplitudes.*

$$A = \sum_j f_j(\mu) \, \exp(i\phi_j) \tag{6.3}$$

It is observed from Eq. (6.3) that maximum wave reinforcement occurs for $\exp(i\phi_j) = 1$, or where $\phi_j = 0$ or 2π; and that the amplitudes diminish for $0 \le \phi \le \pi$. Figure 6.2 illustrates this feature schematically for rotating wave vectors. In Fig. 6.2a the resultant wave phase and amplitude for the general case $0 < \phi < \pi$ is shown, while Fig. 6.2b and c illustrates the diffracted wave profile for maximum reinforcement ($\phi = 0, 2\pi$) and interference ($\phi = \pi$), respectively.

We have considered diffraction from a general space lattice, now we wish to extend this treatment to a crystal lattice array of regular point atoms. Let us therefore consider such a lattice to be located by a reference atom at 0 defining the origin of a unit cell (see Appendix A) and the position the jth atom of the unit cell by a vector r which we define by

$$\underline{r} = u_j\underline{a} + v_j\underline{b} + w_j\underline{c} \tag{6.4}$$

where $\underline{a}$, $\underline{b}$, and $\underline{c}$ are unit vectors defining the unit cell (see Appendix A).

Figure 6.3 illustrates this situation for the diffraction of plane-mono-chromatic electron waves of wavelength λ. The directions of the incident and diffracted waves can be described by unit vectors $\underline{K_0}$ and $\underline{K}$, respectively, while the phase difference between individual diffracted

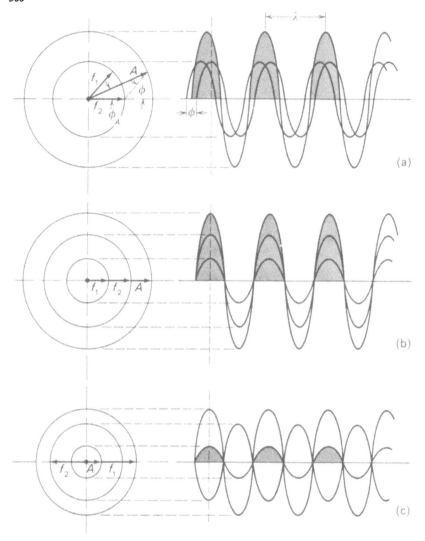

FIG. 6.2 *Diffracted wave (harmonic) phase-amplitude characteristics for constant wavelength* λ.

waves will depend on the relative positions of the jth atoms relative to 0 (at J). The incident and diffracted wave fronts in Fig. 6.3 can be described by $\overline{JM}$ and $\overline{JN}$, respectively, so that the phase difference is defined as the product of $2\pi/\lambda$ and the path difference Δ

$$\phi = \frac{2\pi}{\lambda}\ \Delta$$

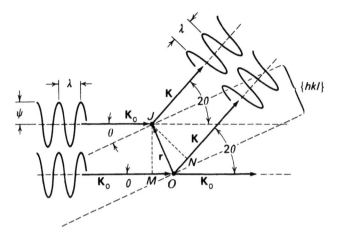

FIG. 6.3 *Phase relation for electron diffraction from a regular crystal lattice unit cell.*

where

$$\Delta = \underline{OM} + \underline{ON} = \underline{r} \cdot (\underline{K} - \underline{K}_0)$$

Thus

$$\phi = \frac{2\pi \underline{r}}{\lambda} \cdot (\underline{K} - \underline{K}_0) \qquad (6.5)$$

We also observe that if atoms O and J in Fig. 6.3 define the reflection
planes {hkℓ}, then the magnitude of r is $d_{hkℓ}$, the interplanar spacing.
Thus we can immediately write

$$\Delta = 2d_{hkℓ} \sin \theta$$

which is equivalent to Eq. (1.9) if $\Delta = \lambda$. We therefore observe that
$\Delta = \lambda$ stipulates the condition for diffraction of electrons, and this
necessitates the condition that $\phi = 2\pi$ as shown in Fig. 6.2b.

6.2.1 RECIPROCAL LATTICE CONVENTION

We will now define a special lattice construction, called the reciprocal
lattice, to aid us in the interpretation of electron diffraction phenomena.
In this lattice, the unit-cell dimensions are reciprocals of the real crystal-lattice unit cells as defined by

$$\underline{a}^* = \frac{1}{V_c} \; (\underline{b} \; \underline{\times} \; \underline{c})$$

$$\underline{b}^* = \frac{1}{V_c} \; (\underline{c} \; \underline{\times} \; \underline{a}) \tag{6.6}$$

$$\underline{c}^* = \frac{1}{V_c} \; (\underline{a} \; \underline{\times} \; \underline{b})$$

where $\underline{a}^*, \; \underline{b}^*, \; \underline{c}^*$ = reciprocal unit-cell vectors

$\underline{a}, \; \underline{b}, \; \underline{c}$ = real (crystal) unit-cell parameters

V_c = crystal unit-cell volume, that is:

$$V_c = \underline{a} \cdot (\underline{b} \; \underline{\times} \; \underline{c}) = \underline{b} \cdot (\underline{c} \; \underline{\times} \; \underline{a}) = \underline{c} \cdot (\underline{a} \; \underline{\times} \; \underline{b})$$

The unique features of such a reciprocal lattice are evident on considering the scalar and vector notation. That is, $\underline{a}^*$ in the reciprocal
lattice is normal to the plane in the real lattice described by $\underline{b}$ and $\underline{c}$;
and dimensions in the reciprocal lattice are fractions of those in the real
crystal. Thus, where (hkℓ) describes a plane in the real crystal, it now
describes a vector in reciprocal space (see Prob. 6.1). Consequently diffraction from a plane in a real crystal can be treated as a common vector
in the reciprocal lattice feeding its intensity into a point. Thus a diffraction intensity point in the reciprocal space corresponds to a plane
(hkℓ) in the real crystal.

We will observe that since

$$\underline{a}^* \cdot \underline{b} = \underline{a}^* \cdot \underline{c} = \underline{b}^* \cdot \underline{a} = \underline{b}^* \cdot \underline{c} = \underline{c}^* \cdot \underline{a} = \underline{c}^* \cdot \underline{b} = 0$$

and

$$\underline{a}^* \cdot \underline{a} = \underline{b}^* \cdot \underline{b} = \underline{c}^* \cdot \underline{c} = 1$$

uniquely characterize the reciprocal lattice as it relates to the real
crystal lattice, we can define a vector in the reciprocal lattice by

$$\underline{r}^* = \underline{g}_{hk\ell} = h\underline{a}^* + k\underline{b}^* + \ell\underline{c} \tag{6.7}$$

Thus dimensions in reciprocal space are reciprocals of those in the crystal, and we observe in Fig. 6.3 that

$$|\underline{g}_{hk\ell}| = \frac{1}{|\underline{r}|} = \frac{1}{d_{hk\ell}} \tag{6.8}$$

We can now express the diffraction condition for electrons as given in Eq. (6.5) by

$$(\underline{K} - \underline{K}_o) = \frac{\lambda}{\underline{r}} = \lambda g_{hk\ell} \tag{6.9}$$

This is essentially the vector form of the Laue [1] condition, which is equivalent to the Bragg reflection conditions as demonstrated previously.

6.2.2 GRAPHICAL REPRESENTATION OF THE DIFFRACTION CONDITION: THE EWALD SPHERE CONSTRUCTION

Equation (6.9) indicates that the condition for diffraction from an (hkℓ) plane is met if the reciprocal lattice vector $g_{hk\ell}$ extends from the origin at 0 (in Fig. 6.3) to reciprocal-lattice point hkℓ. That is, for any adjustment in the direction of $\underline{K}_o$ or the magnitude of λ, there will be a corresponding change in the diffraction condition as described by $g_{hk\ell}$. Let us consider that since Eq. (6.9) describes the diffraction conditions in reciprocal space, that the dimensions of the unit vectors $\underline{K}$ and $\underline{K}_o$ in this space are $1/\lambda$ as given in Eq. (6.9). If we therefore construct a radius vector $\underline{K}_o$ from an origin in reciprocal space of length $1/\lambda$, then any cord from the end of this vector opposite to 0 on the sphere of radius $1/\lambda$ swept out in the reciprocal lattice will describe a reciprocal-lattice point hkℓ for which diffraction will occur, that is, $|\underline{K}| = |\underline{K}_o| = 1/\lambda$ as observed in Eq. (6.9). We can of course arbitrarily multiply both sides of Eq. (6.9) by 2π so that $|\underline{K}| = |\underline{K}_o| = 2\pi/\lambda$. The choice here is a matter of simplicity.

In Fig. 6.4 we have sketched a crystal lattice and its corresponding reciprocal lattice projected in two dimensions. At the point 000 we describe the origin of the crystal lattice and the reciprocal lattice, and arbitrarily extend from this point a radius vector of length $1/\lambda$ depicting the direction of the incident electron wave. If we now describe the sphere (or its great circle projection in two dimensions) as the diffraction condition [Eq. (6.9)], then diffraction will occur for reciprocal-lattice points hkℓ intersected by the sphere so constructed. This is the Ewald sphere. We can observe that if the incident wave vector $\underline{K}_o$ is directed along a real crystal-lattice direction [HKL] (or [uvw]) normal to a plane (hkℓ) in the crystal, then the diffraction condition is satisfied in reciprocal space for intersections of the surface swept out by the radius vector $1/\lambda$ with the reciprocal-lattice points hkℓ, which are equivalent to reflecting planes {hkℓ} in the real crystal.

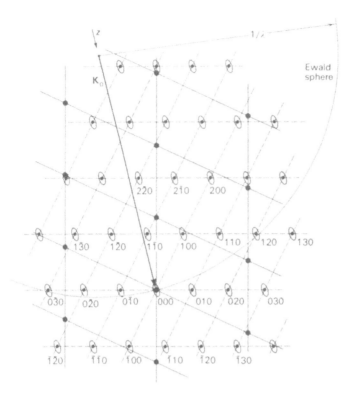

FIG. 6.4 *Ewald sphere in relation to real and recipro-*
cal lattice. Large solid circles joined by solid lines
represent atoms of real lattice. Reciprocal lattice is
represented by smaller solid circles joined by dashed
lines, with ellipses characterizing interference region
at reciprocal-lattice points. Note that sphere is con-
structed tangent to tip of $\underline{K}_O$ at origin in reciprocal
lattice.

It will be observed that the dimensions of a unit cell of real crystals on the order of 2 to 5 Å will reflect a reciprocal lattice unit cell having dimensions of 0.5 to 0.2 Å^{-1}. Consequently, for electron diffraction involving say 100 kV electrons in an electron microscope, the dimension of the Ewald sphere would be described by a radius vector of length 27 Å^{-1}. The Ewald sphere therefore represents essentially a plane section or "cut" of the reciprocal lattice, with deviations from this section appearing only for $|\underline{g}|$ large compared with the reciprocal unit-cell dimensions. Thus, with the incident electron beam normal to this reciprocal-lattice-plane section, and directed along a zone axis [HKL] of the real crystal, diffraction spots are expected in the electron diffraction pattern satisfying the condition

$$Hh + Kk + L\ell = 0 \tag{6.10}$$

If at this point we return to Sec. 1.7, we observe that the electron wave amplitude in a crystal described generally by Eq. (1.44) is essentially of the same form as Eq. (6.3). Thus, where energy discontinuities in a crystal unit cell are characterized by Brillouin zones as discussed in Sec. 1.7, that is, where

$$P \frac{\sin \alpha R}{\alpha R} + \cos \alpha R = \cos |\underset{\sim}{K}_0| R \tag{1.50}$$

the diffraction conditions geometrically expressed in the Ewald sphere of Fig. 6.4 also characterize the Brillouin-zone boundaries in the reciprocal lattice or $\underset{\sim}{K}$ space. In fact the incident electron wave vector $\underset{\sim}{K}_0$ as it describes the Ewald sphere in the reciprocal lattice (Fig. 6.4) terminates on a Brillouin zone. The first Brillouin zone, as depicted in Fig. 1.15 for the boundary conditions $|\underset{\sim}{K}_0| R = \pm \pi$, is constructed in the reciprocal lattice by bisecting the shortest reciprocal lattice vector, $g_{hk\ell}$, on the Ewald sphere by a radius vector that represents the edge of a plane of the Brillouin-zone boundary. The polyhedron generated by the intersections of these planes with the Ewald sphere thus forms the first Brillouin-zone boundary as depicted in Fig. 6.5. In this way the shapes of any order Brillouin-zone boundary can be constructed on considering the intersection of the Ewald sphere with plane sections defined by the reciprocal lattice points satisfying the diffraction conditions.

6.3 ATOMIC CRYSTAL STRUCTURE MODIFICATION OF ELECTRON DIFFRACTION AMPLITUDES: EXTINCTION CONDITIONS

The conditions for the diffraction of electrons, as stipulated in Eq. (6.9) and illustrated graphically in Fig. 6.4, indicate that diffraction can occur when these specific criteria are satisfied. However, the ultimate detection of the diffraction conditions as recorded in the electron diffraction pattern represent the modification or restrictions placed on the general conditions of Eq. (6.9) by the particular crystal structure, that is, the atomic identity influencing the internal potential distributions, and the geometrical arrangement of these atomic species.

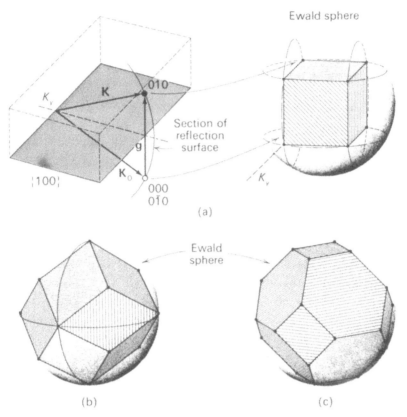

FIG. 6.5 *Construction of Brillouin zones by intersection of Ewald sphere with reciprocal lattice. (a) Generation of the first Brillouin zone for simple cubic crystal by intersection of Ewald sphere with equivalent {100} reflection planes. (b) Second Brillouin zone for simple cubic and equivalent first Brillouin zone for bcc crystals have zone boundaries within Ewald sphere characterized by {110} planes of the crystal. (c) Second bcc Brillouin-zone construction and equivalent first Brillouin zone for fcc crystals. Polyhedron is characterized in bcc and fcc crystals by equivalent {111} and {200} reflection planes.*

Let us now consider a crystal unit cell and denote the diffraction amplitude originating therein by the standard kinematic structure factor, that is,

$$A \equiv F(hk\ell) = \sum_j f_i(\mu) \exp(i\phi_j) \qquad (6.3a)$$

The phase angle ϕ can now be replaced by Eq. (6.5) so that

$$F(hk\ell) = \sum_j f_j(\mu) \exp\left[i \frac{2\mu}{\lambda} \underline{r} \cdot (\underline{K} - \underline{K}_o) \right] \qquad (6.11)$$

and from the diffraction condition specified in Eq. (6.9) we then obtain

$$F(hk\ell) = \sum_j f_j(\mu) \exp(2\pi i \underline{r} \cdot \underline{g}_{hk\ell}) \qquad (6.12)$$

Since $\underline{r}$ and $\underline{g}_{hk\ell}$ describe real and reciprocal lattice vectors as given by Eqs. (6.4) and (6.7), respectively, we can write

$$F(hk\ell) = \sum_j f_j(\mu) \exp[2\pi i(hu_j + kv_j + \ell w_j)] \qquad (6.13)$$

Equation (6.13) now expresses the simple kinematic diffraction amplitude for a unit cell with reflection planes characterized by $\{hk\ell\}$; u_j, v_j, w_j specify the coordinates of the jth atom of a unit cell.

The factor $f_j(\mu)$ is the atomic scattering factor for electrons diffracting from a nuclear potential distribution in the crystal unit cell identified by position coordinates u_j, v_j, w_j. For any atomic species (of atomic number Z) in a unit cell, the atomic scattering factor for electrons can be calculated from the Mott equation [2].

$$[f_{e\ell}(\mu)]_j \equiv f_j(\mu) = \frac{8\pi^2 m e^2}{h^2} \frac{Z - f_x(\mu)}{\mu^2} \qquad (6.14)$$

where $f_x(\mu)$ is the classical atomic scattering factor for x-rays as expressed by

$$f_x(\mu) = \int_0^\infty 4\pi r^2 \rho(r) \frac{\sin \mu r}{\mu r} \, dr \qquad (6.15)$$

and

$$\mu = \frac{4\pi \sin \theta}{\lambda} = 2\pi |\underline{g}| \qquad (6.15a)$$

Thus we can express the atomic scattering factor for electrons as

$$f_{e\ell}(\mu) = \frac{m e^2}{2h^2} \frac{Z - f_x(\mu)}{(\sin \theta/\lambda)^2} \qquad (6.14a)$$

Values of $f_{e\ell}(\mu)$ and $f_x(\mu)$ are tabulated as a function of $\sin \theta/\lambda$, and such tables for electrons are to be found in the books by R. D. Heidenreich [3] and B. K. Vainshtein [4] among many other sources. (Values of $f_{e\ell}(\mu)$ are tabulated in Appendix D.) We observe that while $f_{e\ell}(\mu)$ increases with atomic number Z, the magnitude $f_{e\ell}(\mu)$ also varies with the diffraction angle θ, that is, diffraction amplitudes will be directionally dependent.

6.3.1 GEOMETRICAL STRUCTURE FACTOR

If we now consider the unit cell to consist of atoms of the same species, we can write Eq. (6.13) as

$$F(hk\ell) = f_{e\ell}(\mu) \sum_j e^{2\pi i(hu_j + kv_j + \ell w_j)} \tag{6.16}$$

If we now define

$$\Gamma(hk\ell) = \sum_j e^{2\pi i(hu_j + kv_j + \ell w_j)} \tag{6.17}$$

as the geometrical structure factor, the unit-cell diffraction amplitude for a crystal species is simply

$$F(hk\ell) = f_{e\ell}(\mu)\Gamma(hk\ell) \tag{6.18}$$

and where $f_{e\ell}(\mu)$ is finite and constant for a particular value of μ [Eq. (6.15a)], the amplitude is governed by the geometrical arrangement of the atoms of the unit cell. We can readily observe that Eq. (6.17) is specific for any crystal structure, and therefore modifies the diffraction amplitude for reflections from {hkℓ} planes. We can illustrate this modification by considering crystals with a center of symmetry, for example, those of the cubic system (Appendix A). As an example, the simple cubic lattice can be uniquely specified by u,v,w = 0,0,0 (the center of symmetry or unit-cell origin). The atoms at the corners of this unit cell thus contribute one-eighth of their volume, and the geometrical structure factor for the simple cubic crystal becomes

$$\Gamma(hk\ell)_{\text{simple cubic}} = (1)e^{2\pi(0)} = 1$$

indicating that diffraction is unlimited for any hkℓ. Thus a diffraction pattern for a simple cubic crystal will contain projections of all reciprocal lattice points hkℓ (crystal planes hkℓ) satisfying Eq. (6.10).

In the case of a bcc crystal, the atoms are located by u,v,w = 0,0,0 (corner atoms) and u,v,w = ½,½,½ (body-centered atom). Thus, with 1 atom volume contributed by the corners of the unit cube, and 1 atom volume in the unit-cell center, Eq. (6.7) becomes

$$\Gamma(hk\ell) = 1 + e^{\pi i(h + k + \ell)} \qquad (6.19)$$
$$\text{bcc}$$

It is immediately obvious that

$$(h + k + \ell) \equiv n = 2,4,6, \ldots \text{ even} \rightarrow \Gamma(hk\ell) = 2$$

$$(h + k + \ell) \equiv n = 1,3,5, \ldots \text{ odd} \rightarrow \Gamma(hk\ell) = 0$$

since

$$e^{n\pi i} = \begin{cases} +1 & \text{if } n = \text{even integer} \\ -1 & \text{if } n = \text{odd integer} \end{cases}$$

This means of course that diffraction from the {hkℓ} planes of a bcc crystal will be extinct if (h + k + ℓ) is an odd integer since the intensity for such a reciprocal unit cell point will be zero, that is,

$$|F(hk\ell)|^2 = 0$$

The extinction conditions for an fcc crystal can also be calculated simply by defining the symmetrical array as 000, ½0½, 0½½, ½½0, and substituting for uvw in Eq. (6.17) to obtain

$$\Gamma(hk\ell) = 1 + e^{\pi i(h + \ell)} + e^{\pi i(k + \ell)} + e^{\pi i(h + k)} \qquad (6.20)$$
$$\text{fcc}$$

The extinction conditions for fcc are now observed for

$$\Gamma(hk\ell) = 0$$
$$\text{fcc}$$

to be a mixed condition for even and odd combinations of hkℓ, that is, (h + k), (k + ℓ), (h + ℓ).

In the case of the diamond cubic lattice, a special case of two interpenetrating fcc lattices, it can be shown that (see Prob. 6.2).

$$\Gamma(hk\ell) = 4[1 + e^{\pi i(h + k + \ell)/2}]$$
$$\text{diamond}$$
$$\text{cubic}$$

or

$$\Gamma(hk\ell) = \left[1 + e^{\pi i(h + k + \ell)/2}\right]\Gamma(hk\ell) \qquad (6.21)$$
diamond   fcc
cubic

Similar geometrical structure factors for other crystal unit cells (Bravais lattices) can be derived to show the corresponding extinction conditions. Allowed or expected reflections (hkℓ) in an electron diffraction pattern can then be listed in tabular form as shown in Table 6.1 for the fcc, bcc, diamond cubic, and hcp structures. The order of diffraction planes in Table 6.1 for the respective structures is shown according to increasing magnitude of the reciprocal-lattice vector $|\underset{\sim}{g}_{hk\ell}|$.

TABLE 6.1 *Allowed reflections {hkℓ} for bcc, fcc, diamond cubic, and hcp crystal structures having a single-effective atom species[†]*

bcc{hkℓ}	fcc{hkℓ}	diamond{hkℓ}	hcp {hkℓ}	hcp {hkiℓ}[‡]
110	111	111	100	1010
200	200	220	002	0002
211	220	311	101	1011
220	311	400	102	1012
310	222	331	110	1120
222	400	422	103	1013
321	331	511,333§	200	2020
400	420	440	112	1122
411,330§	422	531	201	2021
420	511,333§	620	004	0004
332	440	533	202	2022
422	531	444	104	1014
510,431§	600,442§	711,551§	203	2023
521	620	642	210	2130
440	533	553	211	2131

[†]*Magnitudes of the reciprocal-lattice vector $|\underset{\sim}{g}|$ for the cubic system are obtained simply from Eq. (6.8); $d_{hk\ell}$ for cubic and hexagonal close-packed structures is given by Eqs. A.1 and A.2, respectively.*
[‡]*Miller-Bravais notation (see Appendix A).*
[§]*Equivalent $|\underset{\sim}{g}|$.*

6.3.2 *DIFFRACTION AMPLITUDES FOR UNIT CELLS OF MIXED ATOMIC SPECIES*

Where the crystal unit cell is composed of a regular arrangement of more than one atomic species (that is, differing Z), the diffraction amplitude is not determined solely on the basis of a geometrical structure factor, but rather the complete form of the unit-cell structure factor must be

considered as given in Eq. (6.13). The atomic scattering factors are therefore considered separately in the individual summations of atoms defining
the unit cell. The result is that where the extinction conditions for
structures having identical atoms cause certain reflections to be absent,
these same structures may show weak or modified reflections where the unit
cell is characterized by more than one type (different Z) of atom. As an
example, let us consider the case of bcc CsCℓ. If we assume the Cℓ atom to
represent the body-centered atom, then the unit cell is characterized by Cs
at 000, and Cℓ at $\frac{1}{2}\frac{1}{2}\frac{1}{2}$. The diffraction amplitude from Eq. (6.13) is then
given by

$$F(hk\ell) = f_{Cs}(\mu) + f_{C\ell}(\mu)e^{\pi i(h + k + \ell)}$$

Thus where (hkℓ) is an odd integer for the CsCℓ lattice, the reflection
will not be extinct as shown in Eq. (6.19) for a simple bcc crystal. A
reflection will occur having an amplitude

$$F(hk\ell) = \left| f_{Cs}(\mu) - f_{C\ell}(\mu) \right|$$

We observe that for crystal structures having different atoms in regular arrangements in the unit cell, the reflections may be missing or weak
depending on the magnitudes of the respective atomic scattering factors.
These conditions must ultimately be considered in the analysis of crystal
structures by electron diffraction, and we shall take up this point again
at a later section of this chapter.

It should be mentioned at this point that structure factors and the
various extinction conditions for most of the known crystal structures and
unit-cell geometries have been tabulated. A complete treatment of these
features can be found in the "International Tables for X-ray Crystallography," originally published in three volumes by Kynock Press, Birmingham,
England, 1962; with subsequent updating and revision.

6.4 GEOMETRICAL NATURE OF ELECTRON DIFFRACTION PATTERNS

Since a crystal is a regular, repeating geometric arrangement of atoms, we
would expect that the resulting diffraction patterns would maintain a similar regularity, even if the diffraction zone is composed of random arrays
of such crystals. As we discussed above, the diffraction condition is met
when the Ewald sphere or sphere of reflection intersects a reciprocal-lat-

tice point corresponding to one tiny, oriented crystal of a polycrystalline aggregate.

It was previously shown that the reciprocal-lattice vectors are so much smaller than the corresponding reflecting sphere radius, that to a good approximation, the Ewald sphere intersects a plane section of the reciprocal lattice. Considering this feature, the geometrical relationship between the reciprocal lattice and the diffraction pattern recorded on the fluorescent screen of an electron microscope or a photographic plate can be established as in Fig. 6.6 through similar triangles. If R is the distance from the center spot 000 to some reciprocal-lattice point hkℓ, as measured from the photographic plane situated an effective distance L from the diffracting crystal or crystals normal to the direction of the incident beam vector $\underline{K}_o$, then we obtain

$$\frac{|\underline{g}_{hk\ell}|}{1/\lambda} = \frac{R}{L} \tag{6.22}$$

or we can write

$$R = \lambda L |\underline{g}_{hk\ell}| \tag{6.22a}$$

where $\underline{g}_{hk\ell}$ is normal to $\underline{K}_o$ and $\underline{s}$, the deviation from the reflecting sphere, is negligible. We observe from Eq. (6.22a) that the radii measured in the electron diffraction pattern are simply the magnitudes of the reciprocal-lattice vectors multiplied by a scale factor. We can, as shown in Fig. 6.6, think of the diffraction pattern simply as a projection of the reciprocal lattice points onto the recording plane (photographic plate). Thus we observe that the reciprocal-lattice point represents the concentration of the diffracted intensity profile, and it is really this profile that is intersected by the reflection sphere. As we shall see later, the intensities associated with the reciprocal-lattice points projected as the diffraction pattern are thus determined finally by the intersection of the intensity profile by the Ewald sphere. The intensity profile at the reciprocal-lattice point depends on the crystal shape, crystallography, and composition. We will treat the theory of diffraction intensity in Sec. 6.5.

By treating the reciprocal-lattice point as an interference region or spatial intensity profile, we can also visualize the concept of geometrical extinction as discussed in Sec. 6.3 for various crystal structures.

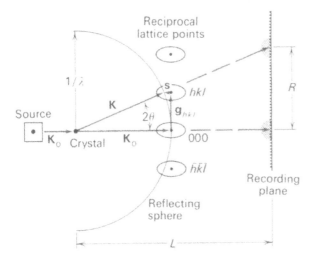

FIG. 6.6 *Reciprocal space intensity projection onto recording plane. Note exaggerated relationship of Ewald sphere to reciprocal-lattice points. In practice, reflection sphere is approximately a plane section as discussed in text.*

That is, where the diffraction intensity is zero (the extinction condition) as determined by the geometrical structure factor, no intensity occurs at the corresponding reciprocal-lattice point, and the reflection is not recorded in the diffraction pattern. Where the crystal is composed of more than one atomic species that independently contribute to the diffraction profile, the intensity at the corresponding reciprocal-lattice points will be proportional to the difference in the magnitudes of respective atomic scattering factors for conditions, that, in the case where the atoms of the unit cell were effectively identical, would demand an intensity extinction at the corresponding reciprocal-lattice point.

6.4.1 MODES OF ELECTRON DIFFRACTION

The modes of formation of electron diffraction patterns can be described by reflection-type diffraction of the electron beam incident on a specimen surface at some glancing angle, or by diffraction of an electron beam transmitted through a thin specimen. In addition to these physical schemes for diffraction of electrons, each mode may employ either low or high magnitudes of accelerating potential; however, transmission-type diffraction, which is the normal mode of selected-area electron diffraction in an elec-

tron microscope, usually requires accelerating potentials exceeding 50 kV
for the penetration of general types of thin-film specimens. Electron diffraction employing generally high-energy electron beams, either in an electron microscope or simple diffraction camera (of the order $V_0 \gtrsim 5$ kV) is
sometimes referred to as high-energy electron diffraction (HEED). High-
energy electron diffraction can be of either the reflection or transmission
mode; and normally produces a diffraction pattern representative of the bulk
of an object or material.

In the case of reflection HEED, the sampling of the object with respect to some distance from the surface can be approximately described again
by the Thomson-Whiddington law [eq. (4.10)]. Where the incident-glancing
angle is small, the depth of penetration will be small (as discussed in
Sec. 5.5.2 in connection with the efficiency of secondary electron emission). Consequently, crystallographic or related structural features associated with a surface condition or specific surface treatment (as, for
example, a reaction zone, surface deformation, or wear, etc.) are elucidated by the HEED reflection mode. Reflection diffraction is particularly
useful for examining surfaces previously observed by replication or scanning electron microscopy. We will explore this feature in describing a
combination of SEM, STM, and micro-scan reflection electron diffraction
(or μ - RHEED) in Section 6.10 at the conclusion of this chapter.

Low-energy electron diffraction (LEED) is primarily useful only in
the reflection mode since at accelerating potentials below 1 kV, the depth
of penetration of electrons is only several atomic layers. This feature,
however, represents the chief advantages of LEED, namely, the ability to
observe diffraction effectively from a single atomic layer composing a
solid surface. We shall treat LEED more extensively in Sec. 6.7, following
the development of diffraction intensity profiles and the character of the
interference region at the reciprocal-lattice points. In this later discussion, LEED patterns can be treated simply as a two-dimensional projection of the surface monolayer, which effectively serves as the reflection
plane. This occurs, as we shall see, by the extension of the interference
region at the reciprocal-lattice points into continuous spikes normal to
the atomic surface monolayer.

6.4.2 *TYPES OF ELECTRON DIFFRACTION PATTERNS*

The type or character of the electron diffraction pattern is of course related to the diffraction mode, but it depends for the most part on the

character of the specimen, that is, on its composition, structure, dimensions, orientation, and the distribution of the various components or polycrystals in the diffraction zone. Basically, electron diffraction patterns are distinguishable as spot patterns resulting from single-crystal diffraction zones or slightly misoriented groups of crystals having a common orientation, ring patterns obtained from randomly oriented crystal aggregates, and various degrees of ring and spot intensities giving rise to texture patterns that indicate the degree of misorientation α of crystal aggregates. We can of course deduce these basic intensity characteristics from our previous treatment of the reciprocal lattice and the diffraction conditions stipulated by the Ewald-sphere construction. That is, for a single crystal, the reciprocal lattice is a three-dimensional array of points, each point characterized by a finite interference region or intensity profile. For a random array of crystal aggregates, effective reciprocal space becomes an interpenetrating array of randomly disposed reciprocal lattices each corresponding to one of the aggregate crystals. The resulting intersections of such a reciprocal-lattice mixture is therefore a continuous arrangement of intensity rings where the Bragg condition is met, and the reflections are allowed by the corresponding structure factor.

Figure 6.7 illustrates the geometry of formation of electron diffraction patterns as it relates to the specimen character depicted. We should caution that in Fig. 6.7 we are considering essentially ideal situations, which we shall later treat as a kinematical approximation when specifically treating the diffraction intensities. Figure 6.7 might generally be considered to represent the high-energy transmission mode of diffraction, such as that which normally occurs in an electron microscope where the specimen sections depicted would be thin foils. The electron diffraction patterns included in Fig. 6.7 are transmission-type patterns as indicated.

We observe that the ring-type diffraction patterns originate as diffraction cones by the intersection of the reflection sphere with the continuous array of reciprocal-lattice points, which can be regarded as concentric spheres in reciprocal space. We can therefore depict the diffraction from polycrystalline aggregates as a successive array of reflection circles on the photographic plane, as depicted in Fig. 6.8a for the transmission mode. Figure 6.8b illustrates the formation of semi-ring patterns for the reflection mode where the specimen edge eliminates a portion of the diffraction cones. Figure 6.9 illustrates typical examples of spot and ring-type reflection electron diffraction patterns obtained in the electron microscope; the character of the patterns as they relate to the geometrical

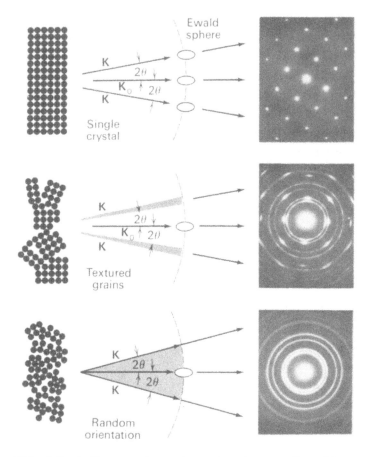

FIG. 6.7 *Influence of specimen structure on formation
of diffraction patterns of Au films.*

arrangement of Fig. 6.8b is apparent. The diffuseness and variations in
intensity profiles result from inelastic scattering and refraction.

Figure 6.10 illustrates transmission-type diffraction patterns for a
number of crystal structures that result from evaporated thin foils of the
materials indicated. The occurrence of the various ring geometries and, to
a first approximation, their respective intensities are determined solely
by the geometrical structure factor $\Gamma(hk\ell)$ discussed previously. As we
observed in Sec. 6.3.1, there are no extinction conditions on $(hk\ell)$ for
simple cubic materials of a common atomic species. Consequently, the electron diffraction pattern for polycrystalline-simple cubic Mn (Fig. 6.10a)
contains reflections for all $\{hk\ell\}$ planes, with the ring radii increasing
with increasing values of $|\underline{g}_{hk\ell}|$ as previously indicated in Eq. (6.22a).

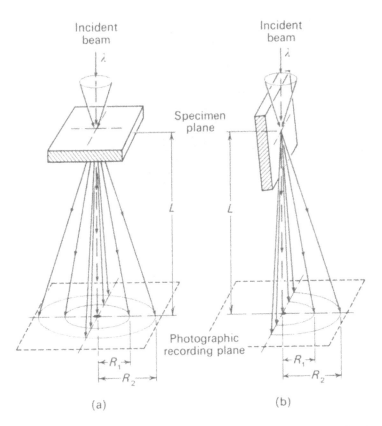

FIG. 6.8 *Simple geometry of formation of (a) transmission and (b) reflection electron diffraction patterns. Ring-pattern formation is specifically depicted while spot patterns result for identical geometries.*

The character of reflections for the bcc, fcc, and hcp extinction conditions can also be observed in Fig. 6.10b, c, and d, respectively, with increasing ring radii corresponding to those allowed reflections listed in Table 6.1.

6.4.3 CALIBRATION OF ELECTRON DIFFRACTION PATTERNS; MEASUREMENT OF λL

We can readily observe that the diffraction geometry indicated in Fig. 6.8 is identical to that shown previously in Fig. 6.6, and we can recall from Eq. (6.22a) that the radii of diffraction rings or spots are proportional to the magnitude of the reciprocal-lattice vector. The constant of proportionality, or camera constant, λL, is determined simply by considering a diffraction ring pattern such as those shown in Fig. 6.10, that is,

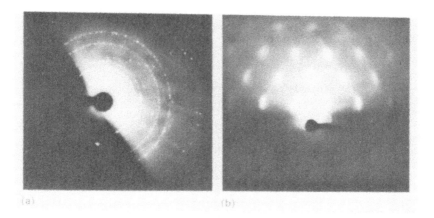

FIG. 6.9 *Reflection electron diffraction patterns. (a) Ring
pattern from Si sputtered onto Si single-crystal substrate;
(b) spot pattern from SnO_2 grown epitaxially on TiO_2 substrate
Orientation is (110). (Courtesy Donald L. Gibbon)*

$$\lambda L = \frac{R}{|g_{hk\ell}|} = Rd_{hk\ell} \qquad (6.23)$$

where $d_{hk\ell}$ is the interplanar spacing for any particular reflecting planes
$\{hk\ell\}$ and R is the particular diffraction ring radius. Equations for $d_{hk\ell}$
for various crystal structures can be found in Appendix A.

Let us consider as an example a cubic crystal. Here we observe that

$$d_{hk\ell} = \frac{a}{\sqrt{h^2 + k^2 + \ell^2}} \qquad (6.24)$$

as a special form of Eq. (A.1), where a is the lattice parameter. We can
now write Eq. (6.22a) in the form

$$\sqrt{h^2 + k^2 + \ell^2} = \frac{a}{\lambda L} R \qquad (6.25)$$

and we observe that by simply plotting this equation of the form y = mx,
with $y = \sqrt{h^2 + k^2 + \ell^2}$, m = a/$\lambda$L, R = x, as shown in Fig. 6.11, the slope
m = a/λL readily determines the camera constant for any known material. We
observe that in Fig. 6.11 the intersection of corresponding values of
$\sqrt{h^2 + k^2 + \ell^2}$ and the ring radii R for the allowed (hkℓ) values depends,
in the particular case of the cubic system, on whether the material is bcc,
fcc, etc. Since the diffraction pattern in Fig. 6.11 is actually that of
an (fcc) Ni thin foil, the values of $\sqrt{h^2 + k^2 + \ell^2}$ as plotted are there-

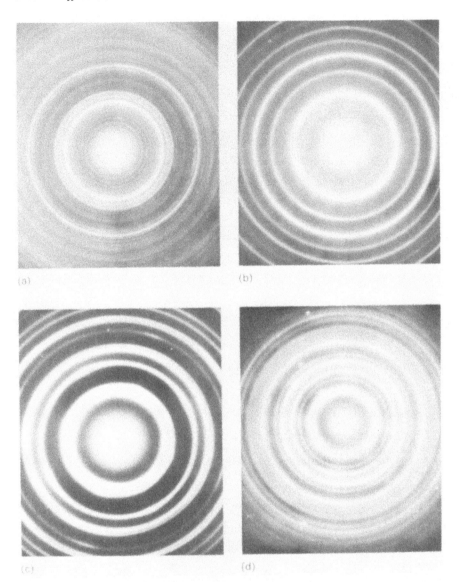

FIG. 6.10 *Transmission-type ring diffraction patterns from polycrystalline thin films (a) β-Mn, simple cubic; (b) Fe, bcc; (c) Rh, fcc, slightly textured; (d) Er, hcp. Extra spots in d originate from neighboring crystals having different composition and structure from Er matrix.*

fore readily computed either by considering the geometrical structure factor of Eq. (6.21) or the allowed {hkℓ} values listed in Table 6.1 and assigned to the rings as shown in Fig. 6.11

We can observe that a similar straight-line relation obtains for any

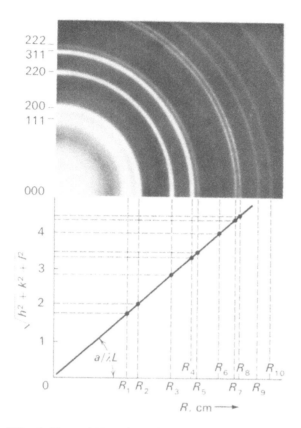

FIG. 6.11 *Calibration of electron diffrac-*
tion patterns. A straight line occurs when
values of $\sqrt{h^2 + k^2 + \ell^2}$ intersect corres-
ponding diffraction ring radii for fcc Ni;
slope yields the camera constant λL.

crystal system simply by considering Eq. (6.23) in the form

$$\frac{1}{d_{hk\ell}} = \frac{1}{\lambda L} R \equiv y = mx \qquad\qquad (6.23a)$$

where $1/d_{hk\ell} = y$, $m = 1/\lambda L$, and $R = x$. The interplanar spacing $d_{hk\ell}$ is
then computed for the particular crystal system, considering the corresponding extinction distances and lattice parameters (see Appendix A for
crystallographic conventions, etc.).

The geometrical nature of the electron diffraction camera constant as
it applies to transmission-type diffraction patterns also applies to reflection electron diffraction. In effect the calibration procedures described apply to both situations depicted schematically in Fig. 6.8.

Having obtained λL for a fixed operating condition, the ring or diffraction spot radii of any other material having an identifiable structure can be related directly to a particular reciprocal-lattice vector; or knowing the reciprocal-lattice vectors, that is, the corresponding reflections {hkℓ}, the lattice parameter can be measured if this is unknown. Expected values of diffraction radii can thus be calculated for any known material if λL remains fixed. This feature is particularly useful for indexing electron diffraction spot patterns, and related crystallographic analysis (discussed later in Sec. 6.8.5). Such calculation are easily arranged in computer programs, and Appendix C illustrates several simple Fortran programs for tabulating expected diffraction radii for fcc, bcc, and hcp materials in which the camera constant is introduced as a single data entry. As the value of λL changes, the data can be recalculated. As noted in Appendix C, similar programs for other crystal structures (for example, tetragonal orthorhombic, etc.) are readily devised simply by inserting the appropriate function (equation) for the interplanar spacing and adjusting the index library for the sequence of allowed reflections {hkℓ}. Alternatively these can be calculated by including appropriate structure factor routines.

The value of λL will of course change with any variation in the accelerating potential, or of the position of the specimen. The variation of λ with accelerating potential has already been presented as Eq. (1.7), and we should realize of course that in the electron microscope, the lens system (intermediate and projector lenses) focus and expand the simple geometry illustrated for a linear diffraction camera in Fig. 6.8. Consequently the value of L measured from diffraction patterns in the electron microscope is really an effective L, or the value of L corresponding to a linear system of Fig. 6.8. So long as the specimen remains fixed in an electron microscope, L will remain fixed; however, if the position of the specimen with respect to the objective pole piece changes, as is frequently the case in conventional instruments employing the various heating, cooling, tilting, etc., stages, the necessity to refocus for these changes will require adjustments in the intermediate-and/or projector-lens currents. In this situation the effective L changes. It is therefore good policy to maintain the projector current of the electron microscope fixed, varying only the objective for object focus and the intermediate-lens current for the subsequent focus of the diffraction pattern, in order to establish some condition of reference. In many contemporary, commercial electron micros-

copes, these features are automatically adjusted by electronic switching
arrangements that establish fixed (reference) lens currents. Figure 6.12
compares the ray paths for conventional electron transmission observations
of an object in the electron microscope, and the observation of the corresponding electron diffraction pattern.

If the effective length from specimen to photographic plane (or plane
of observation, that is, the fluorescent screen) remains fixed, the measurement of ring radii in electron diffraction patterns of a known material
can also serve as an accurate check on the accelerating potential. This
feature is of particular value in electron microscopes operated at very
high voltages, where it is not possible to measure the accelerating voltage
directly. Such a voltage calibration simply involves the recording of a
diffraction pattern of a reference material at a low-measurable accelerating potential in order to calibrate λL. Maintaining L constant as the

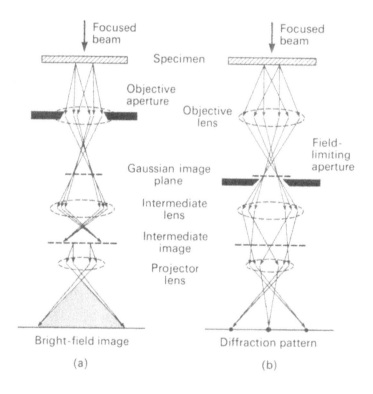

FIG. 6.12 *Comparison of electron ray paths in electron
microscope for (a) conventional electron transmission
microscopy and, (b) selected-area electron diffraction.*

accelerating potential is changed will then allow the magnitude of V_0 to be obtained from Eq. (1.7) by measuring the resulting calibration slope as shown in Fig. 6.11.

6.5 DIFFRACTION INTENSITIES

We have observed in our treatment of electron diffraction thus far that the normal, selected-area diffraction pattern of a structure represents simply the intersection of the reflection sphere (Ewald sphere), becoming effectively a plane section, with the reciprocal-lattice points or with the corresponding reciprocal-structure geometry. The diffraction pattern therefore depicts the image of a portion of the reciprocal lattice as projected normal to the plane section defined by the surface of the Ewald sphere. The scattering amplitude and corresponding intensity associated with the location of intersection of the Ewald sphere with the reciprocal lattice is therefore the reciprocal-space image corresponding to the potential function of the diffraction medium (the crystal specimen).

Since a crystal represents a three-dimensional periodic distribution of scattering material, that is, the atomic array, the scattering process itself can be treated in terms of this characteristic periodicity. Thus, where we have observed in Sec. 1.7 that a one-dimensional solid section can be approximated as a periodic array of rectangular potential wells, the three-dimensional structure can of course be visualized as an extension of this periodicity in the x and y directions in a cartesian space coordinate system. We find in fact the amplitude of scattering over the volume V_r of a specimen to be given by

$$f_{e\ell}(\mu) = \frac{me}{2\pi h^2} \int \phi(\underline{r}) e^{2\pi i \underline{r} \cdot \underline{g}} \, dV_r \tag{6.26}$$

where $\phi(\underline{r})$ is the potential, and $(\underline{r})$ is the vector in the scattering volume as previously treated. Equation (6.26) is essentially of the same form as Eq. (6.15) for x-ray scattering, with the exception that $\phi(\underline{r})$, the atomic potential, now replaces the density factor $\rho(\underline{r})$. We can equivalently write in spherical coordinates

$$f_{e\ell}(\mu) = \frac{2me}{h^2} \int_0^\infty \phi(r) r^2 \frac{\sin \mu r}{\mu r} \, dr \tag{6.26a}$$

where $\phi(r)$ indicates the scalar dependence of potential with r, and μ is simply $2\pi|\underline{g}|$ as defined in Eq. (6.15a).

If we now treat electron diffraction as scattering from a unit cell as previously formulated in Sec. 6.3, we can describe the potential distribution by the superposition of all the component potentials (those associated with the individual atoms). Thus for $\underline{r}_j$, the vector of the center of the jth atom, we can write

$$\phi(\underline{r}) = \sum_j \phi_j(\underline{r} - \underline{r}_j) \tag{6.27}$$

where the complete potential over the whole atomic volume is given by

$$f_{e\ell} \equiv U(\underline{r}) = \int \phi(\underline{r})dV_r = 4\pi \int_0^\infty \phi(r)r^2 dV \tag{6.28}$$

The scattering amplitude for the whole unit cell is then obtained by substituting Eq. (6.27) into Eq. (6.26):

$$F(hk\ell) = \int \sum_j \phi_j(\underline{r} - \underline{r}_j)e^{2\pi i\underline{r} \cdot \underline{g}}dV_r$$

$$= \sum_j \int \phi_j(\underline{r} - \underline{r}_j)e^{2\pi i(\underline{r} - \underline{r}_j) \cdot \underline{g}}dV_r e^{2\pi i\underline{r}_j \cdot \underline{g}}$$

$$= \sum_j f_j(\mu)e^{2\pi i\underline{r} \cdot \underline{g}} \tag{6.12}$$

What we observe is that the scattering amplitude is spherically symmetrical in reciprocal space and depends only on the magnitude $2\pi|\underline{g}|$. We also recognize that Eq. (6.12) is the exponential form of the Fourier series of the general form

$$F = \sum_{n=-\infty}^{\infty} C_n e^{in\phi}$$

where ϕ is the phase angle of the periodic function, that is, $\phi = 2\pi\underline{r} \cdot \underline{g}$, and C_n is the Fourier coefficient, which in the particular case of electron diffraction assumes the identity of the scattering factor or potential. Using the Fourier transform notation we can then write

$$F[\phi(r)] = f_{e\ell}(\mu) \tag{6.26b}$$

and

$$F^{-1}[f_{e\ell}(\mu)] = \phi(\underline{r}) \tag{6.26c}$$

so that in effect Eq. (6.26) has the property of inversion, that is,

$$\phi(\underline{r}) = \frac{\hbar^2}{4\pi^2 me} \int f_{e\ell}(\mu) e^{-2\pi i \underline{r} \cdot \underline{g}} dV_{|\underline{g}|} \tag{6.29}$$

and the combination is usually referred to as a Fourier transform pair in three dimensions. The direct and inverse transformations just discussed are in fact a general property of transforms of the Fourier, Laplace, etc. type. These concepts are discussed routinely in texts dealing with Fourier series or mathematical transform theory. It is recommended that the reader unfamiliar with these concepts consult a reference such as R. Courant and D. Hilbert, *Methods of Mathematical Physics*, Interscience Publishers, Inc., New York, 1953, vol. 2, 1957.

We can show that Eq. (6.12) as derived above is an approximate solution of the time-independent form of the Schrödinger equation (1.16); and therefore F(hkℓ) can be shown to represent the diffracted wave amplitude $\psi(xyz)$; that is, the diffracted wave amplitude can be shown to be identical to the three-dimensional Fourier integral over the scattering power of a specimen. We begin by considering

$$\nabla^2 \psi(xyz) + \frac{2m}{\hbar^2} E\psi(xyz) = 0 \tag{1.42}$$

descriptive of wave motion of the incident electron wave outside the specimen, which, having a plane-monochromatic character, has a solution

$$\psi_o(\underline{r}) = Ae^{2\pi i \underline{K}_o \cdot \underline{r}} \tag{6.30}$$

where

$$|\underline{K}_o| = \frac{1}{\lambda} = \frac{1}{h} \sqrt{2m_o eV_o + \left(\frac{eV_o}{c}\right)^2} \simeq \frac{1}{h} \sqrt{2m_o eV_o}$$

from Eq. (1.7). We then observed that Eq. (1.42) can be written

$$\nabla^2 \psi_o(\underline{r}) + |\underline{K}_o|^2 \psi_o(\underline{r}) = 0$$

which describes the initial wave.

When the electron wave enters the solid (specimen), the wave function must be thought of as composed of the initial wave function and the scattered wave function, that is, we get

$$\psi(\underline{r}) = \psi_0(\underline{r}) + \psi_S(\underline{r}) \tag{6.32}$$

Substituting Eq. (6.32) into Eq. (1.16), with the potential U = U(\underline{r}) described by Eq. (6.28), then results in

$$\nabla^2 \left\{ \psi_0(\underline{r}) + \psi_S(\underline{r}) + \frac{2m}{\hbar^2} [E - U(\underline{r})][\psi_0(\underline{r}) + \psi_S(\underline{r})] \right\} = 0$$

or

$$\nabla^2 \psi_S(\underline{r}) + 4\pi^2 |\underline{K}_0| \psi_S(\underline{r}) = U(\underline{r})[\psi_0(\underline{r}) + \psi_S(\underline{r})] \tag{6.33}$$

This treatment is essentially the approach used originally by H. Bethe, *Ann. Physik, 87:* 55(1928).

6.5.1 KINEMATIC DIFFRACTION

Equation (6.33) has a classic solution, as given by Mott and Massey [5], of the form

$$\psi_S(\underline{r}) = - \frac{1}{4\pi} \int \frac{U(\underline{r})}{|\underline{r} - \underline{r}_j|} [\psi_0(\underline{r}_j) + \psi_S(\underline{r}_j)] e^{2\pi i |\underline{K}|} \cdot |\underline{r} - \underline{r}_j| dV_{r_j}$$

$$\tag{6.34}$$

where $\underline{r}_j$ represents a vector within the scattering volume. We now invoke the basic kinematic assumptions, namely, that the diffracted amplitude is very small and that there is no interaction between incident and scattered waves. This means of course that if we let

$$\psi_S(\underline{r}) \ll \psi_0(\underline{r})$$

then U(r) becomes negligible. Thus for $\underline{r} > \underline{r}_j$ (we stipulate that we are considering the diffracted wave at some distance from the object, a distance that is much greater of course than the dimensions in the atomic volumes) Eq. (6.34) assumes the character of a spherical wave with an amplitude proportional to A of Eq. (6.30) and the Fourier integral of Eq. (6.26) (see Prob. 6.6).

The intensity of the diffracted signal can be related (Sec. 1.3.1) to beam-current flux or beam current per unit area as given generally by

$$J = \frac{\hbar}{2mi} (\psi^* \nabla \psi - \psi \nabla \psi^*) \tag{6.35}$$

where from Eq. (6.30) we obtain

$$J = \frac{\hbar |\underset{\sim}{K}_o|}{m} |A^2| \tag{6.36}$$

for a plane wave propagating in the z direction. Thus we observe that the intensity of the diffracted beam scattered in the direction K at some Bragg angle θ to the incident beam can be expressed simply as

$$\frac{J(\mu)}{J_o} = \frac{|\psi_s|^2}{|\psi_o|^2}$$

so that the flow of electrons in the spherically diffracted beam [Eq. (6.34)] is expressed by

$$J(\mu) = \frac{\hbar |\underset{\sim}{K}_o|}{mr^2} |A|^2 |F(hk\ell)|^2 = \frac{J_o}{r^2} |F(hk\ell)|^2 \tag{6.37}$$

or we can write

$$\frac{J(\mu)}{J_o} = \frac{|F(hk\ell)|^2}{r^2} \tag{6.37a}$$

where r is a radial dimension from the diffracting cell. It must be emphasized that we have dealt above with the unit cell. Equation (6.37a) is therefore the intensity ratio associated with a single unit cell of a specimen material.

Kinematical crystal shape transforms and the intensity profile at reciproca-lattice points Let us now recapitulate in a sense just what it is we are trying to do in this section. First we should be fully aware of the fact that in dealing kinematically with the scattering of electrons from a crystal unit cell, the ensuing diffraction intensity must be imagined to be concentrated, as it were, in the reciprocal-lattice points. This feature is a necessary condition that simply results as a matter of definition of the reciprocal lattice. It is now necessary to expand this treatment to include finite crystal dimensions, that is, to combine the scattering from many unit cells, and to examine the intensity profile that we can mathematically demonstrate to exist in the corresponding reciprocal space. You should be aware of two important features: (1) Since the intensity of diffraction that we stipulate as occurring at reciprocal-lattice points is probabilistic in nature, that is, $|\psi|^2$, we must realize at the outset that the intensity will have a finite distribution at the reciprocal-lattice point. That is, the reciprocal-lattice point will represent an interference region. (2) This interference region will have a definite shape determined solely by the dimensions of the scattering domain.

To demonstrate this let us consider a function $G(\underline{r})$ to be zero outside a crystal and unity inside. If we multiply the periodic potential distribution of an infinite lattice by this function, then $G(\underline{r})$ will define a finite domain having a specific potential distribution. This is essentially the same as placing restrictions on the extent of the periodic array of Fig. 1.12 in three dimensions. Thus the density of scattering matter is simply unity inside the bounds established by $G(\underline{r})$ and zero outside. The crystal scattering amplitude then becomes

$$F[G(\underline{r})] = \int G(\underline{r})e^{2\pi i \underline{r} \cdot \underline{g}}dV_r$$

or

$$F[G(\underline{r})] = S(\underline{g}) = \frac{1}{V_c} \int_{G(\underline{r})=1} e^{2\pi i \underline{r} \cdot \underline{g}}dV_r \qquad (6.38)$$

which is a nonperiodic function appropriately called the crystal shape transform. Since the Fourier transform for an atom is given effectively by Eq. (6.26b), a finite scattering domain described by $G(\underline{r})\phi(\underline{r})$ has a diffraction amplitude described by

$$G[(hk\ell)]_{crystal} = F[G(\underline{r})\phi(\underline{r})] = SF(hk\ell) \qquad (6.39)$$

which in mathematical terms is the convolution of the shape transform and the infinite crystal. The concept of convolution is simply this: if we define two functions, say $\psi_1(|\underline{r}|)$ and $\psi_2(|\underline{r}|)$, to be either periodic or nonperiodic, then an integral of the type

$$g(|\underline{r}|) = F^{-1}F_1F_2 = \int_0^{|r|} \psi_1(|\underline{r}| - \zeta)\psi_2(\zeta)d\zeta \qquad (6.40)$$

where

$$F^{-1}F_1 = \psi_1(|\underline{r}|) \text{ and } F^{-1}F_2 = \psi_2(|\underline{r}|)$$

is the convolution integral. More extensive treatment and proof are to be found in J. Sylie, *Advanced Engineering Mathematics,* p. 188, McGraw-Hill Book Company, New York, 1951 and S. Salvadori and L. Schwartz, *Differential Equations in Engineering Problems,* p. 214, Prentice-Hall, Inc., Englewood Cliffs, N.J., 1954. The property of this integral of particular interest here is that a convolution combines the properties of one function with another in such a way that one is distributed according to the restrictions of the other. This means of course, in the specific treatment of diffraction amplitudes, that the shapes of the interference regions at the reciprocal-lattice points will be described by $S(\underline{g})$ for all points of the reciprocal lattice. The diffraction amplitude corresponding to any point in the reciprocal lattice is then given by

$$S(\underline{g})F(hk\ell)$$

or we can write

$$F^{-1}[S(\underline{g})] = G(\underline{r}) \tag{6.41}$$

Equation (6.41) is simply the inverse of Eq. (6.38), just as Eq. (6.26c) is the inverse of Eq. (6.26b). The significance of these features, particularly Eqs. (6.28) and (6.41), is that while $S(\underline{g})$ refers to the reciprocal-lattice point, $G(\underline{r})$ relates to the real crystal or scattering domain. Thus the size and shape of the scattering domain will be directly related to the intensity profile of the diffraction pattern or specific portions thereof. This means that the character of the scattering domain can be deduced from the diffraction pattern, that is, the specimen structure is manifested in the character of the diffraction pattern.

The intensity distribution in the diffraction pattern for a finite scattering domain of many unit cell dimensions can now be treated by modifying Eq. (6.37a) by the magnitude of the corresponding shape transform:

$$J_g = \frac{J_o}{r^2} \left| F[(hk\ell)]_{crystal} \right|^2$$

or

$$\frac{J_g}{J_o} = \frac{1}{r^2} \left[|F(hk\ell)|^2 |S|^2 \right] \tag{6.42}$$

Diffraction from a general parallelepiped Since the shape of the intensity profile (the interference region) at the reciprocal-lattice points (and the diffraction spots) is determined solely by the shape transform $S(g)$ as defined in Eq. (6.38), we need only evaluate $S(g)$ for any particular scattering domain shape (crystal) in order to determine the corresponding intensity distribution. We will consider the general case of a parallelepiped of dimensions t_1, t_2, t_3 along the corresponding xyz directions (uvw of the unit cell). If g_1, g_2, g_3 represent the components of g at the reciprocal-lattice points (corresponding accordingly to a cartesian coordinate system at the reciprocal-lattice point of x,y,z, that is, $\underline{s}$ in Fig. 6.6) then by establishing an origin at the geometrical center of the parallelepiped, we can expand Eq. (6.38) as

$$S(g_1 g_2 g_3) = \frac{1}{abc} \int_{-t_1/2}^{t_1/2} \int_{-t_2/2}^{t_2/2} \int_{-t_3/2}^{t_3/2} e^{2\pi i (g_1 x + g_2 y + g_3 z)} dx\,dy\,dz \qquad (6.38a)$$

which on integration yields

$$S(g_1 g_2 g_3) = \frac{1}{abc} \frac{\sin \pi t_1 g_1}{\pi g_1} \frac{\sin \pi t_2 g_2}{\pi g_2} \frac{\sin \pi t_3 g_3}{\pi g_3} \qquad (6.43)$$

where a,b,c are the unit-cell dimensions of the specimen material corresponding to the cartesian convention x,y,z, respectively. If we let $a_1, a_2,$ a_3 denote the unit-cell edge lengths a,b,c, respectively, then we can express the intensity distribution in the interference region at a reciprocal-lattice point by

$$|S(g_1 g_2 g_3)|^2 = \prod_{j=1}^{3} \frac{\sin^2 \pi t_j g_j}{(\pi a_j g_j)^2} \qquad (6.44)$$

where $t_j = N_j a_j$

N_j = number of unit cells along domain edge j

g_j = magnitude of deviation vector $\underline{s}$ of Fig. 6.6

The general shape of the intensity profile described in Eq. (6.44) is illustrated in Fig. 6.13; Fig. 6.14 depicts the spatial relationship of the specimen (parallelepiped), and the complete shapes of the first-order interference regions at the reciprocal-lattice points and the projection of

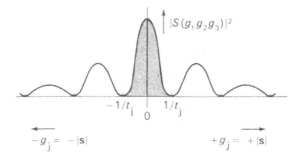

FIG. 6.13 *General appearance of intensity func-*
tion of Eq. (6.44).

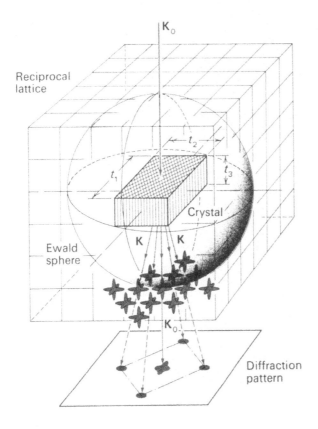

FIG. 6.14 *Phenomenological view of intensity*
profiles at reciprocal-lattice points corres-
ponding to a general parallelepiped. Only
shown are portions of reciprocal-lattice in-
terference regions that contribute directly to
electron diffraction pattern.

the intersection of the reflection sphere with the reciprocal-lattice points to produce the diffraction pattern. The incident wave vector in Fig. 6.14 is directed along the z direction in the real system (crystal and diffraction system) and the diffraction pattern is therefore formed by a projection of the intersection "plane" of the reciprocal space with the xy plane of the real space reference.

We can observe from Fig. 6.14 that, as a general rule, the interference regions appear as extensions (stretching of the intensity profiles) perpendicular to the main faces of the scattering domain or crystal. This feature was first established by von Laue [1,6]; and A. L. Patterson [7] has computed the corresponding shape transforms and interference forms for the sphere and related crystallographic polyhedra.

The various forms of the interference-intensity regions at the reciprocal-lattice points corresponding to the various adjustments of the general parallelepiped can be constructed by simply adjusting the various dimensions in Eq. (6.44), bearing in mind the interference-extension criteria outlined above. In Fig. 6.15 the corresponding intensity profiles for a number of adjustments in the general parallelepiped are sketched with respect to the direction of the initial wave vector $\underline{K}_o$. Again we can imagine the diffraction mode to be of the transmission type as depicted also in Fig. 6.14. It should be observed in Fig. 6.15 that the interference region shown at a reciprocal-lattice point is the same for all reciprocal-lattice points of the orthogonal reciprocal space that results for the orthogonal parallelepiped.

If we direct our attention now to the particular case of the thin plate of Fig. 6.15c, we can extend this form to the special case of a thin foil of effective infinite area in the xy plane, that is, $t_1, t_2 \to \infty$, and t_3 is on the order of $\sim 10^3$ Å. In the electron microscope, t_1 and t_2 must only be greater than the diffraction aperture (Fig. 6.12). The intersection of the reflection (Ewald) sphere with the corresponding reciprocal-lattice points and associated interference regions can then be observed, particularly for the transmission mode, to contribute qualitatively to the intensities of the electron transmission mode, to contribute qualitatively to the intensities of the electron diffraction spot pattern (assuming a single-crystal foil) as illustrated in Fig. 6.16a, while Fig. 6.16b shows an actual electron diffraction pattern for a (100) Pd thin film (having a thickness $t_3 \simeq 800$ Å) recorded in an electron microscope. As observed in Fig. 6.16a, the reflection sphere need not pass directly

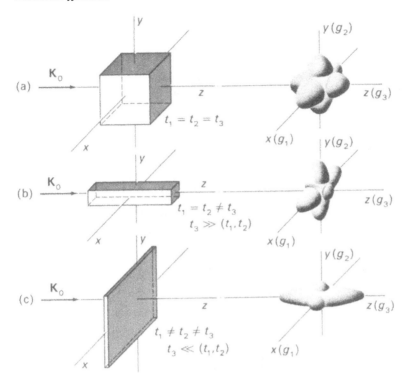

FIG. 6.15 *Kinematic shape transforms and corresponding intensity profiles for principal maxima at reciprocal-lattice points for geometries of scattering domains shown. (a) Cube; (b) needle; (c) thin plate or foil.*

through a reciprocal-lattice point to observe a diffraction spot (hkℓ) in the electron diffraction pattern. The unsymmetrical disposition of diffraction spots in many spot patterns, particularly those obtained in the electron microscope, can now be interpreted, on the basis of Fig. 6.16a, as a tilt of the specimen with respect to the incident beam direction. It is observed in Fig. 6.16 that the spot intensities simply relate to the portion of the interference region intersected by the Ewald sphere that, in one dimension, corresponds to the area of the principal maxima contributing to the intensity. This feature is also readily apparent in the spot pattern of Fig. 6.7, and can be treated as the integrated intensity of a diffraction spot by considering the value of the interference function [Eq. (6.44)] in the cross section.

For the incident wave vector $\underset{\sim}{K}_0$ (Fig. 6.16) directed along the z direction, the integrated intensity can be written

$$I_g(g_3) = \int J_g \, dx \, dy \qquad (6.45)$$

where dx and dy are units of length in the actual electron diffraction pattern of Fig. 6.16b. Substituting for $J(g_{hk\ell})$ from Eq. (6.24) with r = L (the distance from the specimen scattering center as in Figs. 6.6 and 6.8) in Eq. (6.45), we obtain

$$I_g(g_3) = \frac{J_o |F(hk\ell)|^2}{L^2} \int |S(g_1 g_2 g_3)|^2 \, dx \, dy \qquad (6.46)$$

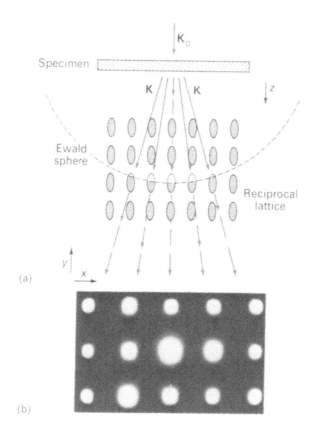

(a)

(b)

FIG. 6.16 *Formation of transmission-type electron diffraction spot patterns from single-crystal thin foils in the electron microscope. (a) Relationship of reciprocal-lattice interference regions and reflection sphere when specimen is normal to incident electron beam. (b) Electron diffraction pattern of Pd foil for 100 kV electrons.*

Since in the context of Fig. 6.16 we have $dx = \lambda L\ dg_1$. and $dy = \lambda L\ dg_2$, we can now write Eq. (6.46) as

$$I_g(g_3) = J_0 \lambda^2 |F(hk\ell)|^2 \frac{t_1 t_2}{a^2 b^2 c^2} \frac{\sin^2 \pi t_3 g_3}{(\pi g_3)^2} \tag{6.46a}$$

At the exact Bragg condition, that is, at the intersection of the Ewald sphere with the reciprocal-lattice point, $g_3 = 0$. This is equivalent to saying $|\underline{s}| = s = 0$ in Fig. 6.6; and since $J_0(t_1, t_2)$ is the incident beam intensity (number of electrons) that reaches the recording plane (fluorescent screen or photographic plate), we can write Eq. (6.46a) as

$$\frac{I_g}{I_0} = \frac{\lambda^2}{V_C^2} |F(hk\ell)|^2 t^2 \tag{6.47}$$

where V_C is the unit-cell volume, and t is the film thickness ($t = t_3$). Equation (6.47) thus relates the diffracted beam intensity to the incident (primary) beam intensity.

It is important to emphasize the very simple and practical implications of Figs. 6.15 and 6.16. If, for example, any of the other diffracting domains in Fig. 6.15 were to replace the specimen geometry in Fig. 6.16, then the corresponding reciprocal-lattice interference regions could be intersected by the Ewald sphere giving rise to the characteristic intensity profiles in the electron diffraction pattern. The significance of this, as we will discuss later in more detail, is that spikes, nodes, and streaks will appear in the diffraction pattern. In addition, you must realize that when the specimen is rotated or if the diffracting volume relative to the electron optical axis is changed, the intensity and shape of the diffraction spots will change. For example if the thin diffracting volume shown in Fig. 6.16 were oriented parallel to the beam, the reciprocal lattice interference zones would be turned 90°. The Ewald sphere cutting these zones would now produce elliptical intensity zones at the diffraction spots rather than the zones shown in Fig. 6.16. Variations on this intensity zone could be achieved at any angle for the thin volume between 0 and 90° of the situation shown in Fig. 6.16. The diffracting volume can in fact contain mixtures of specific shapes which would independently contribute to the diffraction pattern when their reciprocal - lattice interference zones were intersected by the Ewald sphere. This is, as we shall see later, the case with microtwins, precipitates, or other inclusions in a thin film. It is important to remember that each specific diffracting volume maintains a unique reciprocal lattice with regard to spatial geometry and intensity profiles for the reciprocal lattice points. It is the position or geometry of these domains in relation to the optic axis which determines the way the Ewald sphere will cut the corresponding interference zones at the reciprocal lattice points because the optic axis defines the position of the Ewald sphere, as shown in Fig. 6.15.

6.5.2 DYNAMICAL DIFFRACTION

The basic assumption for kinematic diffraction was that the diffracted wave intensities were negligible compared to the incident beam intensity. And, while a phenomenological interpretation for diffraction is obtained on the basis of the kinematical theory, quantitative interpretations in many cases lack the desired accuracy, especially where the diffracted beam or beams possess intensity levels that approximate or equal the primary beam. In such instances, quite clearly, the kinematic assumptions are no longer strictly accurate, and this condition is employed as a criterion for deciding when a transition from kinematical to dynamical treatments is necessary. We can treat this situation by considering that for dynamical diffraction

$$I_g \simeq I_o$$

and Eq. (6.47) becomes

$$\frac{I_g}{I_o} \simeq 1 = \frac{\lambda^2}{V_C^2} \, |F(hk\ell)|^2 t^2 \tag{6.47a}$$

Note that this condition defines dynamical diffraction and differentiates it from kinematical diffraction where $I_g \ll I_o$. The intrinsic relationship of the material as it demands the dynamical considerations be invoked is represented in the critical thickness of Eq. (6.47a), that is,

$$t_{dyn} > \frac{V_C}{\lambda |F(hk\ell)|} \tag{6.48}$$

where Eq. (6.48), as it relates specifically to transmission of electrons through thin films in the electron microscope, indicates the magnitude of film thickness above which the diffraction becomes dynamic in nature. You might recall that a similar criterion was established in our treatment of mass-thickness contrast image interpretations in the electron microscope

[Eq. (5.6)]. In Eq. (6.48) as in Eq. (5.6), the critical thickness decreases for increasing atomic number, that is, in Eq. (6.48) $|F(hk\ell)|$ is increased in proportion to $f_{e\ell}(\mu)$, which is in turn proportional to Z according to the relationship of Eq. (6.14). We also observe from Eq. (6.48) that since λ is proportional to V_0 (the accelerating potential), the kinematical approximation remains valid for thicker foils provided the incident beam energy is high. Figure 6.17 illustrates this feature for a number of fcc metals assuming $f(hk\ell) = 4f_{e\ell}(\mu)$ for reflections (111), and $V_C = a^3$ (cube of the interatomic spacing) in Eq. (6.48).

The physical meaning of dynamical diffraction is that the nature of the diffracted wave field, which results by the interaction of the initial and scattered beams within a crystal, depends on both the crystal geometry (shape) and the Bragg angle (scattering angle). We have treated in our discussion above the two-beam situation, that is, the incident beam and one diffracted beam [a (111) reflection in Fig. 6.17]. The two beams are considered to have equal intensities. We can treat the intensity diffracted for any reflection (for a corresponding $hk\ell$ diffraction spot) for the transmission of electrons through a foil of thickness t as

$$I_g(dyn) = \lambda^2 \left|\frac{F(hk\ell)}{V_C}\right|^2 \frac{\sin^2 t \sqrt{(\pi s)^2 + \lambda^2 \left|\frac{F(hk\ell)}{V_C}\right|^2}}{(\pi s)^2 + \lambda^2 \left|\frac{F(hk\ell)}{V_C}\right|^2} \qquad (6.49)$$

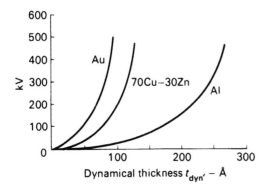

FIG. 6.17 *Dynamical thicknesses for several fcc metal and alloy films considering (111) diffraction as function of accelerating potential (transmission mode). Values of $f_{e\ell}(\mu)$ are from Appendix D.*

with the incident beam intensity unity ($I_O = 1$). The value of s in Eq. (6.49) is equivalent to the distance g_3 in reciprocal space, and is identically the deviation from the Bragg condition as shown in Fig. 6.6, that is $|\underline{s}|$ = s. We therefore observe that for the exact Bragg condition, s = 0; and Eq. (6.49) becomes

$$I_g(dyn) = \sin^2 \lambda \left| \frac{F(hk\ell)}{V_C} \right| t \qquad (6.50)$$

Thus for t very small,

$$\sin^2 \lambda \left| \frac{F(hk\ell)}{V_C} \right| t \simeq \lambda^2 \left| \frac{F(hk\ell)}{V_C} \right|^2 t^2$$

and Eq. (6.50) is observed to be identical to Eq. (6.47a), that is, the diffraction intensity is described kinematically.

We can consider the dynamical theory to apply to the two-beam case discussed above, or to be extended to more than two beams, that is, the n-beam approximation where numerous diffracted beams are observed to possess intensities equivalent to that of the main beam. Figure 6.18 illustrates the dynamical diffraction situations that can be clearly defined solely on the basis of electron diffraction intensities relative to the main beam spot.

It should also be realized that for thicker foils, which necessitate a dynamical treatment of electron diffraction, secondary effects such as diffuse scattering become important, and absorption also influences the diffraction intensities. Thus another indication of dynamical diffraction is found in the detection of such secondary effects, particularly where these give rise to specific diffraction phenomena. These effects also manifest themselves in the diffraction contrast images arising in the normal mode of transmission electron microscopy of crystalline thin-film materials (see Chap. 7).

6.6 DIFFUSE SCATTERING OF ELECTRONS: KIKUCHI DIFFRACTION PATTERNS

In the case of thick single crystals or large-grain areas, the electron transmission diffraction patterns are observed to contain sets of parallel lines, one bright and one dark, having a distinct association with the spots. Where the foils become an order of magnitude or greater in thick-

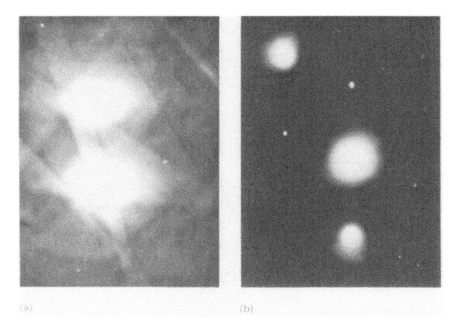

FIG. 6.18 *Dynamical electron diffraction. (a) Two-beam; (b) three-beam diffraction. Diffraction patterns are from thin films of stainless steel observed in electron microscope at 100 kV accelerating potential.*

ness than t_{dyn} of Eq. (6.48) ($t > 10t_{dyn}$), the diffraction pattern consists entirely of arrays of such line pairs. These lines are called Kikuchi lines after their discoverer [8,9], and occur as a result of diffuse scattering of the electron beams (incident and diffracted) as they penetrate the specimen. The basic mechanism of Kikuchi-line formation is depicted in Fig. 6.19. Here we observe a point D some distance below the entrance surface from which the electrons scatter in all directions (with spherical symmetry about D). Those scattered electrons that encounter a set of equivalent crystallographic planes at the Bragg angle θ will therefore diffract, and the diffraction, being symmetrically dependent on the incident beams, will occur as cones, with apexes centered on the net planes. Those electrons scattered at large angles with respect to the incident beam vector (at D) will loose intensity more rapidly than those scattered at small angles, as we observed previously in dealing with the single scattering process (see Sec. 5.2.1). Thus we observe in Fig. 6.19 that diffracted beams at I(e) contain more electrons than those at I(d) because of the difference in the initial scattering angle. The intensities associated with the beams shown are then related by

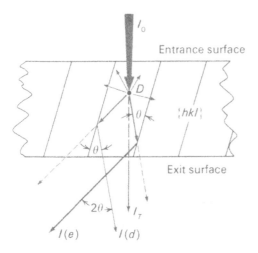

FIG. 6.19 *Mechanism of formation of Kikuchi lines by diffuse scattering of transmitted electrons in thick foils in electron microscope.*

$$I(e) > I_T > I(d)$$

Figure 6.20 indicates the more detailed geometry of the formation of diffraction cones following diffuse scattering in a thin crystalline sample, from which the origin of Kikuchi lines is observed as the intersection of these cones with the recording plane (photographic plate). One cone, corresponding to I(e) of Fig. 6.19, contains an excess of electrons, while the other, corresponding to I(d) has a deficiency of electrons. Because of the smallness of the Bragg angle θ and the large distance from the specimen to the recording plane (L), the intersection hyperbolas are effectively straight lines giving rise to the bright and dark lines (Kikuchi lines) in the photographic plate.

We observe in Fig. 6.20 that the distance between the Kikuchi lines (a bright and dark pair) is

$$x = 2\theta L$$

or we can write

$$x = \lambda L \frac{1}{d_{hk\ell}} = \lambda L \left| g_{hk\ell} \right| \tag{6.22a}$$

Consequently the separation of excess and deficiency lines in a Kikuchi diffraction pattern corresponds to the radii of spots in an electron diffraction spot pattern.

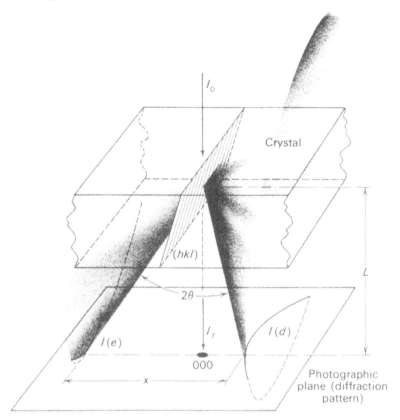

FIG. 6.20 *Geometry of formation of Kikuchi lines in electron diffraction pattern.*

A typical example of Kikuchi diffraction is shown in the electron diffraction pattern from Inconel alloy metal shown in Fig. 6.21. Several diffraction spots also occur, but generally the pattern results by diffraction
of diffuse scattered electrons. The fact that the incident beam is no
longer recognizable as distinct from the diffracted intensities of course
means that the kinematic assumptions no longer apply. This feature can be
used as a check on the applicability of the kinematical assumptions, that
is, when Kikuchi lines begin to appear, the diffraction becomes dynamic in
nature.

We should observe that since the Kikuchi lines correspond to the diffraction spots for any {hkℓ} planes, when a crystal is set at the exact
Bragg condition (where the reciprocal lattice is intersected by the Ewald
sphere), the corresponding Kikuchi lines will pass through the spots, and

FIG. 6.21 *Transmission Kikuchi-line pattern from Inconel (76% Ni, 16% Cr, 7% Fe) in electron microscope. (V$_o$ = 100 kV.)*

the local intensity will be enhanced. It should also be apparent that where the spot diffraction pattern will remain spatially fixed even for a tilt of the specimen or the incident beam, because of the extensions of interference regions perpendicular to the specimen surface (Fig. 6.16), any deviation from the exact Bragg condition will sweep the Kikuchi lines to one side or the other of the corresponding diffraction spot. Thus the angular deviation from the Bragg condition Δθ, as shown in Fig. 6.21, can be determined simply by considering the geometrical adjustment with respect to Fig. 6.20, from which we readily observe that

$$\Delta\theta = \frac{\Delta R}{L}$$

or we can write equivalently

$$\Delta\theta = \frac{\lambda\Delta R}{\lambda L} \qquad\qquad (6.51)$$

in terms of the camera constant λL

In the case where the crystal is exactly oriented (or nearly so) with respect to some crystallographic surface plane, and there is little or no deviation from the exact Bragg conditions, the Kikuchi pattern will be a symmetrical array of excess and deficiency lines with their intensities reinforced where the lines coincide with the spots, or with the interference regions at the reciprocal-lattice points. Figure 6.22 illustrates this condition for a Si crystal (diamond cubic structure) with the zone axis normal to (111).

We shall see in Sec. 6.8 that where several concurrent pairs of Kikuchi lines appear, the intersections can be used to locate zone axes of the crystal; and the crystallographic surface orientation can the be defined. We will also find that where the sign of $\underline{g}$ in an electron diffraction spot

FIG. 6.22 *Three-fold symmetrical Kikuchi diffraction pattern from* [*111*] *Si in the electron microscope.* (*V*$_o$ = *125 kV.*) *Si foil thickness is roughly 7500 Å.*

pattern is arbitrary (in Fig. 6.16, g can be chosen arbitrarily as + or -), the crystallography associated with these patterns maintains a 180° ambiguity. While this is not important in most applications, we will see later that the study of dislocation Burgers vectors and related specific diffraction vectors necessitate the identification of a unique orientation. We can see that Fig. 6.21 is unique as opposed to a symmetrical array of diffraction spots as a simple test of this concept.

It should also be noted that while we have treated Kikuchi lines as the result of a single diffuse scattering process in a single crystal, this condition can arise in any situation that ultimately satisfies the condition illustrated in Fig. 6.19. W. Kossel and G. Mollenstedt [10] have in fact demonstrated this feature where the diffuse scattering was generated in an amorphous surface layer on a thin single crystal, thereby generating Kikuchi lines. It should not be surprizing therefore to observe Kikuchi lines in reflection-mode diffraction patterns, or similar phenomena in situations where diffuse scattering or absorption processes might occur. D. G. Coates [11] observed Kikuchi-type reflections in the scanning electron microscope; these were simultaneously interpreted by G. R. Booker et al [12] to occur as a result of anomalous absorption effects. We have discussed this phenomenon in Sec. 5.5.7 and we will elaborate further in the following section.

Because of the precise demand for crystal symmetry and uniformity in the formation of Kikuchi electron diffraction patterns, the occurrence of Kikuchi lines in an electron diffraction pattern from a thick crystal attests to its degree of perfection. Thus, where a crystal contains numerous defects, the distinction of Kikuchi lines diminshes as a result of the enchanced scattering. It should be realized that ideally more than one scattering incident can occur in a specimen. That is to say, initially scattered electrons may undergo additional scattering. In this way, all of the original beam energy will appear in the diffracted beams, and the transmitted intensity will be zero. We can observe this feature in Fig. 6.22 where no intensity occurs at the center of the pattern.

6.6.1 *ELECTRON CHANNELING PATTERNS*

In Sec. 5.57 we dealt with diffraction effects in the SEM and described the production of electron channeling patterns (ECP's) to be similar in many respects to Kikuchi patterns. Having now described Kikuchi

patterns in some detail, it might be useful to describe the ECP in some
additional detail. The inclusion of ECP discussion in this section is not
to represent that an ECP is strictly the result of diffuse scattering.
Nonetheless, it does possess features which are obviously influenced by
diffuse scattering from the specimen surface region, and has been tradi-
tionally referred to as a pseudo-Kikuchi pattern.

If you will refer back momentarily to Figs. 5.37 and 5.38, you should
note that a parallel beam of electrons incident upon the sample surface is
rocked through a selected angle about a fixed point on the surface. If
this rocking angle is sufficiently large, the Bragg diffraction conditions
are satisfied in a manner phenomenologically similar to Fig. 6.19, producing a series of parallel lines corresponding to the angles at which the
Bragg diffraction conditions are satisfied for various planes exactly like
Kikuchi-line pairs. Furthermore, the line intensities diminish with dimi-
nishing crystal perfection just like Kikuchi lines as noted above. In addition, fine lines appear in the zone center of ECPs which are apparently
the backscattered analogue of those which appear at specific g zones in
convergent beam electron diffraction patterns discussed in detail in Sec.
6.8.7. These lines, shown in Fig. 6.23, correspond to first-order or
higher-order Laue zones and can be referred to corresponding as FOLZ or
HOLZ lines respectively. They can be indexed as shown in Fig. 6.23 and
utilized in making lattice spacing and lattice parameter measurements. It
is believed that lattice strains, both elastic and plastic, can be studied
and quantified by determining not only the variations in the overall ECP
with strain but also examining the HOLZ lines in the pattern centers in
detail. These features have in fact been described recently by D. C. Joy
and D. L. Davidson in *Physical Scanning Electron Microscopy,* Academic
Press, New York, 1982, to which you should refer for additional discussions, and the details for achieving the patterns typified by the examples
shown in Fig. 6.23.

6.7 LOW-ENERGY ELECTRON DIFFRACTION (LEED)

Our treatment of electron diffraction so far has been concerned with high-
energy electrons that is, those electrons accelerated by potentials greater than 50 kV in the electron microscope; and our emphasis has centered
about transmission diffraction from thin films because of the availability
of electron microscopes. The transmission mode also eliminates for the

FIG. 6.23 Selected area electron channeling pattern of (111) silicon single crystal. (a) Solid-state back-scatter detector image taken in a JEOL 35C SEM with a third lens, using a working distance of 9 mm at 35 kV accelerating potential with the beam rocking through 16° and ~ 10 μm diameter spot size. (b) Magnified center portion of zone in (c) showing fine structure formed by diffraction from planes of the first-order Laue zone. (c) Indexing of FOLZ lines in (b) (Courtesy M.C. Madden and J.J. Hren, University of Florida).

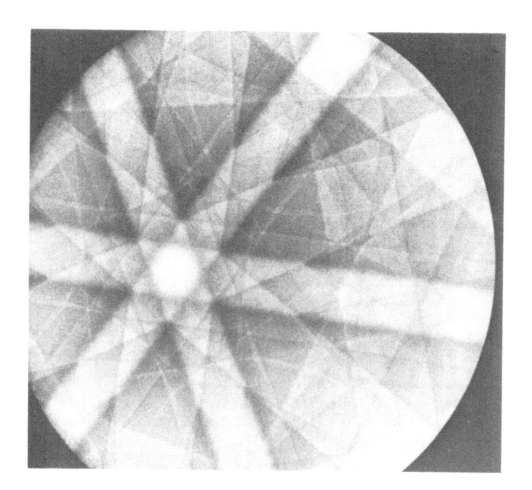

(a)

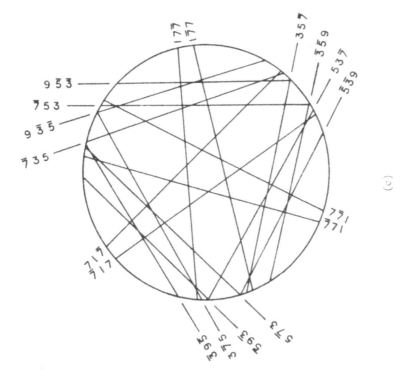

(c)

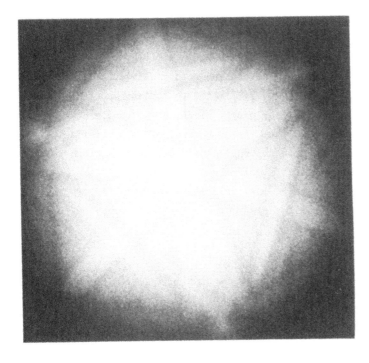

(b)

FIG. 6.23 (continued)

part the need to consider refraction of electrons, which becomes increasingly prominent as the glancing angle decreases [a feature already treated in Chap. 3, especially Eq. (3.1)]. In effect, electrons entering a crystal at an accelerating potential of V_o are further accelerated by the crystal potential, which can be averaged over the volume of the crystal as a mean inner potential. The refractive index is found to be given by

$$r = \sqrt{\frac{V_o + U_{mean}}{V_o}} \qquad\qquad (6.52)$$

where

$$U_{mean} = \frac{48}{V_C} \sum_j U(\underline{r})_j \qquad\qquad (6.53)$$

V_C is the unit cell volume, and $U(\underline{r})_j$ is the potential energy of the jth atom as in (Eq. (6.28). For low-energy electrons, the index of refraction increases; and Eq. (6.52) also illustrates an experimental means for measuring U_{mean} as opposed to the calculation of Eq. (6.53) as it relates to Eq. (6.28). A detailed discussion of the determination of mean inner potentials can be found in the book by Z. G. Pinsker [13]; and values ranging from 5 to 20 volts have been recorded [3].

In the case of low-energy electrons of several hundred volts accelerating potential, we therefore observe several complications. First, as we have mentioned above, refraction becomes appreciable. Second, the penetration of such electrons at low accelerating potentials according to the Thomson-Whiddington law [Eq. (4.10)] is limited to the surface layers of atoms. Since the diffraction of low-energy electrons is confined to the surface layers, certain complications arise because of the potential distribution at the surface, and the absorption of impurities, surface reactions, and related adjustments in the arrangement of surface atoms. While these features contribute to the complications arising for LEED, they also provide a very effective means for studying various structures and surface reactions.

6.7.1 INTERPRETATION OF LEED PATTERNS

Realizing at the outset that LEED can be assumed to occur in most cases from a single surface layer, we can consider the diffraction from a single crystal as a special consideration of kinematical diffraction. That is, we observed in Fig. 6.15 that the interference region at a reciprocal-lattice point for a thin sheet appears as an extension normal to the surface. We can observe from Eq. (6.44) that the length of this extension in the direction normal to the surface plane is

$$g_3 = \frac{1}{t} \tag{6.54}$$

where t is the crystal thickness. Now we suppose that diffraction occurs from a single atomic layer. We then have t = a (where a is the interplanar spacing, and Eq. (6.54) becomes

$$g_3 = \frac{1}{a} \, \overset{\circ}{A}{}^{-1}$$

Under these conditions, it is clear that g_3 of Eq. (6.54) is equivalent to the distance between reciprocal-lattice points. Thus the interference regions become continuous rods through the reciprocal-lattice points, normal to the diffracting layer. The diffraction pattern is then simply a two-dimensional cross grating, with spots identified by the two-index notation {hk}. Figure 6.24a illustrates schematically the formation of this two-dimensional interference feature.

It is to be observed in Fig. 6.24a that where the electron beam is incident normal to the surface of a crystal, we assume that the diffraction can be depicted as shown in Fig. 6.24b as it relates to the identification of the resulting two-dimensional grating pattern. In Fig. 6.23b the path difference between the incident rays is characterized by

$$n\lambda = d' \sin \theta' \tag{6.54a}$$

which we may call the diffraction grating equation; and note its relationship to the Bragg equation (Eq. 1.9).

In Fig. 6.24c the displacement of the Ewald sphere is of no consequence in the formation of LEED surface patterns, and the intensity of spots remains constant overall. Figure 6.25 illustrates schematically the nature of LEED apparatus design and the general nature of LEED patterns for a single crystal surface. [You are urged to read the articles by A. U. MacRae, *Science, 139:* 379(1963); and L. N. Tharp and E. J. Scheibner, *J. Appl. Phys., 38:* 3320(1967).]

While the kinematical assumptions as outlined schematically in Fig. 6.24 provide us with a qualitative interpretation of LEED patterns, many subtle features of certain patterns, which do in fact include intensity

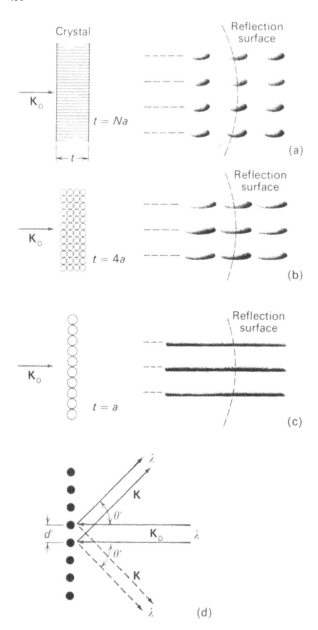

FIG. 6.24 *Development of two-dimensional dif-*
fraction (grating) pattern for LEED from atomic
surface plane. (a) Finite crystal thickness as
in Figs. 6.15c or 6.16. (b) Merging of inter-
ference regions for crystal several atomic layers
thick. (c) Continuous (overlapping) interference
spikes of single atomic layer. (d) Diffraction
geometry for normal electron beam incidence in
LEED from single atomic layer.

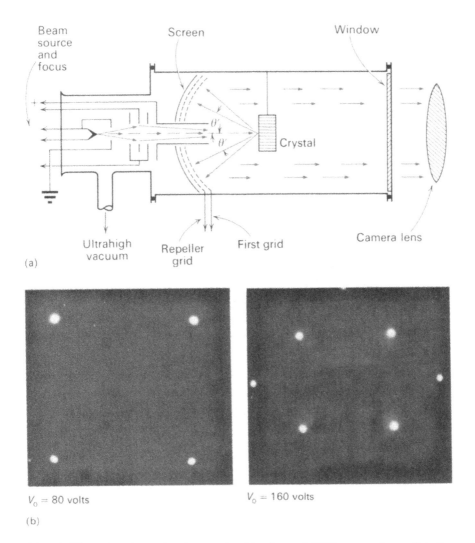

FIG. 6.25 *(a) Schematic (conceptual) view of LEED apparatus. (b) Typical LEED patterns for (100) clean surface of Cu at different potentials. (Courtesy L. K. Jordan, Georgia Institute of Technology.)*

variations of spots, etc., are uninterpretable kinematically. The dynamical treatment of such phenomena must be invoked, and this presents several difficulties. The most serious is of course the fact that a simple two-beam approximation is completely useless because of the continuous two-dimensional nature of the diffraction. A dynamical theory has been outlined by F. Hoffman and H. P. Smith [14] which basically involves the solution of Eq. (6.33) with appropriate adjustments for the surface potential and the provision for treating the n-beam situation.

6.7.2 *APPLICATIONS OF LEED*

The implications are perhaps clear that LEED has had, and will continue to have, important applications in the study of surface structure. It has certain drawbacks, but for studies of metal surfaces it has certain advantages over field-ion emission microscopy discussed in Chap. 2. In the particular case of metals, the problems in characterizing the surface structure mainly involve the assumption that surface atoms preserve, to some extent, their isolated character. That is, the surface-atom electron wave functions are presumed to approximate those for the isolated atom [15,16]. Thus, chemisorbed, etc., atoms arrange themselves in such a way that their outermost orbits can overlap with those of the surface atoms as depicted in the simple sketches of Fig. 2.1. We can realize that by combining LEED analysis with field-ion emission microscopy, a detailed picture of the surface-atomic arrangements and potential distributions will be possible. Z. Knor and E. W. Müller [17] have recently outlined the surface character of metals based primarily on observations by field-ion microscopy; and A. J. Melmed and coworkers [18] have also described a combined LEED-FIM arrangement for surface studies.

Early investigations of atomic surface structure have been pursued by L. H. Germer and associates [19] who looked specifically at the surface atomic arrangement for the (110) Ni surface. Oxygen adsorption on the (111), (100), and (110) surfaces of Cu was studied by L. K. Jordan and E. J. Scheibner [20]. Since about 1960, several thousand LEED studies have been performed with the aim of studying the atomic coincidence of oxygen, nitrogen, and related adsorbates [21] on various metal surfaces and surface crystallographies as illustrated in Fig. 6.26. The technique is not necessarily limited to pure metals or metals in general, and P. W. Palmberg and W. T. Peria [22] have investigated Ge and Na-covered Ge surfaces, while L. Fiermans and J. Vennik [23] have studied the (010) surface of vanadium pentoxide. LEED has become almost routine in detailed surface studies. If you are interested in pursuing LEED applications in more detail, particularly where they relate to surface physics and chemistry, see any current issue of the journal *Surface Science*. You might also consult *Methods of Surface Analysis,* edited by A. W. Czanderna and published by Elsevier, Amsterdam, 1975. Our aim here has been to briefly illustrate the powerful means provided by LEED for the investigation of physisorption, chemisorption, and related phenomena in two-dimensional ordered structures.

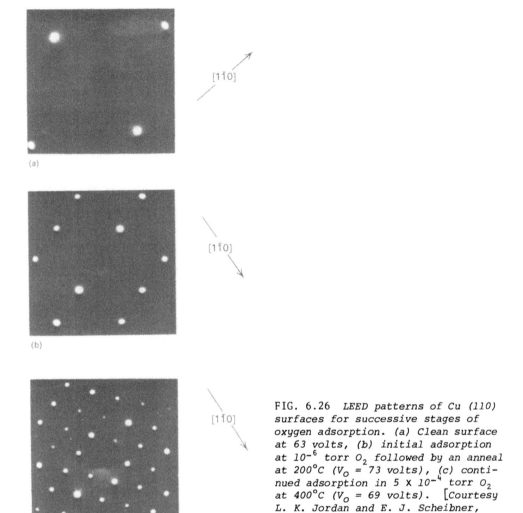

(a)

(b)

[1̄10]

[1̄10]

[1̄10]

(c)

FIG. 6.26 *LEED patterns of Cu (110)*
surfaces for successive stages of
oxygen adsorption. (a) Clean surface
at 63 volts, (b) initial adsorption
at 10^{-6} torr O_2 followed by an anneal
at 200°C (V_O = 73 volts), (c) conti-
nued adsorption in 5 x 10^{-4} torr O_2
at 400°C (V_O = 69 volts). [Courtesy
L. K. Jordan and E. J. Scheibner,
Surface Sci., 10: 373 (1968).]

6.8 APPLICATIONS AND INTERPRETATION OF SELECTED-AREA ELECTRON DIFFRACTION PATTERNS

Our treatment of high-energy electron diffraction, both reflection and
transmission modes, has been concerned primarily with the use of the electron microscope. The reason of course being that the electron microscope
is commercially available and is increasingly found in industrial and university laboratories as a matter of routine. For the remainder of this
chapter, we will deal exclusively with transmission electron diffraction
from thin films in the electron microscope. We will specifically treat the

application and analysis of diffraction patterns that relate to specific thin-foil areas, that is, selected-area electron diffraction patterns.

The unique feature of selected-area electron diffraction patterns is that chemistry and crystallographic features of an object area in a thin foil can be identified with the image observed by direct transmission electron microscopy. Among other features, the selected-area technique allows the various diffracted beams to be isolated; and image signal information specific to these beams is then observed in the dark-field image.

We can illustrate the operational features of selected-area diffraction in the electron microscope by referring to Fig. 6.27. In Fig. 6.27a, a crystalline-foil area is imaged in the electron microscope in the conventional mode of bright-field transmission as earlier illustrated in Fig.5.2. Having then selected an area of the thin section from which a diffraction pattern is desired, the diffraction aperature (field-limiting aperture) is

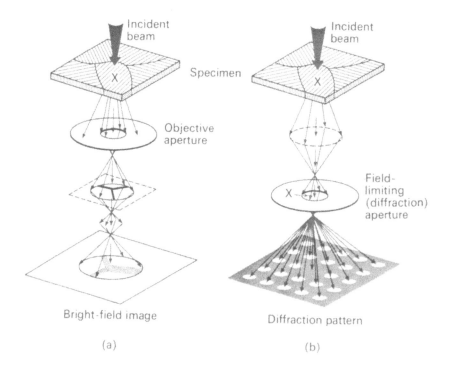

FIG. 6.27 *(a) Bright-field image; (b) selected-area diffraction pattern formation in electron microscope. Grain-boundary junction is specially viewed, with selected-area diffraction being formed in b in area X of thin specimen.*

inserted at the intermediate image planes as shown in Fig. 6.27b. The size
of the field-limiting aperture corresponding to a specific instrument magnification then serves to limit diffraction to particular object dimensions.
The objective aperture is then withdrawn, and the image demagnified by reducing the intermediate-lens current to crossover. This operation focuses
the electron diffraction pattern.

Because of the spherical aberration we observed in Chap. 3, the specimen peripheral areas outside the field-limiting aperture (the selected
area) also contribute to the diffraction pattern as pointed out by A. W.
Agar [24]. A general rule seems to be that the diffraction pattern will
represent an area twice that of the field-limiting aperture. In certain
applications this feature must be carefully considered.

6.8.1 CRYSTALLINE IMAGE AND DIFFRACTION PATTERN ROTATION CALIBRATION

It must be recalled from Chap. 3 (you should especially review Fig. 3.6)
that where a lens-current reduction occurs in the electron microscope, a
corresponding change in the beam rotation occurs. This means of course
that in comparing the image recorded in the situation of Fig. 6.27a (bright-
field image) with the corresponding selected-area electron diffraction pattern of Fig. 6.27b, the relative rotation of the image and diffraction pattern must be known. That is, if a particular crystallographic feature in
the object (for example a slip plane, defect, precipitate habit, etc.) is
to be identified from the diffraction pattern, any rotational ambiguities
must be known. The measurement of the rotation of the image of a crystal
with respect to its diffraction pattern as a function of the intermediate-
lens current thus constitutes a rotation calibration.

Since the rotation is only dependent on the accelerating potential
and the lens currents, rotation curves can be constructed for each operating potential by varying one lens current (the intermediate) while all
other conditions remain fixed. The rotation angle is measured simply by
comparing a known image crystallographic feature with its diffraction pattern. We can therefore use the direction of an annealing twin in a polycrystalline metal foil and compare it with the direction in the diffraction
pattern; or the known habit of a crystal can be used to identify the rotation angle.

A favorite standard is MoO_3, which when heated in a crucible at high
temperatures sublimes into crystalline "smoke" that can be collected on
microscope grids. The flat-elongated pseudo-orthorhombic (very nearly fcc)

crystals orient their long side parallel to [100]. Thus photographing the
image of a crystal edge and superposing the electron diffraction pattern on
the same exposure (as a double exposure) allows the rotation angle to be
measured directly. Figure 6.28 illustrates the technique.

The rotation angle measured directly from Fig. 6.28 is the relative
rotation angle. We observe that the sense of rotation, considering the
image fixed, is that the diffraction pattern rotates counterclockwise. The

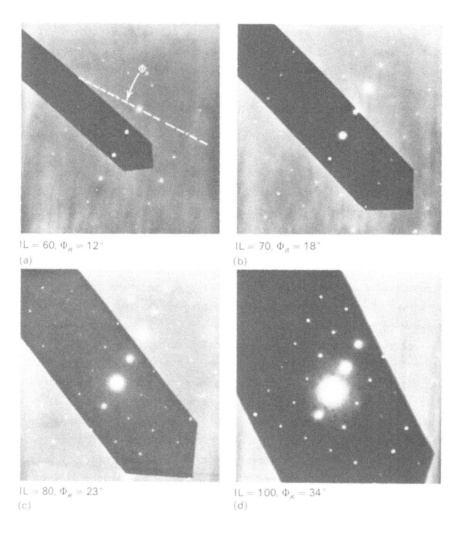

IL = 60, $\Phi_R = 12°$
(a)

IL = 70, $\Phi_R = 18°$
(b)

IL = 80, $\Phi_R = 23°$
(c)

IL = 100, $\Phi_R = 34°$
(d)

FIG. 6.28 *Image-diffraction pattern rotation calibration using* MoO_3
*crystal. Relative rotation angles and corresponding intermediate-lens
currents are indicated. Note sense of rotation. Accelerating potential was 100 kV, and image magnifications correspond to Fig. 5.9.*

rotation sense depends of course on whether the emulsion is up or down, or a similar convention. Thus, for a general image and diffraction pattern taken under identical circumstances, the diffraction pattern must be rotated the required number of degrees clockwise. It should be observed that the diffraction pattern is formed at the intermediate-lens crossover. This implies the elimination of the beam inversion with respect to the magnified image (see Fig. 3.6), and the true rotation angle requires a rotation of 180° in addition to the relative rotation angle calibrated. If we let Φ_A represent the absolute rotation angle and Φ_R the relative rotation angle calibrated as in Fig. 6.28, then

$$\Phi_A = 180° + \Phi_R \tag{6.55}$$

The sense of the rotation is determined by comparing the sequential exposures as in Fig. 6.28.

6.8.2 DIFFRACTION RING (LINE-BROADENING) CALIBRATION AND SPOTTY RING PATTERNS

We have effectively dealt with the indexing of ring diffraction patterns in our treatment of the camera-constant calibration in Sec. 6.4.3 (see Fig. 6.11). The indexing of ring patterns for a known material then simply involves the inverse operation to calibration; that is, knowing λL and the lattice parameter, we simply measure R, the ring radii, and assign the indices according to the appropriate reflection conditions outlined briefly in Sec. 6.3.1.

Where the ring reflections from a thin-film area are very sharp and of continuous intensities, the grain composing the area are randomly oriented and very much smaller than the diffraction aperture. If we consider that all the grains in the selected area are effectively uniform cubes having an edge length $D = t_j$, then from kinematical considerations [Eq. (6.44)], the interference region corresponding to the diffraction ring (as characterized by a continuous intensity ring of the form of Fig. 6.13) will have an angular half-width, W (in radians), of

$$g_j = \frac{W}{2} = \frac{\lambda}{t_j} = \Delta\theta(R)$$

Consequently the average grain size of very small crystals can be approximated from

$$D = \frac{\lambda}{\sqrt{[\theta(R)]^2 - [\theta(R_o)]^2}} \qquad (6.56)$$

where $\theta(R)$ is the intensity half-width (in radians) of the small crystallites having thicknesses or dimensions less than the dynamical thickness of the material, and $\theta(R_o)$ is the corresponding reflection half-width measured for an identical material having a known grain size in excess of the dynamical thickness. This technique is essentially the same method employed in x-ray diffraction measurements of grain size [25]; and it must be cautioned that the accuracy is very doubtful in most cases where the grain size approaches the dimension at which the diffraction becomes dynamical (t_{dyn}) as given in Eq. (6.48). For most materials, assuming a generally random grain orientation, this means that Eq. (6.56) is applicable only where the grain size is less than about 100 to 200 Å. Fairly accurate grain-size measurements can usually be made for $D > 50$ Å using aperture dark-field techniques for a particular reflection ring (where the objective aperture is centered over a portion of a particular ring), and sometimes sensible extrapolations can be made for smaller sizes when patterns can be calibrated for diffraction ring broadening with specific (measured) small grain sizes in a slightly larger range of grain sizes, for example 50 to 100 Å.

We can illustrate the calibration of line broadening in selected-area electron diffraction patterns with reference to Fig. P6.17. The diffraction pattern of Fig. P6.17 illustrates the condition for calibration where $\theta(R_o)$ is measured for any prominent reflection ring. The corresponding reflection line widths for a and b will then allow the corresponding grain sizes to be approximated. Line-broadening measurements can be checked to some extent if the crystal size can be directly observed at the point where the line width reaches a minimum. Regardless of this feature, it should be apparent without undue direction that line broadening as a means of measuring grain size can have serious drawbacks.

Note in Fig. 6.29 that the diffraction rings become spotty when the grain size (in relationship to the diffraction aperture) becomes sufficiently large. This feature can also be utilized as a measure of grain sizes after proper calibration, just as in the case of spotty Debye rings in x-ray diffraction [26]. The reliability of this technique for larger grain sizes is obvious. We should emphasize, for fear you may have been misled,

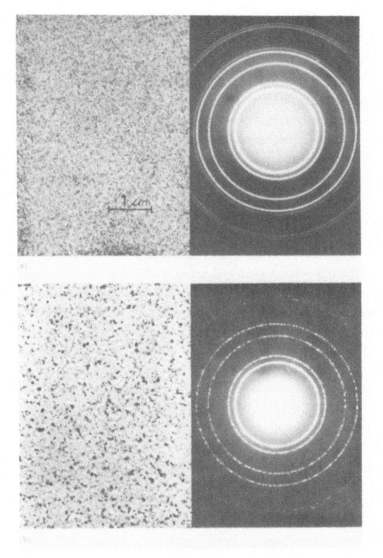

FIG. 6.29 *Diffraction patterns and electron transmis-*
sion images from thin Aℓ foils of randomly oriented
grains. (a) Electron micrograph and selected-area dif-
fraction pattern for polycrystalline Aℓ foil of 200 Å
grain size; (b) Aℓ 600 Å grain size giving rise to
spotty rings. Aperture size constant in (a) and (b).

that particular measurements from diffraction patterns should be considered only as a last resort. Grain sizes can be measured directly from the
transmission images as shown in Fig. 6.29.

6.8.3 *DETECTION OF PREFERRED ORIENTATIONS IN POLYCRYSTALLINE THIN FOILS: TEXTURE PATTERNS*

It was observed in Sec. 6.8.2 that when grains of a polycrystalline material are randomly oriented in thin films, a ring pattern of varying degrees results in the selected-area diffraction pattern. However, if the grains tend to cluster about some particular (preferred) orientation, the rings exhibit intensity maxima in positions corresponding to the prominent reflection planes of the oriented grains. Figure 6.7 has demonstrated this feature qualitatively, and we now wish to pursue, briefly, the quantitative aspects of such textured patterns from selected polycrystalline areas.

We can analyze the origin of texture patterns by considering the reciprocal-lattice conventions outlined previously in this chapter. That is, for thin films containing crystallites (grains) having their crystallographic planes in the plane of the thin section, the texture axis becomes the individual zone axis, and the crystals have a rotational degree of freedom about this axis. For grains having a common orientation, the texture pattern can be a series of discrete rings that become segmented or spotty as the degree of rotational mismatch is reduced. Where the mismatch is everywhere zero, of course the diffracting area is effectively a single crystal, and a spot pattern results. We can therefore imagine such crystals having an orientation [HKL] as a single crystal whose reciprocal-lattice plane (HKL) is rotated about the incident diffraction vector K_0. For a full rotation, the interference regions intersected by the Ewald sphere become rings, while for partial rotations, arcs are formed, and finally spots.

The effective zone axis for this scheme, that is, [HKL], is called the fiber axis. Thus in a thin film containing grains with certain proportions of the selected-area population having fixed orientations, that is, [HKL], more than one texture pattern may be superposed in the selected-area electron diffraction pattern. Some of these concepts will be better understood by considering Fig. 6.30. In Fig. 6.30a we observe the image of grain structures in a polycrystalline, vapor-deposited Aℓ foil. The corresponding selected-area electron diffraction pattern indicates strong texturing on (220). Essentially all of the diffraction originates from {220} planes, with specific concentrations of such planes being disposed by an angular relationship

$$\cos\theta = \frac{h_1h_2 + k_1k_2 + \ell_1\ell_2}{\sqrt{h_1^2 + k_1^2 + \ell_1^2}\ \sqrt{h_2^2 + k_2^2 + \ell_2^2}} \tag{6.57}$$

which is essentially Eq. (A.3) for the fcc Aℓ. The angle θ is taken as the angle between the intensity maxima as measured from the origin 000; $(h_1k_1\ell_1)$ and $(h_2k_2\ell_2)$ are in this case {220} planes, and their respective relationships can be calculated directly. We can also characterize the orientation relationships of Fig. 6.30a as three (100) preferred orientations symmetrically disposed by a rotation of one to the other by an amount θ as given in Eq. (6.57).

Figure 6.30b illustrates a similar but more pronounced (100) orientation containing very tiny included grains having generally random orientations but with some texturing about (220) as in Fig. 6.30a.

Preferred orientations of crystals or preferred directions of orientation in the plane of a thin film can only be completely identified by a rigorous indexing and geometrical investigation of the diffraction pattern. Fortunately in electron microscopy the image is usually available for corroboration or further analysis; and we shall deal further with this feature in Chap. 7.

6.8.4 SINGLE CRYSTAL-SPOT DIFFRACTION PATTERNS AND PATTERN INDEXING

We should be aware of the fact that a spot pattern or a Kikuchi diffraction pattern (in the case of thick sections) results from a single-crystal area, the latter from thick-perfect crystals. In the electron microscope, the thin-foil area need be only as large as the field-limiting (diffraction) aperture as discussed previously. This means that for a polycrystalline thin section, spot patterns can be obtained from individual grains provided the magnified image at the aperture plane is approximately equal to the aperture size. This is one of the most important features of selected-area electron diffraction in the electron microscope, namely, the ability to obtain a diffraction pattern unique to a particular grain. In this way the grain orientation and related crystallographic features can be determined, and the associated and structural characteristics of the transmission image can be identified.

In order to characterize the crystallography of a selected-grain area, the diffraction pattern must be recognizable as being unique to a particu-

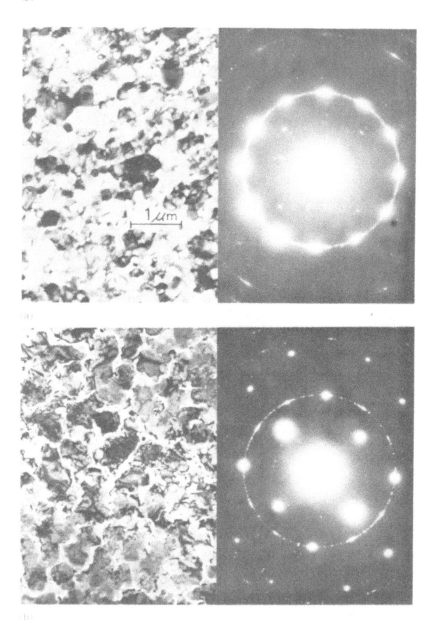

FIG. 6.30 *Texturing in vapor-deposited Aℓ thin films. (a) Electron micrograph and corresponding selected-area diffraction pattern showing strong texturing on (220). Note grain contrast in image. (b) Strongly preferred (100) orientation of grains in Aℓ thin film. Note fairly uniform contrast of grains in transmission image. (100 kV.)*

lar reciprocal-lattice projection. In this way, the crystallographic surface plane of the crystal section can be identified. The basis of the method of identifying a spot diffraction pattern is indexing the spots.

It will be recalled from Sec. 6.2.2 that where the incident electron
beam was directed normal to a reciprocal-lattice plane section representing
a zone axis [HKL] of the real crystal, the resulting projection of the interference regions at the reciprocal-lattice points as diffraction spot
intensities is described by Eq. (6.10):

$$Hh + Kk + L\ell = 0 \qquad\qquad\qquad\qquad (6.10)$$

In indexing a spot pattern, or in generating one, we let the main beam
spot represent the origin (000), and measure the distance to each unique
diffraction spot symmetrically disposed with respect to the main beam spot
as a reference. The distances (radii) for a known crystal structure repre-
sent the magnitude $|g_{hk\ell}|$, as we should now be aware. Since the diffrac-
tion pattern is a reciprocal-lattice net-plane projection, the crystal surface plane (zone axis) can be determined by considering the values of
$|g_{hk\ell}|$ measured as spots in the pattern.

We can construct the diffraction net (a generated pattern) for any
zone axis from Eq. (6.10). Let us consider the three possibilities [H00],
[HK0], and [HKL]. In the first instance we have from Eq. (6.10) h = 0,
thus the diffraction net [H00] would be characterized by spots from planes
{0kℓ} with regard to the extinction conditions for a particular crystal
structure. For [HK0] or any equivalent {hhℓ} combination, the net plane
is characterized by {hhℓ}; and for [HKL] of course the condition is any
{hkℓ}.

Suppose we wish to construct the diffraction net for the [100] zone
axis of an fcc crystal section. We observe from Table 6.1 (or the fcc
structure factor) that the first reflections {0kℓ} which satisfy the zone-
axis criterion [Eq. (6.10)] will be {002}. Thus we have (002), (020),
(00$\bar{2}$), and (0$\bar{2}$0) symmetrically disposed with respect to (000) (the pattern
center). These can simply be placed on a "ring" of radius R_1. The position of these "diffraction spots" on the net are not arbitrary, and will
be determined by satisfying the geometrical relationship shown in Eq.
(6.57). Remember that a crystallographic direction is characterized by a
line from the center of the net through any particular diffraction spot,
so that when one spot is placed on the imaginary diffraction "ring" as
shown dotted in Fig. 6.31(a), other spot positions are determined by cal-

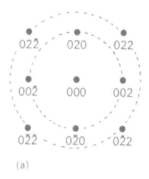

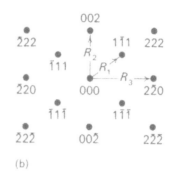

(a)   (b)

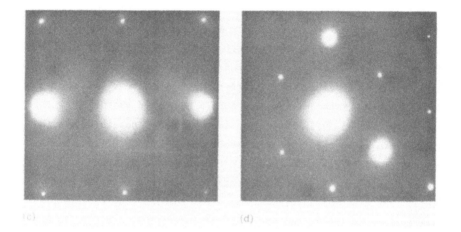

(c)   (d)

FIG. 6.31 *Construction of common fcc diffraction nets and compari-son with equivalent selected-area electron diffraction patterns from fcc stainless steel for 100 kV operation in electron microscope. (a) (100) diffraction net; (b) diffraction net; (c) (100) diffraction pattern; (d) (110) diffraction pattern.*

culating the angle between the respective crystallographic directions according to Eq. (6.57). The first spot placed becomes the pattern reference. So if we place the (002) spot, notice that its negative (00$\bar{2}$) will be diametrically opposite it. Then we use Eq. (6.57) to find the angle between [002] and [020]. We find that θ = 90°. Correspondingly, the (0$\bar{2}$0) spot is diametrically opposite. The next allowed reflections [satisfying Eq. (6.10)] are of the form {022} and we have (022), (0$\bar{2}\bar{2}$), (0$\bar{2}$2), and $(02\bar{2})$. These reflections can be placed on R_2 [dotted in Fig. 6.31(a)] by considering Eq. (6.57), we find the angle between [022] and [002] (our pattern reference) to be 45°. We place its negative $(0\bar{2}\bar{2})$ diametrically opposite, and continue. This scheme continues for higher-order reflections with the resulting construction of a [100] diffraction net as shown in Fig. 6.31a. The [100] zone axis is of course normal to the (100) plane. Consequently the diffraction net can be characterized as the (100) diffraction pattern, that is, the pattern that results for the (100) plane normal to the electron beam. Figure 6.31c. illustrates the corresponding (100) diffraction pattern from fcc stainless steel.

Let us consider the (110) plane in fcc. That is, let us construct the diffraction net for [110] in fcc. We observe again from Table 6.1 (or the fcc structure factor) that the first expected reflections are $\{1\bar{1}1\}$, that is, $(1\bar{1}1)$, $(\bar{1}1\bar{1})$, $(\bar{1}11)$ and $(1\bar{1}\bar{1})$ will satisfy the zone-axis criterion [Eq. (6.10)] zone. The next reflections are $\{002\}$, or (002) and $(00\bar{2})$. Then $\{2\bar{2}0\}$: $(2\bar{2}0)$ and $(\bar{2}20)$, etc. Figure 6.31b illustrates the corresponding (110) diffraction net. Figure 6.31d reproduces an equivalent (110) selected (110) selected-area electron diffraction pattern from fcc stainless steel.

We can continue this scheme for any zone axis construction for any crystal system. We must remember that the crystal system will determine the extinction conditions as given by Eq. (6.17). A number of the common diffraction nets for cubic crystals are given in Appendix A (Figs. A.6 to A.14) along with three orientations in the hcp system (Figs. A.15 to A.17).[†]

It should be apparent that three conditions are considered in the construction of electron diffraction nets (or patterns); the structure factor, the zone-axis criterion, and the calculation of the angle between crystallographic directions through individual spots or points. This system is easily computerized as shown in Appendix C, and diffraction patterns can be generated for any system and at any scale desired by choosing the appropriate parameters.

Having constructed such diffraction nets or equivalent surface-plane projections (HKL) for a particular crystal system, the indexing of selected-area electron diffraction patterns can be performed simply by comparison, as noted in Sec. A.4.1. The indexing can also be facilitated in some cases by comparing the ratios of the magnitudes of the reciprocal-lattice

---

†*It will be instructive at this point to work Prob. 6.9.*

vectors $|\underline{g}|$. This can be illustrated with reference to Fig. 6.31(b) where the radii indicated as R_1, R_2, R_3. etc., correspond to the magnitudes $|g_{hk\ell}|$. Thus, by comparing $R_1/R_2, R_2/R_3, R_3/R_3$, etc., we can compare the indexed spots quite easily due to the fact that the camera constant cancels and only the relative dimensions (radii) must be measured. We observe from Fig. 6.13(b) that

$$\frac{R_1}{R_2} = \frac{\sqrt{3}}{2} \qquad \frac{R_2}{R_3} = \frac{2}{2\sqrt{2}} \qquad \frac{R_1}{R_3} = \frac{\sqrt{3}}{2\sqrt{2}}$$

The ratio method of indexing is then expressed generally by

$$\frac{R_n}{R_m} = \frac{|g_{h_n k_n \ell_n}|}{|g_{h_m k_m \ell_m}|} \tag{6.58}$$

The spot diffraction pattern can also be checked by measuring the angles between spot directions as indicated previously for establishing the preferred orientations of texture patterns. The use of stereographic projections in these investigations also aids in the ultimate identification of a crystal orientation, or in establishing crystallographic directions. For the most part we see that indexing constitutes a system of calculated trials and errors. A method for the exact determination of orientations of thin films has been outlined by H. M. Otte and coworkers [27].

We have been concerned so far with the determination of crystal (grain) surface orientation. We should also be aware that rotation of the diffraction pattern into proper coincidence with the image will also characterize the crystallographic directions in the image. It is obvious, however, that for symmetrical diffraction patterns, there exists a 180° ambiguity in the actual assignment of the indices to the diffraction spots. In addition the diffraction pattern may not represent an exact orientation; this can be noted in some instances by an unsymmetrical disposition of spot intensities about the main beam spot. This is a direct indication that the foil is bent or tilted so that the reciprocal-lattice interference regions are being intersected unsymmetrically by the reflection sphere (Fig. 6.16).

Where the crystal system is unknown, the indexing becomes somewhat more difficult. In these cases the ratio method can be employed to give some indications of g, the reciprocal-lattice vectors. In difficult situations the indexing must be preceded by a structure analysis (Sec. 6.9).

It should also be borne in mind that where some ambiguity may develop in the indexing of spots in the diffraction pattern, it may be helpful to check the angle between directions from a known reference. That is, where one spot is accurately indexed, the angle between the corresponding direction and that for another spot may be measured, and the unknown indices calculated from Eq. (6.57) or Eqs. (A.3) and (A.4). This is also a convenient means to check the degree of tilt of the reflection plane. We have also demonstrated previously that crystallographic orientation can be determined from surface traces by measuring the angles between two or more nonconcurrent traces (review Prob. 5.11).

6.8.5 *CRYSTALLOGRAPHIC DETERMINATION FROM KIKUCHI DIFFRACTION PATTERNS*

We can recall from Sec. 6.6 that the distance between a Kikuchi pair is equivalent to a corresponding $|\underline{g}|$. Thus, from the geometry of their formation (Fig. 6.20), we observe that a parallel line midway between a Kikuchi pair represents the trace of the corresponding diffraction plane (hkℓ) characterized by the dimension $|g_{hk\ell}|$. It is perhaps obvious then that where two concurrent Kikuchi-line pairs occur in a diffraction pattern, their diffraction plane traces $(h_1k_1\ell_1)$ and $(h_2k_2\ell_2)$ will intersect at the corresponding zone axis satisfying the conditions of Eq. (6.10). For an exact case, where the zone axis is parallel to the electron beam, a Kikuchi-line pattern can be constructed in the same way spot patterns (diffraction nets) were constructed in Sec. 6.8.4. We simply consider that the excess and deficiency lines of a Kikuchi pair must lie midway between the reciprocal-lattice spots corresponding to (hkℓ) and ($\bar{h}\bar{k}\bar{\ell}$). In effect, the excess Kikuchi line will correspond to (hkℓ) and the deficiency line to ($\bar{h}\bar{k}\bar{\ell}$). The Kikuchi-line pattern for any orientation is then easily constructed by drawing the corresponding Kikuchi lines perpendicular to the lines joining the origin (000) to the various reciprocal-lattice points or spots (diffraction radii) in the diffraction net, with the lines passing through the midpoints so that each Kikuchi-line pair is symmetrical about (000). Figure 6.32 illustrates the technique for the [111] zone axis for fcc. Note the relationship to Fig. 6.22 (although in Fig. 6.22 the inability to distinguish excess and defiency lines is due to anomalous absorption effects, which are treated briefly in Chap. 7).

The indexing of Kikuchi diffraction patterns thus involves the assignment of indices to the pairs, and the orientation unique to these is then determined by the construction of the parallel (hkℓ) trace midway between

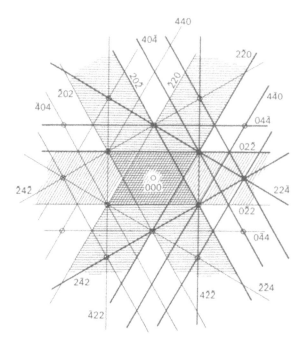

FIG. 6.32 *Kikuchi-line pattern construction for fcc structure with electron beam exactly along [111]. Corresponding diffraction net is shown by open circles. Heavy lines are excess Kikuchi lines and light lines are deficiency Kikuchi lines. Shading accents Kikuchi bands and illustrates three-fold symmetry.*

FIG. 6.33 *Indexing of Kikuchi diffraction patterns. Pairs of Kikuchi lines are chosen as indicated by 1'1, 2'2, and 3'3; dotted center line for each pair represents the corresponding planes $(h_n k_n \ell_n)$. The h,k,ℓ values are calculated by measuring distance between pair n'n as indicated in Eq. (6.22a). Intersection of dotted midplanes corresponds to poles, indicated in the figure as A,B, and C, have indexes $[H_m K_m L_m]$. Indexes of these poles are calculated by using Eq. (6.10). This process is facilitated by knowledge of approximate zone axis and reference to standard stereographic projections (Appendix A). Having located two or more poles in this fashion, pattern center is located as shown. After H, K, and L from Eq. (6.57) angle θ between the Kikuchi poles A,B, and C, and zone axis [HKL] can be directly measured by calibrat-*

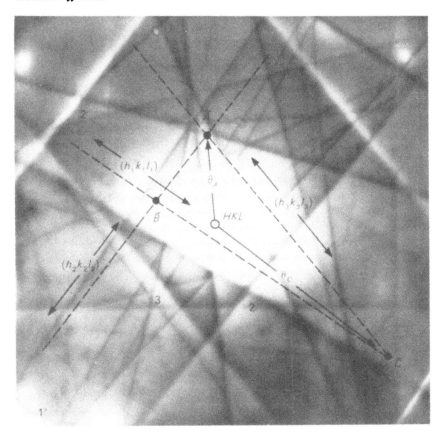

FIG. 6.33 (con't)
ing distance and angle between indexed poles A,B, and
C. Thus, if D_{AB} is distance between A and B, and the
corresponding angle is, from Eq. (6.57),

$$\theta_{AB} = \cos^{-1} \frac{H_A H_B + K_A K_B + L_A L_B}{\sqrt{H_A^2 + K_A^2 + L_A^2} \sqrt{H_B^2 + K_B^2 + L_B^2}}$$

Then the scale factor is simply

$$S_F = \frac{\theta_{AB}}{D_{AB}}$$

in degrees/unit length. Now measure distances between Kikuchi
poles and zone axis to obtain simultaneous equations (6.57) from
which H,K, and L are obtained. This should convince you of the
utility of comparative methods such as Kikuchi map. Pattern
is from stainless steel using 125 kV electrons.

the pair by the location of corresponding zone axes. Figure 6.33 illustrates the indexing of Kikuchi lines and the location of the corresponding
zones. The crystallographic zone axis for the specimen area, that is, its
orientation, is then determined very accurately by using a stereographic
projection, or by direct calculation from the angles subtended by the Kikuchi poles and the main beam location as shown in Fig. 6.33, using Eq. (6.57)
with the distance from the center pole calibrated in degrees as shown.

As we have pointed out with regard to indexing spot diffraction patterns, the crystallography is not unique as a result of the 180° ambiguity
in the assignment of indices as $\pm |\underline{g}|$. The Kikuchi diffraction pattern is,
however, unique in this respect. We should also be aware that the Kikuchi
lines will deviate from the spots or from the exact Bragg condition for any
tilt of the electron beam or specimen as shown in Sec. 6.6. Consequently
orientations determined from Kikuchi patterns are exact, provided three
nonconcurrent Kikuchi-line pairs giving rise to three Kikuchi poles are present in the diffraction pattern [28]; or two Kikuchi poles as discussed in
reference 27.

Levine and co-workers [29] introduced the use of Kikuchi maps in the
solution of Kikuchi diffraction patterns, extending the accuracies of identifying image crystallographies and greatly simplifying the procedures of
identification. The Kikuchi map is a projection of reciprocal space akin
to the stereographic projection. In effect the concept is identical; and
if exact Kikuchi diffraction representation or reciprocal space over a finite angle of tilt from some zone are obtained, the total composition is
the Kikuchi map. The importance of this concept is that Kikuchi diffraction patterns can now be indexed simply by comparison with a standard Kikuchi projection (Kikuchi map)† essentially in the form of Fig. 6.32 for
large deviations from the zone axis. This means of course that any pattern
can be uniquely identified; and it is no longer necessary to have two or
three Kikuchi poles identified in a pattern. Figure 6.22 is in fact the
central portion of a [111] Kikuchi map.

---

†*Standard Kikuchi projections for a [100], [110], and [111] cubic zone
axis are available by order from Technical Information Division, Building
30, Lawrence Berkeley Laboratory, University of California, Berkeley,
Calif., by quoting the following: [100] map; Zn 5270 and MUB 7890. [110]
map; Zn 5274 and MUB [111]; Zn 5279 and MUB 7939.*

Kikuchi maps can be constructed as a montage of Kikuchi patterns within a unit triangle of the stereographic projection. Figures 6.34 and 6.35 illustrate two examples of Kikuchi maps for an fcc and a bcc material respectively. Note in particular the symmetry features associated with specific poles, and especially [111] which can be compared with the Kikuchi pattern for silicon shown in Fig. 6.22, and Fig. 6.32. You might carefully compare the unit triangles of these figures with those apparent in the field-ion micrographs in Chap. 2, and the stereographic projections in Appendix A especially Fig. A.20. A consideration of Prob. 6.13 might be approximate at this point. Also note that the approximate determination of the crystallographic pole in the Kikuchi pattern of Fig. 6.33 simply involves the location of this pattern portion on the Kikuchi map of Fig. 6.34 by inspection. This method provides accuracies on the order of about ±2%. The reader is cautioned that our treatment of indexing of the pole of Kikuchi diffraction patterns and the corresponding beam direction has been rather sketchy, and where more accurate indexing is required, the original references should be consulted for a more detailed treatment of the procedures.

Some applications of Kikuchi diffraction patterns have been outlined by G. Thomas [30] and R. E. Villagrana and Thomas [31] in addition to references already noted. It should be apparent that where Kikuchi lines are available, they can aid in the accuracy of the analysis of a diffraction pattern insofar as determining crystallographic uniqueness.

It must be pointed out, perhaps somewhat prematurely insofar as we have not discussed at length the thin-film image geometry corresponding in the selected-area electron diffraction pattern (either of a spot or Kikuchi type), that identification of a crystallographic surface orientation, no matter how exact, does not ensure coincidence with the true object surface. That is, the crystallographic surface need not be coincidental with the true specimen surface. This can occur because of bending of the foil, irregular object features resulting from the preparation of the foils, and related features. Thus, where accuracies in determining object geometries are required, this fact must be considered. We shall delay until Chap. 7 further details of this feature.

6.8.6 *CONVERGENT BEAM ELECTRON DIFFRACTION (CBED)*

In the formation of selected-area electron diffraction patterns in the transmission electron microscope as illustrated schematically (and very

FIG. 6.34 Kikuchi map of [001] - [011] - [111] unit triangle in fcc type 316 stainless steel (Courtesy of Dr. Nestor J. Zaluzec, Argonne National Laboratory).

FIG. 6.35 *Kikuchi map of* [001] – [0$\bar{1}$1] – [$\bar{1}$1$\bar{1}$] *unit triangle in bcc tungsten. (Courtesy of Dr. Nestor J. Zaluzec, Argonne National Laboratory).*

superficially) in Fig. 6.27 and discussed throughout this chapter, a nearly
parallel beam of electrons illuminates a selected area having a diameter of
roughly 0.5 μm. Systematic diffraction arising from the illuminated area
is then focused onto the back focal plane (or Gaussian image plane) of the
objective lens in the electron microscope, forming the diffraction spot
patterns which have been illustrated above. With the ability to form very
small electron probes (0.5 μm to 0.05 μm and smaller) using high-brightness
field-emission sources, etc., a highly convergent illumination occurs in
the specimen, and the initially well-focused diffraction spots are broadened into discs whose angular dimensions correspond to the incident beam convergence. These features (the formation of a focused electron diffraction
spot pattern and the formation of a convergent-beam induced diffraction
spot-broadened pattern) are illustrated schematically in Fig. 6.36. Since
the convergence angle forming the probe determines the angular dimensions
of the broadened spots in Fig. 6.36, the broadening will increase with an
increasing convergence angle. This is illustrated in Fig. 6.37. Within
broad discs of the CBED pattern (especially at 2.2 x 10^{-2} radians beam
convergence in Fig. 6.37) one can clearly observe the presence of a regular
array of fine higher-order Laue zone lines (or HOLZ lines) similar to those
described in connection with, and illustrated by, Fig. 6.23. These lines
arise as a result of three-dimensional diffraction effects and are, like
Kikuchi lines or Kossel lines, extremely sensitive to specimen orientation.
They are also very sensitive to lattice parameter and lattice parameter
change within a very small probe area, local strain fields and crystal
symmetry. As a consequence, they can be used as very sensitive measures of
these parameters. Thus, while CBED patterns have poor angular resolution,
they contain a substantial amount of information not contained in the SAD
or SAED spot pattern.

The similarities of electron channeling patterns shown in Fig. 6.23
and the convergent beam electron diffraction patterns is illustrated by
comparing Fig. 6.23 with Fig. 6.38 which shows the central portion of a
[111] zone axis pattern from a silicon single crystal foil. The HOLZ lines
of the central zone pattern clearly indicate the three-fold symmetry of the
crystal region illuminated. Information concerning specimen thickness,
crystal potential, structure factor, and the crystallographic space group
can also be obtained by analyzing this diffraction detail in the broadened
disc.

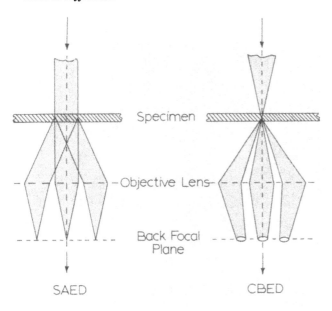

FIG. 6.36 *Simple ray diagrams illustrating conventional, selected-area electron diffraction (SAED) and convergent beam electron diffraction (CBED).*

To accurately analyze pattern symmetry it is necessary to be able to observe one or two dark-field discs as well as the bright-field (000) disc, and this sets a minimum on the angular field of view required, and can vary from roughly 3 to 15°. The higher angles are particularly necessary for viewing the higher order Laue zones and are achieved by short camera lengths.

Convergent beam (CB) patterns or (CBP's) are very simple to achieve if a very small probe can be obtained. One simply goes to diffraction when an area has been illuminated by a sharply focused beam and the discs should be visible. When properly focused by the intermediate lens the shadow image of the specimen (and all real space detail) vanishes, leaving the CBP detail. The probe can be formed in a number of ways (using both condenser lenses or one condenser and the objective lens).

Applications of CBED patterns There are numerous applications of CBED patterns and the technique itself has been known for many years. Kossel and Möllenstedt [32] described the technique as early as 1939, but like many other analytical concepts, instrumental limitations, especially the inability to achieve very small probe sizes in the electron microscope, have

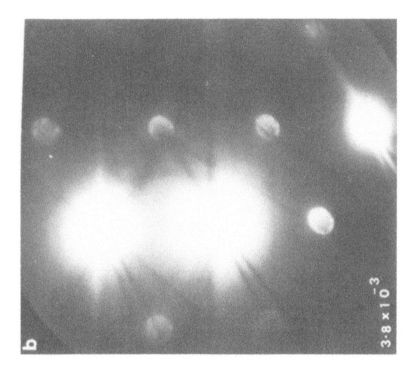

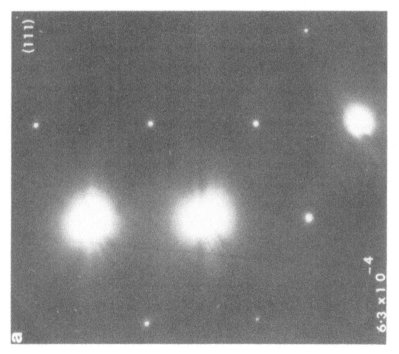

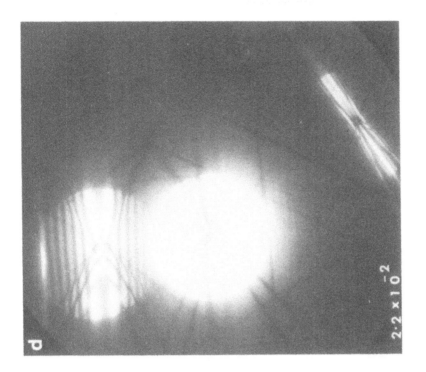

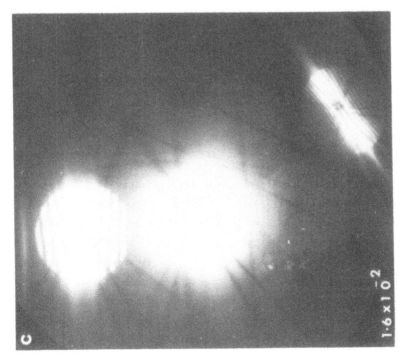

FIG. 6.37 *Comparison of electron diffraction patterns for various values of incident beam convergence indicated (in radians). Conventional spot pattern-to-convergent beam pattern transitions in (111) Si film. (Courtesy Dr. Nestor J. Zaluzec).*

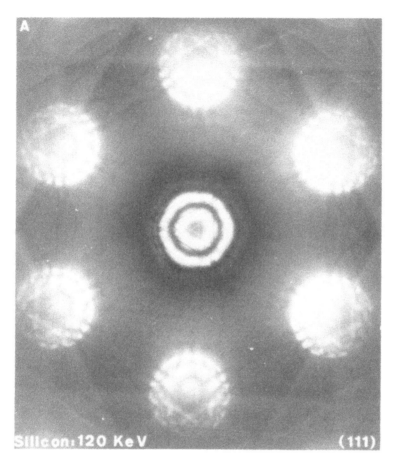

FIG. 6.38 [$\bar{1}11$] convergent beam zone axis patterns from Si
single crystal film showing Kikuchi lines and HOLZ (also
hoLz) lines.(A)   Courtesy Dr. Nestor J. Zaluzec.

limited the applications of the technique. Some applications have been
described by Buxton, et al. [33]. We might mention that, as with all diffraction phenomena, the features recorded in the diffraction pattern occur
by Bragg diffraction from crystal planes. Consequently, there is indeed a
common connection between Kossel line patterns, Kikuchi patterns, convergent beam patterns, channelling patterns and the like, and the similarities in symmetry and line-structure detail would be expected to exhibit
phenomenological similarities as we can indeed observe on comparing Figs.
6.21, 6.22, 6.23, 6.34, and 6.38 for example. We must point out that the
convergent beam electron diffraction patterns are frequently referred to
as Kossel-Möllenstedt patterns [34]. If the pattern is formed by a con-

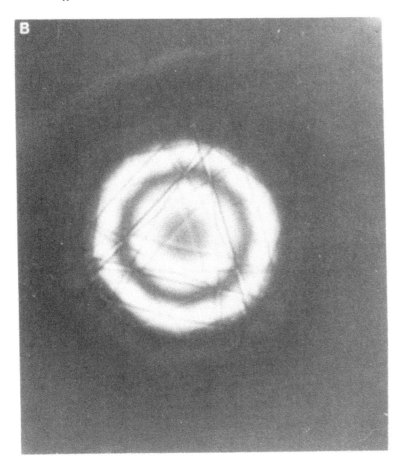

FIG. 6.38 (continued)   (B) *Enlarged view of central zone pattern in* (A).

vergent beam having a very large cone angle (~ 15°), the diameter of the illuminated area is increased and the resulting pattern, resembling the Kikuchi patterns shown in Figs. 6.21 and 6.22, is called a Kossel pattern [34], which, like Kikuchi patterns, displays an area of reciprocal space. Consequently, it is possible to collect the high symmetry, low index zone axes CBED pattern portions and display them on a stereographic projection or zone axis pattern (ZAP) map in a fashion which phenomenologically resembles the Kikuchi maps shown in Figs. 6.34 and 6.35 [35]. These can be useful, as described by Stoter [35], for the identification of precipitate phases in steels for example where the lattice parameters are very close (as shown in Fig. 4.33). CBED can be a very powerful technique in support of EDX analysis and EELS analysis in the analytical electron microscope

having small beam focus capabilities. We will in fact discuss this feature
in more detail in Chap. 7 where the applications of analytical electron
microscopy are described.

   Shaw, et al [36] have recently described the applications of CBED to
the study of precipitate structure, while Kersker, et al [37] have discussed the use of CBED fine structures to measure local strain fields between
coherent precipitates and the matrix. Cowley [38] has also discussed the
applications of microdiffraction in dedicated STEM instruments, and you
should recognize that CBED, as a result of the very small spot size characterizing the focused (convergent) beam, arises from a very small region
(microdomain) of the sample.

6.8.7 SECONDARY EFFECTS IN ELECTRON DIFFRACTION

Our treatment of selected-area electron diffraction spot patterns so far
has concentrated on the primary diffraction features, namely, single diffraction from textured samples or single-crystal grain areas. In practice
electron diffraction patterns frequently exhibit reflections other than the
allowed and expected matrix reflections. The nature of such spots may be
in direct violation of the extinction rules for the matrix, that is, forbidden reflections. Such secondary diffraction effects may arise as a result of a number of features we shall now discuss. Chief among these is
the occurrence of twin spots, double diffraction spots, multiple diffraction effects and the various diffraction effects contributed by oxide
films and other reaction products on the film surface, included phases,
precipitates, crystal ordering, etc., which may constitute the selected
area from which the diffraction signal originates.[†]

Origin and analysis of twin reflections In cubic crystals, a 180° rotation
of the matrix crystallography about some axis (the twin axis) results in a
twinned crystal portion. Another way of putting it is to say that in real
space, twinned regions when compared crystallographically with the matrix
regions are simply mirrored across some {hkℓ} plane. Consequently the
twinned reciprocal lattice can be derived from the matrix reciprocal lattice by a 180° rotation about a corresponding <hkℓ> direction.

   For bcc crystals, the twinning plane is {112} and the rotation axis
is therefore <112>. In the case of fcc crystals, the twinning plane is of
course coincidental with the slip planes or {111}. Consequently for fcc

---

†*In nonprimitive lattices surface layers corresponding to partially filled
unit cells can give rise to forbidden reflections as discussed by W. Krakow,
Surface Sci., 58: 485 (1976).*

the rotation axis is <111>. Since fcc twinning is the more common of the cubic crystals, we shall confine our discussions primarily to fcc crystals.

We can observe in Fig. 6.39 the formation of a twin band in an fcc matrix having an ABC atomic periodicity as indicated. The twin planes are indicated by the arrows, and we might recognize these areas to represent ideally the occurrence of a stacking fault. In effect a twin band in fcc occurs by successive faulting on every {111} plane as shown, thus causing a 180° inversion of stacking order.

Twins in thin fcc and bcc foils can occur as a consequence of annealing, growth accommodation (twin-faults or microtwins), or deformation (mechanical twins). Regardless of the twin character, reflections will occur in the diffraction pattern having a definite relationship to the allowed matrix spots. In fcc crystals, the twin spots generally occur as $1/3 \; g_{hk\ell}$. Numerous schemes are available for directly calculating twin reflections {hkℓ}T for any particular matrix orientation [39,40]. You are urged to consult these references for the details of twin-spot calculations since our discussion here will of necessity be limited to simple qualitative descriptions.

Let us consider the case of twinning in a thin fcc film having a (110) surface orientation (see Fig. A.7). If we assume the twin plane to lie in the [$\bar{1}$12] direction, the twin axis perpendicular to this direction [and perpendicular to the (1$\bar{1}$1) plane] is [$\bar{1}$11] Consequently a rotation of the (110) spot pattern (diffraction net) about the [1$\bar{1}$1] axis by 180° will position the twin spots (T) as shown in Fig. 6.40. We can observe from Fig. 6.40 that the twin spots (T) are characterized by 1/3 <111> if any matrix spot in proximity to the twin spot is taken as the reference 000. For twins along the [1$\bar{1}$2] direction in the fcc (110) orientation, a rotation of the pattern about [$\bar{1}$11] would then produce a similar disposition

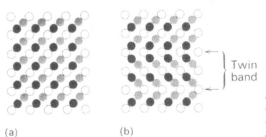

(a)   (b)

Twin band

FIG. 6.39 *Stacking sequences of {111} planes in fcc crystal. (a) Perfect crystal; (b) crystal containing twin band with boundaries indicated by arrows. Stacking sequence is ABC.*

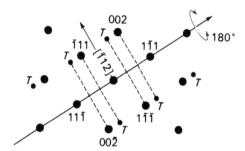

FIG. 6.40 *Location of twin spots (T) for twinning along $[\bar{1}12]$ for (110) fcc surface orientation resulting from rotation of matrix pattern about $[1\bar{1}1]$. (See Fig. A.7.)*

of new twin spots. However, for twinning along $[1\bar{1}0]$ for the (110) orientation in fcc, the twin spot would be coincidental with the matrix spots.

We can illustrate the occurrence of two sets of twin spots by the analysis of thin vapor-deposited Au foils grown epitaxially on (100) rock-salt substrates. In these films, as in many thin-film materials vapor deposited in vacuum, the nucleation and growth processes necessitate the formation of accommodation microtwins. These become randomly disposed on the {111} planes inclined 55° to the (100) surface, making angles of 90° with one another. The resulting twin spots are therefore disposed at right angles about the matrix spots as illustrated in Fig. 6.41a. Figure 6.41b shows the corresponding thin-foil structure with the microtwins and occasional stacking faults imaged by diffraction contrast projection of the {111} plane sections containing them (see Chap. 7). The selected-area diffraction pattern of Fig. 6.41 a has been rotated coincident with the image of b. The twin spots in Fig. 6.41a are identified as 1/3 <110>.

You might find it convenient, in reflecting upon this example, to realize that the indices of a reciprocal lattice point [PQR] for the twinned crystal will be related to the reciprocal lattice point [pqr] of the matrix or original crystal after twinning on the (hkℓ) plane by the following:

$$P = \frac{p(h^2 - k^2 - \ell^2) + q(2hk) + r(2h)}{h^2 + h^2 + \ell^2} \qquad (6.59a)$$

$$Q = \frac{p(2hk) + q(-h^2 + k^2 - \ell^2) + r(2k\ell)}{h^2 + k^2 + \ell^2} \qquad (6.59b)$$

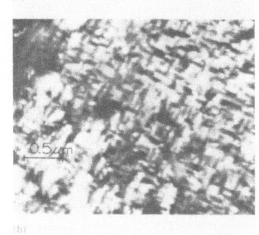

FIG. 6.41 *Microtwins in vapor-deposited (100) Au. (a) Selected-area electron diffraction pattern (100) showing "satellite", twin and double diffraction spots (twin spots are brightest); (b) electron transmission image showing microtwins and {111} faults in projection. (100 kV accelerating voltage)*

$$R = \frac{p(2h\ell) + q(2k\ell) + r(-h^2 - k^2 + \ell^2)}{h^2 + k^2 + \ell^2} \tag{6.59c}$$

For a rotation of $180°$ about the normal to $(hk\ell)$, the twinning plane, we can rewrite this in matrix form as

$$(PQR) = T_{hk\ell} \cdot (pqr) \tag{6.60}$$

or

$$T_{(hk\ell)} = \frac{1}{h^2 + k^2 + \ell^2} \begin{pmatrix} h^2 - k^2 - \ell^2 & 2hk & 2h\ell \\ 2hk & -h^2 + k^2 - \ell^2 & 2k \\ 2h\ell & 2k\ell & -h^2 - k^2 + \ell^2 \end{pmatrix} \tag{6.60a}$$

Equation (6.60a) can be regarded as a general twinning matrix for the cubic
system, and in the fcc system, where twinning occurs on {111} planes, the
indices discussed in connection with the example given in Fig. 6.40 can be
readily obtained. Remember that the indices of the reciprocal lattice
points are the Miller indices of the diffracting planes, and that it is
important in dealing with diffraction patterns to know the indices of the
diffracting plane. For example Eq. (6.60a) will not be the same for twinning on (111) planes and ($\bar{1}$11) planes. Consequently, for twinning on (111)
planes in the (110) fcc surface orientation shown in Fig. 6.40, the twin
spots will coincide with the matrix spots. You might prove this by solving
Eq. (6.40a) for these twinning conditions. Note that the factor 1/3 will
appear regardless of the sign of the indices of the twinning plane, and
that twin spots in the diffraction pattern for fcc either coincide with
the matrix spots or are positioned 1/3 along <111> directions. Similar
features can be observed in bcc and hcp systems, and the reader might consult the paper by Bullough and Wayman [41] or the work of Johari and Thomas
[42].

  We should note in Fig. 6.41a that only two of the spots in association
with a matrix spot for (100) (see Fig. A.6) are twin spots. The opposite
two spots, while identically disposed, occur by double diffraction, which
we shall now treat briefly.

Double diffraction We observed in Sec. 6.6 that where diffuse scattering
occurs (a dynamical consideration), Kikuchi diffraction will result in perfect crystals. In situations where a single diffraction intensity corresponding to a particular reflection $g_{hk\ell}$ is again diffracted, a forbidden
spot or double diffraction spot may occur in the diffraction pattern [43-
45]. That is, a diffracted beam originating by reflection from $\{h_1k_1\ell_1\}$
planes may again diffract from another set of planes $\{h_2k_2\ell_2\}$. The double
diffraction spot that appears in the diffraction pattern will then have
indices $h_1 \pm h_2$, $k_1 \pm k_2$, $\ell_1 \pm \ell_2$.

  We can examine Fig. 6.34a in retrospect with this feature in mind,
and observe that the spots opposite the twin spots result by double diffraction of the twin reflections. Thus the linear deposition of the twin
spots and double diffraction spots corresponds to 2/3 <110>, that is,
$h_1 + h_2$, $k_1 + k_2$, $\ell_1 + \ell_2$.

  Double diffraction spots can occur in a diffraction pattern whenever
the condition outlined above is satisfied. Thus, stacking faults, phases,
precipitates, or other structural features in thin films can give rise to

double diffraction. It is a good policy to check this feature first when encountering forbidden reflections in the selected-area electron diffraction pattern.

Actually, as we imply above, double diffraction can occur in any crystalline regime in which a diffracted beam propagates and is re-diffracted. In fcc and bcc cyrstals which are not twinned, double diffraction does not introduce extra spots because the combinations of any two diffracted beams will only generate allowed reflections for these structures. However when the twin reflection is involved, a double diffraction spot can arise in association with each twin spot. In the case of thin twins inclined to the incident beam direction (and the specimen surfaces) as shown in Fig. 6.41, the intensities of the twin and double diffraction spots can vary as the beams double diffract within the matrix and the twin volume, and are added. Foil buckling also influences the diffraction intensities.

A special case of double diffraction can occur for overlapping crystals, and this is a common phenomenon when foils containing a thick oxide layer or other reaction product are examined in the electron microscope. The double diffraction which occurs from overlapping crystals was treated initially by Bassett, et al. [46] in connection with moire pattern formation associated with images of overlapping crystals, where they demonstrated that the spacing associated with the spacing of the double diffraction spots, with reference to the main beam (000), was equal to the spacing of the moire pattern. We will treat moire images in more detail in Chap. 7.

Obviously if two crystals overlap, each will contribute a primary diffraction spot. But in addition, rediffraction in crystal 2 of a diffracted beam generated in crystal 1 can produce a double diffraction spot in the diffraction pattern. In effect, the double diffraction spot arises from a beam diffracted by both crystals. This process is complicated by rotation of the two crystals and in situations where the overlapping crystals have different compositions, crystals structures, or lattice parameters.

Multiple diffraction effects You can readily appreciate the fact that where two or more crystals overlap, multiple diffraction effects occur as a result of the differences noted above and the relative incidence (rotation and/or translation) of the crystals. This is also true for overlapping layers of polycrystalline materials.

Figure 6.42 offers an example of the diffraction effects which can arise from two simple, overlapping crystal layers. Here a single crystal

FIG. 6.42 *Multiple diffraction effects produced in overlapping*
(001) palladium films. Upper portion shows single Pd layer while
lower portion shows the double layer. The superimposed SAD pat-
terns indicate the associated diffraction phenomena.

film of (001) palladium prepared by vapor deposition onto a (001) NaCℓ
substrate has been folded upon itself. The upper portion shows the single
layer and its (001) fcc diffraction pattern while the lower (darker) region illustrates the bi-layer system. The rotation of the lattices is
readily observed as the angle between any two bright, primary diffraction

spots. Note that in addition to the primary diffraction patterns, the diffracted beams in the first layer act as primary beams for diffraction in
the second layer. Rotation has a very pronounced effect on the diffraction
effects observed for even simple, overlapping crystals, and these effects
are illustrated in Fig. 6.43. Figure 6.44 illustrates the additional diffraction effects characteristic of three overlapping (001) palladium layers
each having a slightly different rotational coincidence, as readily apparent from the displacement of the primary diffraction spots for each layer.

Multiple diffraction effects can also occur, as indicated previously,
by overlapping polycrystalline layers - for example a very small-grain polycrystalline layer overlapping a single crystal layer, or by transformations
and crystal growth within a diffracting volume. These features are illustrated in Fig. 6.45. In Fig. 6.45(a) we observe diffraction spots from a
single crystal of ErO (fcc) acting as primary beams for a thin overlapping
hcp Er foil having a random orientation of fine grains [47]. The diffraction spots (or more accurately the beams associated with them in the sample)
give rise to hcp rings observed only for the more intense hcp reflections.
Such patterns frequently occur for two superimposed materials having different structures, as in the case of electroplated materials or other com-
plexes [48].

Multiple diffraction effects can also occur in electron diffraction
patterns of complex zone transformations. This situation is illustrated
in Fig. 6.45(b) which shows diffraction originating from hcp Er undergoing
oxidation and structural transformation to bcc Er_2O_3 [47]. The profusion
of spots occurs from texturing of the Er grains, double diffraction, twin
reflections, single-crystal spots from transformed areas of Er_2O_3, and
other unidentified reflections. Such diffraction patterns can generally
be characterized by multiple diffraction effects.

Double diffraction and general multiple diffraction effects can also
occur for various superlattice arrangements because each structural element
of the superlattice array will give rise to a discrete diffraction profile,
completely forbidden by the structure factor for the homogeneous alloy. An
excellent treatment of diffraction from superlattices can be found in the
book by C.S. Barrett and T.B. Massalski [49].

It was the aim of this subsection, and especially in presenting the
array of complex diffraction patterns shown in Figs. 6.42 to 6.45, to im-
press upon the reader the rather complex possibilities which need to be
kept in mind when examining thin films in the transmission electron micro-

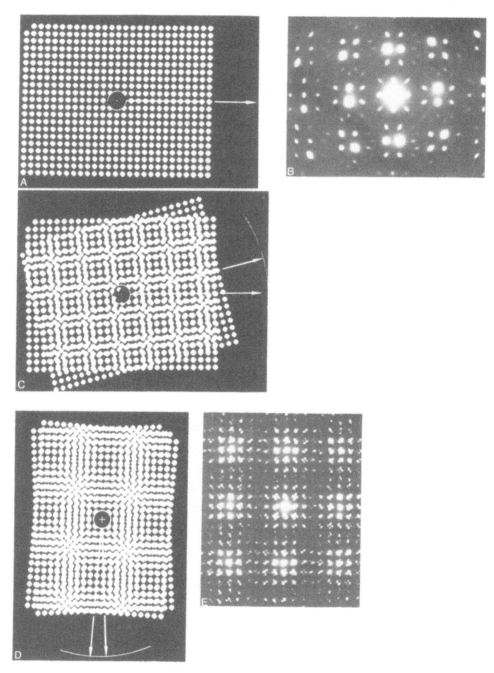

FIG. 6.43 *Simple cubic gratings and diffraction patterns illustrating the origin of multiple diffraction effects in overlapping, rotated, single crystals having the same orientation and lattice parameter. The simple cubic (001) gratings used are compared with (001) palladium overlapping films at approximately the same angle of rotation. The gratings obviously show only primary diffraction from each layer.*

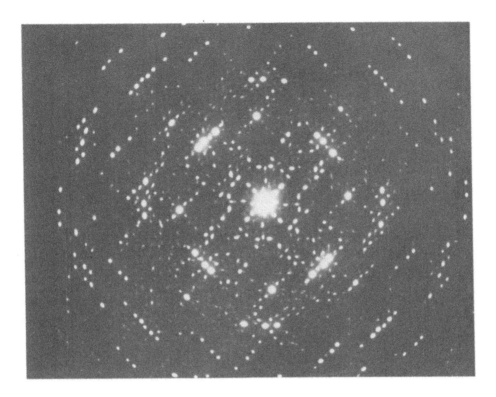

FIG. 6.44 *Selected-area electron diffraction pattern for three overlapping and independently rotated (001) palladium foils.*

scope. This is especially significant when oxide films and other reaction products are present, or where forbidden and unexpected spots appear in the diffraction pattern. Although complex in appearance, many diffraction patterns exhibiting multiple diffraction effects can be modelled in simple ways as illustrated in Fig. 6.43, and geometrical methods for constructing computer-generated electron diffraction patterns can be utilized, allowing for computer studies of oxidation of metals, oxide layer epitaxy, and related phenomena [50].

Thermal diffuse effects In short-range ordered alloys and other short-range ordered solids, superlattice spots can be broadened in specific directions as a result of aging treatments which tend to rearrange the structure. This gives rise to thermal diffuse scattering which appears in the diffraction pattern as broad arcs or what appear to be broad arcs and curved streaks of intensity. These features arise from the Ewald sphere cutting walls of intensity created by diffraction from these ordered

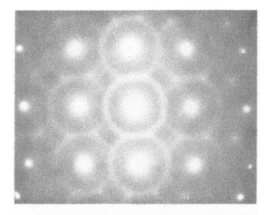

(a)

(b)

FIG. 6.45 *Multiple diffraction effects. (a) Diffracted beams (spots) originating in fcc ErO single crystal act as primary beams for diffraction from hcp Er layers in which ErO is embedded, causing ring reflection for prominent hcp ring intensities. (b) Profusion of multiple diffraction effects including twin and double diffraction spots in thin Er foil undergoing oxidation to bcc Er_2O_3. (100 kV beam.) ((a) from ref. 47)*

regimes, and can be altered in appearance by tilting the foil in the electron microscope, thereby altering the intersection of the intensity zones in reciprocal space with the Ewald sphere. Examples of these effects can be found in the early work of Honjo, et al. [51] and others [52].

Streaking of diffraction spots We can observe in Fig. 6.41a that the twin spots and double diffraction spots appear to be enjoined by faint streaks. This feature is quite common in cases where the thin foil contains effective scattering domains such as closely spaced faults or twins, or regular arrays of precipitates. In the general case, the scattering domains assume the form of thin rectangular sections having a geometrical-crystallographic relationship with the thin-foil section. The streaking of diffraction spots (intensity projections of the reciprocal-lattice interference re-

gions) can be explained qualitatively by the kinematical assumptions outlined in Sec. 6.5.1, with particular reference to Fig. 6.15. In essence we
can treat the regions of closely spaced twins or faults or precipitate
platelets, etc., as finite (very thin) parallelepipeds. The interference
regions for diffraction corresponding to these domains will then exhibit
an intensity spike extended normal to the plane of the domains (the {111}
planes in fcc materials containing faults or twins, or precipitate platelets along {111}). The intersection of the reflection sphere with these
intensity spikes will then exhibit streaking as demonstrated schematically
in Fig. 6.46.

Figure 6.47 shows a transmission image of a thin foil electrolytically
thinned (see Appendix B) from a machined chip of 304 stainless steel and
the corresponding selected-area electron diffraction pattern in the correct
rotation coincidence. Figure 6.47(a) indicates profuse twinning in (1Ī1)
planes normal to the (110) grain surface. This condition effectively
creates alternate and irregular domains of twinned and untwinned parallel-
lepipeds having their thin dimension perpendicular to the electron beam,
and their surfaces normal to [1Ī1] (see Fig. A.7). Streaking of the resulting twin spots, which are disposed as shown in Fig. 6.47 (1/3[1Ī1]), and

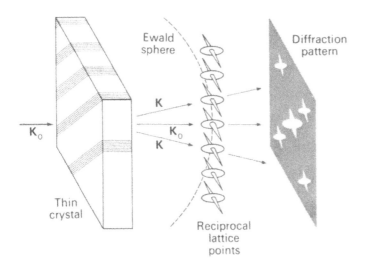

FIG. 6.46 *Origin of intensity streaking in electron
diffraction patterns containing faults, fine twins,
precipitates, or any related scattering domains using
kinematical assumptions.*

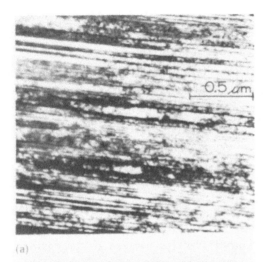

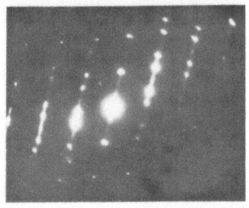

FIG. 6.47 *Prominent streaking and twin reflections observed for machined stainless steel containing dense domains resulting for fine deformation twins. (a) Electron transmission image showing twinned regions with boundaries viewed edge-on; (b) selected-area electron diffraction pattern showing (110) matrix orientation and accompanying streaking and twin reflections. Pattern is rotated into proper coincidence with(a) (100 kV beam.)*

the matrix spots causes a continual overlap of spot intensities at a portion of reciprocal space through which the reflection sphere is most plane.

Diffraction streaks are a common occurrence when specimens containing bundles of different domains are present or even where diffracting domains are grouped together such as in observations of small fiber bundles and the like, or in crystals containing whisker inclusions. Many fibers or microfibrils exhibit streaks perpendicular to the fiber axis because the fibers are composed of linear chains of molecules or of periodic groupings of unit cells which become slightly distorted in their registry as a consequence of the need to accommodate the growth process, and these linear domains being so closely spaced or stacked together produce continues arrays of very long

intensity profiles or rel-rods at the reciprocal lattice points as shown in
Fig. 6.15(b). Regimes of very long streaks, with the streaks spaced according to the dimensions of the domains, can occur in these situations as illustrated in Fig. 6.48. Crystal fibers such as that shown in Fig. 6.48(a)
actually become fluted or corregated along their surfaces as linear facets
form by the arrangement of crystal units as shown in Fig. 6.48(c). Note
that in the diffraction pattern of Fig. 6.48(b) the distance between
streaks can be related to a domain dimension (d) along the whisker axis
since this distance is equal to $\lambda L/d$. Note also the periodicity in the
arrangement or grouping of the streaks in Fig. 6.48(b).

Diffraction from phases and inclusions It should by now be obvious that
forbidden spots or rings in a selected-area electron diffraction pattern
can generally be described by careful consideration of twinning, double and
multiple diffraction effects, etc. It must also be obvious that where a
matrix area has undergone a transformation, structurally, chemically, or
both, the diffraction pattern will reflect this feature in the origination
of corresponding spots or rings. Martensitic transformations thus add
their chracteristic diffraction to the pattern, while precipitates or inclusions occurring in the selected area will also contribute their characteristic diffraction intensities to the selected-area electron diffraction
pattern. We will illustrate many of these features in our treatment of
transmission electron microscopy in Chap. 7.

6.9 CHEMICAL AND STRUCTURAL ANALYSIS BY SELECTED-AREA ELECTRON DIFFRACTION

You may already realize that by considering many of the basic principles
outlined in this chapter, it is possible to work in reverse with respect
to the analysis of an electron diffraction pattern. That is, without a
knowledge of the structure or chemical composition, the diffraction pattern may be analyzed to measure these properties. We can begin by remem-
bering that the diffraction pattern represents the dispostion of the reciprocal-lattice. Thus, by considering the possible reciprocal-lattice con-
structions, the real crystal lattice may be constructed, and the unit cell
identified. With a knowledge of the lattice arrangement, the diffraction
extinctions, and reflection intensities, the disposition and character of
the atomic arrangement can be deduced.

It must be apparent that the diffraction pattern represents only one
projection of the reciprocal lattice. Therefore, in order to reconstruct

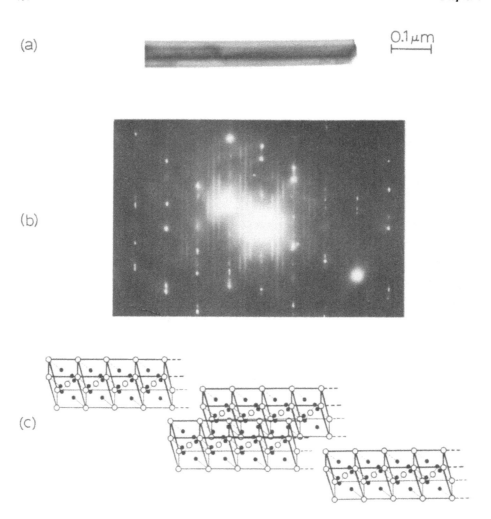

FIG. 6.48 *Diffraction streaks from pyrolucite (MnO$_2$) crystal fiber composed of linear arrays of submicron unit cell stacks along the fiber axis. The fiber grows by adding these arrays in a systematic fashion. (a) Bright-field transmission electron image of the fiber. (b) SAD of fiber section. (c) Schematic diagram illustrating arrays of MnO$_2$ unit cells.*

the three dimensional or true perspective of the reciprocal lattice, two distinct projections or diffraction patterns are required. The tilt stages available for most commercial electron microscopes facilitate to some extent such studies in thin-film materials. Basically, the method of structure analysis involves considerable measurement accuracy and a fair amount of trial and error in attempting a unique identification.

We can illustrate the use of electron diffraction in structure determination with reference to Fig. 6.49 taken from Murr [47]. We observe in Fig. 6.49(a) the selected-area electron diffraction pattern from an Er thin foil undergoing initial stages of oxidation in the electron microscope. Complete indexing of the rings shows several non-hcp reflections as indicated. Figure 6.49(b) shows the indexed pattern resulting after nearly complete oxidation of the area in a. The reflections are immediately observed to correspond to the bcc extinction conditions, and the computed lattice parameter of 10.55 Å identifies the structure as Er_2O_3. The unique feature established in this analysis is the occurrence of reflections for Er_2O_3, which had not been recorded by x-ray diffraction experiments on Er_2O_3. The analysis of the electron diffraction patterns thus allowed a unique space group[†] assignment (T^5) to be made for the Er_2O_3 structure. There are a number of possible reasons for the difference in the nonoccurrence of certain x-ray reflections, but we shall not elaborate on them here, except to point out that the limitation can in many cases be traced to the limited volume of reciprocal space encompassed by the x-ray reflection sphere, that is, to the fact that λ for x-rays is much greater than λ for high-energy electrons.

Since Fourier methods are so important in determining the coordinates of atoms and molecules, such techniques form the basis for much structure analysis. For a detailed treatment of such methods in electron diffraction, you are referred to the book by Vainshtein [3].

We should point out that a complete structure structure or quantitative chemical analysis is difficult with single crystals (spot patterns). This arises mainly because of the two-dimensional cross-grating effect of spot patterns from thin films, and the lack of response to tilt in the electron microscope. It is best to work with ring or texture patterns where possible, since these are responsive to tilt information in their geometrical projections.

Investigations of materials are not limited to thin foils of metal or ceramic sections, but of course can include crystal powders, cleavage sec-

---

[†]*Unfortunately we cannot develop the concepts of space groups here. It is hoped that the reader has already acquired some knowledge of space-group notation. For those who have not, the following references are suggested: M. J. Buerger,* Elementary Crystallography, *John Wiley & Sons, Inc., New York, 1956; and F. A. Cotton,* Applications of Group Theory, *John Wiley & Sons, Inc., New York, 1963.*

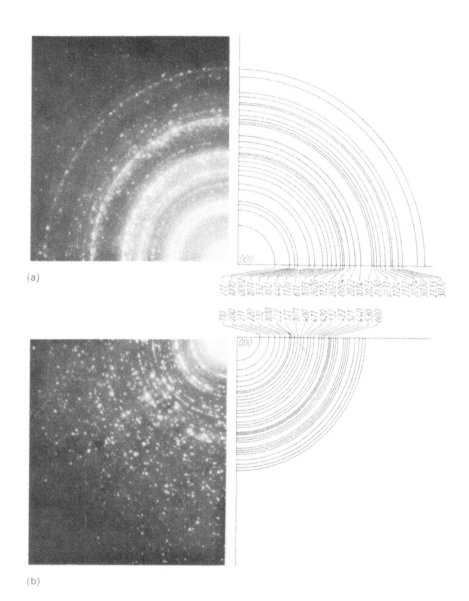

FIG. 6.49 *Electron diffraction structure analysis of* Er-Er$_2$O$_3$ *oxidation transition. (a) Er diffraction pattern at onset of oxidation
showing several weak bcc (Er$_2$O$_3$) reflections; (b) Er$_2$O$_3$ diffraction
showing bcc reflections indexed for a = 10.55 Å. Several Er (hcp)
reflections also remain. (100 kV beam) (From L. E. Murr* [47]).

tions, thin crystals, etc. Organic and inorganic samples are readily amenable to investigation in an electron microscope, and we will reserve for Appendix B some of the details of their preparation. For a lucid account of a number of important structure investigations, see Chap. V of reference 3. A good deal of comtemporary structure analysis by electron diffraction appears in many current issues of *Acta Crystallographica* and related journals.

In conclusion, where accurate measurements of the lattice parameter of a material is required, the electron diffraction pattern can be made more accurate by utilizing high-resolution electron diffraction in the electron microscope. In this facility, the specimen is placed just above the projector lens so that the effective camera distance is influenced only by the projector current. Very accurate calibration is thereby possible.

6.10 OBSERVATIONS OF SURFACES USING MICROPROBE REFLECTION HIGH-ENERGY ELECTRON DIFFRACTION

With the availability of fine-focussed ($<50\mu m$) electron beams in electron optical systems and electron microscopes, it is possible to examine very small (micro-) areas utilizing the integrated modular microanalysis concept illustrated previously in Chap. 4 (Section 4.8). In reality, modern microanalysis systems are the epitomy of this concept. Here we illustrate another example of this approach which utilizes novel combinations of scanning electron microscopy (SEM), reflection electron microscopy (REM), scanning tunneling microscopy/spectroscopy (STM/STS), and microprobe-reflection high energy electron diffraction (μ-RHEED). Figure 6.50(a) illustrates the concept schematically, while Fig. 6.50(b) shows some examples of REM images and corresponding μ-RHEED patterns for gold deposited on a Si(111) surface. The microprobe RHEED is a kind of scanning electron microscopy which uses reflection diffraction spot intensities as an image signal. In Fig. 6.50 (a), the 444 reflection was used to form the REM images. The μ-RHEED technique is actually a method to perform crystallographic analysis of surface micro-areas, which may include surface reaction processes with atomic-layer depth resolution [53,54]. Of course the incorporation of this approach into an STM as shown in Fig. 6.50(a) assures that surface atomic structure can also be related to the crystallographic analysis.

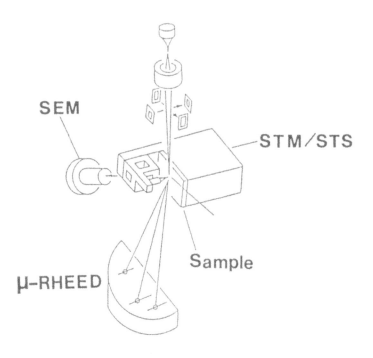

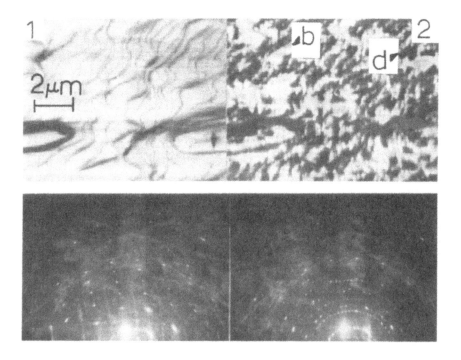

FIG. 6.50 High performance integrated microprobe. (a) Schematic (based upon design of T. Hasegawa, Hitachi Central Research Laboratory). (b) REM images of Si surface before (1) and after (2) gold deposition and RHEED patterns corresponding to dark (d) and bright (b) contrast areas in REM image 2. (Courtesy M. Ichikawa and *Materials Science Reports* [54]).

PROBLEMS

6.1 Prove that $g_{hk\ell}$ is perpendicular to a plane $(hk\ell)$ in the real crystal. Show also that

$$|g_{hk\ell}| = d^*_{hk\ell} = \frac{1}{d_{hk\ell}}$$

6.2 Sketch the diamond cubic unit cell and derive Eq. (6.21).

6.3 Consider a base-centered orthorhombic lattice to be described by unit-cell dimensions a,b,c, that is

$$a = |\underline{a}|, \quad b = |\underline{b}|, \quad c = |\underline{c}|$$

Derive the geometrical structure factor for this lattice assuming the atomic species are identical. Show which of the following planes will diffract electrons; (100),(111),(200),(220) (311),(330),(422),(710).

6.4 Derive an expression for $F(hk\ell)$ for MgO, and indicate the relative intensities of the first six reflections for 100 kV electrons.

6.5 Prove that a bcc real lattice becomes an fcc reciprocal lattice. Sketch the results in three dimensions.

6.6 Show that Eq. (6.34) for kinematical considerations becomes

$$\psi_j(r) = -\frac{1}{4\pi r} e^{2\pi i |\underline{K}| |\underline{r}|} A \int U(\underline{r}_j) e^{2\pi i (\underline{K}_0 - \underline{K}) \cdot \underline{r}_j} dV_{r_j}$$

and show also that $\psi_j(\underline{r})$ is a spherical wave having an amplitude proportional to the incident wave amplitude and the Fourier integral

$$F(hk\ell) = \frac{me}{2\pi\hbar^2} \int \phi(\underline{r}) e^{2\pi i \underline{g} \cdot \underline{r}} dV_{\underline{r}}$$

6.7 By considering a circle of any arbitrary diameter as a unit area of intensity representing the incident electron beam intensity, compare the spot intensities for diffraction from a Ni foil 100 Å in thickness, assuming I_0 unity. The accelerating potential is 100 kV, and the atomic scattering factor for Ni can be assumed to be constant throughout at a value of 4 angstrom units. Sketch the electron transmission diffraction pattern from this foil if the Ewald sphere passes directly through each reciprocal-lattice point (exact Bragg condition) by using circle areas corresponding to the spot intensities.

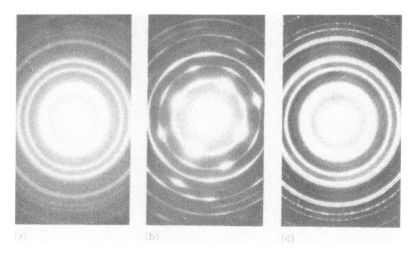

FIG. P6.8

6.8   Of the electron diffraction patterns shown in Fig. P6.8, a is a
      transmission type pattern from a polycrystalline Au foil in the
      electron microscope operated at 100 kV. Calculate the camera con-
      stant from a, assuming a direct scale, and identify the metal foils
      from which b and c originated for the same scale factor as a. (Hint:
      The metallic elements of b and c are among the group Ag, Pd, Rh, Cr,
      Rb, V.)

6.9   The selected-area electron diffraction spot patterns in Fig. P6.9 are
      from fcc stainless steel. Find the surface orientations of the crys-
      tals (grain surface orientations) for the corresponding patterns.

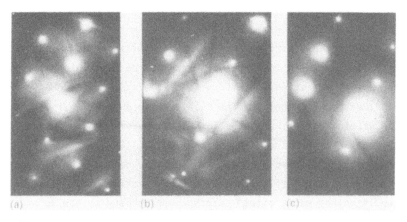

FIG. P6.9

Calculate the deviation from $|\underline{g}| = |31\bar{1}|$ in Fig. P6.9b. Calculate
the camera constant for the patterns (the constant will be the same
for all taken at an accelerating potential of 100 kV). The lattice
parameter for stainless steel (type 304) is 3.56 Å. What can you
say about the disposition of the reflection sphere with respect to
the reciprocal-lattice net plane for a, b, and c? Which pattern
shows evidence for twinning?

6.10 Construct the diffraction nets similar to those of Appendix A (Figs.
A.6 to A.14) for the (101) and (117) crystallographic surface orientations for fcc, bcc, and diamond cubic).

6.11 Construct the (110) diffraction net for a body-centered tetragonal
crystal.

6.12 The electron transmission micrograph of Fig. P6.12a shows the defect
structure of 304 stainless steel following explosive shock loading
at 150 kbars pressure. Figure P6.12b shows the corresponding selected-area electron diffraction pattern. If the relative rotation angle ϕ_R is 20° and the sense of the rotation depicted is clockwise
with respect to the image of a, identify the surface orientation of
a. Index the pattern characteristics observed. The camera constant
in b is $\lambda L = 3.85$ Å cm. What is the nature of the linear defects
observed? Note that b is already rotated into coincidence with a.

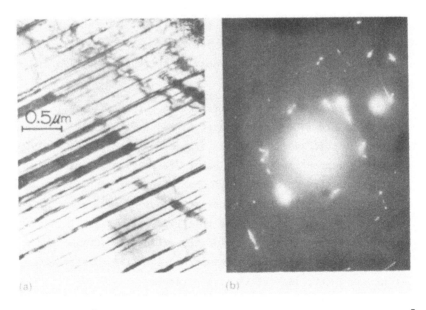

(a)   (b)

FIG. P6.12 [*After L. E. Murr,* Phys. Status Solidi; *19:1 (1967)*]

6.13   Index the Kikuchi diffraction pattern shown in Fig. 6.21 and deter-
       mine the corresponding grain (crystal) surface orientation. Show
       this on the Kikuchi map of Fig. 6.34 by making a xerox copy and
       circling the zone on the copy.

6.14   A thin flake of an unknown origin is to be identified by electron dif-
       fraction in the electron microscope. Outline in detail how you
       would procede if the pattern were composed of rings. What limita-
       would you encounter if the pattern were a spot pattern? What could
       you tell about the flake if a spot pattern were observed? Suppose
       Kikuchi lines were present, what might this tell you? Could you
       identify the crystal if the pattern contained only sharp Kikuchi
       line pairs?

6.15   In a low-energy electron diffraction experiment, a single crystal
       of clean Ni is observed with the electron beam incident normal to
       the surface. Diffraction spots from the (001) surface of the crys-
       tal are observed. Indexing of the pattern such as that of Fig.
       6.24b indicated that for spots corresponding to $\{020\}$ or $\{02\}$, the
       corresponding diffraction grating angle as shown in Fig. 6.24a was
       44.4° for the 50-volt electron beam used. Sketch a section view of
       the surface atoms and their relationship to the bulk atomic arrange-
       ment for (100).

6.16   Plot the critical dynamic film thickness for Mo and Pb as a function
       of accelerating potentials of 50, 100, 150, 200, and 300 kV for
       $\underline{g}_{hkl} = [220]$. Suppose it was desired to examine single-crystal Pb
       films kinematically at 100 kV having thicknesses of 175 Å. How
       might this be accomplished?

6.17   The selected-area electron diffraction patterns shown in Fig. P6.17
       were obtained from thin films of nickel ferrite ($NiFe_2O_4$) reactively
       sputtered onto $NaCl$ substrates at temperatures of -25°C, 25°C, and
       125°C corresponding to a, b, and c, respectively. Recalling that
       the spinel structure of $NiFe_2O_4$ has a reflection plane sequence of
       (111), (220), (311) for increasing values of $|\underline{g}|$, etc., calculate
       the average grain size in the thin films of a and b if the average
       grain size in c is 250 Å. Plot the grain size for the three film
       structures as a function of temperature. (The accelerating poten-
       tial was 60 kV; and the lattice parameter of $NiFe_2O_4$ is a = 8.34 Å.)

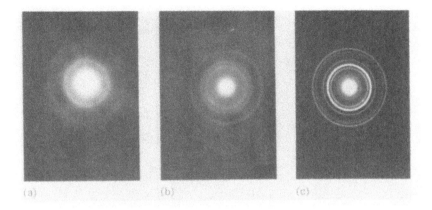

FIG. P6.17 *Diffraction patterns. (Courtesy of R. J. Brook.)*

6.18 Refer to Fig. 6.44 and measure the angles of rotation for the three overlapping (001) Pd layers. Make three xerox transparencies of the single grating shown in Fig. 6.43, and by rotating them into the angular relationships measured from Fig. 6.44, comment on the pattern formed.

6.19 The sequence of selected-area electron diffraction patterns shown in Fig. P6.19 following were made from thin foils prepared from cold-rolled Ni. What information is contained in these patterns? Can you make a plot showing a variation of a microstructural feature with rolling reduction or true strain? What does this tell you about the response to rolling?

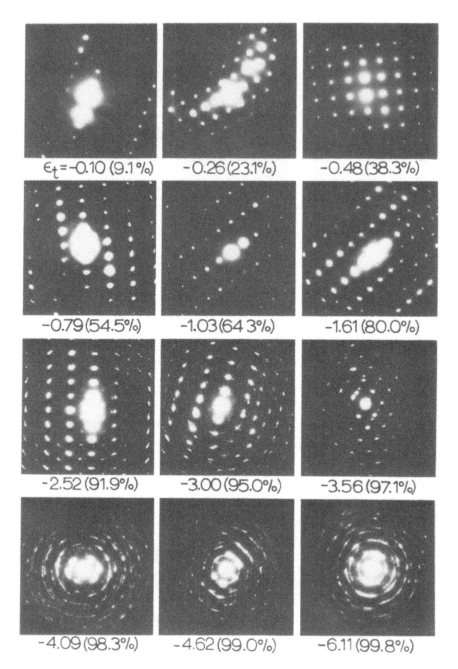

P6.19 [*After W. Zimmer, S. S. Hecker, D. L. Rohr, and L. E. Murr*, Metal Sci. J.; *17:198 (1983)*]

REFERENCES

1. M. von-Laue, *Ann. Physik, 26:* 55(1936).

2. N. F. Mott, *Proc. Roy. Soc. (London), A127:* 658(1930).

3. R. D. Heidenreich, *Fundamentals of Transmission Electron Microscopy,* appendix T, Interscience Publishers, Inc., New York, 1964.

4. B. K. Vainshtein, *Structural Analysis by Electron Diffraction,* appendix III, Pergamon Press, New York, 1964.

5. N. F. Mott and H. S. W. Massey, *The Theory of Atomic Collisions,* Clarendon Press, Oxford, 1949.

6. M. von Laue, *Materiewellen und ihre Interferenzen, Leipzig* (1948).

7. A. L. Patterson, *Phys. Rev., 56:* 972(1939).

8. S. Kikuchi, *Proc. Imp. Acad. Japan, 4:* 271(1928).

9. S. Kikuchi, *Japan. J. Phys., 5:* 83(1928).

10. W. Kossel and G. Mollenstedt, *Ann. Phys., 42:* 287(1943).

11. D. G. Coates, *Phil. Mag., 16:* 1178(1967).

12. G. R. Booker et al., *Phil. Mag., 16:* 1185(1967).

13. Z. G. Pinsker, *Electron Diffraction,* Butterworth Scientific Publications, London, 1953.

14. F. Hoffman and H. P. Smith, *Phys. Rev. Letters, 19:* 1472(1967).

15. G. C. Bond, *Discussions Faraday Soc., 41:* 200(1966).

16. D. A. Bowden, *J. Res. Inst. Catalysis* (Hokaido Univ.), *4:* 1(1966).

17. Z. Knor and E. W. Müller, *Surface Sci., 10:* 21(1968).

18. A. J. Melmed, H. P. Layer, and J. Kruger, *Surface Sci., 4:* 476(1968).

19. L. H. Germer, A. U. MacRae, and C. D. Hartman, *J. Appl. Phys., 32:* 2432(1961).

20. L. K. Jordan and E. J. Scheibner, *Surface Sci., 10:* 373(1968).

21. J. J. Lander, *Surface Sci., 1:* 125(1964).

22. P. W. Palmberg and W. T. Peria, *Surface Sci., 6:* 57(1966).

23. L. Fiermans and J. Vennik, *Surface Sci., 9:* 187(1968).

24. A. W. Agar, *Brit. J. Appl. Phys., 11:* 185(1960).

25. H. P. Klug and L. E. Alexander, *X-ray Diffraction Procedures,* John Wiley & Sons, Inc., New York, 1954.

26. P. B. Hirsch, *Progr. Metal Phys., 6:* 236(1956).

27. H. M. Otte, J. Dash, and H. F. Schaake, *Phys. Status Solidi, 5:* 527 (1964).

28. M. von Heimendahl, W. Bell, and G. Thomas, *J. Appl. Phys., 35:* 614 (1946).

29. E. Levine, W. L. Bell, and G. Thomas, *J. Appl. Phys., 37:* 2141(1966).

30. G. Thomas, *Trans. AIME, 23:* 1608(1965).

31. R. E. Villagrana and G. Thomas, *Phys. Status Solidi, 9:* 499(1965).

32. W. Kossel and G. Möllenstedt, *Ann. Phys.*, *36:* 113(1939).

33. B. F. Buxton, J. A. Eades, J. W. Steeds, and G. M. Rackham, *Phil. Trans.*, *281:* 171(1976).

34. G. Lempfuhl in *Electron Microscopy 1978,* vol. III, State of the Art Symposia, Proc. Ninth Int. Congress on Electron Microscopy, J. M. Sturgess (ed.), The Imperial Press, Ltd., Ontario, Canada, 1978, p. 304.

35. L. P. Stoter, *J. Mater. Sci.*, *16:* 1356(1981).

36. M. P. Shaw, A. J. Porter, R. C. Ecob, and B. Ralph, in *Proc. Electron Microscopy Soc. America,* G. W. Bailey (ed.), Claitor's Publishing Div., Baton Rouge, 1981, p. 352.

37. M. M. Kersker, E. A. Aigeltinger, and J. J. Hren, in *Proc. Electron Microscopy Soc. America,* G. W. Bailey (ed.), Claitor's Publishing Div., Baton Rouge, 1981, p. 362.

38. J. M. Cowley, *Adv. Electronics and Electron Phys.*, *46:* 1(1978).

39. E. S. Meieran and M. H. Richman, *Trans. AIME, 227:* 1044(1963).

40. O. Johari and G. Thomas, *Trans. AIME, 230:* 597(1964).

41. R. Bullough and C. M. Wayman, *Trans. AIME, 236:* 1704(1966).

42. O. Johari and G. Thomas, *Trans. AIME, 230:* 597(1964).

43. R. D. Burbank and R. D. Heidenreich, *Phil. Mag.*, *5:* 373(1960).

44. F. R. C. Schoening and A. Baltz, *J. Appl. Phys.*, *33:* 1442(1962).

45. T. Honma and C. M. Wayman, *J. Appl. Phys.*, *36:* 2791(1965).

46. G. A. Bassett, J. W. Menter, and D. W. Pashley, *Proc. Roy. Soc.*, *A246:* 345(1958).

47. L. E. Murr, *Phys. Status Solidi, 24:* 135(1967).

48. L. E. Murr, *Phys. Status Solidi, 19:* 1(1967).

49. C. S. Barrett and T. B. Massalski, *Structure of Metals,* 3rd ed., McGraw-Hill Book Co., New York, 1966.

50. R. A. Ploc and G. H. Keech, *J. Appl. Cryst.*, *5:* 244(1972).

51. G. Honjo, S. Kodera, and N. Kitamura, *J. Phys. Soc. Japan, 19:* 351 (1964).

52. P. R. Okamoto and G. Thomas, *Acta Met.*, *19:* 825(1971).

53. K. Takayanagi, Y. Tanishiro, K. Kobayashi, K. Akiyama, and K. Yagi, *Japan J. Appl. Phys.*, 26: 957(1987).

54. M. Ichikama, *Materials Sc. Reports*, 4(4): 147(1989).

SUGGESTED SUPPLEMENTARY READING

Andrews, K. W. et al.: *Interpretation of Electron Diffraction Patterns,* Plenum, New York, 1968.

Cohen, J. B.: *Diffraction Methods in Materials Science,* Macmillan Company, New York, 1966.

Cowley, J.: *Diffraction Physics,* 2nd. edition, Elsevier, New York, 1985.

Drits, V. A.: *Electron Diffraction and High Resolution Electron Microscopy of Mineral Structures,* Springer-Verlag, New York, 1987.

Heidenreich, R. D.: *Fundamentals of Transmission Electron Microscopy,* John Wiley & Sons, Inc., New York, 1964.

Hirsch, P. B. et al. (eds.): *Electron Microscopy of Thin Crystals,* Butterworth Scientific Publications, London, 1965.

Larsen, P. K. and J. P. Dobson(eds.): *Reflection High-Energy Electron Diffraction and Reflection Electron Imaging of Structures,* Plenum Publishing Corp., New York, 1988.

Lipson, H. L. and W. Cochran: *The Determination of Crystal Structures: The Crystalline State,* vol. III, G. Bell & Sons, Ltd., London, 1953.

MacRae, A. U.: Low Energy Electron Diffraction, *Science,* vol. 139, p. 379, 1963.

Pinsker, Z. G.: *Electron Diffraction,* Butterworth Scientific Publications, London, 1953.

Reimer, L.: *Transmission Electron Microscopy,* 2nd. edition, Springer-Verlag, New York, 1989.

Thomas, G.: *Transmission Electron Microscopy of Metals,* John Wiley & Sons, Inc., New York, 1962.

Thomson, G. P., and W. Cochran: *Theory and Practice of Electron Diffraction,* Macmillan & Co., Ltd., London, 1939.

Vainshtein, B. K.: *Structure Analysis by Electron Diffraction,* Pergamon Press, New York, 1964.

Von Heimendahl, M.: *Electron Microscopy of Materials,* (translated by Ursula E. Wolff), Academic Press, Inc., New York, 1980.

Von Hove, M. A., et al.: *Low Energy Electron Diffraction,* Springer-Verlag, New York, 1986.

7
TRANSMISSION
ELECTRON MICROSCOPY

7.1 INTRODUCTION

We learned in Chap. 5 that surface morphology of materials is amenable to study by transmission electron microscopy of replicas. In this mode of analysis, the electron transmission image results essentially by replica mass-thickness contrast. Consequently the image interpretations are based solely on these features, and the structure of the replica is itself ignored.

Our interests at this point will now focus, as it were, on the direct imaging of crystal structure, or the intrinsic three-dimensional nature of a thin material under examination by transmission electron microscopy. We have in fact been exposed to image details of crystalline materials in Chap. 6, especially Figs. 6.29, 6.30, 6.41, and 6.47. It is apparent on looking at these examples that electron transmission images of crystalline materials also result by contrast features. However, the mechanism of crystalline materials contrast usually involves primarily diffraction phenomena; specimen mass-thickness simply modifies the degree of diffraction contrast, especially with respect to electron absorption. Depending upon the specific conditions of imaging and the character (composition, thickness, crystal structure, etc.) of the specimen, either mass-thickness or diffraction contrast may dominate, or the contrast features may be mixed or intermixed within the specimen. These features are illustrated in the wide range of examples shown in Figs. 7.1 to 7.3. In Fig. 7.1, a thin aluminum film on a standard copper support grid contains dendritic nuclei of electrodeposited copper. These appear as white specks in the secondary electron image in the SEM, while in the transmission electron microscope (TEM) they appear as well-defined dendritic structures as shown in the insert. The electrodeposited copper dendrite is imaged exclusively by mass-thickness contrast. However the black specks in the background arise by diffraction contrast from the polycrystalline support film. Each black

FIG. 7.1 *Contrast examples in the transmission electron microscope. The larger image portion shows polycrystalline aluminum film containing small copper electrodeposits suspended on a copper support grid and viewed in the SEM. The insert shows a small portion observed in the TEM.*

speck represents an aluminum grain in which the electron beam has been diffracted. This diffracted beam is then blocked by the objective aperture.
Consequently these diffracting grains appear black because of a lack of
signal corresponding to these strongly diffracting volume elements. Figure
7.2(a) illustrates the diffraction contrast which occurs at individual dendrite arms of larger deposits on the aluminum film, and at higher magnification. Figure 7.2(b), on the other hand, illustrates the secondary elec-

FIG. 7.2 *Mass-thickness and diffraction contrast
image features of copper dendrites in the TEM (a)
as compared with secondary electron imaging of
similar dendrites in the SEM (b).*

tron image details of even larger dendrites of electrodeposited copper for comparison. The diffraction contrast in Fig. 7.2(a) is particularly evident as fringe profiles around the dendrite arms, while the thicker dendrite arm segments are clearly imaged by mass-thickness effects.

Figure 7.3 illustrates some additional contrast features involving either mass-thickness or diffraction contrast phenomena. In Fig. 7.3(a) and (b), the hydrocarbon contamination accumulating on the edge of a silicon nitride crystal becomes "fossilized" into Lictenberg figures which grow with time and the accumulation of contamination. This example not only illustrates the mass-thickness contrast features and the in-situ aspects of experimental changes which can be observed in the TEM, but also the realities of specimen contamination which must often be dealt with. Figure 7.3 (c), on the other hand, shows the very distinct nature of mass-thickness (ρt) contrast and diffraction (λ) contrast for the same copper electrodeposit nuclei. Figure 7.3(d) shows very interesting mass-thickness contrast effects for a very tiny chrysotile asbestos crystal. These crystals have a hollow core about which growth occurs, and it is this growth feature and the associated mass-thickness contrast which can provide unambiguous identification or distinction from other asbestiform fiber crystals for environmental considerations for example.

It should be apparent that where image contrast of crystalline materials results by diffraction contrast, the ability of specific (intrinsic) features of a solid to be distinguishable in such an image will depend primarily on the characteristic phase shift imparted to the transmitted electrons. Thus images of grains and grain boundaries as already observed in Figs. 6.29 and 6.30 arise by contrast originating as phase shifts and because electrons encountering various crystallographies diffract at angles determined by Bragg's law as we saw in Chap. 6.

Transmission electron microscopy of crystalline materials therefore affords a direct means of observing the internal structure of solid thin films. We are thus able to distinguish defects as a result of the relative phase shift arising at such distortions with regard to the perfect-crystal portions and thereby giving rise to contrast in the transmitted electron image. Transmission electron microscopy is perhaps the single most effective means for studying solids since it can combine direct image information with selected-area electron diffraction information to completely characterize the physical, chemical, and crystallographic nature of a thin section. The striking feature of transmission electron microscope images

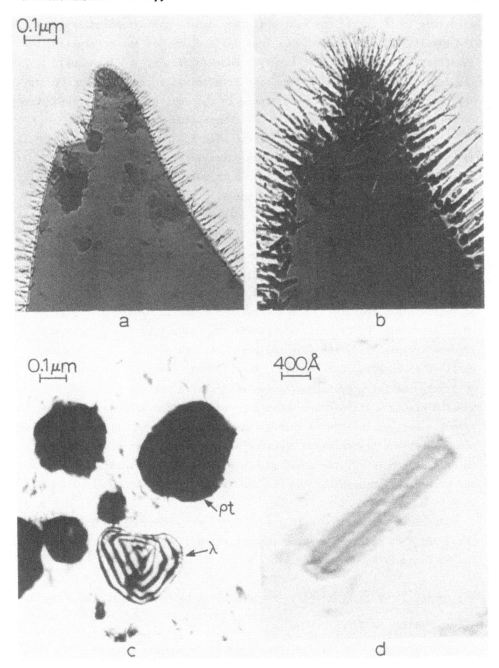

FIG. 7.3 *Contrast examples for materials observed in the TEM. (a) and (b) show amorphous hydrocarbon - based Lictenberg figures growing during observation [see L. E. Murr, Phil. Mag., 25: 721(1971)], (c) shows mass-thickness (ρt) and diffraction (λ) contrast at the same copper nuclei electrodeposited onto a polycrystalline aluminum film as in Figs. 7.1 and 7.2, (d) shows a chrysotile asbestos fiber isolated from drinking water sample imaged by mass-thickness contrast.*

of solids is of course the fact that they can be photographically recorded; the results are therefore visual, and subject only to the interpretations manifested in image-contrast theory. Furthermore, by simultaneously recording characteristic x-ray emission or electron energy loss spectra for any particular region in a thin sample in the TEM, additonal corroboration or actual identification for chemical composition can be obtained.

In this chapter we will deal first with the development of image-(diffraction) contrast theory and the basis for interpreting electron transmission images of crystalline solids, both perfect and imperfect. We will begin, as a logical continuation of the simple diffraction assumptions of Chap. 6, with the kinematical theory. The more sophisticated theory will then be developed as an extension of kinematical approximations at the Bragg condition and in thicker crystals where image contrast is influenced by absorption and multiple beam interactions. We shall then apply the theories of image-contrast formation to the solution of specific problems of interest and concern to the materials scientist. These will include the characterization of crystal defects and the identification of thin solid crystallography using the selected-area electron diffraction techniques outlined in the previous chapter. We will also describe lattice imaging utilizing the selected-area electron diffraction pattern, scanning transmission electron microscopy, and analytical electron microscopy. Numerous examples will be introduced through these treatments to illustrate specific applications of transmission electron microscopy in the investigation of materials properties, the relationship of thin-film structure to physical and mechanical properties, and the general characterization of materials structure.

Specimen preparation will not be dealt with directly in this chapter. The details of specimen preparation for transmission electron microscopy can be found in Appendix B (B.6).

7.2 QUALITATIVE (KINEMATICAL) INTERPRETATION OF CRYSTALLINE IMAGE CONTRAST

7.2.1 *PERFECT CRYSTALS*

We will initially consider contrast arising in perfect crystals. In this treatment, a perfect crystal will be defined by the general absence of defects such as dislocations, stacking faults, or grain boundaries. In effect we shall treat a single-crystal area, that is, the region within a particular crystal grain. This region may have any shape or coincidence with respect to the electron beam, and may also be buckled or elastically strained.

Our starting point will be a consideration of the diffracted beam intensity (the dark-field image) over an area of crystal having a thickness t. As illustrated in Fig. 7.4, the intensity distribution at a point P' on the emergent surface of such a crystal section can be obtained by calculating the amplitude of the electrons scattered by the shaded column in the direction of the diffracted beam. The diffracted wave amplitude is given effectively by Eq. (6.3a) in the form

$$\psi_S \equiv S = \sum_j f_{e\ell}(\mu) e^{2\pi i (\underline{g} + \underline{s}) \cdot \underline{r}_j} \tag{7.1}$$

where s is the deviation (tilt of the crystal section) from the exact Bragg condition as illustrated earlier in Fig. 6.6; we will assume an incident beam amplitude, $|\psi_0|$, of unity.

On examining Eq. (7.1), observe that since the product $\underline{g} \cdot \underline{r}$ is an integer, $e^{2\pi i \underline{g} \cdot \underline{r}_j}$ will be unity, leaving

$$S = \sum_j f_{e\ell}(\mu) e^{2\pi i \underline{s} \cdot \underline{r}_j} \tag{7.2}$$

If we choose an origin at the foil center 0, as shown in Fig. 7.4, the wave amplitude along the column in the z direction can be obtained from

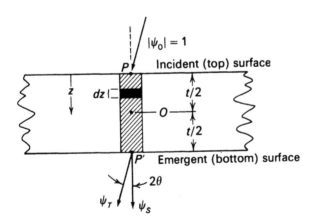

FIG. 7.4 *Section of perfect crystal observed in dark field (tilted illumination).*

$$S = f_{e\ell}(\mu)\int_{-t/2}^{t/2} e^{2\pi isz}dz \qquad (7.3)$$

where s is the magnitude of the deviation from the Bragg condition in the
z direction, and is exactly equivalent to the value of g_3 in our considerations of electron diffraction patterns in Chap. 6 [for example, see Eq.
6.46a)]. Integrating Eq. (7.3), we obtain

$$S \simeq \frac{\sin \pi st}{\pi s} \qquad (7.3a)$$

from which the dark-field image intensity is observed to be proportional to

$$\frac{\sin^2 \pi st}{(\pi s)^2} \qquad (7.3b)$$

We have assumed in our treatment of diffraction from the shaded column
in Fig. 7.1 that the column is composed of unit cells stacked along the z
direction. Since the direction of scattering deviates from the exact Bragg
orientation by an amount s, an element of the column lying between, say,
z and z + dz will have an amplitude proportional to dz [Eq. (7.3)], and the
phase angle ϕ will be $2\pi sz$. It is thus instructive to construct a phase-
amplitude diagram [1] illustrating these features as shown in Fig. 7.5.
The construction of Fig. 7.5 deviates from Fig. 6.2 in that the wave vectors representing individual scattering amplitudes are joined at their tips
with a reference origin at O (Fig. 7.4) and are rotated successively by the
phase angle $\phi = 2\pi nsz$, where n is an integer. Each vector represents an
equivalent chord of a circle whose radius is $1/2\pi s$. The amplitudes scattered by the top and bottom crystal portions are then observed to be $\overline{PO}$ and

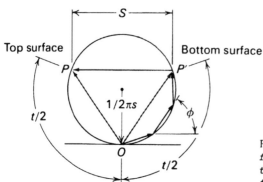

FIG. 7.5 *Phase-amplitude diagram*
for perfect crystal of thickness
t. Origin O is centered on re-
ference column O of Fig. 7.4.

$\overline{QP}'$, respectively, and the resultant diffracted wave amplitude is observed to be S = $|\overline{PP}'|$ [Eq. (7.3a)]. The wave intensity is then $|\overline{PP}'|^2$.

On examining Fig. 7.5 and Eq. (7.3a), we note that the diffracted wave amplitude will vary as

$$\tau = t_o = \frac{1}{s} \tag{7.4}$$

where t_o is the kinematic extinction distance. Thus, for a foil section having a thickness greater than the extinction distance t_o, the diffraction amplitude will oscillate according to Nt_o, where N is an integer. The constant of proportionality in Eq. (7.3a) is π/t_o[†] and Eq. (7.3b) becomes

$$I_S = \left(\frac{\pi}{t_o}\right)^2 \frac{\sin^2 \pi st}{(\pi s)^2} \tag{7.3c}$$

Since the incident beam amplitude was previously specified to be unity, Eq. (7.3c) and Eq. (6.46a) are equivalent, and thus

$$\frac{\pi}{t_o} = \frac{\lambda |F(hk\ell)|}{V_c} \tag{7.5}$$

The value of the extinction distance at the Bragg angle is then given by[*]

$$t_o \equiv t_g \equiv \xi_g = \frac{\pi V_C}{\lambda |F(hk\ell)|} \tag{7.6}$$

and from Eq. (6.48),

$$\xi_g = \pi t_{dyn} \tag{7.7}$$

The wave nature of electrons in a crystal is demonstrated in Fig. 7.6, where the diffracted wave composing the dark-field image is represented by a periodic (sinusoidal) oscillation for a crystal of thickness $N\xi_g$. The corresponding bright-field or transmitted beam intensity oscillations are of the inverted form

---

[†]*This is readily shown in the dynamical treatment of diffraction contrast, and you may wish to glance in advance at Eq. (7.40).*
[*]*The notation t_0, t_g, and ξ_g are sometimes used interchangeably in the literature. We shall henceforth adopt the latter notation to denote the linear measure of the dynamical extinction distance.*

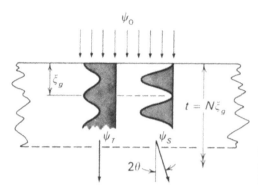

FIG. 7.6 *Intensity oscillations in perfect-crystal section of thickness* $N\xi_g$.

$$\psi_T \equiv T \propto \left(1 - \frac{\sin^2 \pi st}{(\pi s)^2}\right) \qquad (7.8)$$

It is readily observed in Fig. 7.6, and on comparing Eqs. (7.3a) and (7.8), that the bright-field and dark-field intensities are reversed as denoted by these descriptive modes of observation. Thus for a perfect crystal, as shown in Fig. 7.6, the image would consist entirely of an illuminated viewing screen when observed by normal incidence bright-field electron microscopy. Tilting of the incident beam so that only one diffracted beam passes through the objective aperture, or shifting the objective aperture slightly off axis[†] to allow only one strong diffracted beam will then result in some screen intensity. As the thickness varies in such a foil, the exit wave amplitudes intersected by the lower foil surface would be observed to vary. Consequently the screen brightness would change accordingly.

This is ideally the situation that occurs when a depression, etch pit, or a hole is contained in a thin foil. The same situation also exists for a regular taper at the edge of a thin section where the thickness may change regularly in a way analogous to the edge of a razor or knife blade. The resulting screen intensity is therefore a periodic variation in brightness creating fringe patterns corresponding to the tapering contour of holes or etch pits, or an edge. Figure 7.7 illustrates the qualitative features of the bright-field intensity profiles associated with a hole in

---

[†]*We shall refer hereafter to tilted illumination dark field or aperture dark field, respectively, to denote the two operating modes for attaining a dark-field image.*

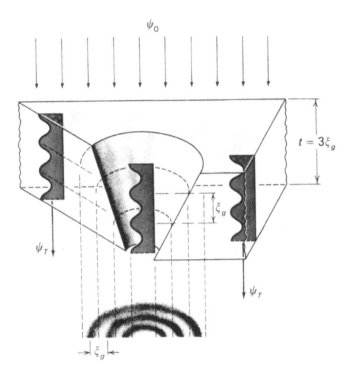

FIG. 7.7 *Geometrical description of formation of fringe patterns at holes or similar thickness irregularities in thin foils. Dark fringes (shaded) correspond to intensity minima in bright field.*

a thin foil; while Fig. 7.9a reproduces an appropriate bright-field example in a stainless steel foil observed at 100 kV.

You should be aware that the spacing of the fringes at holes or similar artifacts is simply the extinction distance ξ_g, as shown geometrically in Fig. 7.7. Since ξ_g corresponds to the extinction distance at a particular Bragg orientation, the number of fringes occurring in an electron microscope image can be related to the linear thickness perpendicular to the specimen surfaces only if $\underline{g}$ is known. This information is contained in the diffraction pattern, where $\underline{g}$, the operating reflection corresponding to ξ_g for the fringes observed, will be recognized as a prominent spot intensity as discussed previously in Chap. 6. Note that the diffracting nucleus marked (λ) in Fig. 7.3(c) exhibits extinction fringes which occur over the symmetrical, cap or pseudohemispherical shape of the nucleus. Here the thickness changes from the center out (decreases), giving rise to concentric extinction fringes. These fringes are the inverse of those occurring for a hole as shown in Fig. 7.9a where the thickness increases from the center out.

We can see from Eq. (7.6) that ξ_g will vary not only with the composition of the material and the reflection vector $\underline{g}$, but also with the electron wavelength. As an example, let us calculate the extinction distance ξ_g for a fringe profile at a hole etched in a pure Ag foil observed in the electron microscope at 100 kV (λ = 0.037 Å) where the operating reflection g is observed from the selected-area electron diffraction pattern to be [200]. The unit-cell volume V_C is readily obtained from

$$V_C = \underline{a} \cdot \underline{b} \times \underline{c} \tag{7.9}$$

For pure fcc Ag, we evaluate $V_C = a^3 = 68.2 \text{ Å}^3$. The structure factor for Ag is observed from Eq. (6.20) to be $4f_{e\ell}(\mu)$; and since

$$\frac{\sin \theta}{\lambda} = (2d_{hk\ell})^{-1} = 0.245 \text{ Å}^{-1}$$

the corrected,[†] interpolated value of $f_{e\ell}(\mu)$ from Table D.3 is found to be 5.41 Å. We then calculate from Eq. (7.6) that ξ_g for Ag at 100 kV is approximately 254 Å. You should confirm this calculation and then try solving Prob. 7.1.

In the treatment of alloy samples or compounds, the value of the atomic scattering factor for electrons $[f_{e\ell}(\mu)]$ is obtained by considering an effective atomic number Z_{eff}. This quantity can be calculated as indicated in Eq. (4.23). The effective atomic or molecular weight can also be calculated for such materials. Because of the importance of a knowledge of ξ_g in the electron microscopy of thin crystalline materials, the extinction distances corresponding to various operating reflections and accelerating potentials commonly encountered are calculated in tabular form in Tables D.4 to D.6. You may now benefit further by working out Prob. 7.2 in addition to Prob. 7.1, or by reexamining Prob. 6.16 in light of Eq. (7.7).

We have observed in the previous discussion that image contrast occurs at holes in perfect crystals because of the intensity oscillations as fringes corresponding to the geometrical taper of the hole or a foil edge.

[†]*The value of $f_{e\ell}(\mu)$ interpolated from Table D.3 (Z = 47) is corrected for relativistic effects by multiplying by $1/\sqrt{1 - (v/c)^2}$. Values of this factor for various accelerating potentials are listed in Table D.2.*

In this particular situation, fringe contrast results by regular variations in the effective specimen thickness, that is, for $t = N\xi_g$, where N is a variable for a fixed $\underline{g}$. Contrast can also arise at extinction contours where a thin crystal is bent or elastically distorted. In this case the intensity oscillations are influenced primarily by variations in the operating reflection g and the deviation from the Bragg condition $\underline{s}$. Thus, in the bright-field image of a bent-foil section, the screen intensity will vary over the area of bending as determined by variations in s in Eq. (7.8), and as the reflection condition for $+\underline{g}$ and $-\underline{g}$ is fulfilled.

Figure 7.8 illustrates geometrically the origin of bend-extinction contours in the bright-field image for a symmetrically buckled (and strained) foil section. The intensity profile for the ideal case is also illustrated, which is the inverted form of Fig. 6.15 for the intensity distribution at a Bragg orientation (or the dark-field image in our present discussion). We can observe from Fig. 7.8 that the symmetry of the bend contours depends on the symmetrical nature of the bending and the coincidence of the section with respect to the electron beam and the Bragg orientation. In practice, thin foils in the electron microscope are strained and bent in irregular ways so that the bend contours consist of lines or related contrast features giving rise to random or symmetrical patterns. Tilting of the foil or the beam or heating the area by focusing the beam intensity to a fine spot will thus cause the bend contours to move as the Bragg conditions vary, that is, as $\underline{s}$ changes ($\pm \underline{s}$). We shall later see that this is one method of distinguishing bend contours from regular defect diffraction contrast features.

A very striking example of bend-contour diffraction contrast from a symmetrical distorted crystal section is reproduced in Fig. 7.9b [2]. The selected-area electron diffraction pattern superposed on Fig. 7.9b is evidence of the symmetrical nature of the crystallographic surface based on our arguments of Chap. 6; and the associated image bend contours are observed to possess a fourfold symmetry. It is instructive to compare the fourfold symmetrical nature of the image of Fig. 7.9b with the Kikuchi diffraction pattern of Fig. 6.22.[†] It should be apparent that the symmetry difference is because of the foil orientation in Fig. 6.22 is (111) while in Fig. 7.9b it is (100).

[†]*An analytical treatment of the crystallographic aspects of bend contours and Kikuchi patterns is given in K. H. G. Ashbee and J. W. Heavens*, Trans. AIME, 239: *1859(1967).*

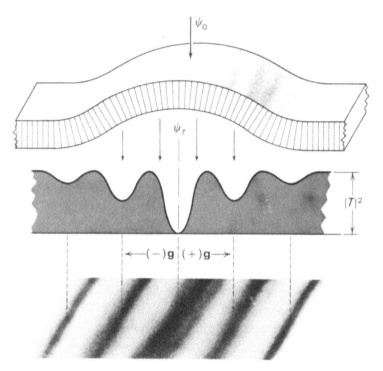

FIG. 7.8 *Geometrical origin of bend-extinction contours for idealized case in thin-foil section observed in bright-field illumination.*

In Fig. 7.10 the diffraction contrast effects, including extinction contours and thickness fringes at tetrahedral etch pits, are observed in a thin Si foil having (111) surface orientation. The intensity reversal of the bright-field image of Fig. 7.10a is strikingly illustrated in the dark-field image of Fig. 7.10b obtained by shifting the objective aperture from the 000 main beam to admit only the strong [220] diffracted beam shown in the superposed selected-area electron diffraction pattern of Fig. 7.10 (aperture dark field). Several interesting features of the electron micro-scope images of Fig. 7.10 can be treated more extensively as demonstrated in Prob. 7.3.

The diffraction contrast effects in uniform thin films were first interpreted by R. D. Heidenreich [3] three decades ago, while these same considerations were not extended to the identification of crystal defects until nearly a decade later [1,4-7]. The extension of the perfect-crystal kinematical column approximation (Fig. 7.4) to the interpretation of lat-tice defects involves simply the consideration of a phase-shifting system

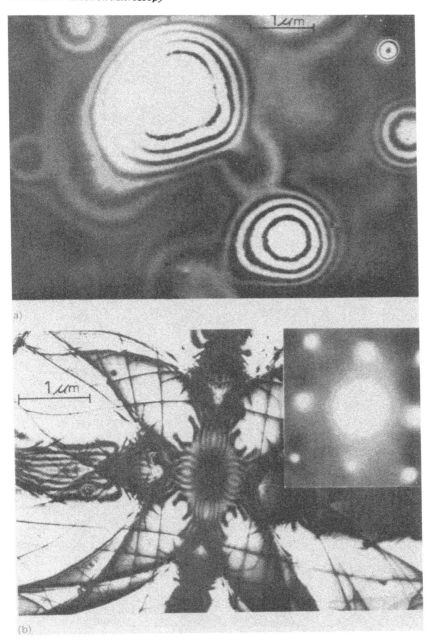

FIG. 7.9 *Examples of bright-field diffraction contrast images of perfect-crystal sections. (100 kV.) (a) Fringe contrast features at holes and etched depressions in stainless steel foil; (b) symmetrical bend-extinction contours resulting from elastically strained bubble in thin foil of Inconel alloy. Symmetrical (100) crystallographic nature of bubble is indicated by superposed selected-area electron diffraction pattern.* [b *from* L. E. Murr, Phys. Status Solidi, 19: *7(1967)*.]

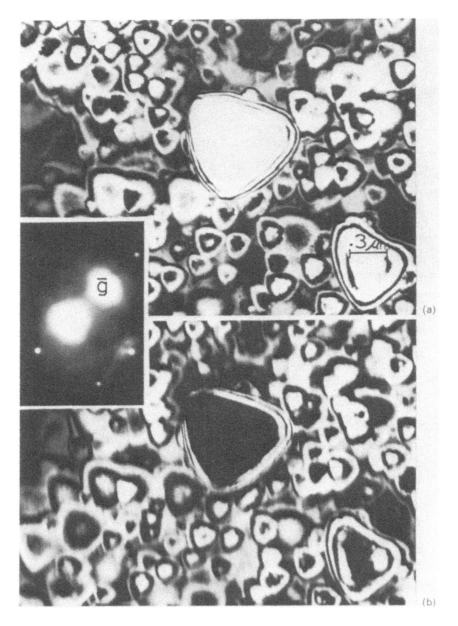

FIG. 7.10 *Bright-field (a) and dark-field (b) electron transmission images of etch pits in (111) Si showing wedge-thickness contrast effects. Superposed selected-area electron diffraction pattern shows strong diffracted beam $g = [2\bar{2}0]$ used to form dark-field image of (b). (125 kV beam).*

that, when defined by a displacement vector, can qualitatively account for defect image contrast.

7.2.2 *DIFFRACTION CONTRAST AT CRYSTAL IMPERFECTIONS*

We have observed in our treatment of perfect-crystal foils that image diffraction contrast conditions can be deduced from simple geometrical arguments. If we now modify these arguments slightly, it can be shown that most common crystal defects, such as grain boundaries, stacking faults, or dislocations, appear in the electron transmission image because of the phase shift they impart to the electron waves.

Let us demonstrate this feature by considering the case of a general phase-shifting plane in an otherwise perfect-crystal section. Figure 7.4 can then be redrawn to depict this situation as shown in Fig. 7.11. In the general case of a grain boundary, for example, a phase shift will arise because of the difference in structure of the adjoining crystals that causes a relative displacement $\underline{R}$ in the location of a unit cell in the diffraction column. As the shaded column in Fig. 7.11 moves to the right, as indicated by the arrow, its point of intersection Q with the phase-shift plane will advance up the plane. At any point of intersection Q, of the column, the unit cell or atomic location will be given by $\underline{r}_j + \underline{R}$, where $\underline{R}$ is a vector characterizing the displacement of the atoms j, from their ideal positions.

The kinematical diffraction amplitude for the crystal section of Fig. 7.11 then becomes

$$\psi_S \equiv S = A \sum e^{2\pi i (\underline{g} + \underline{s}) \cdot (\underline{r}_j + \underline{R})} \tag{7.10}$$

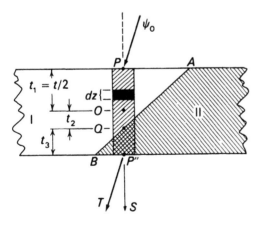

FIG. 7.11 *Column approximation for thin section containing inclined phase-shift plane or crystal interface.*

On examining Eq. (7.10) it will be observed, as pointed out previously, that $e^{2\pi i \underline{g} \cdot \underline{r}j}$ is unity; since both S and R have very small magnitudes (atomic dimensions), $e^{2\pi i \underline{g} \cdot \underline{r}j} \simeq 1$. We are left therefore with the following equation for diffracted beam amplitudes along the column of Fig. 7.11:

$$S = A \int e^{2\pi i s z} e^{2\pi i \underline{g} \cdot \underline{R}} \, dz \qquad (7.11)$$

The second exponential of Eq. (7.11) simply characterizes the interference conditions at the defect or phase-shift plane; and it is obvious that the displacement of atoms within the shaded column by the defect plane of Fig. 7.11 produces a phase angle $\phi = 2\pi \underline{g} \cdot \underline{R}$ in the scattered wave. We can write

$$S = A \int \underbrace{\left[e^{i\phi} \right]}_{\substack{\text{defect} \\ \text{phase factor}}} \underbrace{\left[e^{2\pi i s z} dz \right]}_{\substack{\text{perfect-crystal} \\ \text{phase factor}}} \qquad (7.11a)$$

The characteristic features of the phase-amplitude diagram of Fig. 7.5 are particularly useful at this point in illustrating qualitatively the contrast arising at a crystal interface plane or a stacking fault. We can consider this application generally as shown in Fig. 7.12. Here the broken circle of radius $(2\pi s_1)^{-1}$ represents the perfect-crystal portion (I) of Fig. 7.11. An electron wave entering the crystal at point P scatters regularly within the shaded column to point $Q(t_1 + t_2)$ in the z direction. At this point Q, the column intersects the defect plane and the electron wave undergoes an abrupt phase shift ϕ, equivalent to relocating a new reference circle [of radius $(2\pi s_2)^{-1}$ if $s_1 \neq s_2$][†] at point Q; or else regular scattering of a different nature than in I occurs in II of the thin section of Fig. 7.11 along the phase-shifted or enlarged amplitude-phase circle as shown for a linear distance t_3 of the column. The amplitude from (I) of the thin section of Fig. 7.11 is then characterized by $|\overline{PQ}|$ in Fig. 7.9, while $|\overline{QP''}|$ characterizes the amplitude from (II). The resultant amplitude is therefore $|\overline{PP''}|$ as shown in Fig. 7.12 which is generally different from $|\overline{PP'}|$, the amplitude from the perfect crystal (Fig. 7.1).

[†]*It is important to realize that contrast arising at a grain- or twin-boundary plane results primarily because of the difference in s in I and II of Fig. 7.11, that is, $\Delta s = s_2 - s_1$. We shall deal more specifically with this property in Sec. 7.3.2*

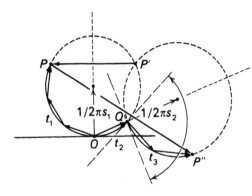

FIG. 7.12 *Phase-amplitude diagram for imperfect crystal containing planar defect such as a stacking fault ($\phi = \pm 2\pi/3$, $s_1 = s_2$), or two crystals (grains) characterized by s_1 and s_2 and separated by interface plane with $\phi \simeq 0$, $s_1 \neq s_2$.*

Since point P is fixed in the convention of Figs. 7.11 and 7.12, PP" varies as the column advances to the right, since point Q varies. Thus, for any two points on the phase-shift plane whose depths differ by ξ_g, point Q will occupy identical positions on the reference (perfect crystal) circle of Fig. 7.12. Under these circumstances, the length t_3 of the column in (II) will be ξ_g and, as a result, P" will be located at an identical position on the (II) circle and $\overline{PP}"$ will have identical magnitudes for these points on the phase-shift plane. Contrast will therefore arise along this plane in the form of fringes parallel to its intersection with the foil surfaces.

We can express the dark-field image-contrast conditions at such a phase shift or interface plane by

$$\frac{\Delta I}{I} = \frac{|PP"|^2 - |PP'|^2}{|PP'|^2} \tag{7.12}$$

with reference to Fig. 7.12; and represent the resulting fringe patterns phenomenologically as depicted in Fig. 7.13. Note again that the bright-field or transmitted image contrast will be the inverse of the dark-field case, so that kinematically, the fringe intensity as shown in Fig. 7.13 will reverse. You should also realize that contrast fringes at grain boundaries arise because only one crystal section in Fig. 7.11 is the reflection position, and the fringes occur at the inclined wedge shown in Fig. 7.13. Figure 7.14 illustrates the contrast fringes that appear in the bright-field image of grain boundaries in a metal, and similar contrast is also observed generally for most materials possessing a regular polycrystalline structure.

In Fig. 7.14 the fringe intensities are not regular as predicted by the kinematic approximation (Fig. 7.12); in some cases the grain-boundary

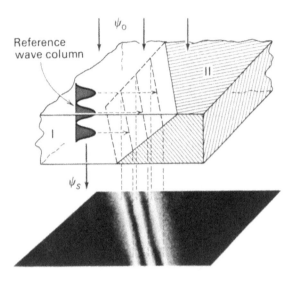

FIG. 7.13 *Geometrical origin of fringe con-*
trast (wedge fringes) at phase-shift or in-
terface plane (grain boundary, twin boundary,
or similar interface) in thin-foil section.
Number of fringes depends on extinction dis-
tance for operating reflection g. If the two
crystal portions, I and II, have a different
crystallographic orientation, fringes will
result either from g(I) or g(II), that is,
from the effective positioning of the refe-
rence wave column. The background intensity
in dark field can be zero as shown, finite,
or alternate between portions I and II.

contrast consists of a continuous intensity change rather than a regular
fringe pattern. The details of these features are accessible only in the
dynamical theory of diffraction contrast (discussed in Sec. 7.3.2). We
will note here that absorption of electrons gives rise to a considerable
portion of these contrast anomalies with regard to a strictly kinematical
approximation.

The contrast at grain-boundary planes in thin polycrystalline materials as observed in electron microscope images can give accurate geometrical information concerning the pole of intersection of individual crystal planes as shown in Fig. 7.14, or of the inclination of the individual
grain-boundary planes with the surfaces, provided the thickness is known
fairly accurately. We can observe in Fig. 7.13 that the angle of inclination of the phase-shift or interface, plane is given by

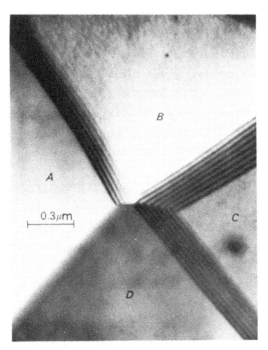

FIG. 7.14 *Bright-field transmission electron microscope image showing fringes at inclined grain boundaries intersecting in polycrystalline section of 304 stainless steel observed at 100 kV. Details of fringe contrast at grain boundaries of discussed in dynamical treatment of Sec. 7.3.2.* [From L. E. Murr, Phys. Status Solidi, 19: 7(1967).]

$$\theta = \tan^{-1} \frac{t}{W} \tag{7.13}$$

Where W is the total width of the projection of the plane as measured directly from the electron transmission image. The crystal thickness t can be obtained from

$$t \simeq N\xi_g \tag{7.14}$$

only if the operating reflection $\underline{g}$ producing the fringe pattern is known, and $\underline{s} \simeq 0$. Thus, where $\underline{g}$ is unknown, or where contrast of the plane occurs as a continuous intensity change rather than oscillating intensity profiles (as shown in the boundary separating grains A and D of Fig. 7.14), t must be obtained in a more direct fashion.

Since the images of coherent boundaries in fcc crystals are coincident with the projection of {111} planes, the angle of inclination of a twin plane, a stacking fault, or a dislocation slip trace in fcc can be calculated directly from a knowledge of the crystallographic surface plane and the direction of the projected slip plane in this surface. This information is obtained from the selected-area electron diffraction pattern as outlined in Chap. 6. We shall defer until later the application of these properties of twin-grain boundary images as they relate to the structural geometry of thin-crystal sections.

As pointed out, the images of twin boundaries represent a crystallographic projection that, in fcc crystals, is defined by a projection of a {111} plane. The mechanism of image-contrast fringe formation at these boundaries is kinematically described by the general situation depicted in Fig. 7.13.

Figure 7.15 shows several examples of twin-boundary contrast in bright-field images of fcc stainless steel. It is instructive to compare the fringe correlation of the twin-boundary planes and thickness contours at an etched depression in Fig. 7.15a with Fig. 7.9a. The geometry of this situation should be quite apparent, particularly if you will now compare Figs. 7.7 and 7.13. Because the fringes of the boundary planes in Fig. 7.15a coincide with the thickness contours of the matrix depression, the operating reflection is associated with the matrix and not the twin-crystal region. The details of the features will become more clear when the dynamical theory is discussed. Figure 7.15b illustrates quite strikingly the attenuation of transmitted signal intensity by absorption for a twin boundary having numerous incoherent (non-{111}) steps. The incident or top surface of the foil can generally be associated with the loss of fringe recognition due to the increased absorption length and other more subtle features, and we shall take up this point in more detail in the discussion of the dynamical theory of image contrast in Sec. 7.3.

Stacking-fault contrast A stacking fault in a crystal presents a special case where, as in Fig. 7.11, the lattice regions on either side of the phase-shift plane (stacking-fault) are identical, that is, where (I) $\equiv$ (II), and $s_1 = s_2$ (Fig. 7.12). A unique feature of the {111} stacking-fault plane in fcc crystals[†] is that the displacement vector $\underline{R}$ is known, and can usually be expressed in the general form

$$\underline{R}_{SF} = \frac{1}{6} <112>$$

[†]*See page 482*

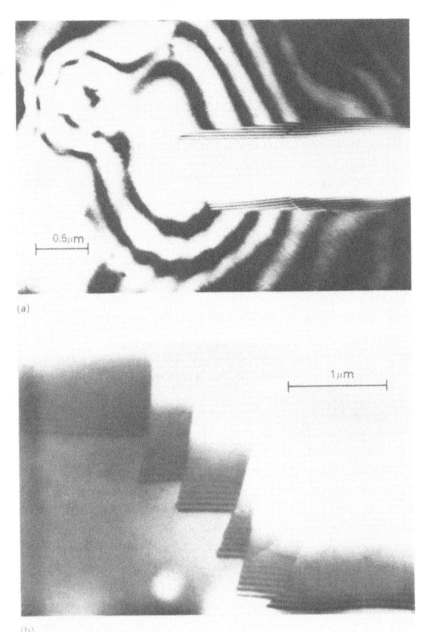

FIG. 7.15 *Bright-field electron transmission micrographs of twin and twin-boundary contrast in 304 stainless steel. (a) Twinned crystal band shown extending into thinned (taper) region of foil. (125 kV.) (b) Incoherently stepped twin-boundary plane showing intensity attenuation by electron absorption (100 kV.)*

Consequently, the phase difference ϕ_{SF} in the wave diffracted from opposite sides of the fault is then

$$\phi_{SF} = \frac{\pi}{3} [h + k + 2\ell] \tag{7.15}$$

where h, k, and ℓ are the Miller indices of the corresponding operating reflection $\underset{\sim}{g}$. The contrast arising at a stacking fault is then kinematically identical with that at a general interface plane as shown in Fig. 7.13; and the mechanism of fringe periodicity can, in this special case, be ideally described by considering $\phi \equiv \phi_{SF} = \pm 120°$, and $s_1 = s_2 = s$ in the phase-amplitude diagram of Fig. 7.12. Figure 7.16a illustrates the nature of stacking-fault contrast[†].

An interesting feature of the stacking-fault phase angle ϕ_{SF} [Eq. (7.15)] in fcc material is that for certain values of $\underset{\sim}{g}_{hk\ell}$, that is [h + k + 2$\ell$] = 6, $\phi_{SF} = 0$, and there is no contrast. Thus tilting a foil area containing a stacking fault can cause the contrast to disappear if the foil is tilted into a reflection position corresponding to, say, $\underset{\sim}{g} = [420]$, [330], etc., for $\underset{\sim}{R} = 1/6[112]$. We shall see shortly that this condition also characterizes dislocations in crystals, and can be used as a means to determine the displacement vector (which, for a dislocation, is its Burgers vector $\underset{\sim}{b}$).

Because of the geometrical nature of an fcc stacking-fault plane, the invisibility criterion discussed above can also occur when three stacking faults occur on adjacent {111} planes in an fcc crystal section. That is, consider a single fault characterized by $\phi_1 = + 120°$. A fault on an adjacent {111} plane will then have $\phi_2 = + 120°$, and the total phase shift across the two fault planes will be $\phi = \phi_1 + \phi_2 = 240°$. Adding a third fault will then produce $\phi = 360°$ or zero phase angle, and the three-fault composite plane will be effectively invisible insofar as image contrast is concerned. This condition is often observed experimentally in thin fcc films where the motion of partial dislocations produces stacking-fault contrast on adjacent {111} planes [7]; and where three such faults overlap, or where two faults are opposite in character (one having $\phi_{SF} = + 120°$, the other having $\phi_{SF} = -120°$), the overlapped or zero-phase portion will show

---

[†]*Stacking faults occur in bcc crystals on the {110}, {112}, and {123} planes. The case of stacking faults in bcc materials occurring on {112} has been treated by P. B. Hirsch and H. M. Otte, Acta Cryst., 10: 447(1957). In hcp materials, stacking faults frequently occur on the (0001) plane.*

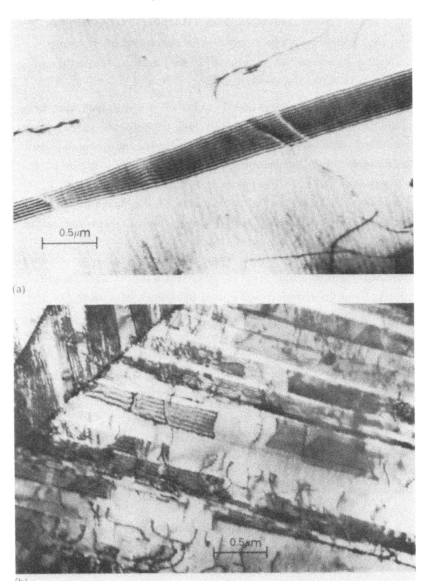

FIG. 7.16 *Stacking-fault contrast in thin metal foils (bright-field images). (a) Characteristic fringes at stacking fault in 304 stainless steel. Note that as foil thickness decreases (to the left), dark fringes in central region of image converge at a point where local thickness decreases by approximately one extinction distance. (b) Stacking faults created in same stainless steel by movement of partial dislocations created during passage of explosive shock wave at pressure of 120 kbars. Note that overlapping faults have created regions of zero phase angle, causing contrast to essentially vanish. (100 kV.)* [*b is after M. C. Inman, L. E. Murr, and M. F. Rose,* ASTM Spec. Tec. Publ., 30: *396(1966).*]

no contrast. Figure 7.16b illustrates this situation for overlapping
faults of the same character, ϕ_{SF} = 120°. The character of stacking
faults is generally described as intrinsic or extrinsic. The nature of
these faults in thin films is treated in Sec. 7.3.

Images of dislocations in thin crystals Images of dislocations were originally observed in the very early thin-film work of Heidenreich [2] , but
were not identified until approximately 5 years later. Dislocations in
thin films were identified independently by W. Bollmann [8] and by P. B.
Hirsch and co-workers [4,9]. It was only a few decades ago that M. J.
Whelan [10] demonstrated that the basic features of dislocation images in
thin foils were in fact experimental proof of the predicted properties of
dislocation types, reactions, and interactions [11-13].

Shortly after the observation and identification of dislocation images,
a kinematical theory for the qualitative interpretation of such images was
advanced [1], followed by the more extensive dynamical treatment [14,15]
discussed in Sec. 7.3. It should be pointed out at the outset, however,
that neither the simple kinematical nor the more quantitative dynamical
treatments provide a completely rigorous interpretation of dislocation
images. Nevertheless, we can unambiguously identify dislocations in thin
films; and in many instances they can be completely characterized.

Let us consider a crystal section (shown in Fig. 7.17) containing a
total dislocation having an edge and screw component. If the diffraction
column corresponding to the original approximation of Fig. 7.4 is taken
through this section along the optic axis (shown dotted), and directed
normal to the dislocation lines (shown by the arrows), an abrupt phase
shift will occur in the diffracted wave. For positions of the column away
from the dislocation lines, however, the phase shift will be a continuous
function of position, with ϕ in Eq. (7.14a) approaching zero in the lattice portions removed from the dislocation line. We also observe from
Figure 7.17 that the displacements of the atoms on opposite sides of the
dislocation line (most apparent for the edge component) are in opposite
directions so that, in effect, $\underline{s}$, the deviation from the reflection position, will be positive on one side and negative on the other. Phenomenologically then, the effect of the dislocation will be to bring one side of
the lattice closer to the reflection position while causing the opposite
side to deviate from the reflection position by an equal amount. The result is that the phase contrast that arises at a dislocation will in rea-

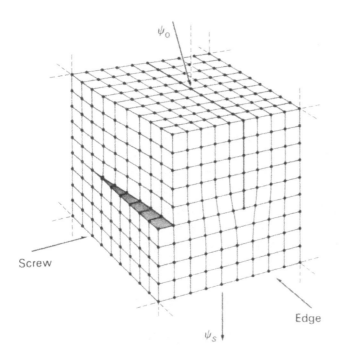

FIG. 7.17 *Total dislocation having edge and screw components in crystal lattice section.*

lie to one side of the actual dislocation line, depending on the coincidence with the reflection position.

We can demonstrate this feature for the screw component of Fig. 7.17 as follows: Consider a screw dislocation line in a crystal section as sketched in Fig. 7.18. The displacement vector is given by

$$\underline{R} = \frac{\underline{b}\zeta}{2\pi}; \quad \zeta = \tan^{-1}\frac{z}{x}$$

where $\underline{b}$, the Burgers vector (parallel to the dislocation line AB), is associated with the distortion of the diffraction column $\overline{PP}'$ into $\overline{DD}'$. The phase angle for the screw dislocation then becomes

$$\phi_{screw} = 2\pi\underline{g} \cdot \underline{R} = \underline{g} \cdot \underline{b} \tan^{-1}\frac{z}{x} \qquad (7.16)$$

where if $\underline{g} \cdot \underline{b} = n$, we can put

$$\phi_{screw} = n \tan^{-1}\frac{z}{x} \qquad (7.16a)$$

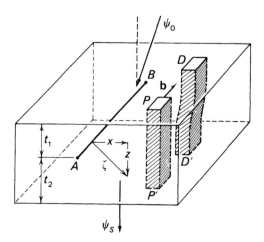

FIG. 7.18 *Schematic view of screw disloca-*
tion in crystal section of Fig. 7.17. Dis-
location line is parallel to foil surfaces
and normal to direction of propagation of
diffracted wave ψ_S.

Since $\underline{g}$ can be positive or negative, n can take on positive, negative, or zero values. We can then substitute Eq. (7.16a) into Eq. (7.11a) to obtain

$$\psi_S = A\int_{-t_2}^{t_1} e^{\,in\cdot\tan^{-1}z/x\,+\,2\pi isz}\,dz \qquad (7.17)$$

as an integral expression of the corresponding phase-amplitude diagram for a screw dislocation. If we now define a parameter $\beta = 2\pi sx$, then for fixed s, β will vary with x, and the center of the dislocation (that is, the dislocation line in its relation to the dark-field image) will correspond to $\beta = 0$.

The phase-amplitude diagram for a perfect screw dislocation can be constructed as shown in Fig. 7.19 for n = 1. The phase-amplitude diagram of Fig. 7.19a for $\beta = -1$ becomes a wound-up spiral for $\beta = +1$ as shown in Fig. 7.19b. In Fig. 7.19c and d the corresponding image intensity profiles are shown for various operating reflections $\underline{g}$. We observe in Fig. 7.19 that the diffraction contrast at screw dislocations will be given generally by

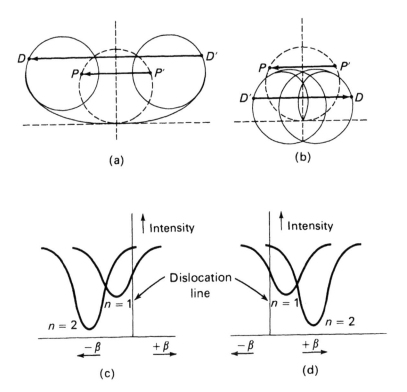

FIG. 7.19 *Representations for screw-dislocation image result-ing for transmission electron microscopy of foil section of Fig. 7.18. (a) Phase-amplitude diagram for screw dislocation for n = 1, β = -1 (unwound spiral). (b) Phase-amplitude dia-gram for screw dislocation for n = 1, β = 1 (wound-up spiral). (c) Bright-field-image intensity profiles of screw dislocations for n = 1,2, and negative deviation of foil area of Fig. 7.18 from reflection sphere, s < 0. (d) Bright-field-image inten-sity profiles of screw dislocations for n = 1,2, and positive deviation of foil area of Fig. 7.18 from reflection sphere, s > 0.*

$$\frac{\Delta I}{I} = \frac{|PP'|^2 - |DD'|^2}{|PP'|^2} \tag{7.18}$$

It is apparent from Fig. 7.19c and d that the dislocation image will be bright in the dark-field mode and dark in the bright-field mode. If the width of the dislocation image as given in the intensity profiles of Fig. 7.19c and d is taken as the intensity half-width, then we can observe that

$$\Delta\beta = 2\pi s \Delta x \cong 2 \tag{7.19}$$

is the maximum half-width for dislocations with n = 2, which is considered common [1]; values of n = 4 seldom occur. The dislocation width Δx is then nominally

$$\Delta x \cong \frac{1}{\pi s} \cong \frac{\xi_g}{\pi} \quad \text{at } n = 2 \tag{7.19a}$$

or roughly twice this value for cases where n = 4. The importance of this feature in thin-film observations by transmission electron microscopy is that values of Δx in Eq. (6.19a) are normally in excess of 100 Å, considerably above the instrument resolution. Consequently resolution is not normally critical in routine thin-film electron microscope studies, and the direct observation of dislocations is limited primarily by contrast conditions as expressed generally in Eq. (7.18). However, it must be recognized in Eq. (7.19a) that dynamical diffraction is tacitly assumed because we have used the dynamical extinction distance. If we do indeed have a situation where the diffraction conditions are strictly kinematical, then the extinction conditions would in principle be governed by the deviation parameter, s. Consequently, for s large, it is possible to reduce Δx. Ideally this condition requires the selective use of a weakly diffracting beam where in fact s will be large.

A pure edge dislocation can of course be similarly treated[†] by considering the appropriate displacement vectors [11-13]

$$\underline{R}_1 = \frac{b}{2\pi} \left[\zeta + \frac{\sin 2\zeta}{4(1 - \nu)} \right] \tag{7.20}$$

$$\underline{R}_2 = - \frac{b}{2\pi} \left[\frac{1 - 2\nu}{2(1 - \nu)} |\underline{n}| |\underline{r}| + \frac{\cos^2 \zeta}{4(1 - \nu)} \right] \tag{7.20a}$$

$$\underline{R}_3 = 0 \tag{7.20b}$$

where $\underline{R}_1$, $\underline{R}_2$, and $\underline{R}_3$ denote displacements parallel to the Burgers vector, normal to the slip plane, and parallel to the dislocation line, respectively [11]. Thus, for metal or alloy foils (ν ≈ 1/3) having an edge dislocation parallel to the surface as in Fig. 7.18 at AB, $\underline{R}_1$ is equal to $\underline{b}/2\pi(\zeta + 3/8 \sin 2\zeta)$, and

---

[†]*Contrast at edge dislocations is treated in a refined kinematical approximation by R. Gevers,* Phil. Mag., 7: *59(1963).*

$$\phi_{edge} = 2\pi g \cdot R_1 = n(\zeta + \frac{3}{8} \sin 2\zeta) \simeq n \tan^{-1} \frac{2z}{x} \qquad (7.21)$$

The diffracted wave amplitude is then given approximately by

$$\psi_S = A \int_{-t_2}^{t_1} e^{in \cdot \tan^{-1} 2z/x + 2\pi isz} dz \qquad (7.22)$$

where we observe on considering Eq. (7.20) in retrospect that ψ_S is the same for an edge and screw dislocation at 2x and x, respectively, for constant s. This means of course that the edge-dislocation image in this particular situation will be roughly twice that for a pure screw dislocation.

We have treated the general case of an edge or screw dislocation in the plane of a thin-foil material. However, you are probably already aware that dislocations will normally be associated with the active slip planes, which are usually inclined to the crystallographic surface. In fcc materials, the slip planes are {111}, while in bcc they can be the {110}, {112}, or {123} planes. Subsequently, dislocations normally occur on inclined planes in association with the planar defects already discussed. The corresponding images of such dislocations will be ideally characterized by the associated contrast features of phase-amplitude diagrams such as those of Fig. 7.19, with some modifications being invoked as a result of the change in the geometical relationships of Fig. 7.18. Phenomenologically, you might imagine dislocations in fcc materials to be a linear portion of the {111} phase-shift plane of Fig. 7.13. In this way the interpretation of the bright-field images of dislocations shown in Fig. 7.20 will be at least comprehensible.

The images of dislocations in Fig. 7.20 represent all of the commonly observed characteristics that are referred to as zig-zag contrast (as in Fig. 7.20a), black-white contrast, and dotted contrast, in addition to the regular or continuous contrast features shown in Fig. 7.20c and described generally by Fig. 7.19. Kinematical considerations can be modified sufficiently to account for some of the image details of Fig. 7.20a and b as shown by Hirsch and associates, [1] but a dynamical treatment becomes increasingly more valuable, particularly where s is small, and where absorption becomes important.

The origin of dotted images is ideally depicted in the modified phase-amplitude diagrams shown in Fig. 7.21, assuming a screw dislocation inclined with the surfaces. Figure 7.21b illustrates the extension of the

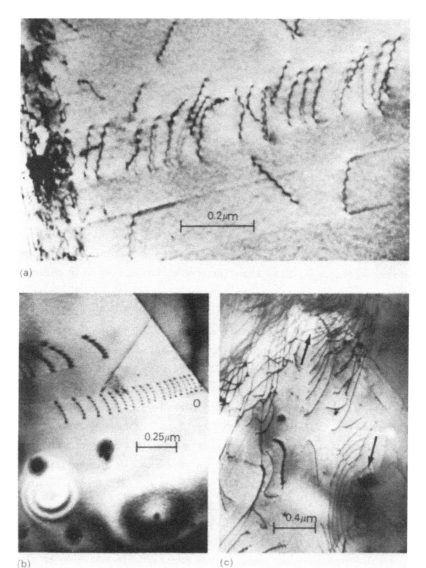

FIG. 7.20 *Characteristic dislocation images in fcc metals.*
(a) Zig-zag and dotted contrast at dislocations associated
with fracture in Inconel foil. Grain surface orientation is
approximately (110). [*From L. E. Murr,* Proc. Penna. Acad.
Sci., *39: 202(1966).*] *(b) Zig-zag and black-white contrast*
effects at a pileup of dislocations at grain boundary in stain-
less steel. (c) Continuous image profiles and zig-zag char-
acteristics of dislocations in stainless steel. Frank-Read
type dislocations (see J. P. Hirth and Jens Lothe, Theory of
Dislocations, *McGraw-Hill, New York, 1968) are observed as*
indicated by arrows. Note zig-zag character of dislocations
near surface. (100 kV.)

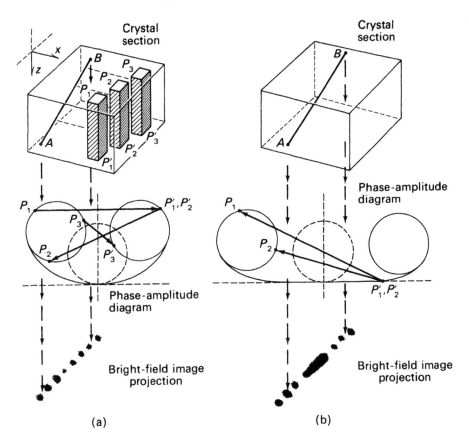

FIG. 7.21 *Modified phase-amplitude diagrams for inclined screw dislocation to qualitatively depict origin of dotted contrast and dotted or zig-zag characteristic images associated with dislocation segments near foil surface as shown for example in Fig. 7.20c. (a) n = 1, β = -1; (b) n = 2, β = -1.*

phase-amplitude circles to qualitatively account for dotted or zig-zag type features associated with dislocation images close to the surfaces as shown in Fig. 7.20c. For a more extensive treatment of phase-amplitude diagrams as they relate to dislocation contrast, see Hirsch and associates [1] and Whelan [6].

We have observed that while the image characteristic (zig-zag, dotted, etc.) does not relate to the intrinsic nature of the dislocation, the geometrical features of the dislocations nevertheless do. In this sense the dislocation pileup of Fig. 7.20b and the curved nature of dislocations emanating from sources in Fig. 7.20c do reflect their theoretically predicted properties [11-13].

Bowing of dislocation images is a result of a normal stress acting in slip plane. The bowing normally occurs in association with the direction of motion. In this respect, the sense of motion is apparent in many electron micrographs as a result of the bowing of dislocations and their characteristic contrast left in the wake of their change in position along a particular slip plane. This feature is particularly noticeable in Fig. 7.20a where the ends of the dislocation images terminating on the surfaces are streaked, thus showing the geometrical width of the slip plane by diffraction contrast. The contrast at slip traces has been treated by Howie and Whelan [14].

The geometrical features of dislocations in a thin-foil slip plane imaged in the electron microscope are shown in Fig. 7.22. Here we observe that the true radius of curvature R can be deduced directly from the projected image [2] by considering

$$R = R' \frac{(\cos^2\phi + \sin^2\phi\cos^2\theta)^{3/2}}{\cos^2\theta} \qquad (7.23)$$

where R' is the radius of curvature measured directly from the electron micrograph as shown in Fig. 7.22. The thickness of the section is also directly amenable in Fig. 7.22 by inserting the measured value of W and θ in Eq. (7.13).

Recalling that the stress on a dislocation is given generally by [11-13]

$$\sigma = \frac{Gb}{2R} \qquad (7.24)$$

where G is the shear modulus, and b is the magnitude of the Burgers vector ($|\underline{b}|$), the normal stress in the slip plane can be calculated directly from the image detail provided b is determined. We shall treat the determination of b in Sec. 7.3.2.

In the case of the dislocation pileup shown in Fig. 7.20b, we can consider the total force on the ith dislocation to be given by [13]

$$F_i = \frac{Gb^2}{2\pi(1-\nu)} \sum_{\substack{j=0 \\ j\neq i}}^{n} \left(\frac{1}{X_i - X_j}\right) - \sigma b = 0 \qquad (7.25)$$

where ν is Poisson's ratio, and n is the number of dislocations piled up behind the lead dislocation (shown as 0 in Fig. 7.20b). Consequently,

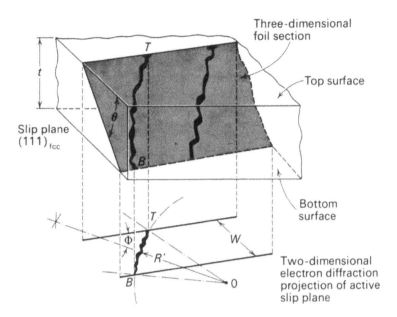

FIG. 7.22 *Schematic section showing dislocations on inclin-*
ed plane in thin foil, and corresponding two-dimensional
electron transmission microscope image projection. [*From*
L. E. Murr, Phys. Status Solidi, 19: 7(1967).]

having found σ in Eq. (7.24) by measuring R directly from Fig. 7.20b, the
dislocation force can be calculated.[†] It is found in the solution of Eq.
(7.25) that

$$\sigma^* = n\sigma \qquad (7.26)$$

where σ^* is the stress on the lead dislocation (at 0 in Fig. 7.20b) due to
the n dislocations piled up behind it. Since the spacing of the pileup has
a geometrical relationship to the acting stresses, we can evaluate Eq.
(7.26) approximately from

$$\sigma \cong \frac{nGb}{\pi L} \qquad (7.27)$$

where L is the length measured from the lead dislocation at 0 to the ith
dislocation measuring n in the pileup as shown in Fig. 7.20b.

It is also possible to measure stacking-fault energies from similar

---

[†]*For an example of the consequence of dislocation pile-ups in their rela-*
tionship to materials properties, see J. D. Meakin and H. G. F. Wilsdorf,
Trans. AIME, 218: 445(1960).

direct observations of dislocation nodes and related features, particularly
in fcc materials of low stacking-fault energy. We will treat such operations in more detail in a later section.

We should also point out that the various reactions and interactions
of dislocations can be observed directly within the electron microscope.
The motion of dislocations as a result of straining thin foils directly
within the electron microscope can be observed [16-20] and it is also possible to observe similar motion by stressing dislocations by focusing the
electron beam onto a small area of the foil. In this way, thermal and contamination stresses will cause dislocations to move [9,21,22] as shown in
the sequence of electron micrographs in Figs. 7.23 and 7.24. In Fig. 7.24,
dislocations apparently emitted from the grain boundary are caused to bow
and move under the action of the electron beam. This action might be compared with the dislocations apparently piling up at a grain boundary as
shown in Fig. 7.20c. Indeed, Murr [23,24] has demonstrated that grain
boundaries are principal sources of dislocations, especially during the
early stages of deformation. Some interesting possibilities exist for the
direct observation of such phenomena in thicker materials and at correspondingly higher accelerating potentials. These will be described briefly in
the context of in-situ high-voltage electron microscopy in Chap. 8.

Weak-Beam TEM As we noted earlier in connection with Eq. (7.19a), dislocation image widths are governed by the dynamical extinction distance for
strictly dynamical diffraction, or by the deviation parameter, s, for
strictly kinematical diffraction. You might recall from Chap. 6 (6.5.2)
that dynamical diffraction was defined as the condition where $I_o \cong I_g$,
while in kinematical diffraction $I_g \ll I_o$. In the two-beam dynamical case
for thick films where $I_o \cong I_g$, $s = 0$. If the Ewald sphere is shifted by
tilting the beam or if the specimen is tilted, it is possible to move the
Ewald sphere away from a particular intensity rel-rod, reducing the brightness, and creating a situation where $I_g < I_o$. In doing this, $s \neq 0$, and
as I_g is reduced, the magnitude of s increases (either + or -); producing
a phenomenological kinematical condition. These conditions for diffraction can, obviously, be selected by placing the objective aperture over
particular diffraction spots to form the dark-field image. For the conditions where s is large enough to invoke a kinematical-like response while
Io and I_g are dynamically coupled, the image width of any diffracting regime will be reduced. For crystal defects such as dislocations, the Bragg

condition for the weak beam might still be achieved as a result of plane bending near the dislocation core. This will contribute significantly to the local signal strength and provide a very narrow signal distribution confined to the plane bending region or dislocation core. Under optimum conditions, this reduction can be very significant, allowing detail as small as a few tens of Angstroms to be resolved. It has been determined that a beam is "weak" enough to provide this imaging advantage if $|s|$ > 2×10^{-2} Å^{-1} [25].

The concept of a weak-beam method of electron microscopy was introduced in 1969 by Cockayne, et al [26] and has been widely used during the last decade to study dislocation separations for stacking-fault energy measurements, small loops and precipitates, grain boundary structure and second phase particle interfaces, precipitation on dislocation lines, and a host of other applications [25-29]. There are certainly useful prospects for the technique in the study of heavily deformed materials where image overlap of closely spaced dislocations and dislocation loops can obscure any real understanding of the residual microstructure.

Although it is possible to select weak beams by simply placing the aperture over them, or deflecting them into the optic axis by beam tilt as shown in Fig. 5.2, not all reflections yield a sufficiently large $\underline{s}$, and not all are "dynamically coupled" to the regime of interest. Figure 7.25 illustrates the simplest of the weak-beam concepts where the weak - $\underline{g}$ reflection is used to image closely-spaced dislocations. In general it is possible to select a weak-beam for some specific reflection at the Bragg condition. For example in Fig. 7.25 $\underline{g}$ is at the Bragg condition and -$\underline{g}$ is the weak beam. A more effective weak beam condition might occur if $2\underline{g}$ were at the Bragg condition, etc. Since the aperture is used in Fig. 7.25 to select dark-field image reflections, chromatic aberration distortion contributes significantly to the dark-field image features. Weak-beam (kinematical) dark-field images as well as any dynamical dark-field image can of course be clarified by tilting the beam so that a particular reflection is on the optic axis, thereby eliminating the chromatic image distortion.

To optimize the weak-beam mode, a Bragg reflection $\underline{g}$ may be selected, and the specimen oriented to be far from satisfying the Bragg condition for this reflection. Under these conditions, a dark-field image of the perfect crystal using the reflection $\underline{g}$ would show very weak intensity because in this particular orientation into which the crystal has been

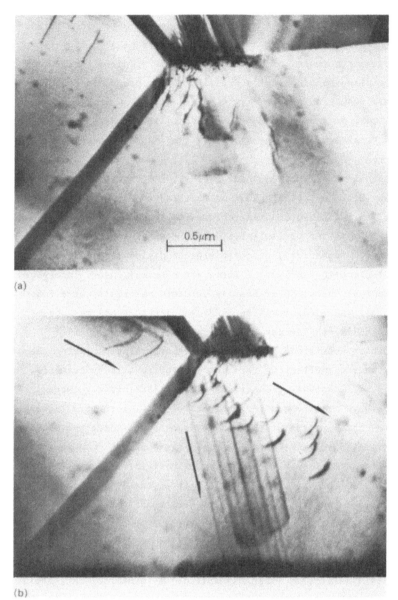

(a)

(b)

FIG. 7.23 *Direct observation of dislocation motion in thin*
stainless steel foil. (a) *Area showing junction of three*
grains contains numerous dislocations. Electron beam is then
focused onto area, causing local heating and buildup of con-
tamination that combine to cause complicated stress fields to
act on the dislocations. (b) *Same area of a is shown following*
several seconds of beam heating. Dislocations have moved along
several slip systems as indicated by arrows. Note severe bow-
ing of dislocations resulting from normal stress component
resolved in slip planes. (100 kV.)

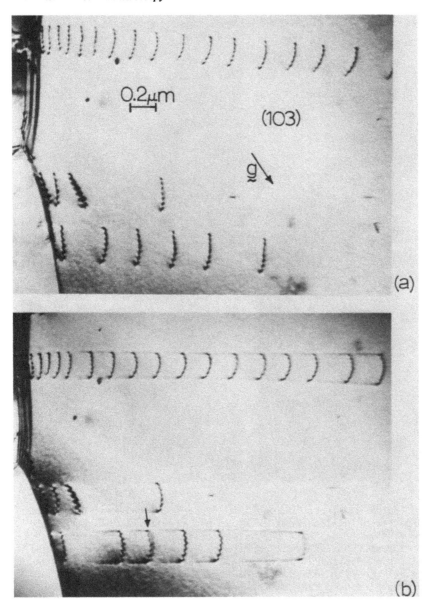

FIG. 7.24 *Electron beam-induced motion of dislocations emanating from a grain boundary in a thin stainless steel foil (a) prior to beam heating, (b) after beam heating. The operating reflection, g, is [020]. The arrow in (b) emphasizes the splitting of one of the total dislocations into two separated partial dislocations.*

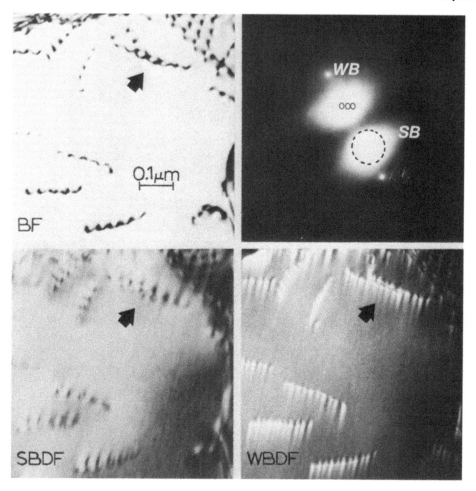

FIG. 7.25 *Aperture-selected strong-beam (dynamical) and weak-beam (kine-matical) imaging of dislocations in stainless steel. The bright-field (BF) image is shown along with the selected-area electron diffraction pattern. With the aperture (dotted circle) over SB, $\underline{g}$, the strong-beam dark-field image results (SBDF). With the aperture shifted over the weak beam (WB), $-\underline{g}$, the weak-beam dark-field image results (WBDF). Note espe-cially the separation of fine dislocation image widths denoted by the arrow. (200 kV, $\underline{g} = [\bar{1}11]$).*

tilted, it scatters only weakly into $\underline{g}$. However a dislocation in this volume of crystal could give rise to local lattice plane bending which could satisfy the Bragg condition for $\underline{g}$, producing a localized increase in the image intensity. You might refer to Fig. 7.17 to gain some addi-tional insight. Imagine that the incident beam (ψ_0) satisfies the Bragg condition for the perfect lattice planes and produces a strong diffraction spot. If this crystal is tilted, the diffracted beam will decrease in intensity but on tilting the Bragg condition might still be maintained near the core of the edge dislocation by diffraction from the "bent" lattice planes. Thus, although the background dark-field intensity from the lattice is weak, a very prominent but local intensity can arise at the dislocation (near the dislocation core) where the lattice plane bending is most severe . Indeed, it is this very local satisfaction of the Bragg condition rather than a strict satisfaction of the conditions for kinematical diffraction which provides the real advantage of the weak-beam method, although it is phenomenologically a kinematical situation since $I_g \ll I_0$. This has the same effect on the image width, Δx, as assuming a kinematical extinction condition in Eq. (7.19a), but it localizes the conditions to specific defects. This condition can also be achieved by using higher order reflections which are brought onto the optic axis through incident beam tilt, leaving the specimen fixed. In the former scheme, a beam is selected and placed on the optic axis through beam tilt, and the crystal is then also tilted to reduce the strength of $\underline{g}$ (and increase $\underline{s}$). Consequently both beam tilt and specimen tilt are required in that mode. Figure 7.26 illustrates the optimized weak-beam imaging technique utilizing beam tilt of higher-order reflections. The advantages in terms of image clarity (resolution) are clearly evident in the WBDF image as compared to the BF image.

7.3 DYNAMICAL THEORY OF THE INTERPRETATION OF THE IMAGES OF CRYSTAL LATTICE IMPERFECTIONS

We have demonstrated in Sec. 7.2 that the phase relationships among electron waves scattered by individual atoms of a crystal are important in determining the diffracted beam intensity at the recording plane of an electron microscope and, conversely, the transmitted signal contributing to the bright-field image. We have shown in effect that phenomenologically the visibility of defects in transmitted electron images is due to diffraction contrast. Treated kinematically, assuming some finite deviation from the reflection position, the qualitative aspects of extinction contours, fault or phase-shift planes, and dislocations have been demonstrated.

However, two important aspects of image formation were neglected in the kinematical approximation, namely, that in many real situations in the electron microscope, $\underline{s} = 0$; and that inelastic scattering or energy absorption of the scattered electrons was neglected. We can observe that

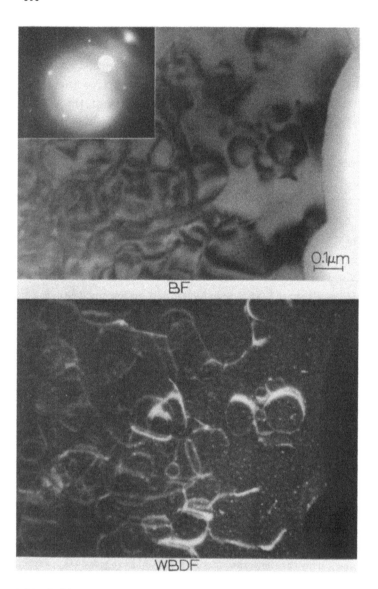

FIG. 7.26 *Optimized (tilted) weak-beam dark-field imaging of dislocations, loops, and voids in neutron irradiated α-Al$_2$O$_3$. The position of the aperture over the weak beam is indicated by the exposed circle. The loops are interstitial and initially faulted on (0001) or {10$\bar{1}$0}, but at a critical size unfault to form b = 1/3<10$\bar{1}$1> loops which intersect to form the dislocation network shown. These details are only evident from the WBDF image. The establishment of networks provides an unsaturable interstitial sink, with the corresponding growth of voids from vacancy condensation. The voids lie in rows along [0001] but randomly in the (0001) basal plane, so columns of voids are seen in this [0001] projection. (Courtesy L. W. Hobbs, from F. W. Clinard, Jr., L. W. Hobbs and G. F. Hurley in J. Nucl. Mater., 1982.)*

as s = 0, the extinction condition as defined in Eq. (7.5) becomes kinematically unmanageable, and the approximations of Eqs. (7.3c) and (7.8) become meaningless. A paradox arises since $t_0 = \xi_g$, a finite value for any particular operating reflection $\underline{g}$.

It will also be recalled from our treatment of electron diffraction in Chap. 6, that for foil materials of any appreciable thickness, the scattering of electrons becomes dynamical, and the scattered intensity, either as a single diffracted beam or many beams, appears equal in magnitude to that of the transmitted beam intensity. In general, dynamical diffraction is given by Eq. (6.49) if we treat only the spot intensity for a two-beam situation as represented ideally in Fig. 6.18a.

Let us now reconsider the electron wave function of Eq. (1.42) as treated in Sec. 6.5. We will choose as a starting point Eq. (6.33) and write

$$U(r)[\psi_0(\underline{r}) + \psi_s(\underline{r})] = U\psi_T + U\psi_S \tag{7.28}$$

We then consider that

$$T*T + S*S = 1 \tag{7.29}$$

or

$$|T|^2 + |S|^2 \equiv I_T + I_S = 1 \tag{7.29a}$$

where T and S are the transmitted and diffracted wave amplitudes, respectively. We will now consider Eq. (7.28) fully, and consider the following system of differential equations for the two-beam case.[†]

$$\frac{d\psi_T}{dz} = \frac{i\pi}{\xi'_g} \psi_S$$

$$\frac{d\psi_S}{dz} = 2\pi i s\psi_S + \frac{i\pi}{\xi'_g} \psi_T \tag{7.30}$$

where the amplitudes of the transmitted and diffracted waves are given by

$$T = e^{-i\pi sz}\psi_T$$

$$S = e^{-i\pi sz}\psi_S \tag{7.31}$$

[†]*We use ξ'_g in place of the extinction distance ξ_g to denote the inclusion of absorption phenomena.*

Substitution of Eqs. (7.31) for the wave functions of Eqs. (7.30) then results in

$$\frac{dT}{dz} = - i\pi sT + \frac{i\pi}{\xi'_g} S$$

$$\frac{dS}{dz} = i\pi sS + \frac{i\pi}{\xi'_g} T \tag{7.32}$$

as the dynamical expressions for the transmitted and diffracted wave amplitudes originally derived by A. Howie and Whelan [14,30] for a perfect-crystal section. In matrix notation, this system of equations can be written

$$\frac{d}{dz} \begin{pmatrix} T \\ S \end{pmatrix} = \pi i \begin{pmatrix} -s & \frac{1}{\xi'_g} \\ \frac{1}{\xi'_g} & s \end{pmatrix} \begin{pmatrix} T \\ S \end{pmatrix} \tag{7.32a}$$

The so-called anomalous absorption is expressed in the previous equations through

$$\frac{1}{\xi'_g} = \frac{1}{\xi_g} + \frac{i}{\delta_g} \tag{7.33}$$

where δ_g is the absorption length, given by the notation τ_o or τ_g in other treatments [14,33,35], that is:

$$\delta_g \equiv \tau_o \equiv \tau_g \simeq 10\xi_g \tag{7.34}$$

Equation (7.33) is essentially equivalent to replacing the lattice potential of Eq. (6.28) by a complex potential of the form

$$U_i = U(\underset{\sim}{r}) + iV(\underset{\sim}{r}) \tag{7.35}$$

where

$$V(\underset{\sim}{r}) = V_T + \sum_{\underset{\sim}{g}} V_S e^{2\pi i \underset{\sim}{g} \cdot \underset{\sim}{r}} \tag{7.36}$$

as originally proposed by K. Moliere [31] and H. Yoshioka [32].

You can readily show that, with reference to Prob. 7.8, Eq. (7.32) has solutions

$$T = \cos \pi\sigma z - i\left(\frac{\pm s}{\sigma}\right) \sin \pi\sigma z$$

$$S = \frac{i}{\sigma\xi_g'} \sin \pi\sigma z \tag{7.37}$$

where σ is defined by

$$\sigma = \frac{1}{\xi_g} \sqrt{1 + (s\xi_g)^2} = \sigma_r + i\sigma_i \tag{7.38}$$

σ_r and σ_i denoting real and imaginary components, respectively, given approximately by

$$\sigma_r = \frac{1}{\xi_g} \sqrt{1 + (s\xi_g)^2} \simeq \frac{1}{\xi_g} \; ; \; \sigma_i = \frac{1}{\delta_g \sqrt{1 + (s\xi_g)^2}} \simeq \frac{1}{\delta_g} \tag{7.39}$$

If absorption is ignored, that is, if we let $\delta_g = \infty$, then the diffracted wave intensity for the dynamical dark-field image of a perfect crystal will be given by

$$I_S = S^*S = \left(\frac{\pi}{\xi_g}\right)^2 \frac{\sin^2 \pi\sigma_r z}{(\pi\sigma_r)^2} \tag{7.40}$$

from which we observe that at the exact Bragg condition, $s = 0$, Eq. (7.40) is qualitatively the same as Eq. (7.3c).

In the general case of a crystal containing a phase-shift plane or a dislocation as shown in Fig. 7.18, we can write [30]

$$\frac{dT}{dz} = i\pi\sigma Se^{i\phi}; \; \frac{dS}{dz} = i\pi\sigma Te^{-i\phi} \tag{7.41}$$

where ϕ is the phase angle associated with the defect as given previously by $\phi = 2\pi g \cdot \underline{R}$. Howie and Whelan [30] have solved equations of this type by using computer techniques, as have H. Hashimoto and associates [33,34]; we shall not become involved here in the rigors of the mathematical details except for brief expositions of the important implications. In addition to the original treatments by Howie and Whelan [10,30], and Hashimoto [33,34], you will find extensive dynamical treatments of defect images by R. Gevers [35,36] Gevers et al. [37,38], Art et al. [39], and in the book by Amelinckx [15]. Computed micrographs for crystal defects are treated in the book by Head, et al. (*Computed Electron Micrographs and Defect Identification*, North-Holland, London, 1973.)

*7.3.1 PHENOMENOLOGICAL EFFECTS OF ELECTRON ABSORPTION ON DEFECT
IMAGE CONTRAST*

With the inclusion of rigorous mathematical proofs, we can demonstrate graphically the phenomenological effects of electron absorption on the images
of stacking faults, dislocations, and similar phase-shifting interfaces,
for example, twin or grain boundaries. Let us consider the case of a thin
foil of thickness $t = 10\xi_g$. If the absorption length is taken as $\delta_g \approx t$,
then the image profile as a function of the distance along the phase-shift
plane as shown in Fig. 7.11 (z/ξ_g) will appear generally as shown in the
curves of Fig. 7.27a and b, which result for the solution of Eq. (7.41)
[14] taking the origin at $z = 0$ (the top surface) in Fig. 7.11. The curves
of Fig. 7.27a and b pertain specifically to defects having a displacement
vector $\underline{R}$, which can be defined in the lattice of I or II in Fig. 7.11.
Consequently, we can consider Fig. 7.27 to represent ideally a dislocation   .
or stacking fault.

There are two other characteristic intensity profiles that result dynamically when absorption becomes appreciable, and when the reflection
conditions are suitably adjusted. These can be represented graphically by
various degrees of bright-field intensity attenuation near the top of the
foil; and, as shown in Fig. 7.27c and d, can be identified generally as
type 2 or type 3 absorption profiles, respectively. Despite the fact that
we have not given a rigorous mathematical proof of the image absorption and
intensity profiles sketched in Fig. 7.27, their applications in the interpretation of electron microscope images of crystalline defects is clearly
illustrated in Fig. 7.28. If you are unhappy with this qualitative approach, you will find solace in the works of Howie and Whelan [14,30],
Hashimoto and coworkers [33,34], and Gevers et al. [37,38].

Two important features should be noted in the intensity profiles of
the type 1 and 2 bright-field images for planar defects. That is, in the
type 1 image for stacking faults, the first and last fringes are the same
(Fig. 7.27a), while in the dark-field image, the fringes are different.
In type 2 images of grain boundaries and twin boundaries in particular,
the first and last fringes in the bright-field image are different (Fig.
7.27c), while conversely the dark-field-image fringes are similar. Thus
by comparing the bright- and dark-field images of planar defects, it is
possible to determine the respective foil surface. We will show in Sec.
7.3.2 that examination of the fringe patterns can be employed in the char-

characterization of the nature of the defect, that is, we can decide whether it is a stacking fault, grain boundary, or twin boundary; and can also
determine the type (intrinsic or extrinsic) of stacking fault.

7.3.2 *CHARACTERIZATION OF CRYSTAL IMPERFECTIONS*

Gevers [35] has shown that the amplitudes of the transmitted and diffracted
waves in a crystal section such as that shown in Fig. 7.11 can be expressed
in the form

$$\begin{pmatrix} T \\ S \end{pmatrix} \quad \begin{pmatrix} T_1 T_2 + S_1 S_{\bar{2}} e^{i\phi} \\ S_1 S_{\bar{2}} + T_1 S_2 e^{-i\phi} \end{pmatrix} \tag{7.42}$$

where

$$T_j S_j (s) = T_j S_j (-s) \quad j = 1,2 \tag{7.43}$$

Thus, that for a crystal characterized by part 1 and 2 corresponding to (I)
and (II), respectively, of Fig. 7.11, the wave amplitudes will be influenced by the phase shift ϕ, for $\phi \neq 0$, and the value of the deviation s. Equation (7.43) therefore defines the sign of s in Eq. (7.42); and the respective values for S and T of Eq. (7.42) are the same as those presented in
Eq. (7.37) for the appropriate crystal portions, that is, for part 1 or 2.

If we limit our treatment here to relatively thick crystals, the approximation of Eq. (7.39) will be valid, and we can write

$$\sin \pi\sigma z \simeq \frac{i}{2} e^{-\pi i \sigma z} \quad \cos \pi\sigma z \simeq \frac{1}{2} e^{-\pi i \sigma z} \tag{7.44}$$

Equations (7.37) can then be expressed more generally as

$$T_j = \frac{1}{2} \left(1 + \frac{s_j}{\sigma_j} \right) e^{-\pi i \sigma_j z_j} \tag{7.37a}$$

$$S_j = - \frac{1}{2\sigma_j \xi_g} e^{-\pi i \sigma_j z_j}$$

where j = 1,2 of Eq. (7.42) with reference to Fig. 11.

Nature of grain-boundary images Let us consider again a thin foil as depicted originally in Fig. 7.11. We direct our attention to the column
passing from P to P' perpendicular to the foil surfaces, and assume in
the particular case of a grain boundary that parts I and II are different

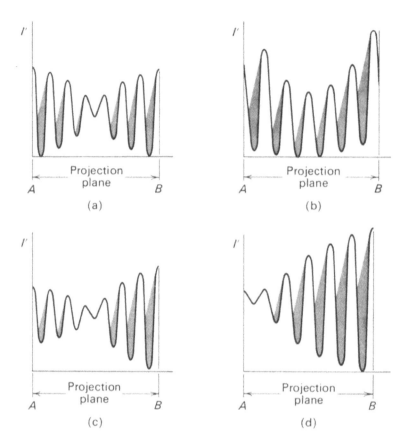

FIG. 7.27 *Graphical representation of various profiles resulting for solution of dynamical equations. Absorption conditions depicted phenomenologically for various bright-field images at phase-shift plane with φ generally negative. Intensity shown normalized as I' = I_T/I_O; assumption in each case is that t ≃ 7ξ_g. Exact solutions are found in A. Howie and M. J. Whelan, Proc. Roy. Soc. (London), A263: 217(1961) and R. Gevers et al., Phys. Status Solidi, 4: 838(1964). Shading of wave profiles depicts origin of dark fringes in the image with reference to Fig. 7.11. A denotes entrance or top surface, and B exit or bottom surface as shown in Fig. 7.11. Projection is inverted 180°. (a) Type 1 bright-field-image profile. (b) Type 1 dark-field-image profile shown for comparison. Note that fringe nature is opposite at exit or bottom surface. Image profile is shown is also approximately bright-field image for φ = 0 at a grain boundary. (c) Type 2 bright-field-image profile. (d) Type 3 bright-field-image profile.*

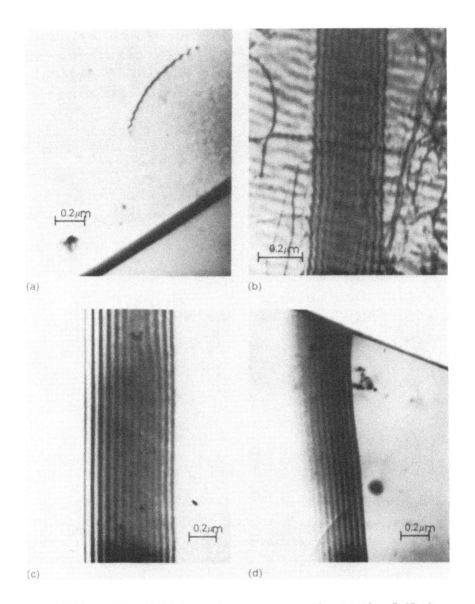

FIG. 7.28 *Bright-field image types corresponding to Fig. 7.27 observed in fcc 304 stainless steel. (a) Dislocation image showing type 1 contrast of Fig. 7.27a, 125 kV; (b) stacking-fault contrast (type 1) of Fig. 7.27a, 100 kV; (c) twin boundary type 2 contrast of Fig. 7.27c, 125 kV; (d) twin boundary type 3 contrast of Fig. 7.27d, 125 kV. Grain surface orientation in all cases is close to (110), and all images are oriented so that <110> is parallel and vertical with page. Foil thickness in all cases $> 10\xi_g$.*

crystals of the same material. Let g_1 define the set of lattice plans near the exact Bragg orientation for reflection in I; the family of active reflecting planes in II will be defined by the reciprocal-lattice vector $g_2 = g_1 + \Delta g$. The deviation of the exact Bragg orientation in each part is measured by the so-called excitation error s_1 or s_2, corresponding to parts I or II, that is,

$$s_j = \frac{|K_0|^2 - |K_j|^2}{2|K_0|^2} \qquad j = 1,2 \qquad (7.45)$$

In effect we can then specify the diffraction conditions in I by $g_1 s_1$, and in II by $g_2 s_2$; and for the associated excitation errors s_1, s_2, we can write

$$\Delta s = s_1 - s_2 = |(\Delta g)_n| \qquad (7.46)$$

where $(\Delta g)_n$ is the component of Δg normal to the foil surfaces.

Where the inclined plane in Fig. 7.11 is a grain boundary, the phase angle ϕ is zero. Equation (7.42) then appears as

$$\begin{pmatrix} T \\ S \end{pmatrix} = \begin{pmatrix} T_1 T_2 + S_1 S_{\bar 2} \\ S_1 T_{\bar 2} + T_1 S_2 \end{pmatrix} \qquad (7.47)$$

or

$$\begin{pmatrix} T \\ S \end{pmatrix} = \begin{pmatrix} T_2 & S_{\bar 2} \\ S_2 & T_{\bar 2} \end{pmatrix} \begin{pmatrix} T_1 \\ S_1 \end{pmatrix} \qquad (7.47a)$$

For contrast to occur at the boundary, it is required that either one or both parts of Fig. 7.11 contribute to the diffraction; however, it usually occurs that only one part is near any Bragg orientation, particularly where a two-beam situation is observed. A very simple situation then appears.

Let us assume that I in Fig. 7.11 is far from any exact Bragg orientation for reflection. We then have $S_1 \simeq 0$, and $T_1 \simeq e^{-i\pi s_1 z_1}$; and

$$T_2 \cong T_2 e^{i\pi s_2 z_2} \qquad S_2 \cong S_2 e^{i\pi s_2 z_2} \qquad (7.48)$$

insofar as the diffraction column of Fig. 7.11 is concerned. The distance of wave propagation through part I is specified as $z_1 = t_1 + t_2$, and for part II, $z_2 = t_3$. Similarly we could just as easily have taken I as being in the reflection position, and stipulated that $S_2 \simeq 0$, etc. The resulting fringe patterns for these two situations are then seen to occur simply as wedge fringes as shown in Fig. 7.13. However, since absorption is phenomenologically included in the present treatment, the resulting image intensity profiles (with bright-field illumination) will appear qualitatively as depicted in Fig. 7.29; assuming type 3 absorption (Fig. 7.27d). The resulting fringe patterns at grain boundaries as predicted by dynamical two-beam considerations and depicted in Fig. 7.29 can be treated simply by considering the crystal portion not in the reflection position as effectively a vacuum, as pointed out by Gevers [36].

The characteristic fringe profiles occurring at grain boundaries as shown in Fig. 7.14 should now be phenomenologically clear. We have of course neglected the many special cases, where, for instance, g_1 and g_2 may not be very different, etc. It should be obvious that where both parts of the bicrystal reflect nearly identically, the contrast at the interface will in some cases appear slightly below that of any one crystal part, and in some cases be devoid of fringes. The interface between grains A and D of Fig. 7.14 illustrates such a situation.

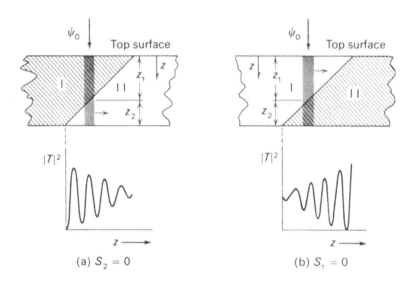

FIG. 7.29 *Dynamical nature of grain- or phase-boundary images. (a) Part 1 of bicrystal section is in reflecting position;* $|g_1| > |g_2|$, $s < s_2$. *(b) Part II is in reflecting position;* $|g_1| > |g_2|$, $s > s_2$. *Crystal portion not in reflecting position acts as vacuum layer insofar as diffraction is concerned.*

While it is not obvious from our present rather qualitative treatment of grain boundaries, Gevers et al [37] have shown in an exact theory that the nature of the first and last fringes in a grain boundary bright-field image are different, while those of the dark-field image are the same. Compare this with the fringe nature of stacking faults where in bright-field illumination, the fringes are the same (top and bottom), while in dark-field illumination, the fringes are different. This is shown in Figs. 7.27 and 7.28. We shall use this feature later to distinguish faults from crystal interfaces, and to determine the foil surface.

Nature of twin images and related coherent boundaries When the contact plane separating two crystal phases in a crystallographic plane for both crystals, the interface is coherent. Coherent twin boundaries in fcc materials are then observed to coincide with the {111} planes, and are easily defined crystallographically. We can, following the notation of Gevers [36], describe a coherent twin boundary generally by letting $\underline{a}$ and $\underline{b}$ represent lattice vectors of the boundary plane of Fig. 7.11, and then describing a vector $\underline{c}$ not in the boundary plane for part I. The corresponding vector in II will then be $\underline{c} + \Delta\underline{c}$ defined such that $|\Delta\underline{c}| \ll |\underline{c}|$. Consequently, for any vector g_1 of part I, there corresponds a reciprocal-lattice vector in part II of $g_1 + \Delta\underline{g}$ with the same Miller indices such that $|\Delta\underline{g}| \ll |\underline{g}_1|$. This is similar in effect to the deviation of operating reflections in the two parts as outlined in Sec. 7.3.2., and we can generally represent $\Delta\underline{g}$ as being normal to the coherent boundary plane.

The same reflection conditions occurring for grain boundaries or related noncrystallographic interfaces as described in Sec. 7.3.2 occur in the case of coherent twin boundaries. In the general situation, $s_1 \neq s_2$, $\phi = 0$, and only one part, either the matrix or its twin, can be near the reflecting position. We can then consider again that where one side of the coherent boundary deviates appreciably from the reflecting position, the section functions as a vacuum layer as described above.

For coherent twin boundaries in fcc materials, the interfacial planes separating a twin band from the matrix are parallel.[†] It is of interest therefore to consider the contrast arising at such boundaries. We can

---

[†]*Twins of this character are of the first order. These are the rules in fcc materials. Twin crystals bounded by planes with opposite slopes are possible and are called second-order twins, but there are no reported cases for fcc materials.*

depict the geometrical situation by attaching another part, having a parallel-coherent plane, to the section of Fig. 7.11. Two distinct reflection conditions are apparent, and two boundary fringe characteristics are expected. These occur if we can consider the matrix as a vacuum (part I) or the twin crystal as a vacuum (part II). The situation is depicted schematically in Fig. 7.30.

The resulting boundary fringe patterns in the two cases shown in Fig. 7.30 occur at each of separate wedges, and are mirror images because of the characteristic excitation errors shown. Again, we are depicting a type 3 absorption in Fig. 7.30. The nature of the extreme fringes is dependent on the sign of Eq. (7.46) as outlined in detail by Gevers et al [37].

The characteristic twin-boundary images depicted in Fig. 7.30 are profoundly illustrated in the bright-field images of coherent twin boundaries in fcc metals shown in Fig. 7.31. The images represent several reflection conditions in addition to those described generally in Fig. 7.30. In particular, the situation where both the twin and matrix are

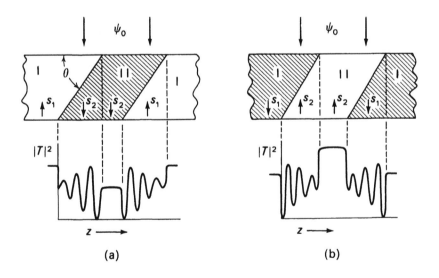

(a)   (b)

FIG. 7.30 *Phenomenological origin of characteristic bright-field fringe patterns at coherent twin boundaries and general domain wedge fringes. (a) $s_1 \simeq 0$, and part 1 acts as vacuum; (b) $s_2 \simeq 0$, and part II acts as vacuum. Reflecting crystal is shown crosshatched (type 3 absorption). Sense but not magnitude of excitation error is shown by $\underline{s}_1, \underline{s}_2$.*

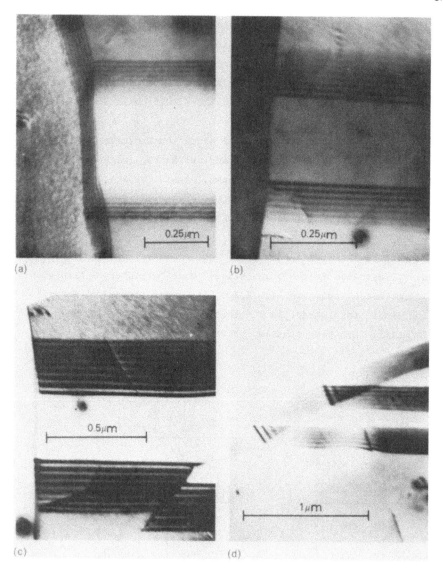

FIG. 7.31 *Dynamical wedge fringes at coherently twinned fcc cry-*
stals (bright-field images). (a) Intersection of twin band with
grain boundary in Inconel alloy showing wedge-absorption profiles
characteristic of Fig. 7.30b. (100 kV.) (b) Twin band in stainless
steel showing mirror-image reversal of twin boundaries (compared
with a) characteristic of Fig. 7.30a. Note contrast change between
twin and matrix. (c) Twin band in stainless steel showing type 2
contrast compared with type 3 of a. (125 kV.) (d) Twin band in
stainless steel where $s_1 \simeq s_2$; both twin and matrix act as dif-
fraction vacuum. (125 kV).

nonactive is shown in Fig. 7.31d. The twin boundaries are distinguished
by reflection conditions very similar to those that produce stacking-fault
contrast for $\phi = \pm 2\pi/3$. The situation can of course become inverted where
both crystal portions are strongly diffracting. The absorption-intensity
profile of Fig. 7.31d is a type 2.

Electron transmission theory for coherent twin- or domain-boundary
images has been formulated very thoroughly in the work of Gevers [36,37].
You will find the conclusions of the exact theory formulated as Eq. 63 of
reference 37.

We should point out that the images of deformation twins in fcc materials result in a manner identical to that shown in Fig. 7.30 with the
exception of thick foils where the boundary-phase images overlap (for a
microtwin). Contrast arising at microtwins observed in the electron microscope is treated in the dynamical thoery by J. Van Landuyt and coworkers
[40], Gevers and coworkers [41], and by Remaut et al [42].

Distinguishing stacking-fault, twin-, and grain-boundary images in the
electron microscope We should be immediately aware that twin- and grain-
boundary images result through the formation of wedge fringes where $\phi = 0$
and $s_1 \neq s_2$. By comparison, stacking-fault contrast results under the
condition that $\phi \neq 0$ and $s_1 = s_2$. Such features, when appropriately accounted for in the two-beam dynamical theory [Eq. (7.42)] serve to distinguish these defects by their characteristic fringe patterns along the
supplemental information, obtained from selected-area electron diffraction
patterns, concerning the crystallography of the parts separated by the
interface, and the image geometry.

Many of the distinguishing features of these defects have already
been discussed separately in the foregoing sections, particularly with
respect to the corresponding bright-field images. And, on looking back
at Figs. 7.28 and 7.29, you will readily see the distinctive nature of
twin boundaries as opposed to single stacking faults. Additional distinction is afforded by the corresponding dark-field image; and by comparing
the bright- and dark-field image for any planar interface, it is possible
to distinguish the top from the bottom surface with respect to the entrance and exit surfaces of the electron waves. The characteristic features are summarized in Table 7.1. Note that the top surface of the thin
section is associated with the side of the projected image, which is the
same in bright and dark field.

TABLE 7.1 *Planar-defect-image characteristics in the electron microscope*

Characteristic	Stacking fault	Twin boundary	Grain boundary (high angle)
Phase angle Reflection cond. Extinction error	$\phi = 0, \pm 2\pi/3, \pm\pi$ $g_1 = g_2$ $s_1 = s_2$	$\phi = 0$ $g_1 \neq g_2$ $s_1 \neq s_2$	$\phi = 0$ $g_1 \neq g_2$ $s_1 \neq s_2$
Usual bright-field image type (see Fig. 7.27)	1	2, 3	2, 3
Projected image profiles† (constant thickness)	BF DF	BF DF	BF DF
Nature of first and last fringe in bright field	same	different	different
Nature of first and last fringe in dark field	different	same	same
Nature of fringe addition with increasing foil thickness† (Bright field) Thickness increase to the left ←			
Orientation of parts I and II in Fig. 7.29	same	twin	different

†*Solid heavy lines are dark fringes, dotted lines or open spaces are bright fringes.*

Another distinguishing feature associated with the fringe patterns is the change of fringes with section thickness. In the case of a stacking fault, the fringes add at the center of the image for $\phi = \pm 2\pi/3$, and where wedge fringes are associated with a stacking fault near the edge of a foil, the wedge fringes pass through the fault-plane image at the center portion where a new fringe is added [13]. When $\phi = \pm\pi$ for stacking faults, fringes add at the surface [44]. This is, however, rare in fcc materials.* Twin boundaries add fringes in a manner somewhat similar to that of a stacking fault, primarily by the addition of fringes along one side (surface) of

---

*When this occurs for stacking faults or similar domains, the extinction distance is $\xi_g/2$ and not ξ_g. See Prob. 7.25

the image, each fringe adding abruptly as the thickness changes by approximately ξ_g. When twins add fringes in the central portion of the image, the fringes usually add off-center where the image is of type 2(Fig. 7.21). The addition of fringes in the central portion of a twin-plane image is distinct from a stacking fault as shown in Table 7.1. Grain boundaries add fringes primarily along one edge as shown in Table 7.1.

Figure 7.32 reproduces representative images of planar interfaces in fcc materials (stacking-fault and twin-boundary images) demonstrating the characteristics outlined generally in Table 7.1. You should also compare the criteria of Table 7.1 with the image characteristics of Figs 7.14, 7.15, 7.28, and 7.29.

It is sometimes difficult to distinguish twin boundaries from high-angle grain boundaries, particularly where no junctions of these are in the field of view. The only check of this distinction is the analysis of the adjoining grain orientations, or simply the observation of the selected-area electron diffraction pattern. As pointed out in Chap. 6, a twin crystal will produce twin spots in the diffraction pattern which, if distinct from the matrix reflections, can be easily identified. An obvious feature of grain boundaries will be, in the case of high-angle boundaries, distinct orientation changes or pattern rotations on crossing the image plane from one part to the next. You are perhaps already quite aware of these characteristics.

Images of low-angle grain boundaries are also distinguishable from high-angle boundaries, and have been treated by the two-beam dynamical theory by Gevers and coworkers [41]. These images show a distinct mosiac contrast associated with the boundary-plane projection; and we shall treat these boundaries in more detail in Sec. 7.4.2. You might glance in advance at Fig. 7.53 in order to be convinced of the distinctions.

Determination of stacking-fault nature in fcc materials The nature of a stacking fault in fcc materials is intrinsic or extrinsic. Intrinsic faults are characterized in a lattice array by the occurrence of a missing plane portion, the missing segment bounded by partial dislocations. In the case of extrinsic faults, a lattice region contains an inserted segment again bounded by partial dislocations. We can illustrate this feature for fcc materials having a lattice periodicity ABCABC, etc., in a perfect-crystal section as shown in Fig. 7.33.

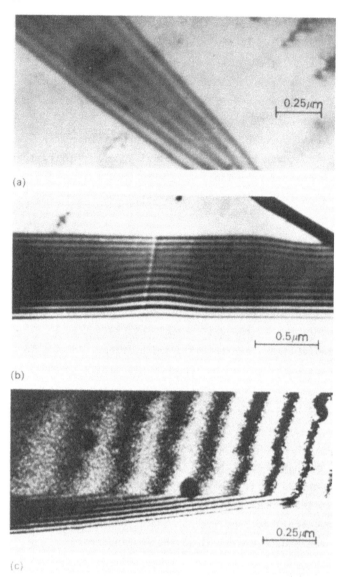

FIG. 7.32 *Image characteristics of stacking faults, and twin boundaries with changes in foil thickness. (a) Stacking fault in stainless steel showing relationship of wedge fringes to projected type 1 image. (100 kV.) (b) Twin boundary in stainless steel showing fringe properties similar to a but differing as indicated in Table 7.1. (125 kV.) (c) Twin boundary in Inconel 600 showing relationship of wedge fringes to projected type 2 boundary image. (100 kV).*

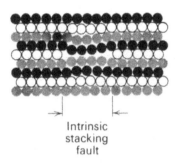 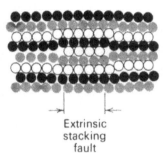

Intrinsic
stacking
fault

Extrinsic
stacking
fault

FIG. 7.33 *Intrinsic and extrinsic stacking-fault character in fcc lattice having atomic plane periodicity ABC.*

Because of the atomic features associated with each type of stacking fault as shown in Fig. 7.33, the diffraction contrast image of the faulted region, when observed in the electron microscope, will be different as a result of the phase difference the electron wave experiences in each situation.

Intrinsic stacking faults on the fcc (111) plane are formed by displacement vectors

$$\underline{R}_1 = \frac{a}{6}\,[\bar{1}\bar{1}2] \quad \underline{R}_2 = \frac{a}{6}\,[\bar{1}2\bar{1}] \quad \text{and } \underline{R}_3 = \frac{a}{6}\,[2\bar{1}\bar{1}]$$

while extrinsic faults are simply the negatives:

$$\underline{R}_{1e} = \frac{a}{6}\,[11\bar{2}] \quad \underline{R}_{2e} = \frac{a}{6}\,[1\bar{2}1] \quad \text{and } \underline{R}_{3e} = \frac{a}{6}\,[\bar{2}11]$$

We can also show that

$$\underline{R}_{1e} = \underline{R}_2 + \underline{R}_3 \quad \text{or} \quad -\underline{R}_1 = \underline{R}_2 + \underline{R}_3$$

and so we can observe that the phase angle ϕ for these two fault characteristics will be opposite in sign. The nature of the fault is then uniquely identified when the sign ϕ is determined.

As a convenience in determining the sign of ϕ associated with a stacking-fault image, let us consider the edge of a faulted region that is a Frank sessile dislocation (a partial) having a vector $a/3\,[111]$. The shear vector of the stacking fault characterizing the displacement bringing the perfect lattice into coincidence with the faulted position is then $a/3\,[111]$, which is equivalent to $a/6\,[11\bar{2}]$ since

$$\frac{a}{6}\,[11\bar{2}] + \frac{a}{3}\,[111] \longrightarrow \frac{a}{2}\,[110]$$

which is a lattice vector. If we let $\underline{D}_i$ denote the displacement vector for an intrinsic stacking fault, and $\underline{D}_e$ an extrinsic fault, we observe that

$$\underline{D}_e = \frac{a}{3}\,[111], \quad \underline{D}_i = -\frac{a}{3}\,[111] = -\,\underline{D}_e$$

Hence for an operating reflection $\underline{g}$, ϕ will simply change sign for one fault type or the other. And, since the images of stacking faults have been shown (Table 7.1) to be determined by ϕ, we can find ϕ from the images. In fact, Gevers and coworkers [43] have shown that really only the dark-field image is required.

If we let

$$\phi' = 2\pi \underline{g} \cdot \underline{D} \qquad \phi' = \phi + 2\pi$$

the we observe in Table 7.2 that

$$\phi' = 2\pi |\underline{g}| \cdot |\underline{D}| \cos \beta \tag{7.49}$$

where β is the angle between the displacement vector and the reciprocal-lattice vector, with $\underline{g}$ always pointing to the right as shown. All possible situations are shown in Table 7.2 along with the resulting bright- and dark-field images. Only the first and last fringes are shown as dark (solid) and bright (dotted) following the notation of Table 7.1. We denote the images as class A and B following the notation of reference 43, in essence defining the diffraction vectors. The fcc diffraction vectors producing stacking-fault contrast can now be considered in Eq. (7.49) with the results indicated in Table 7.3. Tables 7.2 and 7.3 now fully characterize the images of stacking faults in fcc materials, and the results of these tables can be summarized in the following procedure for characterizing a stacking fault in the electron microscope:

1. Observe the fault image in bright field, tilt the specimen for maximum contrast (ensuring a strong two-beam situation).

2. Photograph the bright-field image.

3. Reduce the intermediate-lens strength to obtain the selected-area electron diffraction pattern in the immediate area of the stacking fault. Enclose the fault in the field-limiting aperture (Fig. 6.27).

4. Record the diffraction pattern.

5. Tilt the operating reflection into the optic axis or place the objective aperture over the spot and produce the dark-field image.

6. Record the dark-field image.

7. Develop the recorded images and diffraction pattern, and rotate the images and the selected-area electron diffraction pattern into the coincidence prescribed by the rotation calibration described in Sec. 6.8.1.

8. Identify the operating reflection $g_{hk\ell}$, and rotate the dark-field image and the diffraction pattern (in coicidence) so that $\underline{g}$ points to the right as in Table 7.2. Knowing $\underline{g}$, the class of fault, A or B in Table 7.3, is identified.

9. Note the fringe profile of the dark-field image and compare first and last fringes with that for the characteristic fault class in Table 7.2.

10. The fault is then readily and unambiguously identified. As a final check, compare the bright- and dark-field images so as to determine the foil surface (Table 7.1) and the characteristic geometry. This can be checked with the situations depicted in Table 7.2.

TABLE 7.2 *Fringe character for intrinsic and extrinsic stacking-fault images (first and last fringes shown as heavy solid and dotted lines corresponding to dark and bright fringes, respectively)*[†]

[†]*After R. Gevers, A. Art, and S. Amelinckx*, Phys. Status Solidi, 31: 563 (1963).

TABLE 7.3 *Reflection characteristics for fcc stacking faults.*

Operating reflection $\underset{\sim}{g}$	ϕ'	ϕ	Sign	Class
[200]	$4\pi/3$	$-2\pi/3$	-	A
[222]	$4\pi/3$	$-2\pi/3$	-	A
[440]	$16\pi/3$	$-2\pi/3$	-	A
[400]	$8\pi/3$	$2\pi/3$	+	B
[111]	$2\pi/3$	$2\pi/3$	+	B
[220]	$8\pi/3$	$2\pi/3$	+	B

It has been shown by Art and coworkers [39] that the condition for similarity of fringes at the foil surface on comparing the bright- and dark-field images is[†]

$$t \gtrsim 0.4 \; \delta_g = 4\xi_g \quad \text{if} \quad \delta_g \cong 10\xi_g$$

where t is the thickness of the thin foil. This means that where fewer than four dark fringes are observed in the bright-field image, the nature of the fault may not be amenable to analysis by the above procedure. Figure 7.34 illustrates the analysis of stacking faults produced in a thin foil where the nature of the bottom fringe in the dark-field image is readily observable above the background. (Note that, as indicated above, generally four fringes are required to make a real distinction of the fringe nature, although if you look closely at Fig. 7.34 the fringe distinction is apparent even for 2 fringes. By following the general procedure outlined above, we observe in Fig. 7.34 that the fault nature is intrinsic.

You should also be aware of the importance of rotation calibration as discussed in Chap. 6. Without a knowledge of lens rotation, the position of $\underset{\sim}{g}$ as it relates to the image cannot be determined unambiguously. The procedure outlined above pertains primarily to fcc and diamond cubic materials. Stacking faults in fcc materials are observed generally to

---

[†]*The nature of stacking faults can be determined in some cases using only the bright-field image, see J. Van Landuyt, R. Gevers, and S. Amelinckx, Phys. Status Solidi, 18: 167(1966).*

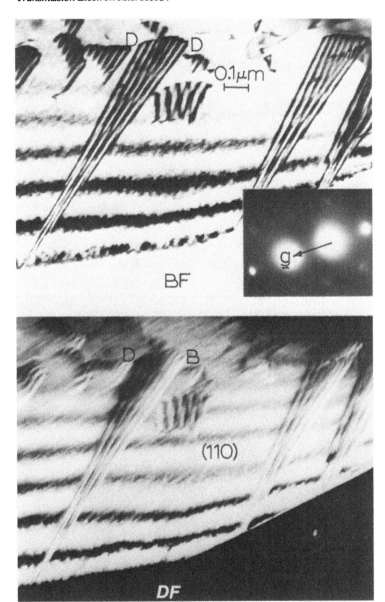

FIG. 7.34 *Determination of stacking fault nature in a stainless steel foil. The upper bright-field (BF) image shows stacking faults emanating from the foil edges. Note the correspondence of the wedge fringes with the stacking fault fringes. The selected-area electron diffraction pattern insert shows the properly oriented operating reflection: $g = [1\bar{1}1]$. The lower dark-field (DF) image is obtained with the aperture over g. Note that from Table 7.3 the faults are Class B faults. Turning the book $180°$ to have g point in the direction shown (to the right) in Table 7.2 will readily indicate the faults to be intrinsic. (200 kV)*

be intrinsic in nature, although in heavily deformed materials overlapping faults can give rise to extrinsic regions. Extrinsic faults are also 1 - layer twins in fcc and if the twin reflection is also coincident with $\underset{\sim}{g}$, special dark-field effects can occur.

A similar method for characterizing the nature of stacking faults in hcp crystals has been described by Blank et al [45].

Nature of partial dislocations in fcc materials As already pointed out, partial dislocations in fcc materials are characterized by a Burgers vector of the form a/6 <112>. This fact leads to the condition that $\underset{\sim}{g} \cdot \underset{\sim}{b}$ has values of $\pm 1/2$, $\pm 2/3$, $\pm 4/3$ in association with stacking faults in these materials in addition to values of n = 0,1,2, etc. Image profiles for partial dislocations will differ for the fractional values of $\underset{\sim}{g} \cdot \underset{\sim}{b}$ as shown in Fig. 7.35, in contrast to Fig. 7.19 for a total dislocation. We can observe in Fig. 7.35 that for $\underset{\sim}{g} \cdot \underset{\sim}{b} < 1/3$, partial dislocations tend toward invisibility ($\underset{\sim}{g} \cdot \underset{\sim}{b} = 0$).

Let us illustrate the nature of partial dislocations as follows: We might initially consider the dislocation reaction

$$\frac{a}{2} [01\bar{1}] \longrightarrow \frac{a}{6} [\bar{1}2\bar{1}] + \frac{a}{6} [11\bar{2}]$$

which depicts the dissociation of a total dislocation on the left into two

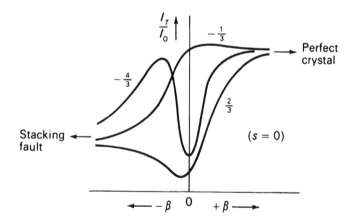

FIG. 7.35 *Image profiles for partial dislocations in fcc materials (bright field).*

partials on the right. Let us suppose that $\underline{g}$ = 1/a [200]. Then we have for the complete reaction

$$\underline{g} \cdot \underline{b} = \frac{1}{a} [200] \cdot \frac{a}{2} [01\bar{1}] = 0 \qquad \text{the total dislocation is invisible}$$

$$\underline{g} \cdot \underline{b} = \frac{1}{a} [200] \cdot \frac{a}{6} [\bar{1}2\bar{1}] = -\frac{1}{3} \qquad \text{this partial is effectively invisible}$$

$$\underline{g} \cdot \underline{b} = \frac{1}{a} [200] \cdot \frac{a}{6} [11\bar{2}] = +\frac{1}{3} \qquad \text{this partial is effectively invisible}$$

This reaction is therefore unobservable in the electron microscope. Let us now suppose the operating reflection to be $\underline{g}$ = 1/a [020]. We then obtain for the complete reaction

$$\underline{g} \cdot \underline{b} = \frac{1}{a} [020] \cdot \frac{a}{2} [01\bar{1}] = +1 \qquad \text{the total dislocation is visible}$$

$$\underline{g} \cdot \underline{b} = \frac{1}{a} [020] \cdot \frac{a}{6} [\bar{1}2\bar{1}] = +\frac{2}{3} \qquad \text{this partial is visible}$$

$$\underline{g} \cdot \underline{b} = \frac{1}{a} [020] \cdot \frac{a}{6} [11\bar{2}] = +\frac{1}{3} \qquad \text{this dislocation is effectively invisible}$$

Numerous additional examples can be cited to illustrate the characteristic features. Namely when both partials are invisible, the total dislocation is invisible. Conversely, when the total is visible, one or both of the partials will be visible. You can further demonstrate to yourself that conditions may arise when two partials, visible in the image, collapse to form an invisible total dislocation (try working Prob. 7.13).

Figure 7.36 illustrates a number of the cases discussed above for partial dislocations in 304 stainless steel. The nature of partial dislocations is particularly prominent in fcc materials of low stacking-fault energy inasmuch as they serve to define the faulted regions. You should note in Fig. 7.30 that the "invisible" overlap region is in fact not visible. This occurs because in reality the phase shift is not abrupt but occurs gradually. This feature is treated in detail by P. Humble, *Phys. Status Solidi, 30:* 183(1968).

Identification of deformation microtwins and measurement of their density in thin-film materials Images of deformation twins in thin films can be constructed phenomenologically by considering the overlapping wedge-fringe contrast that occurs when region II of Fig. 7.30 is made so small that the projected intensity profiles overlap. This situation has been considered

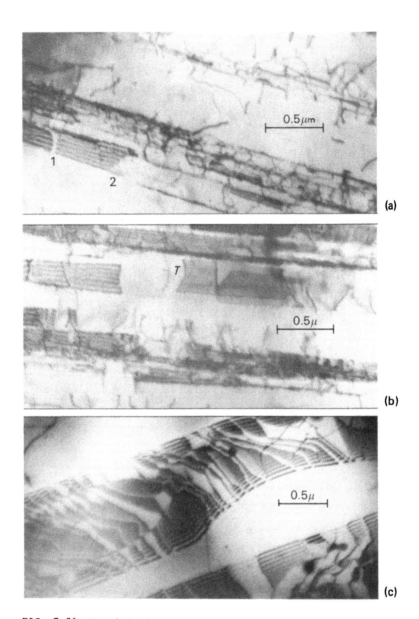

FIG. 7.36 *Partial dislocations in stainless steel. (a) Both partials at 1 and 2 separated by ribbon of stacking fault, are invisible. (b) Bounding partial at stacking-fault ribbon is visible at T. (c) Complicated arrays of partial dislocations and phase extinctions caused by overlapping fault segments show cases where one or both partials are invisible.*

in detail by Van Landuyt and coworkers [40], and we will not burden you with the theoretical development. In essence, a fringe contrast arises as a result of diffraction conditions outlined in Sec. 7.3.2 (illustrated in Fig. 7.30).

In fcc materials, deformation twins occur along the {111} planes, and can be distinguished in some cases by their characteristic contrast. However, where the distance between twin planes is very small (of the order of a few interplanar spacings), the resulting fringe profiles cannot be distinguished from simple stacking faults. Overlapping stacking faults in fcc materials also produce bundles of hcp material forming ε-phase platelets that, aside from the characteristic hcp reflections in the selected-area electron diffraction patterns, cannot be distinguished from single or overlapping stacking faults. Consequently, where the character of the faults is of a twinned nature, twin spots will occur in the selected-area electron diffraction pattern. If the twin spots or the hcp ε-phase spots can be identified in the diffraction pattern in the electron microscope, then tilting the corresponding diffraction spot into the optic axis (or centering the objective aperture over the spot) will cause the fault contrast to reverse. Thus where the faults appear as dark bands in the bright-field image, they will appear bright in the dark-field image. We can unambiguously identify the fault character if the nature of the diffraction spot can be identified.

We illustrate the technique with an example of deformation twins observed in machined stainless steel as shown in Fig. 7.37. In Fig. 7.37a the bright-field images shows linear defect contrast along the <112> directions in the (110) surface of a thin section of a machined chip. The {111} fault planes are normal to the crystallographic surface, and this information is readily available from the inserted selected-area electron diffraction pattern. In addition to the (110) fcc matrix reflections, extra spots and diffraction streaks are observed. The pattern can be compared with Fig. 6.33 so you can identify the extra spots as the corresponding twin reflections for the two sets of linear-defect images. The origin of the streaks should also now be apparent (if this is not the case, you should reread Sec. 6.8.7). The dark-field image of Fig. 7.37b, with the aperture centered over the twin spot shown circled, unambiguously clarifies the image identity.

Because the twin boundaries in Fig. 7.37 are normal to the (110) surface, we can estimate the average twin width (the width of the twin

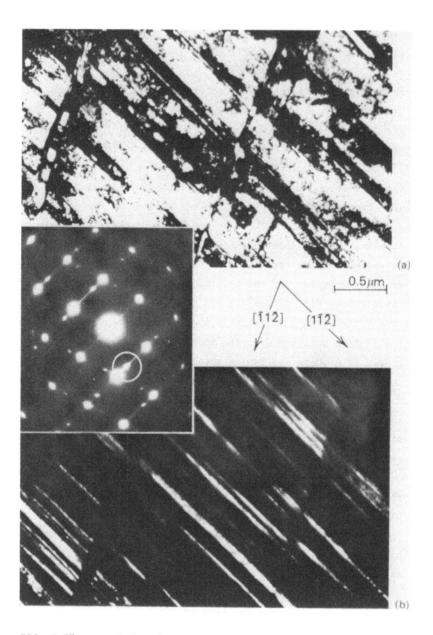

FIG. 7.37 *Identification of deformation twins in thin section*
of machined chip (corresponding to depth of cut of 0.005 in.) of
stainless steel. (a) Bright-field image. (b) Dark-field image
with objective aperture centered over twin spot (1/3 [111]) shown
circled in superposed selected-area electron diffraction pattern.
The grain surface normal is observed to be nearly exactly [110].

band) directly. We can observe that in Fig. 7.37a the angle between the twin directions does not deviate measureably from that which we should expect in the ideal case. Consequently we may conclude that the (110) crystallographic surface is approximately coincident with the true specimen surface. We shall deal more extensively with this feature in Sec. 7.4.3, but it is instructive to realize this now. The width is then obtained by averaging the measurements in the dark-field image (Fig. 7.37b).

Suppose that in the general case the twins are inclined to the foil surfaces. The width is then not distinguishable from the region of overlap. However, having measured the twin width as demonstrated above, the total twin volume over a volume of thin foil can be approximated from the geometrical situation depicted in Fig. 7.38. We observe that to a good approximation for w_T (the twin width) small compared with the section thickness t, the average twin volume is given by

$$V_T = tw_T \sum_j \frac{L_j}{\sin \theta_j} \qquad w_T \ll t \qquad (7.50)$$

where L_j = total length of twinned material in jth direction or associated with particular crystallographic planes

θ_j = inclination angle as shown in Fig. 7.38.

The volume percent of twinned material in the matrix of the thin section is then calculated from

$$\text{Volume percent} = \frac{V_T}{tA} 10^2$$

where A is the area of the section over which the twin lengths are measured. We then observe simply that

$$\text{Volume percent} = \frac{w_T 10^2}{A} \sum_j \frac{L_j}{\sin \theta_j} \qquad (7.51)$$

where it is obvious that the section thickness need not be known or calculated in determining the volume percent of twins. It should be apparent that the inclination θ is determined from Eq. (6.57) by knowing the indexes of the twin plane and the surface orientation.

Equation (7.51) will apply to calculations of any linear defect possessing a crystallographic relationship (or a noncrystallographic relationship) to the matrix. Thus phase-volume percentages can be calculated using

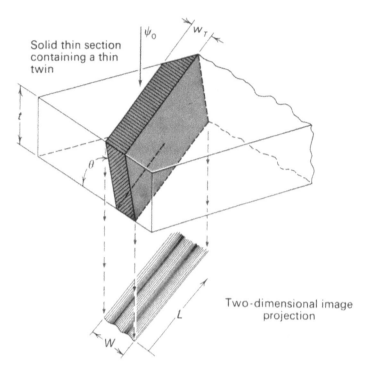

Solid thin section
containing a thin
twin

Two-dimensional image
projection

FIG. 7.38 *Geometrical features of microtwins or twin faults
in thin-foil materials observed in electron microscope.*

the foregoing outlined. In addition, the percentage of, for example, martensite or similar phases can be measured directly from the micrograph by
estimating its area in relation with the field of view. You must remember
that this is only strictly true when the thickness of the foil is small
compared to the size (or volume) of individual phases or inclusions. Indeed, if the boundaries of phases or inclusions are not inclined (θ_j = 90°),
or if the phases are very large relative to the boundaries, and if the
thickness of the foil is small, then the volume fraction can be accurately
determined by simply measuring the area of the phase or inclusion in the
image. To avoid ambiguity, this should be done in the dark-field image.
In fact, for very irregular inclusions or phases where this condition is
met, the volume fraction can be rapidly and accurately determined by first
weighing the print of the dark-field image and then cutting out all the
dark-field areas and re-weighing the photograph for the segments cut out).
This technique can apply to a wide range of such measurement requirements
in the electron microscope.

It might be useful to illustrate not only the difficulties in accurately determining certain phase volume fractions, but also the ability to uniquely distinguish and isolate specific phase regimes utilizing dark-field electron microscopy based upon the identification of specific reflections in the selected-area electron diffraction pattern. Figure 7.39 shows a region in a deformed 304 stainless steel foil where α' martensite is nucleated at the intersection of twin-faults (which are intermixed bundles of twins, stacking faults, and ε-phase; all created by regular and irregular overlapping of intrinsic faults) [46,47]. Although the orientation is similar to that in Fig. 7.37, the twin-faults along the $[\bar{1}1\bar{2}]$ direction are tilted off their normal angle (relative to the specimen surface) and are observed in some projection. When the twin reflection is used to form the dark-field image, the strongly diffracting twin volume segments are observed. When the α' (bcc) reflection is utilized, the martensitic regions formed at the intersection of twin-faults along $[\bar{1}10]$ (which are generally out of contrast) with those along $[\bar{1}1\bar{2}]$ are readily apparent. These martensitic volumes are inclined to the foil surface by roughly 35°, although since the foil is obviously tilted this angle is uncertain. To determine the volume fraction of martensite here requires a knowledge of the twin-fault thickness and the angle of inclination of the transformed regions. The twin-fault thickness can be determined by tilting the foil until the beam is parallel to twin-faults along $[\bar{1}1\bar{2}]$. This is accomplished by measuring from the tilted dark-field image as in Fig. 7.37.

Determining the nature of the Burgers vector of a dislocation In Sec. 7.2.2 we learned that the phase factor characteristic of a dislocation is

$$\underline{g} \cdot \underline{b} = n$$

In Sec. 7.3.2 we noted that, in the case of partial dislocations, $n = m/3$ where $m = 0,1,2$, etc. Consequently for $m = 0,1$, there is ideally no visible contrast at partial dislocations.

In the case of a total dislocation, from Sec. 7.2.2, $n = 1,2,3$, etc. The corresponding condition for no contrast was then observed to be $n = 0$, that is $\underline{g} \cdot \underline{b} = 0$. In order for n to be zero, b must be perpendicular to $\underline{g}$. This feature points up a rather direct method for determining the direction of a Burgers vector. That is, the dislocation or a dislocation segment is observed in bright field. The foil is then tilted until the diffraction pattern is observed. Then, following tilting, the pattern is again observed. A prominent reflection will normally occur on tilting

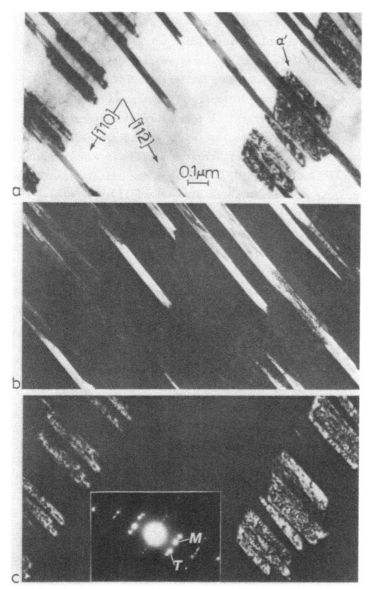

FIG. 7.39 *Selective dark-field microscopy illustrating the formation of
strain-induced α' martensite at twin-fault intersections for a foil area
in 304 stainless steel tilted off (110). (a) Bright-field image. (b) Dark-
field image using the twin reflection (T) indicated in the selected-area
electron diffraction insert in (c). (c) Dark-field image using the α' martensite reflection (M) indicated in the selected-area electron diffraction
pattern insert. The faint spots and streaks near the center spot and particularly within the central intensity distribution are ε (hcp) reflections.
(Courtesy Dr. K. P. Staudhammer, Los Alamos National Laboratory.)*

the foil for no contrast at the dislocation. This is the operating reflection g for the invisible segment. The selected-area electron diffraction pattern is then rotated into the proper coincidence with the image. The Burgers vector direction is then drawn normal to g on the image. Figure 7.40 illustrates this procedure for undissociated[†] dislocation networks in a thin stainless steel.

In order to establish the direction of b unique to the foil crystallography, we need to observe two distinct situations where the dislocation is out of contrast. We then find g_1 and g_2. The Burgers vector is then parallel to $g_1 \times g_2$.

The sense of the Burgers vector (the sign of the corresponding dislocation) can be described by knowing the image side of the dislocation and the sign of s when the direction has been established. The sign of s can be immediately deduced from the diffraction pattern if Kikuchi lines are present as shown in Fig. 7.41a. The image side is determined by comparing bright-field images taken with two different signs of s for the same g; or with two different signs of the same g vector with the same s. The latter procedure is usually the most reliable, and for good contrast we normally find s positive (as in Fig. 7.40). Figure 7.41b illustrates the concept qualitatively.

The actual magnitude of the Burgers vector (its length) can be determined by trial and error in most cases, and from observations of the image character. We already observed in Sec. 7.2.2 that the image profile depends on the value of n. This feature is particularly pronounced where an extinction contour crosses the dislocation image; we illustrate the expected image alteration as shown in Fig. 7.41c.

With a knowledge of the visible image, we can then satisfy $g \cdot b = n$. As an example, suppose $s = +$. Then, from the image character, n = 2. Consequently we have $g \cdot b = 2$. The operating reflection for contrast can then be observed from the selected-area electron diffraction pattern. Let us suppose that $g = [200]$. Then we have, for $g \cdot b = 2$, $b = a/2 <110>$ or $a = <100>$ for a total dislocation in fcc. We might then tilt the foil so that another g operates, and again determine the image character as shown in

---

[†]*The meaning of dissociated and undissociated networks or dislocation nodes is clarified in Sec. 7.4.2. We will find that dislocations that split into partials are said to dissociate. The region between such dislocations is a region of stacking fault.*

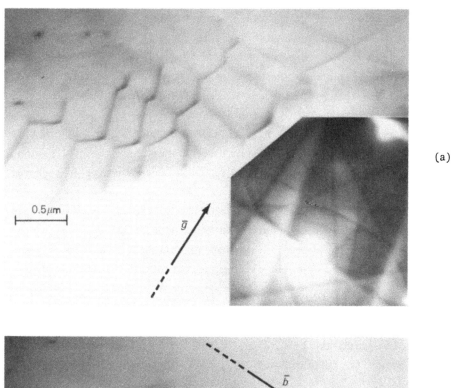

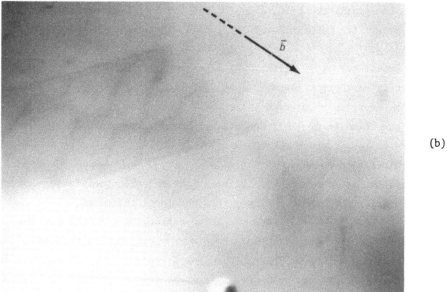

FIG. 7.40 *Determination of dislocation Burgers vector. (a) Network of dislocations in a (111) plane in 304 stainless steel observed in bright-field illumination at 125 kV. Foil is then tilted until a dislocation or segment of dislocations becomes invisible. (b) Selected-area electron diffraction pattern is then observed to possess a prominent reflection corresponding to $g \cdot b = 0$. Burgers vector of invisible segment is then approximately normal to g as sketched in b. By noting diffraction condition g, where dislocations are in contrast, and by repeated tilt experiments, Burgers vectors for dislocations can be uniquely determined.*

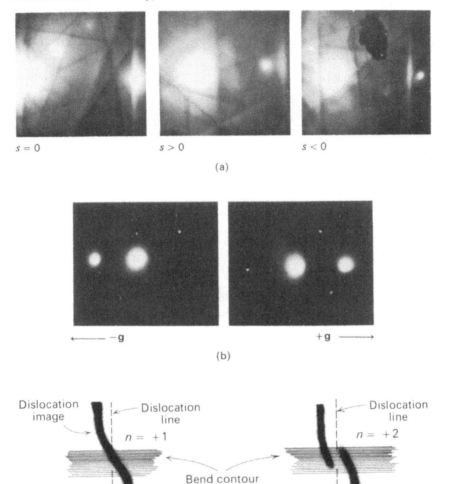

FIG. 7.41 *Diffraction characteristics for dislocation images. (a) Determination of $\underset{\sim}{s}$; (b) sign of $\underset{\sim}{g}$; (c) image features for $\underset{\sim}{g} \cdot \underset{\sim}{b} = n$.*

Fig. 7.34c. The resulting two equations: $\underset{\sim}{g}_1 \cdot \underset{\sim}{b} = n_1$ and $\underset{\sim}{g}_2 \cdot \underset{\sim}{b} = n_2$ can now be solved simultaneously for $\underset{\sim}{b}$.

The determination of the nature of $\underset{\sim}{b}$ for dislocations in thin films has been outlined in a number of studies of various materials including fcc [48,49], bcc [50], diamond cubic [51,52], and hcp [53]. Even dislocation networks in polyethylene single crystals have been investigated using the methods outlined above [54]. The reader might also refer to the book by Edington [55]. We should mention that although it may not always be possible or even necessary to determine the exact nature of the Burgers vector, it is sometimes useful to determine the character of the dislocation, or the distribution of character in an area. This can be done in an apparent way by simply measuring the angle between the dislocation lines and the Burgers vector direction determined from the $\underset{\sim}{g} \cdot \underset{\sim}{b} = 0$ condition. For pure edge character, this angle is 90°. Correspondingly, for pure screw dislocations it is zero. Although dislocations vary in character, a plot of the distribution can sometimes give some indication of specific microstructural phenomena. You might refer to Prob. 7.27.

Measurement of dislocation density   The measurement of dislocation density in thin films where each dislocation is individually recognizable is certainly straightforward and you can readily demonstrate this feature with reference to Fig. 7.17a and Prob. 7.15. In situations where the dislocation density is high and very complicated arrangements present, their measurement becomes effectively a statistical estimate. Several methods have been employed in measuring dislocation densities [56-58].

J. Bailey and P. B. Hirsch measure the toal projected length of dislocation images in an A that is presumably representative of the density distribution within a foil, assuming the dislocations to be randomly oriented. If $\Sigma\ell$ is the total length of dislocations over an area A of foil thickness t, the density is given by

$$\rho = \frac{4\Sigma\ell}{\pi A t} \tag{7.52}$$

R. K. Ham draws a random net of lines on a transparent overlay. The total length of the lines is measured as L. The overlay is then placed over the micrograph to be investigated, and the number of intersections of dislocations with grid lines is counted. If the number of such intersections is N, the average dislocation density is then expressed as

$$\rho = \frac{2N}{Lt} \tag{7.53}$$

This expression follows directly from Eq. (7.52) since $\Sigma\ell = \pi N A/2L$.

The alternate technique proposed by A. S. Keh is one of the more popular since it is somewhat more regular. In this method, regular nets of lines are constructed with sets of regular but different spacings perpendicular to one anotehr as a transparent overlay. If the total length of lines in one direction is L_1 and that in the normal direction L_2, then the density is given by

$$\rho = \left(\frac{N_1}{L_1} + \frac{N_2}{L_2} \right) \frac{1}{t} \tag{7.54}$$

where N_1 and N_2 are the number of intersections along the two respective sets of grid lines.

You must bear in mind that these methods reveal estimates, and only where individual dislocations are readily observed can accurate densities be established. This implies that the operating reflection giving rise to dislocation contrast must be known and maintained consistent for the comparison of dislocation densities, or corrections made for that fraction of dislocations invisible as a result of a particular $\underset{\sim}{g}$ as illustrated in Table 7.4 for fcc materials. It should also be borne in mind that as Ham [59] has pointed out, up to 50 percent of the dislocations in a material may be lost during electropolishing, if this is the method of thin-foil preparation. Consequently the dislocation densities actually measured are usually somewhat lower than those to be found in bulk solids. It must also be cautioned that where the dislocation density becomes so great that intersections are not clearly distinguishable, or where the overlap of individual dislocations produces continuous contrast effects, the estimates of density lose significance. As a rule of thumb, it is usually found that meaningful estimates of dislocation density cannot be attained where ρ exceeds about 10^{12}cm^{-2}. Much of the difficulty in dealing with very high dislocation densities arises because of the image overlap in normal 2-beam situations. Consequently, dislocations spaced less than an extinction distance cannot be distinguished. This difficulty can of course be overcome in certain cases by utilizing the weak-beam method described previously.

Despite the limitations on the measurement of dislocation density, it is sometimes useful to observe the trend of dislocation increase or decrease as a function of some variable such as stress, temperature, etc. In this respect the errors in measurement are usually consistent, and the trend established is indeed representative. This treatment is particularly revealing in the study of mechanical properties of materials and related areas. We illustrate this feature in Fig. 7.42, which shows the increase in dislocation density for cold-rolled and shock-loaded α brass (70 percent Cu and 30 percent Zn) using Keh's method to estimate the dislocation densities.

TABLE 7.4 *Proportion of Invisible Total Dislocations for Various Operating Reflections, g, in FCC Materials*[†]

Operating Reflection, g	Proportion of Dislocations Invisible
111	1/2
200	1/3
220	1/6
311	1/6
222	1/2
400	1/3
331	1/6
420	0 (all visible)

[†]*Based upon data described in P. B. Hirsch, et al.* Electron Microscopy of Thin Crystals, *Butterworths Scientific Publishers, London, 1977, p. 423 (Second Edition).*

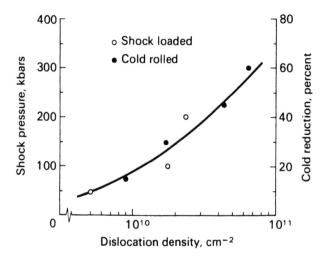

FIG. 7.42 *Estimates of dislocations in α brass after explosive shock loading and cold reduction.* [*Data from L. E. Murr and F. I. Grace,* Trans. AIME: 2225 (1969).]

Characterization of dislocation loops If vacancies or interstitials precipitate over a single crystallographic plane of a material, the conditions
exist for collapse or expansion of the lattice region. In the case of a
vacancy disk in fcc materials on the (111) planes, the collapse of this
region results in a sessile dislocation loop with a Burgers vector a/3
[111], and with an intrinsic stacking-fault region within the bounds of
the loop. The situation is ideally represented in Fig. 7.33, which also
shows the extrinsic stacking-fault character for the interstitial loop.

Because of the geometrical nature of such dislocation loops, a simple
method exists for the characterization of their images as vacancy or interstitial. The method involves an intuitive identification of loop geometry,
and considers the loop to lie on an inclined plane of a crystal matrix.
Table 7.5 illustrates the essential features of the method, which incorporates the proposals of D. J. Mazey and associates [60], and B. Edmundson
and G. K. Williamson [61]. We observe in Table 7.5 that if the operating
reflection producing contrast of a loop is to the right of the tilt axis,
then a clockwise tilt will pass the foil through the Bragg position as indicated. On passing through the reflection position, the image side will
be observed to change, becoming "outside" for interstitial loops and "inside" for vacancy loops. We observe that as we tilt the foil clockwise
along the loop plane, the image will continuously grow in size for an interstitial character, and will first grow then decrease for a loop having
a vacancy character on going through the reflection position. Careful comparison of the image and diffraction pattern will usually allow the loop to
be characterized in this manner.

Loops occur in a large number of materials for numerous reasons, chief
among them being some form of thermal quenching. Loops have been described
in fcc materials by R. M. Cotterill [62], in bcc materials (for example,
Mo) by J. D. Meakin and associates [63], and in hcp Mg by P. B. Price [64]
to present but a few examples. Loops can generally be described as pris-
matic or regular, depending on their image character, and can of course result for various modes of quench of a material (point defects produced by
radiation damage can also coalesce as described in Sec. 8.5). Figure 7.43
illustrates the appearance of loop images.

Diffraction contrast at precipitates Precipitates or inclusions of a general nature will be observed in electron microscope images of thin sections
because of a difference in the contrast originating at such particles. In

TABLE 7.5 *Characterizing dislocation loops in the electron microscope*

Reflection condition	**n** Tilt	**n** **s** $s > 0$	$s < 0$ **n** **s**
Vacancy loops	( )	( )	( )
Interstitial loops	( )	( )	( )

the general case, the particle will be chemically different from the matrix. Consequently Z will be different and the local scattering factor (electron phase shift) will be different from the matrix. For a precipitate with an effective Z very much greater than the matrix, transmission through the particle may be very small. The particle will then give rise to contrast simply as a result of the shadow it projects. A type of mass-thickness contrast may then result where the matrix region about the inclusion is relatively undisturbed, that is, the local strain field is small. Such particles are of the incoherent type.

For coherent precipitates, the strain field resulting from the lattice accommodation extends some distance from the particle-matrix interface. This lattice distortion gives rise to a local phase shift, which in turn influences the image contrast. As a consequence, a precipitate may show an intrinsic structural contrast in addition to an "isostrain" contrast.

We should not be confused that coherent precipitates show diffraction contrast while inchoherent ones do not. Both can give rise to a characteristic diffraction contrast.[†] However, only the coherent particles will have isostrain contours associated with the image. The appearance of these contours will be governed chiefly by the foil thickness and the operating reflection producing contrast. At the origin of the strain

---

[†]*Diffraction contrast at incoherent inclusions is treated briefly in a paper by R. J. Horylev and L. E. Murr,* Proc. Electron Microscopy Soc., *ed. C. J. Arceneaux, p. 168, Claitor's Publishing Div., Baton Rouge, La., 1969.*

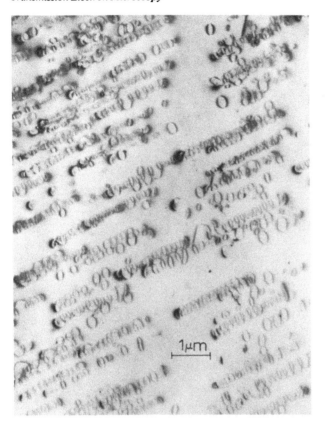

FIG. 7.43 *Dislocation loops and helical disloca-*
tions in Al-Mg alloy observed in high-voltage
electron microscope at accelerating potential of
750 kV as discussed in Chap. 8. (Courtesy Hita-
chi, Ltd., Tokyo).

field, a neutral plane exists so that in the image, the strain contour
that effectively surrounds the particle will split at this plane along a
line of "no contrast." This line of no contrast will be perpendicular to
g. These effects were originally observed by V. A. Phillips and J. D.
Livingston [65] for coherency strains about Co particles in a Cu matrix.
You should readily grasp the concept with reference to Fig. 7.44, which
shows the images at fine precipitates (carbide) in Inconel 600 alloy
films.

The contrast arising at the elastic-strain fields about second-phase
particles is not so well understood in terms of the diffraction contrast.
These features have been discussed by Howie and Whelan [66], whose treat-

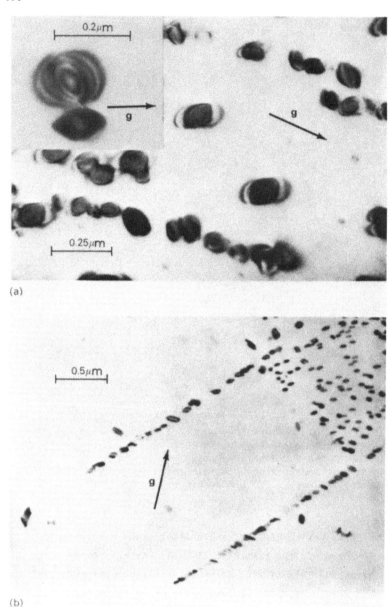

FIG. 7.44 *Images of precipitates in thin films of Inconel*
600 alloy. (a) Isostrain contours arising from elastic dis-
tortions of matrix, coherently accommodating precipitate lat-
tice, are particularly prominent. Insert shows details of
diffraction contrast effects. Line of no contrast is essen-
tially perpendicular to g, operating reflection. [*After L. E.*
Murr, Phys. Status Solidi, 19: 7(1967).] *(b) Similar preci-*
pitates under slightly different contrast conditions and pris-
matic dislocation loops. The loops and precipitates are, for
the most part, indistinguishable from the precipitate contrast
without tilting. Many precipitates are aligned in slip traces.

ment was expanded upon by M. F. Ashby and L. M. Brown [67]. Heidenreich
[68] also presents a treatment of the intensity profiles at isostrain contours using a somewhat different approach than that of Ashby and Brown.
The important consideration is, however, that the image character of a coherent precipitate can be related to the radial-strain field surrounding
the particle.

Vacancy clusters, voids, and color centers In the quenching of metals and
in other treatments producing large numbers of vacancies these can coalesce
forming large discs bounded by dislocation loops or form other void structures having large lattice volumes. Metals and other materials subjected
to radiation damage can also produce large voids by coalescence phenomena
which can assume polyhedral (crystallographic) shapes as a result of minimum energy configurations, etc. Voids accumulating in a crystal can also
establish a specialized arrangement as a result of energy considerations
and these arrangements frequently result in a void lattice. In many solids,
particularly ionic solids, radiation-induced cation vacancies can also
create special color centers when the site is charge compensated by the
trapping of an electron in the site. Coalescence of these sites produces
large charge-compensated voids called F-center aggregates or color-center
aggregates. Figure 7.45 illustrates large vacancy loops created from collapsed vacancy discs in quenched nickel and color-center aggregates in
electron-irradiated CaF_2. The color-center aggregates are organized in an
equilibrium void lattice array in the (111) planes, giving rise to the
close-packed symmetry observed [69]. Similar void lattices have been observed in a variety of irradiated materials [70,71]. As shown in Fig.
7.45 the contrast arising at vacancy discs, voids, void lattices, bubbles,
etc. can be a mixture of diffraction effects arising from crystallographic
features or local strains as well as mass-thickness effects.

7.4 STRUCTURAL GEOMETRY IN THIN-FILM MATERIALS

One of the most unique features of transmission electron microscopy is the
ability to view solid sections in perspective, and to analyze the geometry
of an image using spherical and solid-geometrical theory. Several of these
features have already been discussed in our earlier treatments of dislocation images in thin films (Sec. 7.2.2), the measurement of foil thickness,
and the estimation of defect densities (Sec. 7.3.2). It is instructive at
this point to elaborate on some of the more fundamental geometrical fea-

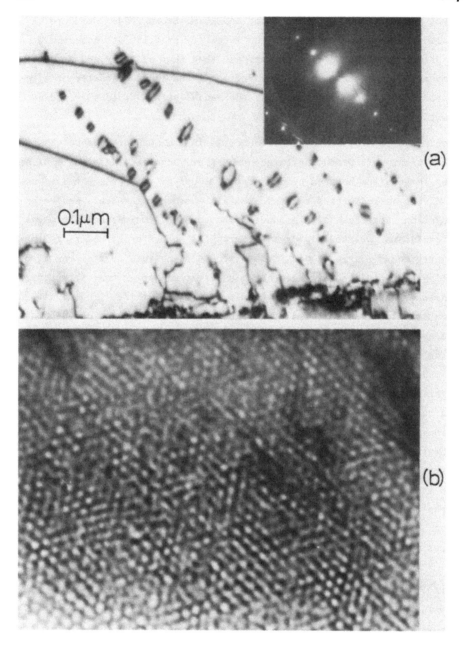

FIG. 7.45 *Collapsed vacancy discs giving rise to large dislocation loops in quenched Ni (a); and color-center aggregates giving rise to void lattice structure in electron-irradiated (111) CaF₂ (b). Magnification of (a) and (b) is indicated in (a).*

tures of thin films, namely, the geometry of grains, phases, and various crystal interfaces. In essence we will simply add the third dimension to our treatment of two-dimensional image projections of grain and phase boundaries discussed in Sec. 2.4.3, thereby enhancing not only our precision of measurement, but also our perspective of thin solid matter.

7.4.1 CHARACTERIZATION OF GRAIN-BOUNDARY STRUCTURE

Let us consider the geometry associated with the junction of three homogeneous grains or phases as shown in Fig. 7.46. While we can readily measure the intersection angles in the image in a manner identical to that attainable by surface microscopy, the electron transmission image gives us a projection of the three-dimensional solid section. By geometrical means, it is then possible to reconstruct the object shape, and to determine rather accurately the inclinations of the boundary planes and the true dihedral angles of intersection of the grains using the techniques outlined by Murr [2].

The geometrical situation for the image of Fig. 7.46 is sketched in profile in Fig. 7.47. Utilizing the conventions of Fig. 7.47, we first measure the foil thickness in the area of the junction by using Eq. (7.13) in the form

$$t = W_T \tan \theta_T$$

where W_T is the projected width of the twin boundary measured directly in Fig. 7.48, and θ_T is the inclination angle as determined from Eq. (6.57). Having determined the thickness, the inclinations of the grain boundaries can be determined from

$$\theta_A = \tan^{-1} \frac{t}{W_A}$$

$$\theta_B = \tan^{-1} \frac{t}{W_B} \tag{7.55}$$

$$\theta_C = \tan^{-1} \frac{t}{W_C}$$

The true dihedral angles between the intersecting grain-boundary planes can then be calculated directly from

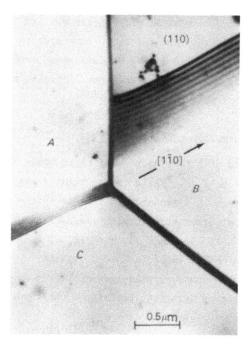

FIG. 7.46 *Bright-field image of grain-
boundary triple junction in stainless
steel. Junction area (in grain B) contains crystallographic twin boundary from
which local foil thickness can be measured directly, allowing associated boundary geometries at junction to be uniquely determined. (125 kV).*

$$\left.\begin{array}{l}
\cos\Omega_A = \cos\theta_B\cos\theta_A + \cos\omega_A\sin\theta_B\sin\theta_A \\[2mm]
\cos\Omega_B = \cos\theta_C\cos\theta_B + \cos\omega_B\sin\theta_C\sin\theta_B \\[2mm]
\cos\Omega_C + -\cos\Omega_A\cos\theta_C + \cos\omega_C\sin\theta_A\sin\theta_C
\end{array}\right\} \qquad (7.56)$$

where Ω_A, Ω_B, Ω_C are the true dihedral angles, and ω_A, ω_B, ω_C are the angles measured directly from the image of Fig. 7.46 as shown in Fig. 7.47.

In many observations of thin-film geometry, no crystallographic traces
are present in the image. Thus no unique means exists for the determination
of the local thickness t, and we cannot analyze the object geometry. However, in most cases it is possible to initiate a crystallographic trace by
focusing the electron beam onto the junction or foil area of interest,
thereby causing the initiation of dislocation motion. A slip or stacking-
fault trace may then result as shown, for example, in the sequences of Figs.
7.23 and 7.24. This technique has been used in the analysis of grain-
boundary junctions [72,73].

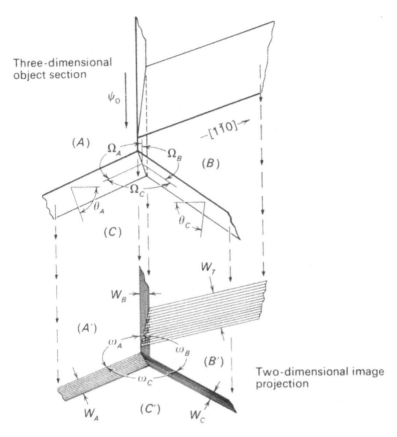

FIG. 7.47 *Schematic representation of object section and pro-jected image geometry of Fig. 7.46.*

We can employ a similar technique to study grain growth and related phenomena directly within the electron microscope. That is, by heating a film section with the focused electron beam it is possible to raise the local temperature sufficiently to initiate grain growth or phase transformations in much the same way as we indicated in the discussions of the applications of thermionic electron emission microscopy in Sec. 2.4.3. However, in the present situation we can measure true radii of curvature by using Eq. (7.23) and our knowledge of the boundary-plane inclinations. In this way it is possible to study grain-growth kinetics, phase-transformation rates, and related solid-state phenomena. Figure 7.48 illustrates the application to studies of grain growth in a thin ceramic foil.

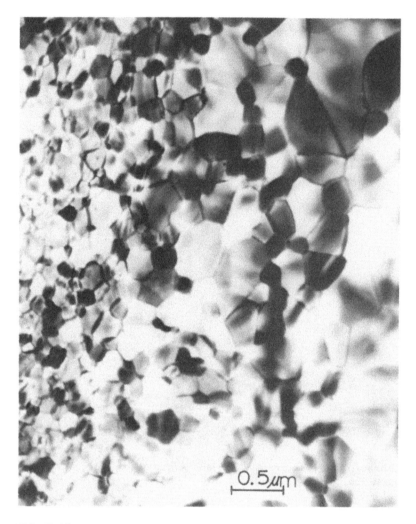

FIG. 7.48 *Direct observation (bright field) of grain growth
in Er$_2$O$_3$ thin foils heated in electron beam. Direction of
growth is to left. Foil thickness is roughly 800 Å. (100 kV.)*

You must bear in mind that while very direct and useful information
can be obtained from grain-growth studies, etc., directly within the electron microscope, such processes in thin films may not and often do not represent entirely the bulk characteristics, and this must be carefully considered in interpreting the results. The representation is apparently applicable to bulk solids when the foils are sufficiently thick; and for certain applications, the use of a specimen heating stage in a high-voltage
electron microscope would be the most desirable method of investigation.

In the case of electropolished foils, grain boundaries are perhaps the only manifestations of the microstructure that are relatively stable.

7.4.2 *DIRECT MEASUREMENT OF SOLID INTERFACIAL ENERGETICS*[†]

Associated with any thermodynamic environment in a solid is a conditional equilibrium or a pseudoequilibrium. What we mean to illustrate is that for a fixed temperature, grain growth will cease in a polycrystalline solid when the surface tensions at the boundary junctions are balanced. If we make the assumption that at sufficiently high temperature, the surface tensions and surface free energies are essentially equivalent, then we can write generally that

$$\frac{\gamma_3}{\sin\Omega_A} = \frac{\gamma_1}{\sin\Omega_B} = \frac{\gamma_2}{\sin\Omega_C} \qquad (7.57)$$

where γ_1, γ_2, and γ_3 are the free energies resolved within the boundaries opposite grains A, B, and C of Figs. 7.46 and 7.47. Equilibrium of the junction is usually described for $\Omega_A = \Omega_B = \Omega_C = 120°$.

Measurement of the average free energy for grain boundaries and interfaces

The average high-angle (high-energy) grain-boundary free energy of a solid can be measured directly in a combined measurement of the surface free energy using an experimental method called zero creep [23,74,75]. This method is generally more direct than the field-ion microscopy technique outlined in Sec. 2.5.2. In the usual zero-creep method, very thin wires having diameters on the order of 0.001 to 0.004 in. and having various weights of the same material suspended from them are heated at some fixed temperature for several hundred hours in vacuum or inert atmosphere. The wires with heavier weights extend and those with weights unable to balance the surface tension contract. A plot is then made of load versus strain rate from which the load at zero strain rate (balance of surface tension) is readily calculated from

$$w = \pi r \left(F_s - \gamma_{gb} \frac{r}{\ell} \right) \qquad (7.58)$$

[†]*We will use interchangeably throughout the notation interfacial energy or interfacial free energy. You are cautioned that we are not characterizing the free energy per unit area of surface, but rather the reversible work of formation of the unit area of boundary.*

where w = weight at balance

     r = wire radius

    F_s = surface tension; γ_{gb} = grain-boundary tension

    ℓ = average grain length

The average grain length is readily observed from the bamboo-type character the wire develops after either long-time creep or high-temperature anneal because of the thermal grooves formed where the grain-boundary planes meet the wire surface. The grain boundaries align themselves normal to the wire axis.

We can consider the equilibrium of the surface- and grain-boundary free energies of the wire to be resolved as shown in Fig. 7.49. This situation is equivalent to considering that in Eq. (7.57), $\gamma_2 = \gamma_3 = F_s$, $\gamma_1 = \gamma_{gb}$, and $\Omega_A = \Omega_C$. We can then write

$$\gamma_{gb} = 2F_s \cos \frac{\Omega_B}{2} \qquad\qquad (7.59)$$

It is therefore necessary to measure only Ω_B, the dihedral groove angle. This is done at high magnification in the electron microscope by making a shadowgraph of the wire groove [23].

We now have, on combining Eqs. (7.58) and (7.59), two equations with two unknowns, γ_{gb} and F_s, from which a unique solution can be obtained.

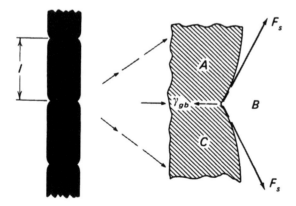

FIG. 7.49 *Thermal grooving (bamboo structure) in equilibrated solid wires.*

The values of both F_s and γ_{gb} must be recognized as average values. Nonetheless, the measurements have been direct, and are representative of the equilibrium conditions at the corresponding test temperature.

Measurement of twin-boundary free energies in fcc materials With a knowledge of the grain-boundary free energy in an fcc material, the average twin-boundary free energy can be measured directly as shown by Murr [23]. Here we must modify our previous considerations of an equilibrated triple grain-boundary junction, and consider a system of two interconnected junctions describing the intersection of a coherent twin band with a grain boundary as depicted in Fig. 7.50. The equilibrium equations for the system then become [+],

$$\gamma_{tb} + \gamma_{AB}\cos\Omega_1 + \gamma_{T_AB}\cos\Omega_2 + \Sigma M_I = 0$$

$$\gamma_{tb} + {}_{T_AB}\cos\Omega_3 + \gamma_{AB}\cos\Omega_4 + \Sigma M_{II} = 0 \qquad (7.60)$$

where ΣMs are the associated torque terms. The simultaneous solution of these equations then yields the following two ratios

$$\frac{\gamma_{tb} + EM}{\gamma_{gb}} = \frac{\cos\Omega_2\cos\Omega_4 - \cos\Omega_1\cos\Omega_3}{\cos\Omega_3 - \cos\Omega_2} \equiv C_{AB}$$

$$\frac{\gamma_{tb} - EM}{\gamma_{gb}} = \frac{\cos\Omega_2\cos\Omega_4 - \cos\Omega_1\cos\Omega_3}{\cos\Omega_1 - \cos\Omega_4} \equiv C_{T_AB} \qquad (7.61)$$

where γ_{tb} is the resolved twin-boundary free energy, and $\gamma_{gb} = \gamma_{AB} = \gamma_{T_AB}$ is the average high-angle grain-boundary free energy.

This technique has been applied to a number of metals and alloys [23] by observing the geometries of twin-grain intersections as shown typically in Fig. 7.51, and resolving the features using the conventions sketched in Fig. 7.50. The results have been rather interesting as shown in the distributions of the data for Inconel and stainless steel reproduced in Fig. 7.52.

Measurement of low-angle grain-boundary free energies You should already have concluded that the geometrical resolution of interfacial energies can

---

[+]*The origin and measurement of torques has been treated by L. E. Murr, R. J. Horylev, and W. N. Lin, Phil. Mag., 20: 1245(1969).*

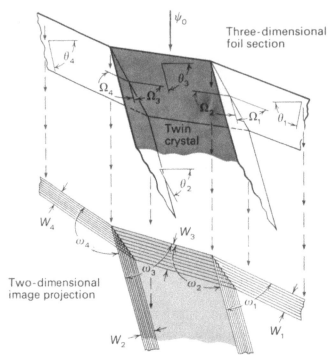

FIG. 7.50 *Crystal section and projection view of*
twin-boundary-grain-boundary system in equilibrium.

be generally extended to any equilibrium situation. Thus the related properties and structure of phase boundaries in metals and ceramics are amenable to direct study, as are any class of boundary structure. Of particular interest to the solid-state scientist is the energy properties of low-
angle grain boundaries, that is, those boundaries composed of discrete
dislocation arrays.

It should hopefully be apparent that with the capability to resolve
boundary-interface details with high contrast and resolution, as well as
being able to directly observe the associated grain or phase crystallographies by selected-area electron diffraction, the study of thin sections in
the electron microscope is an unprecedented approach to solid-state investigations. We can demonstrate this feature by considering the junction of
recognizable low-angle[†] grain boundaries with high-angle grain boundaries
or twin boundaries as outlined previously.

[†]*In such analysis we must arbitrarily set special conditions for specific*
boundary identification such as the resolution of the dislocation charac-
ter in the boundary and a crystallographic misorientation across the boun-
dary of not more than a few degrees.

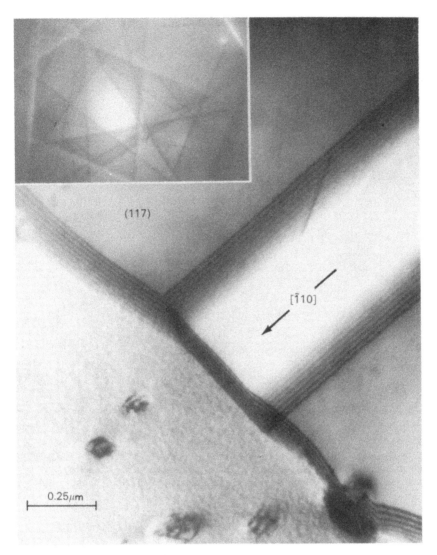

FIG. 7.51 *Coherent (11Ī) twin-boundary-grain-boundary intersection in thin Inconel film. Geometrical features correspond to
Fig. 7.50 Note precipitate in lower-right grain-boundary image.
(100 kV).*

Let us initially consider the two cases depicted in Fig. 7.52. We
observe that for Fig. 7.52a, a low-angle boundary is considered to be in
equilibrium at a junction with two grain boundaries, while in b the junctions are composed of a twin band intersecting a low-angle grain boundary.
In Fig. 7.52a Eq. (7.57) can be appropriately written

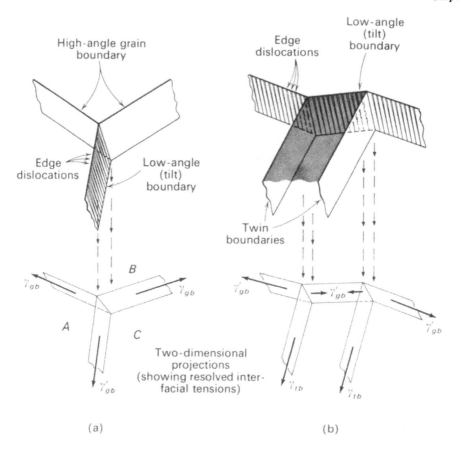

FIG. 7.52 *Low-angle grain-boundary or other special interphase junc-*
tions. (a) Equilibration of low-angle grain boundary with high-angle
boundaries; (b) equilibrium of low-angle boundary with twin boundaries.

$$\frac{\gamma_{gb}}{\sin\Omega_A} = \frac{\gamma'_{gb}}{\sin\Omega_B} = \frac{\gamma_{gb}}{\sin\Omega_C} \tag{7.57a}$$

where γ_{gb} is the average high-angle grain-boundary free energy, and γ'_{gb}
is the average low-angle grain-boundary free energy.

We can similarly consider Fig. 7.52b to simply involve replacing γ_{gb}
in Eq. (7.61) with γ'_{gb}. Thus, having previously evaluated γ_{tb}/γ_{gb}, and
knowing γ_{gb}, the average value of γ'_{gb} can be obtained statistically. This
concept can also be applied to other special boundaries and interphase
boundaries which are designated γ'_{gb} [23].

Low-angle grain boundaries are generally of two types, tilt or twist
[11 to 13]. Depending on the contrast and diffraction conditions giving

rise to an image, these types can be readily characterized by the regular array of dislocations in the case of the tilt boundary, and the cross-grid network associated with a boundary of the twist type. We can illustrate the details of a tilt boundary with reference to Fig. 7.53. Here the dislocations composing the boundary are readily apparent.

Using selected-area electron diffraction, we can measure the misorientation of the lattice spearated by the low-angle grain boundary in Fig. 7.53. The magnitude of the Burgers vector is then obtained from [13]

$$\Theta_m \cong \frac{|b|}{d} \tag{7.62}$$

where Θ_m is the angle of misorientation, and d is the spacing between dislocations (edge type) in the boundary. Then, since the total energy per unit area of a tilt boundary is given by [12,13].

$$\gamma'_{gb}(\text{tilt}) \cong \frac{G|\underline{b}|}{4\pi(1-\nu)} \left[\frac{4\pi(1-\nu)\gamma_C}{G|\underline{b}|^2} - \ln \right] \tag{7.63}$$

where G is the shear modulus of the material, and ν is Poisson's ratio; the approximate core energy of the individual dislocations composing the boundary, γ_C, can be calculated directly (see Prob. 7.19) [23].

Measurement of stacking-fault free energy We have seen that dislocations in a material can lower their energy splitting into two partials. When the splitting occurs over several lattice spacings, a visibly faulted region occurs. The width of this faulted region (stacking fault) will depend on the energy of formation of the fault interface. In essence the ribbon of fault will be appreciably extended in materials of low stacking-fault energy, with the width decreasing with increasing stacking-fault free energy. It is therefore possible to measure stacking-fault energy directly by simply considering the equilibrium spacing of partial dislocations.

If the material can be considered to be elastically isotropic, the stacking-fault energy is given by

$$\gamma_{SF} = \frac{G\underline{B}_2 \cdot \underline{b}_2}{8\pi d_o} \frac{2-\nu}{1-\nu} \tag{7.64}$$

where $\underline{b}_1$ and $\underline{b}_2$ are the Burgers vectors of the partials ($\underline{b}_1 + \underline{b}_2 = \underline{b}$), and d_o is a factor relating to the width of the partials. Considering Shockley partials in fcc materials (Sec. 7.3.2), we have

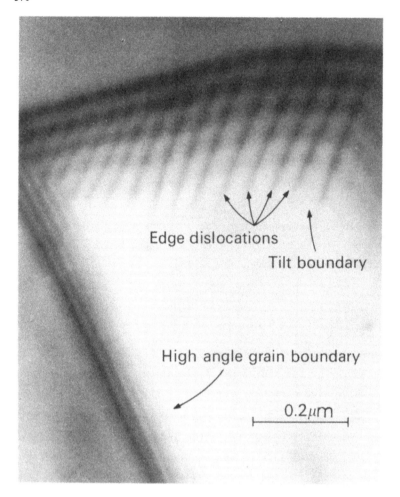

FIG. 7.53 *Tilt-boundary structure in stainless steel showing its composition by regularly spaced edge dislocations.*

$$d_o = d \frac{2 - \nu}{2 - \nu - 2\nu\cos2\phi} \qquad (7.65)$$

where d is the actual separation of the partials (the length of stacking fault), and ϕ is the angle between the direction of the fault ribbon and the Burgers vector $\underline{b}$. Since both Shockley partials of the type a/6 <112> have equal magnitudes, we can rewrite Eq. (7.64) as

$$\gamma_{SF} = \frac{Gb^2(2 - \nu - 2\nu\cos2\phi)}{8\pi d(1 - \nu)} \qquad (7.66)$$

One very troublesome feature associated with Eq. (7.66) is that the equilibrium spacing of the partials must relate to thermodynamic equilibrium, and should be free of the effects of surface stresses in thin films, and of local temperatures.[†] This can hardly be assured. An attempt has been made in thin stainless steel foils to measure stacking-fault energy by Eq. (7.66) assuming all partials to be pure edge in character and having a Burgers vector in the fault plane normal to the direction of the ribbon (which was parallel to the dislocation lines), so that $\phi = 90°$ in Eq. (7.66). The results are plotted in the distribution of Fig. 7.54 from which the mean stacking-fault free energy is observed to be 22 ergs/cm^2 (22 mJ/m^2). The spread in the data presumably reflects the variation in local equilibrium conditions, and it cannot be stated with any assurance what the equilibrium temperature was although the material was annealed at 1000°C. The character of faults measured corresponded somewhat to those shown in Fig. 7.36c, but with a more regular separation d.

Stacking-fault energies can be measured for bcc and hcp materials by observing the separation of partials, and by considering the appropriate dislocation reactions. T. E. Mitchell [76] has outlined the dislocation reactions and the generation of stacking faults in bcc crystals.

One of the more popular methods for measuring stacking-fault energy in fcc materials has been to determine the radius of curvature at extended partial dislocation nodes. A simple relationship then describes the stacking-fault energy as [49].

$$\gamma_{SF} = \frac{Gb^2}{2R} \tag{7.67}$$

where R is the true radius of curvature of the node. You should immediately realize that where such node radii R' are measured in an electron transmission image, and the dislocation networks themselves occur on an inclined {111} plane, the true radii of curvature are given by Eq. (7.23), following the corresponding notation shown in Fig. 7.22.

In fcc materials we can consider two types of dislocation reactions involving partial dislocations separated by a ribbon of stacking fault. In the first, we can consider two sets of Shockley partial dislocations, each on a concurrent {111} plane. The front partial of each is considered

---

[†]*It should be emphasized that all stacking-fault energy measurements in thin films will be susceptible to surface effects that render dynamic features unrepresentative of bulk characteristics.*

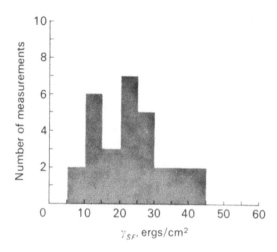

FIG. 7.54 *Measurement of stack-ing-fault energy in 304 stain-less steel foils.*

mobile and each is observed to move in the respective {111} planes, fur-ther extending the stacking fault. If these planes intersect, the mobile partials on meeting at the junction will form a dislocation called a stair rod [13]. This stair rod at the junction, and the two end partials (one on each slip plane), all connected by stacking faults, form an immobile dislocation triad called a Cottrell-Lomer lock. We can visualize one pos-sibility with respect to Figs. 7.55a and b as follows:

$$\frac{a}{6} \, [2\bar{1}1] + \frac{a}{6} \, [\bar{1}2\bar{1}] \longrightarrow \frac{a}{6} \, [110]$$

Because these triads are immobile, they act as obstacles to the movement of dislocations on the $(11\bar{1})$ and (111) planes.

In the second type of reaction, we can assume that two sets of par-tials separated by a region of stacking fault are mobile on the same (111) plane. If one of the partials of each set has a common partial, on ap-proaching one another reaction will occur in which the two separate stack-ing faults will fuse to form an isolated extended node. Figure 7.55c and d illustrate schematically the formation of a dislocation node. The radii of curvature are also apparent in Fig. 7.55d.

Figure 7.56 shows actual images of Cottrell-Lomer locks and networks of dislocation nodes observed in stainless steel foils. It should be rea-lized of course that these reactions are most likely where γ_{SF} is small, that is, where dislocations are measurably extended.

We can now appreciate the fact that Eq. (7.67) expresses the equili-brium curvature of an isolated partial dislocation. The curvature is

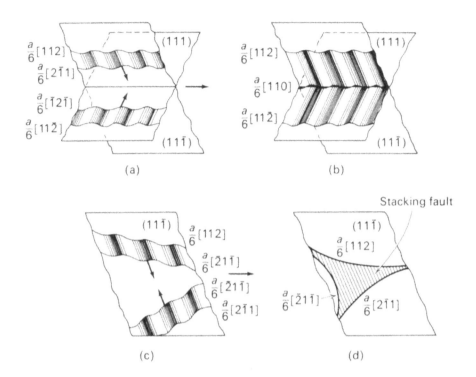

FIG. 7.55 *Dislocation reactions on fcc {111} planes. (a) and (b) show generation of Cottrell-Lomer lock. (c) and (d) show creation of dislocation nodes extended in (11$\bar{1}$) plane.*

influenced on the one hand by the effective shear stress exerted by the stacking fault within the node, and on the other hand by the line tension associated with the partial dislocation. Because of these features, the correct form of Eq. (7.67) will depend to some extent on whether the ribbons are edge or screw in character.

Numerical solutions of the node problem have been given by Brown and others [67]. Their results indicate that on considering a general node, that is, simply measuring the radii of curvature without determining the node character as edge or screw, the average stacking-fault free energy is actually twice the value given in Eq. (7.67). The general equation is then

$$\gamma_{SF} \simeq \frac{Ga^2}{6R} \tag{7.68}$$

where a is the lattice parameter of the material (see Prob. 7.20).

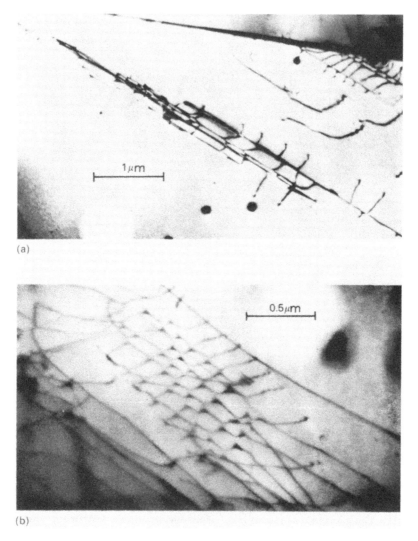

(a)

(b)

FIG. 7.56 *(a) Cottrell-Lomer locks and (b) extended dislocation nodes in 304 stainless steel foils. (100 kV.)*

A number of stacking-fault energy measurements have been made using Eq. (7.67) in metals and alloys where it was observed that alloying lowers the stacking-fault energy [23,77-81]. This is of course a means for measuring γ_{SF} for metals of high stacking-fault energy, that is, by extrapolation. Figure 7.57 illustrates this particular feature for Cu-Aℓ alloys from which the value of γ_{SF} for pure Cu is extrapolated to be approximately 70 ergs/cm^2 (see Prob. 7.26).

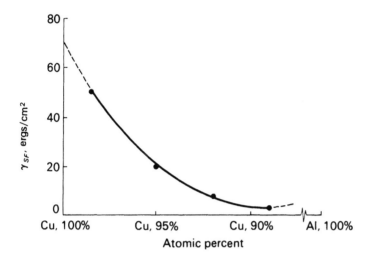

FIG. 7.57 *Measurement of stacking-fault energy from extended dislocation nodes in Cu-Al alloys. Points plotted represent averages calculated from rigorous form of Eq. (7.68) as discussed by L. E. Murr,* Thin Solid Films, 4: no. 6, 389(1969).

We should also point out that where mobile stacking faults intersect on all four active slip systems in fcc, a stacking-fault tetrahedron can result. The stacking-fault energy can be measured from the equilibrium size of such tetrahedra as described by J. Silcox and Hirsch [82], R. E. Smallman and K. H. Westmacott [83] and T. Jossang and J. P. Hirth [84].

7.4.3 SURFACE COINCIDENCE AND TRUE OBJECT GEOMETRY IN THIN SECTIONS

In our geometrical analyses of thin-film structure in the foregoing sections, we obtained the local film thickness and crystallography from the selected-area electron diffraction pattern. The object features were then reconstructed on the basis that the crystallographic surface plane was the true specimen plane. This is not generally true. We must therefore point out that in order to unambiguously define the true object geometry in a thin-film section, it is necessary to define the thin-film surfaces with respect to the electron beam. Ideally we would like to ensure that the crystallographic orientation coincides nearly exactly with the true specimen surfaces.

Let us consider this situation as depicted in Fig. 7.58. In Fig. 7.58a a foil section has its surface randomly disposed with respect to the

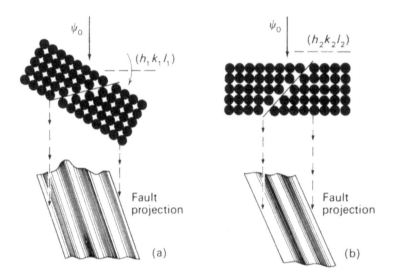

FIG. 7.58 *Disposition of surface plane and crystallographic*
orientation of thin film in electron microscope. (a) Crystal-
lographic surface $(h_1 k_1 \ell_1)$ is not coincident with object sur-
face plane. Inclination angle of fault plane measured from
image does not then represent true angle of inclination with
object surfaces. (b) Coincidence of crystallographic $(h_2 k_2 \ell_2)$
and true specimen surfaces.

electron beam. Nonetheless a crystallographic orientation $(h_1 k_1 \ell_1)$ is observed. In Fig. 7.58b the specimen surface plane is normal to the electron
beam and coincident with the crystallographic surface plane $(h_2 k_2 \ell_2)$. The
projection width of the fault plane changes appreciably, and as a consequence the geometrical reconstruction of the object features will be significantly different from Fig. 7.58a to b.

The only means to assure that the image of a crystal section meets the
exact case of coincidence is to compare the deviation of crystallographic
trace directions in the image with those deduced from the selected-area
electron diffraction pattern. This is illustrated for a thin-foil section
of stainless steel following explosive shock loading in Fig. 7.59. The
traces of twins and stacking faults along the {111} planes are exactly
coincident with those directions observed in the selected-area electron
diffraction pattern when properly rotated with respect to the image. The
crystallographic surface and the specimen surface are then unambiguously

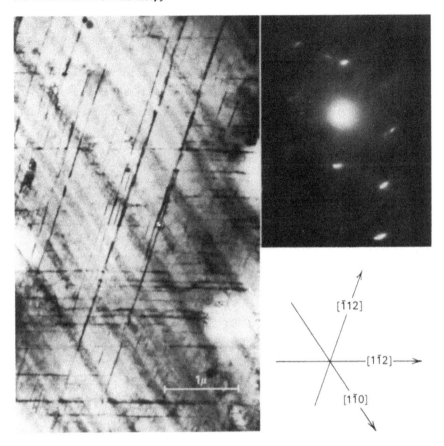

FIG. 7.59 *Coincidence of specimen surface and the (110) crystallo-graphic surface in thin stainless steel foil following shock load-ing to 425 kbars.* [*After L. E. Murr,* Phys. Status Solidi, 19: 7(1967)].

shown to be almost exactly coincident. This feature is also evident in Fig. 7.37.

Murr [72] has shown that measurements of interfacial free energies of electropolished stainless steel in the electron microscope neglecting the coincidence of the specimen and crystallographic surfaces must be correct-ed by a factor of about 0.7. Consequently, where exact solutions for ob-ject geometry and related structural features are desired in transmission electron microscopy, one must be careful to consider this important pro-perty.

7.5 HIGH-RESOLUTION ELECTRON MICROSCOPY

7.5.1 LATTICE IMAGES AND MOIRE' PATTERNS

We have already observed in Chap. 3 that when the resolution of an electron
microscope is sufficiently small, it is possible to directly observe planar
arrays of atoms in sufficiently thin foils by transmission electron microscope (see Fig. 3.16).

The mechanism by which such images are obtained lies in considering
the Abbe' theory[†] of optical image formation, which essentially states that
diffracted amplitudes are formed of the object at the back focal plane
(BFP) of the objective lens (see Fig. 3.12). The interference or interaction between the transmitted and diffracted beam or beams then produces
a lattice fringe image. In effect, for a 2-beam lattice fringe image, the
objective aperture must be positioned to pass the transmitted and one diffracted beam. This can be accomplished by shifting the aperture or tilting
the illumination. The fringes formed are normal to the diffraction vector
(g) of the diffracted beam admitted, and they ideally possess the spacing
of the corresponding lattice plane ($d_{hk\ell}$). Figure 7.60 illustrates these
features for a thin specimen of platinum phthalocyanine where the objective
aperture is positioned to admit two diffracted beams, g_1 and g_2 which produce two separate fringe images having a spacing d_1 and d_2 respectively as
shown.

It is apparent from Fig. 7.60 that specimen parameters such as thickness and orientation have a profound effect on the fringe appearance and
contrast. Using a two-beam theory of dynamical electron diffraction
Hashimoto, et al [85] and Amelinckx [86] have treated these effects in
detail as implicit in the following equation for the intensity of the resultant electron wave function:

$$|\psi|^2 = A + B\sin(2\pi gx - \pi st) \tag{7.69}$$

where

$$A = 1 + \left(\frac{\pi}{\xi_g}\frac{\sin\pi st}{\pi s}\right)^2 ; \quad B = -\frac{2\pi}{\xi_g}\frac{\sin\pi st}{\pi s}$$

We see from Eq. (7.69) that for a given specimen thickness, t, and devia-

---

[†]*See E. Abbe*, Arch. Mikr. Anal., *9, 413(1837).*

tion from the Bragg condition, $\underline{s}$, a modulated sinusoidal intensity occurs in the x direction (in the plane of the specimen) with periodicity d (the interplanar spacing). From this expression we see that the visibility of the fringes depends upon both thickness and deviation from the Bragg condition, and if $\pi st = n\pi$ there are no fringes. In fact we see that both the amplitude and the periodicity of the fringes will depend on t and $|\underline{s}|$. Consequently, there are situations when the fringes do not accurately represent the lattice spacing. These features must be carefully considered when variations in fringe spacing and intensity are to be related to variations in chemical composition for example [87]. Because of absorption and other interference effects, foils should be as thin as possible and the thickness should be uniform. The maximum two-beam fringe visibility occurs at the exact Bragg condition so the diffracted beam used should correspond to $|\underline{s}| = 0$. Consequently tilting the foil can be useful in achieving sharp fringe images when $|\underline{s}| = 0$ can be achieved.

If we consider the path difference at the gaussian image plane between the paraxial ray and the "aberration" ray (as in Fig. 3.13e), the phase of the aberration ray will be given by[†]

$$X = -\frac{2\pi}{\lambda}(C_s\alpha^4 - 1/2\Delta f\alpha^2) \tag{7.70}$$

where C_s is the spherical aberration coefficient of the objective lens, and Δf is the amount of image defocus. The maximum intensity change in the grating image will then be observed to occur at some point of defocus. We see therefore that fringe contrast of crystal lattices appear at an out-of-focus condition by phase contrast.

Figure 7.61 illustrates the images of a pyrophyllite crystal along its c axis. The image show the effect of overfocus and underfocus (a and b, respectively) on the reversal of fringe contrast using the [020] diffraction vector. Figure 7.61c and d shows overlapping crystals of pyrophyllite exhibiting moire patterns. The contrast reversal due to overfocus and underfocus is again exhibited in a' and b' respectively.

We observe in Fig. 7.61c and d that a distinct pattern characteristic, moire pattern, results for overlapping crystal films. The mechanism of formation of moiré fringes is analogous to the superposition of two opti-

---

[†]*See chaps. V and X, and Appendix B of R. D. Heidenreich,* Fundamentals of Transmission Electron Microscopy, *Interscience Publishers, Inc., New York, 1964.*

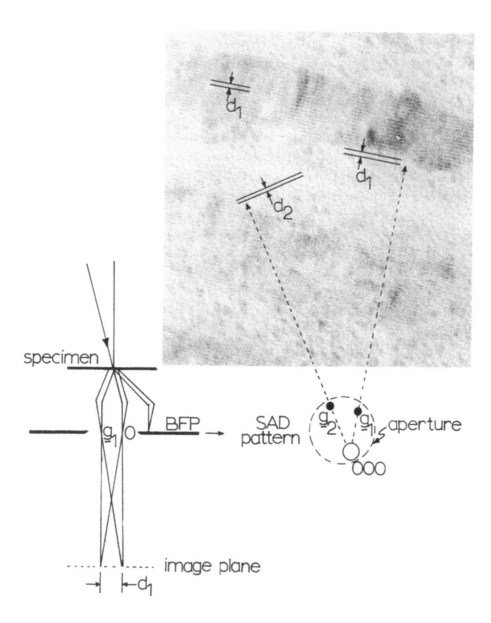

FIG. 7.60 *Formation of lattice fringe images in platinum phthalocyanine corresponding to lattice plane spacing, d_1 and d_2. The objective aperture is placed at the back focal plane (BFP) in either a beam tilted or untilted mode (200 kV).*

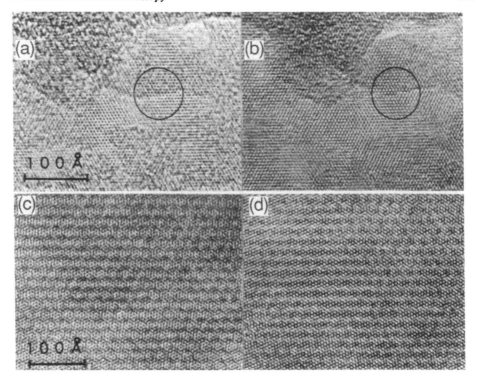

FIG. 7.61 *Phase-contrast images of pyrophyllite crystal planes,*
$a = 4.57$ Å. *(a) and (b) show same crystal portion at different*
focuses to produce contrast reversal. (c) and (d) show moire
patterns for overlapping crystals observed at different focuses
to produce a contrast reversal as in a and b. (Courtesy T. Komoda,
Hitachi Central Research Laboratory.)

cal gratings differing in line spacing or in the angular coincidence of
superposition.

A distinct advantage in imaging lattice features by the moire' technique is the enhanced fringe spacing. That is, for two crystals, each with
a line spacing d_1 and d_2, respectively, the moire' fringe spacing for parallel superposition (without rotation) is given by

$$D_p = \frac{d_1 d_2}{d_1 - d_2} \qquad (7.71)$$

If these crystals are rotated, the fringe spacing of the moire' pattern is
given by

$$D \simeq \frac{d_1 d_2}{\sqrt{(d_1 - d_2)^2 + d_2 d_1 \theta^2}} \qquad (7.72)$$

where θ is the (small) rotation angle. It is therefore possible to image crystal detail by superposition of moire' patterns even if the normal crystal line spacing is too small to resolve by phase contrast. Consequently, a moire pattern can be regarded as a magnified image of the reflecting planes, and patterns form if a beam diffracted by both crystals (a double diffraction spot) is allowed to recombine with the transmitted beam to form the initial image. Double diffraction is therefore an essential prerequisite for moire' pattern formation, and the aperture must be positioned to allow one or more double diffraction spots corresponding to the moire' pattern fringes to be generated in the image. The fringes run perpendicular to the operative vectors for the two overlapping crystals for parallel moire fringes and parallel for rotation moire' fringes. Thus parallel and rotational moire' patterns can be distinguished. Optical analogues demonstrating the formation of parallel and rotational moire' patterns are easily constructed from two overlapping line gratings or by lattice gratings as shown in Fig. 6.43.

It should be apparent that, as demonstrated originally by Bassett, et al. [88], if one lattice in thin, overlapping crystalline foils contains a dislocation, the dislocation can be observed by the image-plane magnification afforded by the formation of a moire' pattern. This feature along with the general nature of moire' fringe pattern appearance is demonstrated in Fig. 7.62 for overlapping thin crystals of $PrCo_5$ prepared by splat cooling. Bassett and co-workers [88] were the first to show an edge dislocation in a crystal section using the moire' fringe technique.

The dynamical theory of moire' pattern formation has been formulated by Hashimoto, et al [85]. In practice, however, the features implicit in Fig. 7.62 can be easily obtained and readily understood on the basis of the phenomenological concepts for two-beam fringe formation shown in Fig. 7.60, with $\underline{g}$, or $\underline{g}_2$ representing double diffraction spots in the selected-area electron diffraction (SAD) pattern.

7.5.2 *MULTI-BEAM IMAGING AND BLOCK STRUCTURES*

Although Fig. 7.60 illustrates the use of three beams in forming two distinct sets of lattice fringes, many studies of fringe images utilize only

FIG. 7.62 *Moire fringe patterns from overlapping crystals of PrCo₅ pre-pared by splat cooling. Note dislocation images particularly prominent within circled region (125 kV electron beam).*

two beams (000 + g) for small lattice periodicities. However multi-beam lattice fringe images produced by a number of reflections in a systematic row are more useful when the lattice periodicities are large. In general, the information available in fringe images is somewhat limited since only one set of lattice planes is visible. Multi-beam images utilizing all or many of the available diffracted beams should ideally provide structure images able to resolve the atom positions, but the spherical aberration effect limits the resolution to about 3.5 Å [89]. Further distortions of the image can also occur from beam divergence, astigmatism, etc. as dis-cussed in Chap. 3.

For materials with sufficiently large atomic or molecular spacings, the multi-beam lattice image can produce a projected charge density (PCD) picture as described originally by Cowley and Iijima [90]. Subsequently, theoretical treatments have been developed which, along with computer simulation techniques, have provided excellent agreement with the experimental observations [91,92].

Some of the first structures to be studied intensively by multi-beam lattice imaging were oxides, particularly niobium oxides. These systems, built up from a combination of corner and edge-sharing octahedra, produce monoclinic "block" structures with two long cell parameters (a_0 and c_0) and one short axis (b_0) of 3.8 Å. Projections along this axis reveal the $a_0 c_0$ superstructure, and the block structure images clearly show the accommodation of complex stoichiometry and non-stoichiometry through block size periodicities and "defect blocks". Figure 7.63 illustrates the perfect block structure periodicity (2 x 2 + 2 x 3 block rows) accommodating $Nb_{22}O_{54}$. In examining specimens as illustrated in Fig. 7.63, they must be uniformly thin (< 100 Å for 100 kV electrons) and oriented close to a zero-axis orientation [($h0\ell$) in the case of Fig. 7.63]. The objective lens defocus should be chosen to transfer the majority of the intersecting beams with similar amplitude and phase (the so-called Scherzer defocus). This is typically about - 900 Å from the Gaussian focus [93]. Figure 7.63 illustrates the potential for multi-beam structure imaging applications which have been made and continue to be made in the general area of inorganic crystal chemistry.

Lattice structure images can also be improved by higher voltage because of the greater resolution which can be attained, the differences in beam interaction, and diffraction producing many strong reflections (and a broader spectrum in the transfer function[†] or the main transfer interval) which can be utilized in forming the structure image. Figure 7.64 illustrates this feature in the examination of the crystal structure of tetragonal (a = 7.6 Å and c = 4.9 Å) $Si_3N_4 \cdot Y_2O_3$ growing into a glassy phase. The specimen in Fig. 7.64 was oriented with the electron bean nearly parallel to the c-axis [normal to (010)], and the image was obtained at an underfocus of about 700 Å. It should be pointed out in Fig. 7.64 that the heavy (Y) cations are resolved separately as dark spots while the bright spots correspond to the Si-O-N tetrahedron whose two edges are parallel to [001]. Furthermore, the boundary separating the amorphous or glassy phase from the crystalline $Si_3N_4 \cdot Y_2O_3$ is essentially parallel to [$1\bar{1}0$][94].

[†]*For a discussion of transfer functions and the role they play in image formation you should peruse the article by J. M. Cowley in* Introduction to Analytical Electron Microscopy, *edited by J. J. Hren, et al. Plenum Press, New York, Chap. 1, 1979, pp. 1 - 42.*

FIG. 7.63 *Underfocus, multi-beam lattice image showing octahedral "block" structure of $Nb_{22}O_{54}$ taken with 100 kV electrons (Courtesy F. Nagata, Hitachi Central Research Laboratory).*

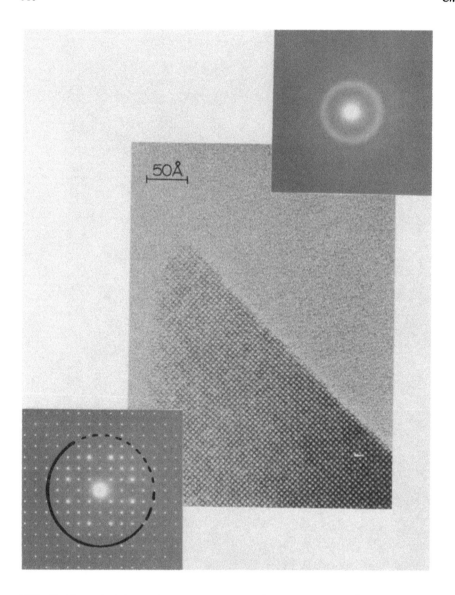

FIG. 7.64 Multi-beam (1 MV) structure image of $Si_3N_4 \cdot Y_2O_3$ growing in a glassy phase. The corresponding SAD patterns are shown
superposed, and the objective aperture size defining the beams
admitted for image formation is indicated. The dotted portion of
the aperture implies that a portion of the amorphous pattern diffraction intensity was also simultaneously admitted in forming the
.glassy phase structure image. Although the glassy phase structure
is too small to "resolve", there is evidence of short range image
order [94]. (Courtesy Dr. Shigeo Horiuchi.)

Mineral structures are particularly amenable to study by lattice imaging techniques because many mineral specimens can be cleaved or grown as uniformly thin platelets, and many have fairly large structural units. Growth phenomena and structural transformations can be studied in great detail using high-resolution transmission electron microscopy (HRTEM) as illustrated in part in Fig. 7.64, and as shown in more detail in Fig. 7.65.

Figure 7.65 in some respects summarizes the concepts of both two-beam lattice fringe imaging and multi-beam images discussed in this section. It also provides an excellent example of the applications of these techniques in detailed structure analysis work.

For metals, alloys, and other materials with small lattice structures, multi-beam images taken with a limited number of reflections (as illustrated in Fig. 7.65b) can be particularly useful as pointed out recently by Rez and Krivanek [95]. However the key to obtaining higher spatial resolution seems to hinge upon the defocus condition, and Hashimoto, et al [96] have described an aberration-free focus (AFF) condition which brings additional, higher-order reflections into the image plane in phase to achieve resolutions on the order of 1 Å. Calculations have also demonstrated that different elements possess different fine structure in such images, and Hashimoto, et al.[97] have demonstrated the feasibility of distinguishing a row of aluminum atoms in a thin gold crystal. Consequently, lattice (structure) imaging not only holds the prospect for providing information about local composition through fringe spacing-interplanar spacing-composition relationships, but also single particle resolution approaching the atom probe mass spectrometer described in Chap. 4.

In studying defects (including grain and interphase boundary structure) by structure imaging techniques, it is important to be sure the foils are very thin, and that interfaces in particular are parallel to the electron beam direction. Otherwise interference effects produce image artifacts as a result of the inclination. In general, the following conditions must be borne in mind when attempting to effectively apply the structure image techniques utilizing high-resolution transmission electron microscopy:[†]

[†]*See also Lars Kihlborg (ed.),* Direct Imaging of Atoms in Crystals and Molecules, *The Royal Swedish Academy of Sciences, Stockholm, 1980.*

Chapter 7

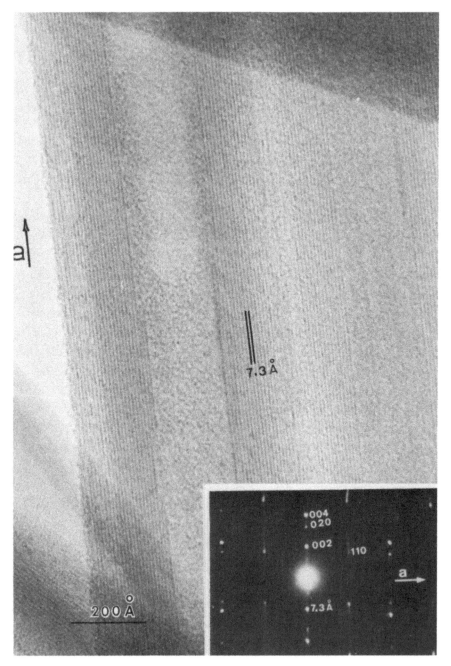

FIG. 7.65(a) *2-beam lattice fringe images corresponding to the (002) planes parallel to the fibril length in a typical clinochrysotile. The SAD pattern insert must be rotated into proper coincidence with reference to "a" (Courtesy Dr. Helena de Souza Santos, Universidade de São Paulo).*

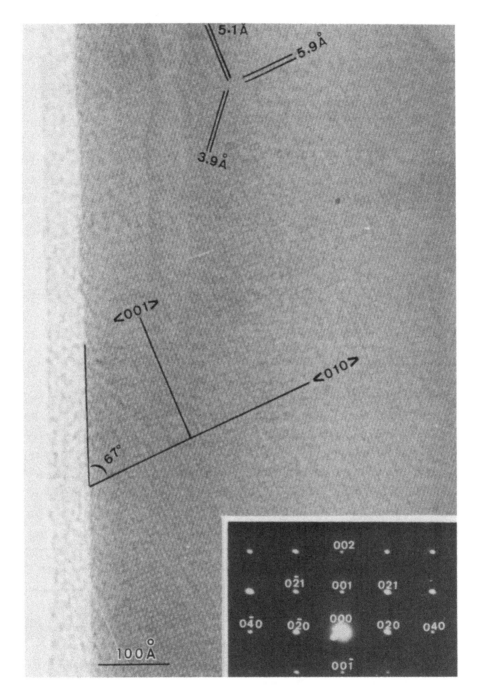

FIG. 7.65(b) *4-beam lattice image showing three fringe systems of
3.9 Å, 5.14 Å, and 5.9 Å corresponding to the (021), (020), and (001)
planes of forsterite after heating a clinochrysotile fibril to 800°C.
The 5.9 Å fringe system makes a 67° angle with the original fiber
length (a-axis of chrysotile). The SAD pattern insert corresponds to
the (001) forsterite orientation and must be rotated to correspond with
the [010] direction indicated. After Souza Santos and Yada [98].
(Courtesy of Dr. Helena de Souza Santos, Universidade de São Paulo.)*

1.) High resolution must be routinely obtainable with the instrument and there must be exceptional mechanical and electrical stability.

2.) For a 2-dimensional lattice image using 100 kV electrons, the crystal thickness cannot exceed 100 Å. For a 1-dimensional systematic image in which reflections along only one row are utilized, the thickness may be as large as 300 Å. Multi-beam imaging is also generally limited to this thickness range at 100 kV.

3.) Accurate defocusing of the objective lens by 500 to 1000 Å is required for optimum image contrast. This requires high magnifications to be used on the order of 200,000 X so that defocus changes and fringe details can be directly observed.

4.) For good structure images many beams must be used.

5.) Specimen tilting is important and must be utilized in attaining optimum structure image contrast and detail.

6.) There must be a direction in the structure along which equivalent units line up to give the projection geometry of the heavy atom positions.

7.6 SCANNING TRANSMISSION ELECTRON MICROSCOPY (STEM)

As one might expect, the major difference between the scanning transmission electron microscope (STEM) and the conventional transmission electron microscope (CTEM) described previously in this chapter and Chap. 3 is the beam scan. This is of course an oversimplification, but indeed in the STEM the image is formed by scanning a focused beam across a specimen in a raster fashion while detecting the transmitted, elastically scattered electrons with an annular detector located at the beam exit side of the specimen as shown schematically in Fig. 7.66. The electron current that strikes the annular detector is used to modulate the intensity of a synchronously scanned CRT or video display. In practice, the image is viewed by utilizing a scan converter and some type of storage device. In addition to the annular detector signal, signals resulting from no-loss and energy-loss electrons passed through the detector can be simultaneously recorded as implicit in Fig. 7.66. It is in fact possible to form an image by combining these three signals in a variety of ways. In addition, the corresponding signals can be manipulated, filtered, or otherwise processed by computer and other electronic techniques to provide specific signal manipulation.

A significant and distinguishing feature of the STEM is the formation of a very fine electron probe. As shown in Fig. 7.66, the electron probe

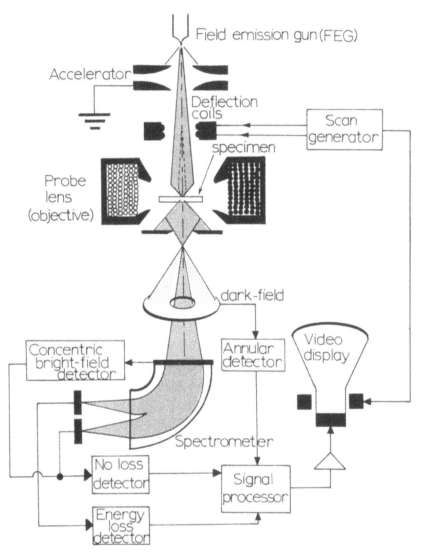

FIG. 7.66 *Schematic diagram of a scanning transmission electron microscope (STEM). Only the objective or probe lens is shown. In practice several additional lenses may be included in the beam forming processes before and after the specimen, respectively.*

in the STEM is formed by optically demagnifying the electron source which is either a LaB_6 thermionic-field emitter or a tungsten field-emission source. Some of the main features of very fine-focused electron probes have been discussed in our general treatment in Chap. 3 and in specific applications described in Chap. 4, and will not be further developed here.

However, probe sizes in the STEM have been reduced to 5 Å and below using
a field emission gun (FEG), and this capability provides an opportunity to
visualize single atoms [99].

7.6.1 *STEM IMAGING MODES*

The most common imaging mode in the STEM is the dark-field mode implicit
in Fig. 7.66 where the annular detector collects nearly all the electrons
scattered outside the central beam. This mode is much more efficient than
dark-field imaging in the CTEM because of the magnitude of the scattered
signal. However when the annular detector is used to collect all of the
scattered radiation, the information contained in the diffraction pattern
is lost.

Although it is not completely unambiguous from Fig. 7.66, the electron
beam incident upon the specimen has a convergence defined by the objective
aperture. Consequently, for every position of the incident beam a convergent beam electron diffraction (CBED) pattern is produced on the detector
plane as described in Sec. 6.8.6. Each diffraction spot in the CBED pattern forms a circular disc (Fig. 6.36), and the intensity distribution
within the disc will change with specimen thickness. When a small detector aperture is placed in the middle of the central spot of the CBED pattern, the imaging conditions will be identical to the usual bright-field
mode in the CTEM. The signal strength will of course increase with an
increased detector diameter, pass through a maximum, and then go to zero
as the detector diameter approaches the objective aperture diameter. In
forming the CBED pattern the objective aperture is withdrawn and even a
field-limiting aperture is unnecessary because the fine-focussed beam defines the diffracting volume. This microdiffraction feature is also a
unique aspect of the STEM in its specific applications especially in the
minerals and materials sciences. We will discuss microdiffraction in this
context in Sec. 7.7.

Reciprocity in STEM and CTEM In many respects the STEM is similar to the
CTEM while in others it is distinctly and even uniquely different. Indeed,
as Cowley [100] has pointed out "it is easy to yield to the temptation of
thinking about scanning transmission electron microscopy as an incoherent
imaging process since one can visualize in terms of geometric optics and
simple scattering theory how a narrow beam is scanned across the specimen
and some fraction of the transmitted or scattered intensity is detected to

form the image signal. This viewpoint can lead to serious errors". Cowley [101] has discussed the relationship between STEM and CTEM in terms of the reciprocity principle which was originally demonstrated by Helmholtz for the case of sound wave propagation.

Figure 7.67 illustrates the principle of reciprocity applied to CTEM and STEM imaging. The essence of this principle is that the electron wave amplitude at a point B due to a source at A is the same as that at A due to a point source at B. This does not imply the reversibility of the geometric-optics notion of ray paths implicit in Fig. 7.67. Indeed reciprocity holds only when transmission through the system involves scalar fields and elastic scattering processes. If magnetic field distortions or inelastic scattering becomes appreciable, reciprocity does not apply. In other words for rather idealized imaging conditions in relatively thin crystalline films ($\sim$ 0.1 µm at 100 kV), the diffraction contrast features outlined for the CTEM in the preceding portions of this chapter will likely be duplicated in STEM images. But for increasing specimen thickness reciprocity becomes increasingly inappropriate. If reciprocity is not applicable it is then necessary to calculate or evaluate STEM images from first principles. Joy, et al [102] have also pointed out that extinction fringes and related diffraction phenomena may not appear the same in a STEM image not because reciprocity fails, but because of the limiting effects of the signal-to-noise ratio. The signal/noise ratio is related to the exit angle in the STEM ($2\beta'$ in Fig. 7.67), which is also related to the image quality. A STEM detector angle $2\beta'$ is equivalent to an incident CTEM beam from an incoherent source with a range of angles of incidence corresponding to $\beta'|g|\xi_g$, which is related to the dimensionless deviation parameter, $(s\xi_g)$, in Eq. (7.39) for example. Thus, for specific diffration conditions the image will be influenced by β'. Maher and Joy [103] have shown that for $\beta'|g|\xi_g > 1$, the STEM image is often markedly different from the corresponding CTEM image owing to the averaging out of dynamical effects. This averaging is most severe for thickness and bend contours, less severe for stacking faults, and is sometimes negligible for dislocations. In order for STEM images to closely resemble CTEM diffraction contrast images it is necessary for $(\beta'|g|\xi_g)$ to be less than about 0.3. As an example, for 100 kV electrons incident on a thin copper film where $g = [200]$, this corresponds to $2\beta' > 3 \times 10^{-4}$ radians. This will yield a very low signal/noise ratio, and in most instances some compromise must be reached to optimize the imaging conditions.

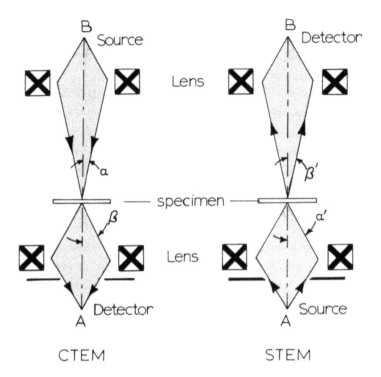

CTEM   STEM

FIG. 7.67 *Schematic illustration of reciprocity prin-*
ciple applied to CTEM and STEM. The STEM diagram is
obtained by applying time reversal to the CTEM system.
In the STEM the objective lens is the probe lens.

Lattice Imaging in the STEM As we noted in Sec. 7.5, at least two beams
must pass through the objective aperture and interact coherently to resolve
the lattice periodicity of a crystalline film in CTEM. In STEM, essentially the same is true: two or more convergent beam discs (Fig. 6.36) must
overlap and interact coherently. This is accomplished by allowing the
detector aperture to include the region of overlap. Coherent interaction
of the beams necessitates the use of a field-emission electron source.
Spence and Cowley [104] have recently discussed the theory of STEM lattice
imaging. In principle, it is possible to obtain instrument resolutions of
better than 2 Å, especially at high voltage, and a few scanning transmission electron microscopes have been designed to operate at 1 MV.

Crewe [105] has recently discussed the prospects for STEM applications, and you might peruse this reference for some additional, brief in-
sights. However it is obvious that the STEM in principle has some unique
analytical characteristics which will make it particularly useful in the

context of microanalysis. We will explore these concepts in the following description of analytical electron microscopy.

7.7 ANALYTICAL ELECTRON MICROSCOPY (AEM)

In Chap. 4 (Sec. 4.8) we described a concept referred to as integrated modular microanalysis systems. In this concept, numerous microanalysis systems are integrated in a design to promote an analytical synergism having unique advantages over any single, even dedicated instrument. The analytical electron microscope (AEM) is simply an extension of this concept. Beginning with the very small probe available in a basic STEM design, modifications are made to allow the system to function as both a STEM and a CTEM, with modules attached for rapid conversion for SEM, EDX spectrometry, and electron energy loss spectrometry (EELS). This concept is illustrated schematically in Fig. 7.68, and Fig. 7.69 shows a commercially available SEM system which incorporates this modular concept into an analytical facility.

As we noted in Sec. 4.8, when combining analytical techniques there is always the necessity to compromise between the requirements for an optimum analysis in a dedicated system and those in the integrated modular system. This is especially true when STEM and CTEM are to be combined with surface imaging and microanalysis (SEM and EDS). Generally a wide lens gap or large pole piece bore is one of the principle design considerations in order to provide for sufficient excitation in the various modes and to allow an x-ray detector to be positioned as close as possible to the specimen in order to achieve a high detector yield. This large lens bore is often achieved at the expense of some loss of resolution. In order to effectively combine CTEM and STEM, an objective lens capable of very high excitation is required. In the STEM excitation state an electron probe is focused onto the object plane while in the CTEM excitation state the object plane is actually imaged in the intermediate lens. With certain innovations in lens design, it is possible to reduce the electron optical column designs of contemporary instruments. Column designs similar to Fig. 7.69 can accommodate operating voltages of 200 kV.

Commercially available analytical electron microscope systems such as that shown in Fig. 7.69 employ four lenses in the imaging lens system below the specimen and up to three condenser lenses before the specimen.

The imaging lens system (objective, intermediate, and two projector lenses) can be optimized for STEM, SEM, EDS, energy analysis (EA) utilizing

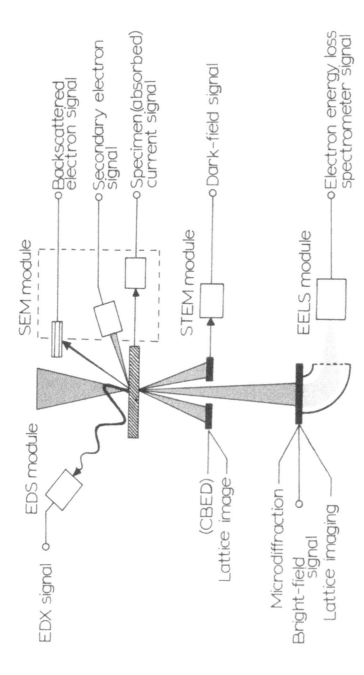

FIG. 7.68 *Schematic diagram depicting the AEM concept. Signal and image information commonly available is noted. Other source or detector modules could conceivably provide additional signal information. Note that operation in the SEM, STEM, or other beam rocking modes, etc. requires a system of deflection/scan coils above the specimen which is not shown.*

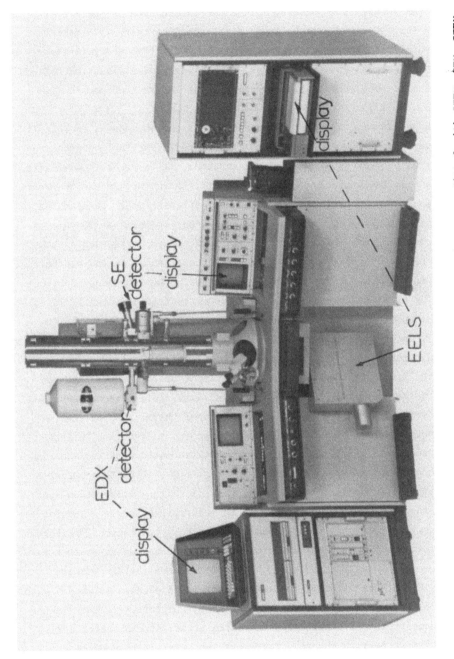

FIG. 7.69 Commercial AEM, Hitachi H-600 analytical electron microscope fitted with STEM, SEM, CTEM, EDX and EELS system modules. The secondary electron detector (SE) and EDX displays are noted. (Courtesy of Hitachi, Ltd., Tokyo, Japan).

EELS, microdiffraction, and beam rocking modes to produce CBED patterns
similar to the rocking-beam electron channeling patterns (ECP) in the SEM
as described in Chapters 5 and 6. Since the information in the STEM is
collected serially, the various scanning attachments to the TEM make the
integrated microanalysis schemes particularly attractive and relatively
simple to achieve in practice. In addition, image and diffraction pattern
signal data can be altered by energy filtration in the EELS module (See
Fig. 7.68).

Although we have briefly discussed some of the advantages of the STEM
in the previous section in connection with microdiffraction phenomena resulting from the very small beam focus attained, it is probably useful to
reiterate this in connection with the AEM. It is in fact in the AEM where
applications of CBED patterns can have a significant impact on microanalysis capabilities, and microdiffraction from areas as small as 100 Å on a
side is possible even in commercially available instruments such as that
shown in Fig. 7.69. There are several techniques available for microdiffraction in the AEM. These include the focused probe technique identical
to that for STEM imaging, the focused aperture technique (which is identical to the conditions for selected-area electron diffraction in the TEM
(or CTEM), and the rocking-beam method developed by Van Oostrum, et al
[106] and Geiss [107]. In the latter method, the deflection (or scanning)
coils (Fig. 7.68) are used to pivot the incident beam over the specimen
surface in such a manner that the bright and dark-field images appear sequentially over the STEM detector at the base of the microscope (concentric detector in Fig. 7.68). The detector records an increase in intensity as each dark-field image is formed. With repeated pivoting, the increase in electrons striking the detector result in a momentary increase
in the signal sent to the STEM video display, forming a diffraction spot
whenever the incident beam is tilted at the appropriate angle. The resulting diffraction pattern appears similar to the more conventional selected-
area electron diffraction patterns formed without beam rocking and at reduced probe focus in the CTEM mode of operation. For the SAD pattern
formed by CTEM the portion of the image forming the pattern is chosen with
the size of the diffraction (or field-limiting) aperture. In the rocking
beam method, it is the diameter of the solid-state detector which determines the image portion chosen. The energy analyzer (EELS module) can
also provide energy filtered patterns and other signal-modified information.

In most commercially available AEM systems, the diameter of the probe is varied by varying the first condenser lens strength, while the second condenser lens is used in conjunction with the condenser aperture to control the beam convergence. Although the consequence is at least an order of magnitude greater than the TEM mode, the spot sizes achieved can be as small as 5 Å with a field-emission gun. However, as a result of this small convergent probe, the diffraction spots tend to be discs, as discussed previously. By measuring the disc diameter and the distance between any spot and the center spot (R), the convergence angle, α' can be measured from

$$\alpha' = \theta\left(\frac{\text{disc diameter}}{R}\right) \tag{7.73}$$

where θ is the Bragg angle for the particular spot characterized by R. You might refer back to Figs. 6.37 and 6.38 for some of the details of the CBED patterns.

As we noted previously, lattice imaging can be accomplished both in the CTEM and STEM modes. This can provide unprecedented microanalytical capabilities when very minor variations in chemistry or stoichiometry occur which might be detectable in the CBED pattern information, and requires confirmation at the sub-microscopic level. Of course good lattice images require rather thin films as noted previously, and this requirement can place severe limitations on the utility of this mode of microanalysis.

Less stringent microanalysis requirements are placed on the energy-dispersive x-ray microanalysis capability in the AEM, and this is also generally true for the electron energy analysis (EELS) mode of analysis as well. By comparing the EDX and EELS signals it is possible to corroborate specific elemental microanalyses and accurately identify chemically unique zones (such as precipitates) as small as 500 Å on a side or smaller. In addition, EELS analysis can compliment EDX analysis in the range of elements below oxygen (Z = 1 to Z = 8) Figure 7.70 illustrates some of the unique applications of EDS and EELS modes of microanalysis in the AEM in the analysis of very small precipitates. Figure 7.70(a) also illustrates the use of extraction replication in the analysis of precipitates and inclusions as discussed in Chap. 5.

You should appreciate the concept and applications of the AEM implicit in the illustrations provided in Fig. 7.70. Indeed, Fig. 7.70 also illustrates the experimental synergism involving techniques of specimen preparation, imaging, and microanalysis which is becoming ever more crucial

(a)

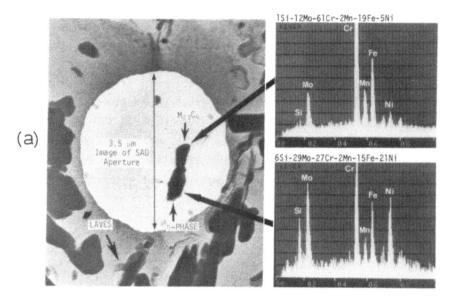

(b)

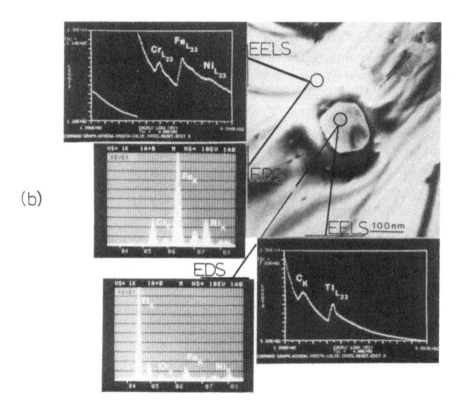

in providing unambiguous solutions to the increasingly complex problems modern materials technology faces, particularly at the microscale.

It seems unnecessary to pursue the details of AEM manipulations further because contemporary instrument designs are rather logical and simplified to the point where even a casual user familiar with electron optical systems can perform the diverse functions through mode switching and related operations. One of the unique features of many commercial AEM systems in particular is the electronic lens control coupled with microprocessor memory functions which automatically adjust to pre-set conditions or make automatic variations based upon simple push-button commands.[†]

7.8 IMAGES OF FERROELECTRIC AND FERROMAGNETIC DOMAINS

In materials that are ferromagnetic (or antiferromagnetic or ferroelectric or antiferroelectric) the occurrence of local magnetic or electric fields leads to refraction of the incident electrons. These materials contain domains, characterized, respectively, as ferromagnetic or ferroelectric domains that are in effect separated by an interface of domain boundary. The domain structure, like the grain structure, can then be observed in the corresponding deflected electron image by diffraction contrast or Lorentz contrast.

7.8.1 *FERROELECTRIC DOMAINS*

The most common of the ferroelectric materials possess the perovskite structure. Of these, barium titanate is perhaps the best known, and will serve as an example of domain boundary images since a considerable number of investigators have specifically observed domain structures in it. Ferroelectric domain boundaries in barium titanate originate by diffraction

---

[†]*You should consult* Introduction to Analytical Electron Microscopy, *J. J. Hren, et al (eds.), Plenum Press, New York, 1979 for additional details of AEM and some descriptions of operating procedures for many related techniques.*

◀ FIG. 7.70 *EDS and EELS analysis in the AEM. (a) Carbon extraction replica of 20% cold-worked 316 stainless steel aged 10,000 h at 600°C (showing two morphologically similar but chemically distinct precipitates using EDS (Courtesy Dr. P. J. Maziasz); (b) EELS and EDS spectra from TiC in a 316 stainless steel matrix. The carbon is only observed in the EELS spectrum. The image is a CTEM bright-field image of the precipitate following aging of the stainless steel for 24 h at 900°C. (Courtesy of N. J. Zaluzec and P. J. Maziasz).*

contrast in the same way that grain boundaries, etc., are imaged. A fringe
pattern results that is similar to those shown in Fig. 7.32. Ferroelectric
domain-boundary fringes are distinguishable from regular crystal boundaries
by the fact that the extreme fringes in the bright-field image differ, and
are the same in the dark-field image. Recall that normally the images of
crystal interfaces have extreme fringes that are just the opposite in
bright and dark field (see Table 7.1) [30].

In the ABO_3 perovskite structure, ferroelectric polarization occurs
by a displacement of the B atoms from the centers of the oxygen tetrahedra.
In $BaTiO_3$, the Ti ions are displaced along a cube direction that becomes
the c axis of the tetragonal form. The domain structure originates by the
simultaneous polarization in many parts of a crystal as it is cooled
through the transition temperature; each part having a characteristic polarization direction as shown in Fig. 7.71.

Two types of domain boundaries are observed to occur as a consequence
of the domain c axes on either side of the boundary being mutually perpendicular or antiparallel as shown in Fig. 7.71. For the 90° boundaries indicated in Fig. 7.71a, you should note the schematic similarity with Fig.
6.39. The lattice of Fig. 7.71a is in fact twinned. However, the twinning
vector is very small. Nonetheless, the diffraction vectors on either side
of the boundary are different, and the diffraction conditions as discussed

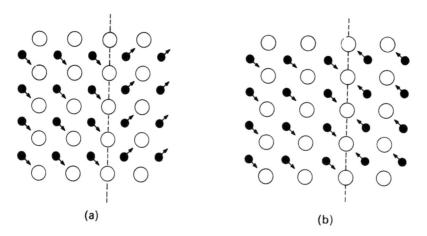

(a)   (b)

FIG. 7.71 *Ferroelectric domain-boundary formation in barium tita-*
nate. (a) Displacements of Ti atoms are mutually perpendicular;
(b) displacements of Ti atoms are antiparallel. Only a section
view of the oxygen tetradedra is shown by large circles with Ti
atoms indicated by small solid circles.

previously in Sec. 7.3.2 apply here if the domain boundary in a thin section is inclined to the surfaces. For the 180° boundaries depicted in Fig. 7.71b, the diffraction conditions approximate those for a grain boundary as discussed in Sec. 7.3.2. Wedge fringe images occur when the boundary is inclined to the electron beam [108].

7.8.2 FERROMAGNETIC DOMAINS: LORENTZ ELECTRON MICROSCOPY

The force exerted on a moving electron by an electric field is simply eE, where E is the field strength. The direction of the force is along the direction of the field. In a magnetic field, a similar force is exerted on the electron, given by

$$\underline{F} = -\frac{e}{c}\,\underline{v}\times\underline{H}$$

where v describes the electron velocity and H the magnetic field strength. The direction of this force, called the Lorentz force, is at right angles to the moving electrons and to the field direction. When an electron beam passes through a ferromagnetic thin film, the magnetic fields in the individual domains will exert a force on the electrons. Since the direction of magnetization is different on opposite sides of a domain boundary (or wall), the sense of deflection will differ, and the domain will be imaged because of the convergence or divergence of electrons in this area causing enhanced or retarded brightness with respect to the background intensity. Contrast [109] arises when the focal plane of the thin section is located in front of or behind the specimen for walls causing convergence or divergence of the electrons, because an excess or deficiency of electrons will result in the image. It is therefore necessary to either under or overfocus the image in order to observe ferromagnetic domain walls. A pattern of bright and dark lines bounding the domain walls will then occur as shown in Fig. 7.72. Domains in ferromagnetic materials are characterized by 90 or 180° walls originating by polarizations analogous to those shown in Fig. 7.71 for ferroelectric materials. You should be aware of the fact that in order to perform magnetic domain observations the electron microscope must be fitted with a special objective pole piece, and a special Lorentz attachment to the electron microscope capable of applying external in-plane magnetic fields to the specimen [110]. You might also peruse the use of an electrostatic microscope for Lorentz microscopy as recently outlined by H. Karamon, G. F. Rempfer, J. Dash, and M. Takeo [*J. Appl. Phys.*, 5: 2771 and 2774(1982)].

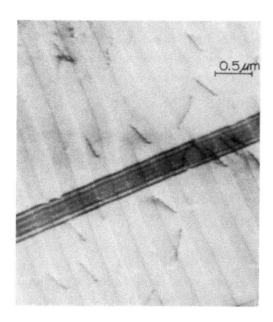

FIG. 7.72 *Domain walls crossing stack-*
ing fault in thin cobalt crystal. (100
kV.) [Courtesy J. Silcox, Phil. Mag.,
8: 7(1963)].

J. P. Jakubovics [111] and M. Wilkens [112] have shown that ferromagnetic domains can also be imaged by diffraction contrast arising from
the deflected electrons. In effect the scattered and transmitted electron
intensities act in a manner similar to diffraction from a deformed crystal
in dark-field illumination. Diffraction contrast is not apparent in the
bright-field images, and the domain character is only imaged by Lorentz
contrast as discussed above.

    Since magnetic properties of materials can be determined by the reac-
tion of domain configurations to some external magnetic field, studies of
thin ferromagnetic films can be carried out directly with the electron
microscope. The ability of the walls to be impeded by defects and/or a reaction with defects can conceivably be studied and related to other materials properties or to some deformation variable. Some work has been done
in attempts to establish a correlation of thin-film magnetic properties and
structure with limited success. E. M. Hale and co-workers [113] and H. W.
Fuller and Hale [114,115] have also treated the intensity distributions in
domain walls in considerable detail. Fidler and Shalicky [116] have also

recently studied the changes in the magnetic domain structure during the
annealing of amorphous $Fe_{40}Ni_{40}B_{20}$ where cross-tie walls are observed.

7.9 TRANSMISSION ION MICROSCOPY

Since the theme of this book includes ion microscopy, this chapter would
not be complete without at least mentioning, briefly, the concept of transmission ion microscopy, and more specifically the scanning transmission ion
microscope. As we noted initially in Chap. 1 in our brief discussion of
ion properties, ions, even small ions such as hydrogen, possess intrinsically unique advantages over electrons with respect to penetrating power
and ideal resolution as a result of the enormous mass difference between
even a hydrogen ion (or proton) and an electron; resulting in the ion's
much shorter de Broglie wavelength. Although there are undesirable effects
as a result of the interaction of fast ions with the specimen, such interactions can also provide unique opportunities to probe matter not available
through electron transmission. These include charge exchange and molecular
ion dissociation processes which can provide contrast mechanisms not available in the TEM or STEM.

Levi-Setti [117] has discussed both the feasibility and the promise
of the proton scanning microscope and Escovitz, et al [118] have also described a scanning transmission ion microscope (STIM) with a field-ion
source designed to eventually evolve into a high-resolution STIM. Figure
7.73 illustrates the STIM design utilized by Escovitz, et al [118] at 65
kV accelerating potential. In this design, a field-ion source provided
hydrogen ions with a specific brightness near 10^3 protons $s^{-1}Å^{-2}sr^{-1}V^{-1}$.
However the prototype instrument described by Escovitz et al [118] had a
limited resolution of between 0.1 and 0.2 µm, and the image quality, even
for biological specimens, was extremely poor. More recent work by Levi-
Setti [119] has shown some improvement in resolution (<0.05 µm), and some
interesting ion-solid interaction phenomena utilizing a liquid metal (Ga^+)
ion (LMI) source [120] and a magnetic-sector spectrometer for ion microanalysis of primary or secondary ions. This analytical feature involves
incorporating a magnetic-sector spectrometer at the detector plane of Fig.
7.73 in a configuration similar to the EELS module in the STEM or AEM
(Fig. 7.66 and 7.68).

In a conceptual sense, an analytical ion microscope (AIM) might also
be constructed having unique capabilities similar to the AEM. Indeed it
is possible to provide a scanning ion microscope module and SE detector

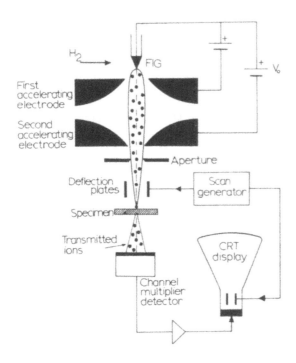

FIG. 7.73 *Schematic diagram of a scanning transmission ion microscope utilizing a field-ion gun (FIG) and a two-electrode Butler lens.*

for surface image formation in the STIM as well as a SIMS module. Other module functions could also be added to a basic STIM design as shown in Fig. 7.73 providing an integrated modular microanalysis system as described conceptually in Sec. 4.8. Indeed, this concept may eventually lead to a combined AEM/AIM incorporating both electron and ion source modules, electrostatic and magnetic-sector spectrometers, and related microanalysis modules in a single optical column employing lower accelerating voltages and utilizing electrostatic lenses, or a dual column system. Even more complex imaging synergism could be envisioned employing electron-acoustic microscopy modules and the like [121]. Such systems may have important applications in the analysis of laminated materials systems such as VLSI circuit arrays and other thin film arrays and composites of various materials such as metal-ceramic, ceramic-semiconductor, or metal-ceramic-semiconductor film overlays.

7.10 APPLICATIONS OF TRANSMISSION ELECTRON MICROSCOPY AND RELATED TECH-
NIQUES: STRUCTURE - PROPERTY RELATIONSHIPS

For the reader who is already a practicing research scientist, it will be
completely unnecessary to point out that one of the most fruitful applications of transmission electron microscopy and related techniques is in the
characterization of materials defects, and the correlation of the defect
nature with the material's residual properties. For a student having pursued this chapter unfailingly, this fact will perhaps be equally apparent.
It will therefore be our aim in the present section simply to present several additional convincing examples of the powerful approach to materials
characterization provided by transmission (thin-film) electron microscopy.
These will be supplemented appropriately with a few references to the current research literature.

Numerous examples can be cited to illustrate that the electrical properties of vapor-deposited solid-state devices vary appreciably depending on
the evaporation parameters, for example, substrate temperature, pressure,
rate of evaporation, etc. (see Appendix B for the details of vapor deposition). It is often both possible and necessary to inspect individual
thin-film devices in order to determine whether a recognizable structural
characteristic is responsible for the observed behavior.

As it turns out, the structure of thin films prepared by vapor deposition depends significantly on the degree of vacuum as well as the substrate
temperature, the crystallinity of the substrate, the rate of evaporation,
and other features. Figure 7.74 illustrates the typical structure of vapor
deposited Ni foils having similar thickness and identical substrate temperature, but evaporated at successively decreasing pressures. The transmission images and selected-area electron diffraction patterns clearly indicate the grain size to increase with decreasing pressure. Figure 7.75 reproduces similar data for Al for various substrate temperatures. Resistivity data for the specimens of Fig. 7.75 were observed to have a similar
but inverted shape with respect to the curves of Fig. 7.75 and this response is shown in Fig. 7.76. The implications of the data of Figs. 7.75
and 7.76 are readily amenable from the direct observations of thin-film
structure. The electrical properties are observably deteriorated by the
introduction of lattice defects.

Some of the most significant advances in the fabrication of new materials, particularly high-strength metals and alloys, has come about because of the information gleaned from thin-film observations in the elec-

FIG. 7.74 *Electron diffraction patterns and transmission im-*
ages of vapor-deposited Ni foils. Foils were all approximate-
ly 1000 Å thick and evaporated onto NaCl (100) substrates at
$300°C$ at a rate of about 1000 Å/sec. (a) 2×10^{-2} torr; (b)
2×10^{-5} torr; (c) 2×10^{-9} torr. [From L. E. Murr and M. C.
Inman, Phil. Mag., 14: 135(1966).]

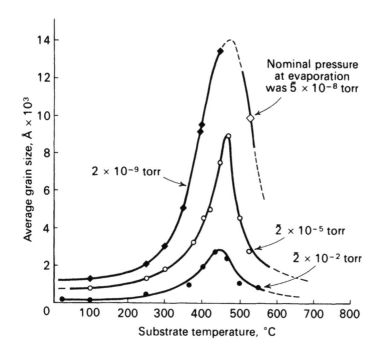

FIG. 7.75 *Variation of grain size of pure vapor-deposited Al on NaCl (100) substrates as function of pressure and substrate temperature. Evaporation rate was always about 1000 Å/sec, and thin thickness ranged from 1000 to 2000 A.* [*From L. E. Murr and M. C. Inman,* Phil. Mag., 14: *135(1966).*]

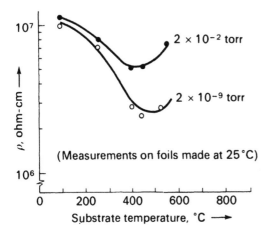

FIG. 7.76 *Resistivity data corresponding to vapor-deposited Al films in Fig. 7.75.*

tron microscope. The ability of dislocations to be pinned or their motion impeded by the presence of precipitates or similar dispersoids can be veiwed directly in the electron microscope as shown in Fig. 7.77. From such direct information, it is possible to devise new materials with special properties. We can cite, for example, the fiber-reinforced materials and the dispersion-hardened metals and alloys of which Fig. 7.77b is a typical example. Microstructural changes occurring in response to simple uniaxial tensile straining of twin sheets of type 304 stainless steel is illustrated in Fig. 7.78, and more complicated stress-strain and stress-state and/or strain-rate phenomena can also be studied in a very direct way utilizing the electron microscope [46]. High-strain phenomena, which is becoming increasingly important in metal forming operations is also beginning to be understood through detailed studies involving TEM. Indeed the evolution of shear bands at very high strains in many materials, particularly low stacking-fault free-energy alloys, is only amenable to detailed understanding by TEM and AEM utilizing microdiffraction.

Selected-area electron diffraction, and where available microdiffraction, will become increasingly important in defining microstructural and microchemical changes in modern materials developments. This approach, coupled with EDX and EELS analysis in the AEM will provide insight into increasingly smaller features in microcircuit fabrication, laser microwelding applications, and other emerging developments of international high-technology. Figure 7.79 illustrates the use of stationary microdiffraction to examine lath structure in Zr-15% Nb. These laths alternate between β and $\beta + \omega$ phase regions with an occasional hcp region being detected, and EDX analysis in the AEM has indicated no segregation between the phase regions. Although the regions examined in Fig. 7.79 are of the order of <500 Å on a side, the use of a field-emission gun in the AEM could theoretically allow microdiffraction to be recorded from areas as small as 15 Å on a side from 1000 Å thick films. This could allow for analytical capabilities approaching the smallest conceivable microcircuit components in VLSI systems and the like [122].

We could continue to cite examples of the direct observation of materials structures and the correlation with residual electrical, physical, and mechanical properties. However, you should by now have gained at least an appreciation for the advantages in transmission electron microscopy. A glance at the current research literature, particularly the metallurigcal, ceramics, and materials science journals, will readily amplify

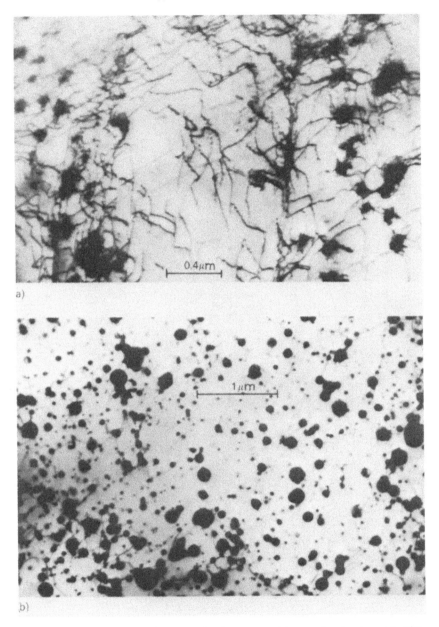

FIG. 7.77 *Strengthening matrix by generation and pinning of dislocations at precipitates and dispersed particles. (a) Dislocations
associated with fine carbide precipitates in a 76 Ni-16 Cr-Fe alloy
following explosive-shock deformation at 200 kbars. Note dense
tangles and loops surrounding precipitates. (125 kV). (b) Dislocations pinned by ThO_2 particles disbursed in Ni matrix (100 kV).*

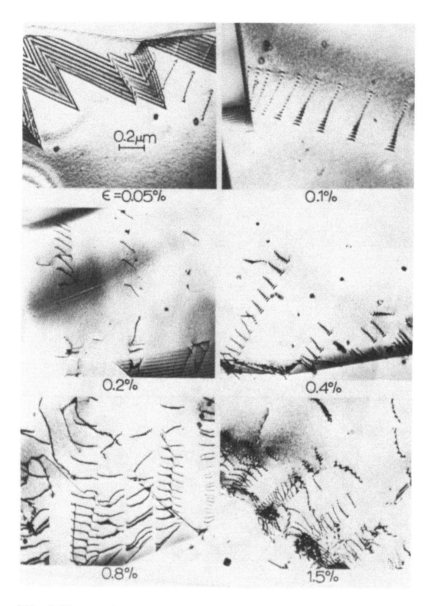

FIG. 7.78 *TEM images of equivalent low-strain regime in the stress-strain diagram for type 304 stainless steel. Dislocation emission from the grain boundaries is clearly observed to play a prominent if not controlling role in low-strain deformation. [After L. E. Murr and S-H. Wang, Res Mechanica, 4: 237(1982).]*

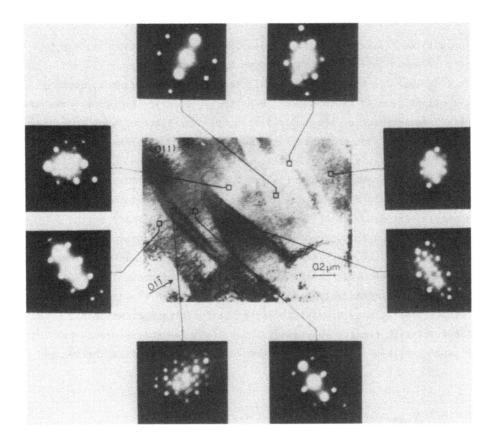

FIG. 7.79 *Analysis of lath structure in Zr - 15% Nb using stationary microdiffraction technique in the AEM (courtesy of Dr. N. J. Zaluzec).*

these impressions. The limitations of transmission electron microscopy in the study of thin-film properties are governed chiefly by the inexperience of the scientist or his or her lack of imagination. You need to page through this chapter a few times, in retrospect reflect upon its enormous range of implications and applications, peruse the suggested supplementary reading list and journal literature in your library, and spend a little time reflecting upon and mulling over the details; and re-reading or clarifying some of the vague or confusing points. The TEM can truly provide a unique analytical experience and often affords an opportunity to probe where no one has probed before, and to see in many instances what no one has ever seen.

7.10.1 TRANSMISSION ELECTRON HOLOGRAPHY

We have already discussed the principal features of electron holography
in our treatment of reflection electron holography in Chap. 5 (5.4.1).
Here we alluded to the fact that while conventional electron microscopy
deals primarily with electron wave amplitudes, electron holography enables
explicit measurement of the phase distribution of electron wave functions.
Transmission electron hologram formation is illustrated schematically in
Fig. 7.80, which is essentially the same arrangement shown in Fig. 5.22
(b). In the arrangement shown in Fig. 7.80, a coherent electron wave is
emitted from a cold field-emission (FE) electron gun [Fig. 5.22(a)], and
transmitted through a thin film specimen material. An image of the
specimen is formed through the objective lens as in standard or conventional electron microscopy except that the electron beam is overfocused to
obtain a collimated beam having a divergent angle of about 10^{-7} rad, and
the specimen is located in half of the specimen plane to allow for a
reference beam to pass through the other half of the specimen plane. The
Möllenstadt-type electron biprism is biased to make the reference beam and
the image overlap to form an interference pattern which is magnified and

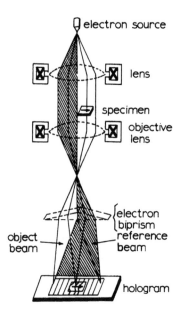

FIG. 7.80 *Schematic representation of hologram formation by interfering
electron beams in transmission mode.*

recorded on film as an electron hologram. This hologram is then reconstructed optically as shown in Fig. 5.23(b).

Electron holography was first described by Gabor [124] and more recent
applications have been summarized by Tonomura [123]. The coherent electron
beam required for successful electron holography also provides unprecedented lattice and atomic imaging by reducing the chromatic aberration and
providing increased beam brightness. Spherical aberrations can also be
corrected by optical reconstructions [Fig. 5.23(b)]. The possibilities
for observing precipitates and microstructural details in thin films by
electron holography have not been extensively explored but open a potentially exciting era for electron microscopy. Electron holography combined
with stereological observations by tilting thin specimens to create
electron stereograms or stereographic electron interferograms pose unprecedented observational opportunities in thin film materials.

An electron beam passed through electromagnetic potentials contains
information on those potentials that can be expressed in a phase-distrib-
ution image observed by interference microscopy. Domain structures in thin

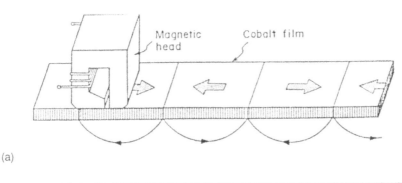

(a)

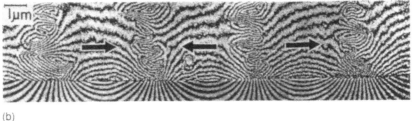

(b)

FIG. 7.81 *Interference micrograph of magnetically-recorded cobalt thin
film. (a) Schematic view of recording. (b) Contour map showing magnetic
streams of magnetization which collide to form a vortex structure (Courtesy
of A. Tonomura [124]).*

films can be observed in this manner as illustrated in Fig. 7.81 which shows a thin cobalt film on which magnetic recording was carried with a moving magnetic head illustrated schematically. These observations show detailed magnetic field distributions. You might compare these images with those for domain walls in cobalt shown in Fig. 7.72.

7.10.2 ENERGY - FILTERED IMAGING

In conventional transmission electron microscopy at accelerating voltages of 100kV, inelastically scattered electrons cause image blurring due to chromatic aberration of the objective lens when the fraction of the inelastically scattered electrons becomes comparable with the zero-loss electrons. This is particularly prominent in Fig. 7.25. This is also true for electron diffraction patterns as well. As we will illustrate in Chap. 8, increasing the voltage significantly will decrease the chromatic

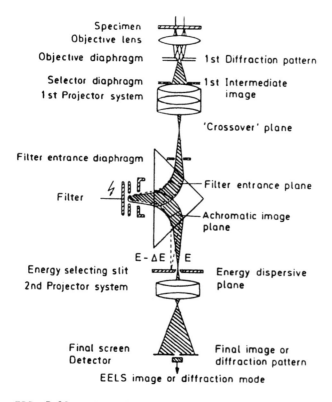

FIG. 7.82 *Schematic diagram for an energy-filtering electron microscope. (After reference 127).*

aberration and improve image clarity. However at lower accelerating
voltages image clarity can be improved by energy filtering to separate
inelastically and elastically scattered electrons, and also to filter
electrons within an energy-loss window. These processes are often referred to as energy-filtering electron microscopy (EFEM). An imaging energy
filter originally developed by Castaing and Henry [125] and attached to a
TEM column by Ottensmeyer [126] is illustrated schematically in Fig. 7.82.
In this arrangement, inelastically scattered electrons in the image
produced at the filter entrance are removed so that only the unscattered
electrons of the primary beam and very narrowly scattered electrons contribute to the final image. This zero-loss filtering is illustrated in
the images of microtwins in vapor-deposited silver films shown reproduced
in Fig. 7.83.

For further details and examples of energy-loss imaging and related
electron spectroscopic imaging and diffraction, the reader is referred to
Reimer, et al. [127].

7.10.3 PROGRESS AND APPLICATIONS IN LATTICE AND ATOMIC IMAGING

While we have demonstrated numerous examples of lattice and atomic images
in Section 7.5.2, more recent developments warrant additional comment, and
will serve to illustrate progress being made in applications of lattice
and atomic imaging, particularly in the materials sciences.

Figure 7.84 provides an excellent example of the progress being made
in high-resolution atomic imaging and its utility in the clarification of
atomic structure issues. The structure of grain boundaries has been an
area of interest in metallurgy and the materials sciences for decades [23],
and attempts to simulate or model the atomic nature of grain boundaries has
more recently been complimented by high-resolution imaging in the TEM.
Figure 7.84(a) shows a routine lattice image of a grain boundary in (111)
gold observed at 200kV while Fig. 7.84(b) shows the details of the atomic
nature of the Gibbs dividing surface [23] and atomic arrangements which
accommodate the misorientation in aluminum observed at 800kV. While the
crystal misorientation is certainly evident in Fig. 7.84(a), the atomic
nature of the interface (or Gibbs dividing surface) is not.

While Fig. 7.84 provides a vivid comparison of lattice and atomic
imaging, and the utility of atomic resolution in understanding the details
of atomic structural issues, recent discoveries of ceramic superconductors

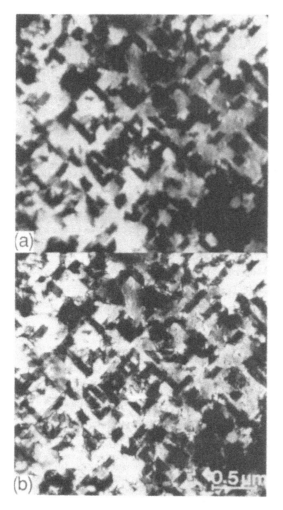

FIG. 7.83 *Microtwins in a vapor-deposited silver film imaged in (a) the unfiltered and (b) the zero-loss filtered mode by conventional transmission (bright-field) electron microscopy. [Courtesy of L. Reimer and* EMSA Bulletin, *20(1): 75 (1990)].*

provide even more compelling examples of high-resolution atomic imaging in providing structural information relating to materials technology issues.

Figure 7.85 shows a sequence of TEM images for superconducting $YBa_2Cu_3O_7$. Discovered in early 1987 [128] and based on the pioneering work of J. G. Bednorz and K. A. Müller [130] (for which they received the Nobel Prize in Physics in 1987), Y-Ba-Cu-O superconductors have attained superconducting transition temperatures routinely around 90K. Early

FIG. 7.84 *Comparison of lattice (a) and atomic (b) structure images of grain boundaries. (a) 65° misorientation of crystal grain lattices in gold observed at 200kV. (b) 21° misorientation of atomic crystal grains in aluminum observed in the atomic resolution microscope (ARM) at the University of California, Berkeley at 800kV. (Courtesy of R. Gronsky).*

research suggested a prominent role of the coherent permutation interfaces illustrated in the diffraction contrast image shown in Fig. 7.85(a) and the atomic image of Fig. 7.85(b). But aside from flux pinning, these interfaces do not seem to play a significant role in the mechanism of superconductivity in this ceramic system. On the other hand, crystal

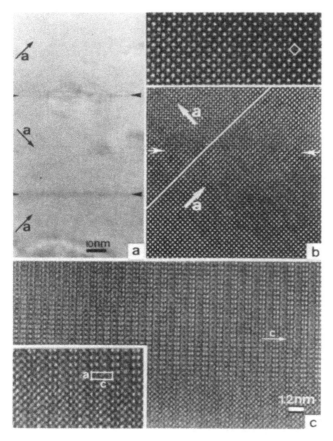

FIG. 7.85 *High-resolution images of YBa$_2$Cu$_3$O$_7$. (a) [001] zone showing coherent twin boundaries along <110>. The unit cell a-lattice directions are shown. (b) High magnification of (a). (c) [100] zone showing long period (c-axis) (see insert). From reference [129]; Courtesy of G. Van Tendaloo.*

defects such as those illustrated in Fig. 7.86 can trap flux and alter the superconductivity as it is reflected in the resistance-temperature signatures shown in Fig. 7.87 [131,132].

Figure 7.86, when compared with Fig. 7.85 [in particular Fig. 7.85 (c)] provides a range of examples of fundamental principles involving electron microscopy and diffraction. For example, the lattice (and atomic) periodicity illustrated in Fig. 7.85(c) is clarified to some extent by comparison with the lattice images and unit cell schematic shown in Fig. 7.86. Furthermore, the diffraction conditions illustrated in the selected-area electron diffraction pattern insert in Fig. 7.86 give rise to

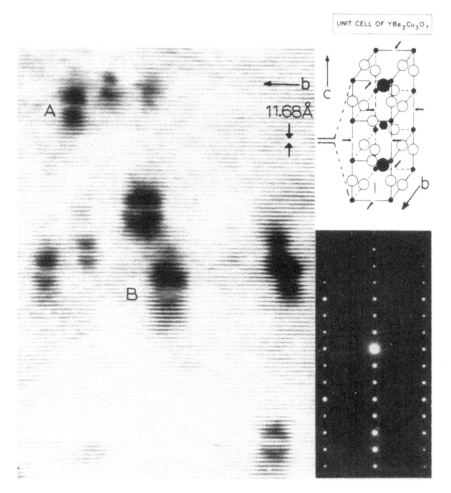

FIG. 7.86 *[100] zone (viewed along the a-axis) of YBa₂Cu₃O₇ after shock loading at 7GPa peak pressure. The photograph was printed with reduced contrast to show fringe splitting in defect zones marked A and B. The black regions are strain field (diffraction) contrast regions perpendicular to the diffraction vectors illustrated by the selected-area diffraction pattern insert. In the unit cell insert the small solid circles are copper, the open circles are oxygen. The arrows illustrate oxygen vacancies. (From reference 131).*

strong strain contrast images associated with the precipitate-like defects, and these strain-induced diffraction contrast images are superimposed upon the lattice fringe image. This image superposition provides direct information about the nature of the defects which would not be apparent from the lattice fringe image alone. While Fig. 7.85(c) provides some

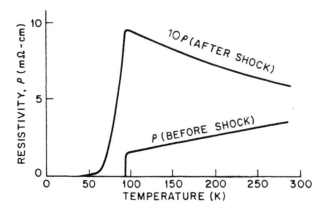

FIG. 7.87 *Examples of resistance-temperature (R-T) signatures for sintered YBa$_2$Cu$_3$O$_7$ before and after shock loading at 6 GPa. During explosive fabrication or shock deformation the shock wave produces defects (as in Fig. 7.86) which can be altered and eliminated with annealing at elevated temperatures in oxygen (After reference [132]).*

additional clarification of the lattice fringe images in Fig. 7.86, the resolution of specific atomic structure is not unambiguous, and requires even better resolution combined with image simulation to provide detailed structure information.

Figures 7.88 and 7.89 are an attempt to illustrate these features for a Tl$_2$Ba$_2$Ca^{1}Cu^{2}O$_8$ (2212) ceramic superconductor first described by Sheng and Hermann in late 1988 [133], and maintaining the highest consistent super-conducting start [T$_c$ (start)] temperature (of 125K) into the end of 1990. This compares with a T$_c$ (start) of slightly above 90K shown in the fully annealed Y-Ba-Cu-O system in Fig. 7.87; and the Bi-Sr-Ca-Cu-O and Bi-Pb-Sr-Ca-Cu-O systems (described by Maeda, et al. [134] and Murr and Niou [135] respectively) where by comparison the T$_c$ (start) is nominally between 100 and 115K. The ideal unit cell shown in Fig. 7.89(b) is based on X-ray diffraction analysis reported initially by Subramanian, et al. [136,137]. In the calculated image shown in Fig. 7.89(d), the Tl, Ba, Ca, and Cu atoms are represented as dark spots with different contrasts depending upon their atomic numbers; where oxygen atoms located between the heavy atoms are not imaged with the simulated image resolution. The simulated images are also shown as a function of crystal thickness (t = 3.855 Å), and the contrast and brightness is observed to become stronger with increasing thickness because of dynamical scattering. Oxygen vacancies (O$_v$) on the two Ca layers are represented by brighter spots in

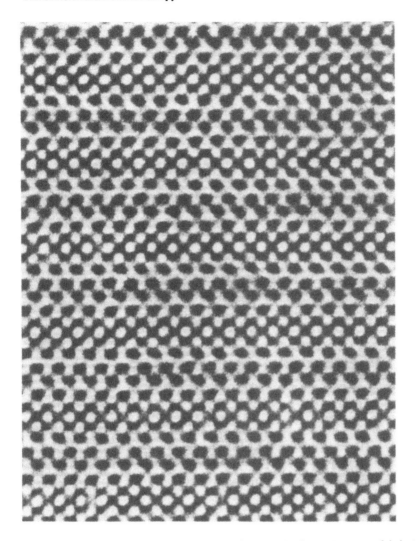

FIG. 7.88 *High-resolution atomic image of $Tl_2Ba_2CaCu_2O_8$ high-temperature superconductor taken at 400kV accelerating potential. (Courtesy of K. Hiraga, Tohoku University).*

the calculated image [Fig. 7.89(d)]. Generally good agreement is observed for the calculated images in Fig. 7.89(d) and the atomic resolution images obtained in the TEM (Fig. 7.88).

High-resolution image simulation Figure 7.89 illustrates the prospects for determining ordered arrangements of oxygen atoms in the superconducting oxides by computer image calculations. Computer generated image simulation

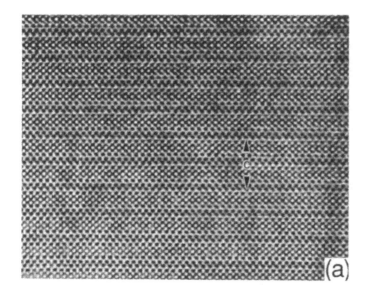

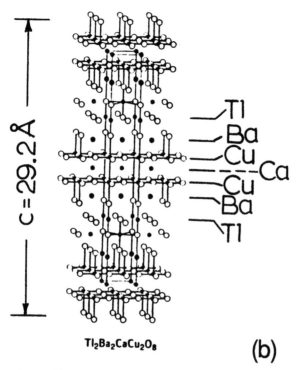

Tl₂Ba₂CaCu₂O₈   **(b)**

FIG. 7.89 $Tl_2Ba_2CaCu_2O_8$ superconductor image analysis and simulation
showing demagnified images of Fig. 7.88 [(a) and (c)] and simulated image
(d) compared with idealized unit cell schematic (b).   (Images and
simulation courtesy of K. Hiraga, Tohoku University).

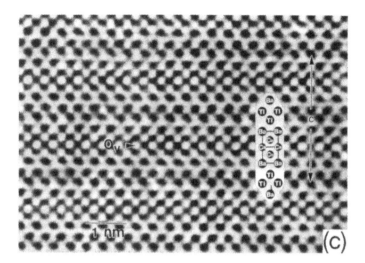

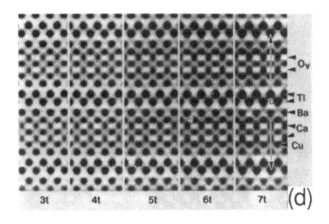

is becoming increasingly important in high-resolution electron microscopy, and several computer software packages have been developed [138,139]: NUMIS (Northwestern University Multislice Imaging System) and Synoptics in England (SEMPER); which can be integrated into computer systems compatible with commercially available transmission electron microscopes and AEM systems. These multislice image computation packages are based on the original multislice image computations described by Cowley and Moodie [140], and Goodman and Moodie [141]. Applications to structure analysis of superconductors as illustrated in Fig. 7.89 have also been described by Horiuchi, et al. [142]. Image simulation is also accomplished with a CEMPAS program [143] with image processing carried out using the SEMPER program.

It should be remembered that atomic image features (atom positions) will reverse in brightness depending upon the specific conditions of under or over focus (defocus conditions) (see Fig. 7.61). In the overfocused condition bright spots represent projections of atomic potentials. To carry out image computations the details of image formation are required, including: accelerating potential, the spherical aberration coefficient, and the amount of defocus [C_s and Δf respectively in Eq. (7.70)]; and the crystal structure and orientation (zone axis). The focus spread and focus step is also important. These conditions are often referred to as the Scherzer focus.

High-Resolution Cross-Sectional TEM (XTEM) While we are discussing multi-slice image simulation and high-resolution atomic imaging in the TEM, it might be appropriate to mention cross-sectional TEM or XTEM which is described in more detail in Appendix B.

In XTEM, various interfacial regions of interest are prepared for viewing in the transmission electron microscope by sandwiching the regions into a rigid structure which can be sliced, mounted, ground and polished, dimpled, and thinned selectively to electron transparency. Some of the most dramatic examples of the use of these techniques have involved the study of interfacial regimes in a variety of semiconductor materials systems and devices as discussed in the review by Al-Jassim, et al. [144]. In these techniques [145-147], devices or device sections to be viewed are sliced along the areas to be examined in cross-section and glued together or in stacks using superglue or fast-setting epoxies. Standard discs (3mm) or small specimens are then cut from the stack to include the inter-facial regions of interest. Mechanical dimpling to accelerate the thinning of areas of interest is then followed by ion milling to produce electron transparent thin sections which can be routinely examined at lattice or atomic resolutions.

Figure 7.90 illustrates the practical features of XTEM in the atomic image examination of the SiO_2/Silicon interface in a portion of an integrated circuit. The circuit region is built up using epoxy and silicon wafer blanks from which a cross-section is cut and a disc cut as shown dotted. The disc is then ion-milled to produce a thin electron transparent region between the SiO_2/Silicon interface. This technique can be used to examine ion bombardment or surface damage, depth profiles, and other features in cross-section in addition to device interfaces. You should examine Appendix B which illustrates the XTEM specimen preparation techniques in more detail.

As device technology becomes more sophisticated in the development of device arrays and ULSI circuitry, XTEM will be a very necessary technique even for routine diagnostics in order to examine interfacial reactions and precipitation on the atomic scale. Several IC facilities in fact utilize analytical, high-resolution transmission electron microscopes in the manufacturing areas to understand irregularities in device operations which can be traced to atomic-level, interfacial phenomena.

7.10.4 COMPUTED IMAGES AND COMPUTER PROCESSING OF ELECTRON MICROSCOPE IMAGES

Techniques have evolved over the past two decades, somewhat in consonance with advances in computer and computing technology, for theoretically developing and ideally simulating crystal defect and crystal structural images [139,148,149]. Such images, which can be printed out in intricate grey-scale plots by computer-driven graphic displays, (as illustrated for example in Fig. 7.89(d)) can be altered to match actual TEM images through the systematic variation of such parameters as defocus and chromatic aberration in the case of lattice or atomic images; or Burgers vector (b), operating reflection (g), deviation from the Bragg condition (s), $g \cdot b$, absorption, etc. in the case of crystal defects. In doing such computer-image matching, it is possible to understand the specific nature of crystal defects or other structural phenomena which influence the image features through a comparison of actual and computed micrographs.

These features are illustrated in Fig. 7.91 for intrinsic stacking faults in an $Al_{67}Ni_8Ti$ intermetallic alloy [150]. Computer-simulated

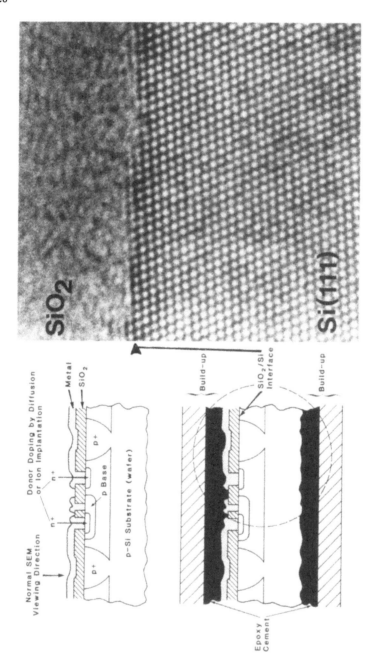

FIG. 7.90 *Schematic illustration of cross-section developed from built up device region where a standard TEM disc is cut and ion-milled to electron transparency in order to view the oxide/semiconductor interface. The atomic image shows the (111) silicon/SiO2 interface. (High-resolution image is courtesy of John Bravman, Stanford University).*

electron micrographs shown in Fig. 7.91(c) were instrumental in determining the displacement vector associated with the faults to be approximately $0.4^a{}_o{<}001{>}$.

In addition to the ability to generate electron micrographs by computer, the computer (microprocessor) can also manipulate image formation on-line or reconstruct image details (even three-dimensionally), correlate image data, and process images as in the signal mode associated with the STEM. Software capabilities for reconstructing images or for simulating images through simultaneous video display are becoming available commercially as special features of transmission electron microscope operation. Many capabilities are being developed jointly by university and commercial laboratory scientists, and many researchers will make these software innovations available on request. This is especially true of national laboratories and laboratories supported by NSF, DOE, NIH, and other federal agencies in the United States. These include Arizona State University, Argonne National Laboratory, and the University of California, Berkeley, (Electron Microscopy Center for Materials Research, Argonne National Laboratory, Materials Science Division, 9700 S. Cass Ave., Argonne, IL 60439; National Center for Electron Microscopy, Lawrence Berkeley Laboratory, 1 Cyclotron Rd., Berkeley, CA 94720). Electron microscopy software and hardware information is also available through the Electron Microscopy Society of America. The interested reader might consult the *EMSA Bulletin*. (See for example *EMSA Bulletin, 20(1)*: 1990).

7.10.5 NOVEL CONCEPTS AND APPLICATIONS

There are novel imaging techniques which can be applied to the actual image-forming process in the TEM, or to image interpretation or display following image recording in the TEM. Some of these are implicit in the short description of computer-aided diagnosis described above, and others include novel specimen manipulation or the use of optical techniques outside the TEM, including holography and 3-dimensional viewing. Three-dimensional imaging and viewing usually involves the simultaneous recording images which have been tilted (as a result of actual specimen tilt) by at least 12°. Such images, when viewed simultaneously as a stereo pair through a 3-D optical (stereo) viewer, can give a rather striking 3-D perspective, allowing crystal defects or inclusions to be viewed in their true perspective with respect to other defects, interfaces,

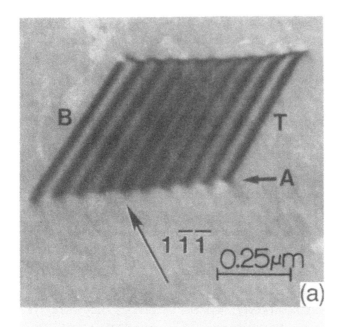

(a)

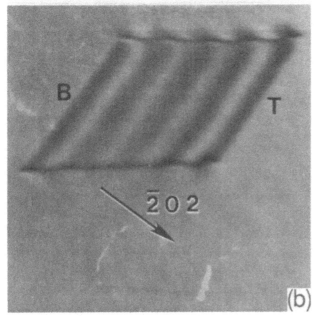

(b)

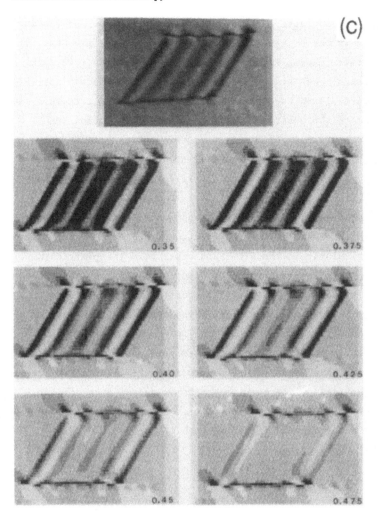

FIG. 7.91 *Bright-field TEM micrographs and computed images of intrinsic stacking faults in an intermetallic Al-Ni-Ti alloy. (a) and (b) show bright-field images with $\underline{g} = [1\bar{1}\bar{1}]$ and $\underline{g} = [\bar{2}02]$ respectively. T and B indicate the top and bottom of the foil. (c) Comparison of experimental bright-field image in (b) with simulated images produced using a range of magnitudes of $\left|\underline{R}\right| = (n a_o)$; values of n are shown in the simulated images. ($\underline{g} = [\bar{2}02]$, $\left|\underline{s}\right| = 0.05$; beam direction along $[10\ \bar{1}\ 10]$). The computations were based on methods described by Head, et al. [148]. (Courtesy C. D. Turner from Reference [150]).*

or the actual specimen surfaces. This is particularly useful for viewing thick film sections in the high-voltage electron microscope (see Chap. 8).

In an alternative imaging technique, called $2\frac{1}{2}$-D imaging, two dark-field images are taken, each at a different objective lens focus [151,152]. When these images are viewed as a stereo pair, the height of a feature can be directly related to the position of its associated reflection in reciprocal space. This allows diffraction patterns which contain very closely-spaced spots corresponding to closely-spaced images to be resolved, and is particularly applicable to phase identification in complex alloys [150].

Finally, we might mention that color-coded graphic display systems are being developed for a variety of the electron microscope signal information obtained particularly in the STEM, and other color-display schemes. For example, Murr [153] has produced color electron micrographs based on a technique originally devised by Müller [154] for the investigation of point defects in field-ion emission images (see Chap. 2). This particular technique has potentially useful applications in the study of dynamic phenomena in thin foils because of the enhanced perception rendered by the color acuity. It consists basically in superposing the transmitted images of sequential bright-field electron micrographs by use of color filters and a beam-splitter prism [153]. Color addition using combinations of yellow-and-blue or red-and-green filter schemes will produce a color spectrum of several thousand angstroms.

PROBLEMS

7.1   Show that the extinction distance for Ag($\underline{g}$ = [200]) for observations in the electron microscope at 100 kV is 254 Å. What is the percentage variation in ξ_g for observations at 500 kV? Calculate the extinction distances for Au and Ag at 1 MeV where $\underline{g}$ = [20$\bar{2}$].

7.2   Calculate the extinction distance for an 80/20 Cu-Zn alloy and $A\ell_2O_3$ assuming $\underline{g}$ = [002] and a 300 kV electron beam.

7.3   Calculate the approximate foil thickness in Fig. 7.9a if the operating reflection $\underline{g}$ is [$\bar{1}$11]. What would happen to the fringes in Fig. 7.9a if the accelerating voltage were tripled? Examine Fig. 7.10 and calculate the average thickness of the Si foil based on the extinction distance. Show that by strictly geometrical arguments, and assuming the (111) Si surface to be exactly normal to the electron optic axis, the thickness of the foil can be calculated. Compare the thickness obtained in the geometrical case with that obtained on considering the extinction distance.

7.4   Sketch to scale the foil section shown in the image of Fig. 7.15a if the angle between the surface plane and the boundary planes is 35°. What is the most probable operating reflection $\underline{g}$ in Fig. 7.15a? (Assume the twin boundary planes have parallel slopes.)

7.5   Draw the amplitude-phase diagrams for two overlapping stacking faults in an fcc material having an operating reflection $\underline{g} = [\bar{1}11]$. Show that three such faults will produce no contrast. (Draw the phase-amplitude diagrams to some constant scale.) Using $\perp$ to indicate a partial dislocation, show the arrangement of dislocations necessary to produce the contrast at the invisible fault portion of Fig. 7.16b.

7.6   Calculate the average stress acting on the dislocations shown in Fig. 7.20a. Make a histogram of the stresses for 20 dislocations in the (111) planes. Observe that the two dislocations in the top part of Fig. 7.23a bow appreciably when they move as shown in b. Assume the normal stress in the slip planes is constant. Prove that the trailing dislocation (a) has split into two partials (b).

7.7   Calculate the force on the lead dislocation in the pile-up shown in Fig. 7.20b. Find the total stress on this dislocation if the Burgers vectors are a/2 $[10\bar{1}]$. State any assumptions made and outline in detail the analysis you use.

7.8   If we differentiate Eq. (7.32) with respect to z, and substitute back for the first derivative, we can obtain two second-order linear differential equations. Show that the solution of this equation in T and S results in the forms given in Eq. (7.37).

7.9   Consider the matrix of Eq. (7.47a). If we wish to treat a grain section separated by an interface into parts I and II, show that if II is tilted away from any diffracting condition, the resulting image is described generally by

$$T \simeq T_1 e^{i\pi s_1 z_1} \qquad S \simeq S_1 e^{i\pi s_1 z_1}$$

Make a sketch of the object section and show the projection of the interface image. What will happen to the image if the specimen is tilted so that I is in the diffracting position and II is not? What will be the appearance of the interface if the crystal section is bent so that both I and II are in a diffracting position? If neither I nor II are in diffracting positions, what will be the appearance of the interface? What will be the approximate value of T and S?

7.10 Let us consider an idealized situation for two stacking faults parallel to the foil surfaces in a thin specimen. Using kinematical assumptions, namely, that the amplitude of diffraction is given by

$$A = \int e^{2\pi i s z} dz$$

derive an expression for the image intensity. Show that when a third fault is added to the transmission area, the diffraction intensity I = A*A = 0, and that the faults are invisible. Is there a dependence of the image on the foil thickness and the disposition or spacing of the faults? If so, show this in the equation you derive.

7.11 The motion of a dislocation in Aℓ is observed in the electron microscope. In the initial observations, the Burgers vector is determined to be a/2 [01$\bar{1}$]. When the dislocation is observed in bright-field illumination, the reflection producing contrast is observed to be 1/a [$\bar{2}$20], and the dislocation is initially observed to lie in a (11$\bar{1}$) plane. Under continued heating of the area by the electron beam, this dislocation is observed to move. Prove that this dislocation will be visible, and that its motion will be observable in the electron microscope image. At one point, the dislocation cross-slips onto a (111) plane, and in this section of foil both the (11$\bar{1}$) and the (111) planes are inclined to the surface. What will happen to the image if the operating reflection remains the same, and the Burgers vector is now a/2 [$\bar{1}$$\bar{1}$0]? Indicate two possible surface orientations for this situation and indicate the inclination angles of the (111) and (11$\bar{1}$) slip planes for each orientation.

7.12 Consider the images of Fig. 7.34. If the diffraction pattern shown were rotated 180° what would be the fault character? If the dark-field image of Fig. 7.34 and the diffraction pattern were each rotated 180°, what would be the stacking-fault character? Suppose only b is rotated 180°, what is the character? Now consider the stacking-

fault images and selected-area electron diffraction pattern for a
Cu-5 atomic percent Aℓ alloy in Fig. P7.12. Assuming the proper rotations have been made, find the fault nature. Indicate the upper
and lower foil surfaces and calculate the foil thickness.

7.13 A mobile Shockley partial dislocation in a stainless steel foil
having a Burgers vector a/6 [$\bar{2}1\bar{1}$] is observed to approach a Frank
dislocation having a Burgers vector of a/3 [11$\bar{1}$]. On meeting, they
disappear. On observing the diffraction pattern, a four-beam dynamical situation occurs, with the three reflection spots indexed as
[111], [200], and [02$\bar{2}$]. Which of these spots dominantly characterizes the operating reflections? The specimen is now tilted several
degrees and the total dislocation is observed to move and then to
split into two mobile partials forming a ribbon of stacking fault
between them. Only one partial is visible, however. If the operating reflection is observed to be [020], what are the possible Burgers
vectors of the Shockley partials?

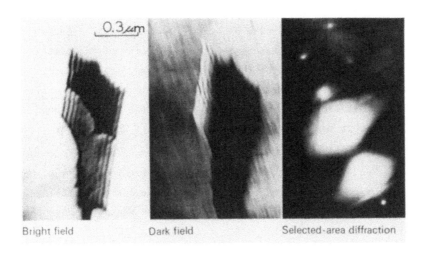

Bright field   Dark field   Selected-area diffraction

FIG. P7.12

7.14 If we assume in Fig. 6.41b that the images of faults are all microtwins having an average thickness of 25 Å, calculate the average
twin-fault density (volume percent) in the area of specimen shown.

7.15   Estimate the dislocation density in the foil area shown in Fig. 7.20a.
       How does this value compare with those quoted in the literature from
       etch-pit measurements? Indicate your source or sources of reference.
       What is the dislocation density in Fig. 7.20c? Compare the results
       of measurement using the methods of Ham [57] and Keh [58].

7.16   If the foil thickness in Fig. 7.44a is 1100 Å, estimate the volume
       percent of precipitate in the area shown assuming the precipitates
       to be approximately spherical particles. Outline your method of
       analysis or any assumptions you make.

7.17   Calculate the true dihedral angles of intersection of the twin boun-
       daries with the grain boundary in Fig. 7.51, and find the ratio of
       twin-boundary energy to grain-boundary energy $(C_{AB} + C_{T_{AB}})/2$.

7.18   In Fig. 7.53 find the true dihedral angle of intersection and calcu-
       late the average low-angle grain-boundary free energy if the operat-
       ing reflection producing the high-angle grain-boundary fringes is
       $[1\bar{1}1]$. What is the ratio of low-angle (tilt boundary) free energy
       to grain boundary free energy?

7.19   Calculate the average dislocation core energy in stainless steel by
       assuming the image for a low-angle boundary shown in Fig. 7.53 is
       typically representative. Assume the Burgers vector of the tilt
       boundary dislocations to be of the general form $a/2 <110>$. Obtain
       the foil thickness from measurement of the image width of the dislo-
       cations, and estimate the actual inclination of the low-angle boun-
       dary with respect to the specimen surfaces. (Hint: The necessary
       feature is the extinction distance.)

7.20   Assume that in Fig. 7.56b the dislocation nodes lie in a (111) plane
       perpendicular to the electron beam. Estimate the average stacking-
       fault energy of the material. Assume the Burgers vector to have the
       general form $a/6 <112>$.

7.21   In Fig. 7.77 make a histogram of the distribution of dispersoid
       sizes and find the mean particle size. Assuming the particles es-
       sentially spherical, estimate the volume percent of ThO_2 if the
       thickness is 1000 Å. What is the average dislocation density in b?
       What is the dislocation density in a if the foil thickness is the
       same? Are dislocation dipoles observed in a? If so, outline their
       mechanism of formation.

7.22 Copper is known to be embrittled by small additions of Sb. If the Sb is thought to segregate to the grain boundaries, devise an experiment to indirectly support this contention. We must assume here that the Sb cannot be directly observed. (Hint: Consider the basic stress equation for brittle-fracture characterization, the Griffith equation.) Outline the experiments you would perform bearing in mind that the measurements must involve transmission electron microscopy.

7.23 Magnetic hystersis in Armco iron is observed to vary with pressure in the range above 50 kbars. Outline an experiment where the domain structure of such a magnetic material could be directly related to pressure-induced hysterisis changes.

7.24 A semiconducting material is observed to change its electrical properties drastically when the temperature is raised above a certain critical level. Indicate how you could determine whether a structural or chemical change were involved by using the electron microscope.

7.25 The electron micrographs in Fig. P7.25 illustrate the characteristic diffraction contrast for stacking faults having $\phi = -\pi$ for s = 0. The thin-foil material was pure Ni. Considering that for stacking faults with $\phi = -\pi$, the extinction distance is $\xi_g/2$ and not ξ_g, discuss whether the images represent stacking faults. Note that Fig. P7.25 shows the bright- and dark-field images, respectively. The (110) selected-area electron diffraction pattern in c has already been rotated by the absolute amount, taking into account relative rotations and lens inversions. Compare your conclusions with those of the author in *Thin Solid Films*, 4: No. 6, 389(1969).

7.26 In the electron micrograph in Fig. P7.26, extended dislocation nodes are observed to have formed in a Cu-5 atomic%Aℓ alloy as typically shown circled. Following the conventions outlined in *Thin Solid Films*, 4: No. 6, 389(1969), calculate the stacking-fault free energy by measuring both the diameter of a circle inscribed within the nodes, and the radii of curvature at the nodes. The character angle (angle between the dislocation arm and the Burgers vector) can be measured by considering only the invisible arm connecting extended and unextended nodes. The slip plane shown is the (111) plane, and

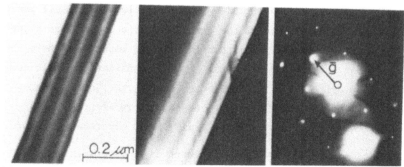

FIG. P7.25

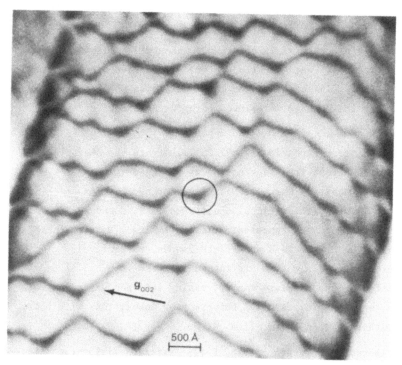

FIG. P7.26

the grain surface is tilted several degrees off (110). Compare your results with those in Fig. 7.57.

7.27   Figure P7.27 shows dislocations and dislocation loops in molybdenum following explosive shock deformation. Vacancy loops, identified by tilting experiments outlined in general in Table 7.5, are indicated by "V". In one of the tilted views shown, the general field of defects appears essentially invisible. If we assume this satisfies the $g \cdot b = 0$ criterion, make a histogram of dislocation character and comment on the average character for the explosively generated dislocations.

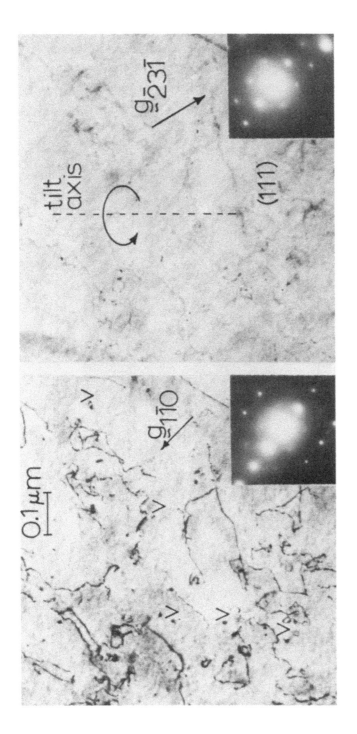

FIG. P7.27

7.28   In Fig. P7.28 is shown a bright-field, dark-field image sequence
       illustrating a chromium oxide bicrystal on a thin film nickel sub-
       strate as part of a fundamental study of oxide solar absorber coat-
       ing growth. Determine the nature (composition) of these oxide grains
       and find their apparent crystallographic misorientation.

7.29   What crystal planes are imaged in Fig. 7.60? What is the magnifi-
       cation of the lattice image?

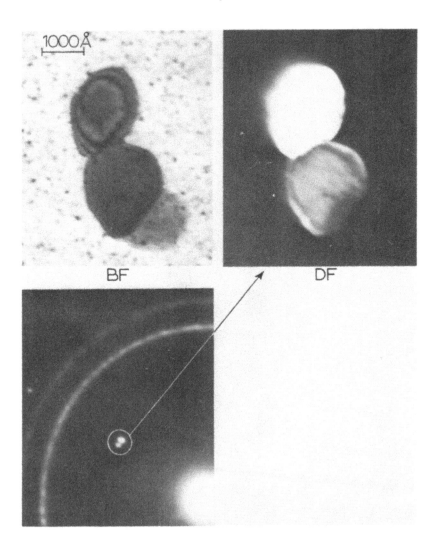

FIG. P7.28

7.30 Discuss the image of Fig. 7.62 in terms of the rotation of overlapping of PrCo$_5$. Make a sketch of the diffraction pattern and show the position of the aperture (as a dashed circle) to form this image.

7.31 Figure P7.31 shows a {200} lattice image for a vapor deposited [001] palladium film. The atomic plane contrast is good enough to allow for enlargement of the page to do a more detailed inspection of the image. What is the total magnification of your image? Place circles around dislocations in your image. How many dislocations are present? What is the equivalent dislocation density? Compare this value with those shown for other metals in Fig. 7.42. Refer to L. E. Murr, *Scripta Metall. et Material., 24*: 783 (1990) and discuss the implications of these image features and the prospects for promoting electron tunneling-induced deuterium fusion in a palladium electrode in a cold-fusion cell.

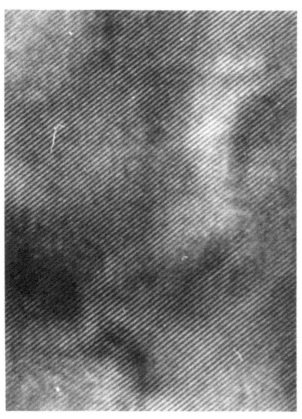

FIG. P7.31

REFERENCES

1. P. B. Hirsch, A. Howie, and M. J. Whelan, *Phil. Trans. Roy. Soc. (London), A252:* 499(1960).

2. L. E. Murr, *Phys. Status Solidi, 19:* 7(1967).

3. R. D. Heidenreich, *J. Appl. Phys., 20:* 993(1949); and *Fundamentals of Transmission Electron Microscopy,* chap. 6, Interscience Publishers Inc., New York, 1964.

4. M. J. Whelan *et al., Proc. Roy. Soc. (London), A240:* 524(1957).

5. M. J. Whelan and P. B. Hirsch, *Phil. Mag., 2:* 1121,1303(1957).

6. M. J. Whelan, *J. Inst. Metals, 87:* 392(1959).

7. W. Bollman, *Elec. Microscopy: Proc. Stockholm Conf.* 1956, p. 316, Almquist and Wiksell, Stockholm, 1957.

8. W. Bollmann, *Phys. Rev., 103:* 1588(1956).

9. P. B. Hirsch, R. W. Horne, and M. J. Whelan, *Phil. Mag., 1:* 677(1956).

10. M. J. Whelan, *Phil. Mag., A149:* 144(1959).

11. W. T. Read, *Dislocations in Crystals,* McGraw-Hill Book Company, New York, 1953.

12. J. P. Hirth and Jens Lothe, *Theory of Dislocations,* McGraw-Hill Book Company, New York, 1968.

13. Johannes and Julia Weertman, *Elementary Dislocation Theory,* The Macmillan Company, New York, 1964.

14. A. Howie and M. J. Whelan, *Proc. Roy. Soc. (London), A263:* 217(1961).

15. S. Amelinckx, *Direct Observation of Dislocations,* Academic Press, Inc., New York, 1964.

16. L. E. Murr, *Proc. Penna. Acad. Sci., 38:* 126(1964).

17. R. M. Fisher, *Rev. Sci. Instr., 30:* 925(1959).

18. H. G. F. Wilsdorf, ASTM *Spec. Tech. Publ., 245:* 43(1958) and *Structure and Properties of Thin Films,* p. 151, John Wiley & Sons, Inc., New York, 1959.

19. P. J. E. Forsythe and R. N. Wilson, *J. Sci. Instr., 37:* 37(1960).

20. L. M. Howe, J. L. Whitton, and J. F. McGurn, *Acta Met., 10:* 773(1962).

21. D. W. Pashley and A. E. B. Presland, *Phil Mag., 7:* 1407(1962).

22. L. E. Murr, *Proc. Penna. Acad. Sci., 39:* 202(1966).

23. L. E. Murr, *Interfacial Phenomena in Metals and Alloys,* Addison-Wesley, Reading, Mass., 1975. (Reprinted by Tech Books, Herndon, VA in 1990).

24. L. E. Murr, and S-H. Wang, *Res Mechanica, 4:* 237(1982); see also L. E. Murr, *Mater. Sci. Engr., 51:* 71(1981).

25. D. J. H. Cockayne, *Z. Natur., A27:* 452(1972).

26. D. J. H. Cockayne, I. L. F. Ray, and M. J. Whelan, *Phil Mag., 20:* 1265(1969).

27. E. L. Hall and J. B. Vander Sande, *Phil. Mag., 37:* 137(1978).

28. J. B. Vander Sande in *Introduction to Analytical Electron Microscopy* J. J. Hren, et al (eds.), Chap. 20, Plenum Press, New York, 1979, p. 535.

29. D. J. H. Cockayne in *Ann. Rev. Mater. Sci., 11:* 75(1981).

30. A. Howie and M. J. Whelan, *Proc. Roy. Soc. (London), A267:* 206(1962).

31. K. Moliere, *Ann. Physik, 34:* 461(1939).

32. H. Yoshioka, *J. Phys. Soc. Japan, 12:* 618 (1957).

33. H. Hashimoto, A. Howie, and M. J. Whelan, *Phil. Mag., 5:* 967(1960).

34. H. Hashimoto, A. Howie, and M. J. Whelan, *Proc. Roy. Soc. (London), A269:* 80(1962).

35. R. Gevers, *Phys. Status Solidi, 3:* 415(1963).

36. R. Gevers, *Phys. Status Solidi, 3:* 1672(1963).

37. R. Gevers, et al., *Phys. Status Solidi, 4:* 383(1964).

38. R. Gevers et al., *Phys. Status Solidi, 5:* 595(1964).

39. A. Art, R. Gevers, and S. Amelinckx, *Phys. Status Solidi, 3:* 697 (1963).

40. J. Van Landuyt, R. Gevers, and S. Amelinckx, *Phys. Status Solidi, 9:* 135(1965).

41. R. Gevers, J. Van Landuyt, and S. Amelinckx, *Phys. Status Solidi, 18:* 325(1966).

42. G. Remaut et al., *Phys. Status Solidi, 13:* 125(1966).

43. R. Gevers, A. Art, and S. Amelinckx, *Phys. Status Solidi, 13:* 563 (1963).

44. J. Van Landuyt, R. Gevers, and S. Amelinckx, *Phys. Status Solidi, 7:* 519(1964).

45. H. Blank, et al., *Phys. Status Solidi, 7:* 747(1964).

46. L. E. Murr, K. P. Staudhammer, and S. S. Hecker, *Met. Trans., 13A:* 627(1982).

47. K. P. Staudhammer, L. E. Murr, and S. S. Hecker, *Acta Met., 31:* 267 (1983).

48. J. Washburn et al., *Phil. Mag., 85:* 991(1960).

49. M. J. Whelan, *Proc. Roy. Soc. (London), A249:* 114(1959).

50. W. Carrington, K. Hale and D. McLean, *Proc. Roy. Soc. (London), A259:* 1960).

51. E. Aerts et al., *J. Appl. Phys., 38:* 81(1967).

52. E. Levine, J. Washburn, and G. Thomas, *J. Appl. Phys., 38:* 81(1967).

53. E. Ruedl and S. Amelinckx, *J. Nucl. Mater., 9:* 116(1963).

54. V. F. Holland et al., *Phys. Status Solidi, 10:* 543(1965).

55. J. W. Edington, *Practical Electron Microscopy*, VanNostrand, New York, 1975.

56. J. Bailey and P. B. Hirsch, *Phil. Mag., 5:* 485(1960).

57. R. K. Ham, *Phil. Mag., 6:* 1183(1961).

58. A. S. Keh, *J. Appl. Phys., 31:* 1501(1960).

59. R. K. Ham, *Phil Mag., 7:* 1177(1962).

60. D. J. Mazey, R. S. Barnes, and A. Howie, *Phil. Mag., 7:* 1861(1962).

61. B. Edmundson and G. K. Williamson, *Phil. Mag., 9:* 277(1964).

62. R. M. Cotterill, *Phil. Mag., 6:* 1351(1961).

63. J. D. Meakin, A. Lawley, and R. C. Koo, *Appl. Phys. Letters, 5:* 133 (1964).

64. P. B. Price in G. Thomas and J. Washburn (eds.), *Electron Microscopy and the Strength of Crystals,* p. 41, Interscience Publishers, Inc., New York, 1963.

65. V. A. Phillips and J. D. Livingston, *Phil. Mag., 7:* 969(1962).

66. A. Howie and M. J. Whelan, *Proc. Roy. Soc. (London), A267:* 206(1962).

67. M. F. Ashby and L. M. Brown, *Phil. Mag., 8:* 1083(1963).

68. R. D. Heindenreich, *Fundamentals of Transmission Electron Microscopy,* Interscience Publishers, Inc., New York, 1964.

69. L. E. Murr, *Phys. Stat. Sol. (a), 22:* 239(1974).

70. J. L. Brimhall and G. L. Kulcinski, *Rad. Effects, 20:* 25(1973).

71. J. Gittus, *Irradiation Effects in Crystalline Solids,* Applied Science Publishers, London, 1978.

72. L. E. Murr, *Acta. Met., 16:* 1127(1968).

73. L. E. Murr, *J. Appl. Phys., 31:* 5557 (1968).

74. H. Udin, A. J. Shaler, and J. Wulff, *Trans. AIME, 185:* 186(1949).

75. A. P. Greenough, *Appl. Mater. Res., 4:* 25(1965).

76. T. E. Mitchell, *Phil. Mag., 17:* 1169(1968).

77. L. M. Brown, *Phil. Mag., 10:* 441(1964); J. P. Hirth and J. Lothe, *Acta Met., 13:* 279(1964); and L. E. Murr, *Thin Solid Films 4:* No. 6, 389(1969).

78. A. Howie and P. R. Swann, *Phil. Mag., 6:* 1215(1961).

79. G. Thomas, *J. Australian Inst. Metals, 8:* 80(1963).

80. P. C. J. Gallagher, *J. Appl. Phys., 37:* 170(1966).

81. F. Haussermann and M. Wilkens, *Phys. Status Solidi, 18:* 609(1966).

82. J. Silcox and P. B. Hirsch, *Phil. Mag., 4:* 72(1959).

83. R. E. Smallman and K. H. Westmacott, *J. Appl. Phys., 30:* 603(1959).

84. T. Jossang and J. P. Hirth, *Phil. Mag., 13:* 675(1966).

85. H. Hashimoto, M. Mannami, and T. Naiki, *Phil. Trans. Roy. Soc., 253:* 459(1961).

86. S. Amelinckx, *The Direct Observation of Dislocations,* Academic Press, New York, 1964, p. 164, p. 406.

87. R. Sinclair in *Introduction to Analytical Electron Microscopy,* J. Hren, et al (eds.), Chap. 19, Plenum Press, New York, 1979, p. 507.

88. G. A. Bassett, J. W. Menter, and D. W. Pashley, *Proc. Roy. Soc., A246:* 345(1958).

89. J. M. Cowley and S. Iijima, *Physics Today, 30(2)*: 32(1977).

90. J. M. Cowley and S. Iijima, *Z. Naturforsch. 27a:* 445(1972).

91. M. A. O'Keefe, *Acta Cryst., A29:* 389(1973).

92. P. L. Fejes, *Acta Cryst., A33:* 109(1977).

93. S. Iijima, *Acta Cryst., A29:* 18(1973).

94. S. Horiuchi and M. Mitomo, *J. Mater. Sci., 14:* 2543(1979).

95. P. Rez and O. L. Krivanek, Proc. 9th Int'l Congress on Electron Microscopy, *Electron Microscopy 1978,* The Microscopical Society of Canada, Toronto, Vol. I, 1978, p. 288.

96. H. Hashimoto, Y. Sugimoto, Y. Takai, and H. Endoh, Proc. 9th Int'l Congress on Electron Microscopy, *Electron Microscopy 1978,* The Microscopical Society of Canada, Toronto, Vol. I, 1978, p. 284.

97. H. Hashimoto, A. Kumao, and H. Endoh, Proc. 9th Int'l Congress on Electron Microscopy, *Electron Microscopy 1978,* The Microscopical Society of Canada, Toronto, Vol. III, 1978, p. 244.

98. Helena de Souza Santos and Keiji Yada, *Clays and Clay Minerals, 27:* 161(1979).

99. A. V. Crewe, *Quart. Rev. Biophysics, 31:* 137(1970).

100. J. M. Cowley in *Introduction to Analytical Electron Microscopy,* J. J. Hren, et al (eds.), Plenum Press, New York, Chap. 1, 1979, p. 1.

101. J. M. Cowley, *Appl. Phys. Letters, 15:* 58(1969).

102. D. C. Joy, D. M. Maher, and A. G. Cullis, *J. Microscopy, 108:* 185 (1976).

103. D. M. Maher and D. C. Joy, *Ultramicroscopy, 1:* 239(1976).

104. J. C. H. Spence and J. M. Cowley, *Optik, 50:* 129(1978).

105. A. V. Crewe in Proc. 9th Int'l Congress on Electron Microscopy, *Electron Microscopy 1978, Vol. III,* J. M. Sturgess (ed.), Microscopical Society of Canada, Toronto, 1978, p. 197.

106. K. J. Van Oostrum, A. Lienhouts, and A. Jore, *Appl. Phys. Letters, 23:* 283(1973).

107. R. H. Geiss, *Appl. Phys. Letters, 27:* 174(1975).

108. H. Blank and S. Amelinckx, *Appl. Phys. Letters, 2:* 140(1963).

109. H. W. Fuller and M. E. Hale, *J. Appl. Phys., 31:* 238(1960).

110. J. Silcox, *Phil. Mag., 8:* 7(1963).

111. J. P. Jakubovics, *Phil. Mag., 10:* 277(1964).

112. M. Wilkens, *Phys. Status Solidi, 9:* 255(1965).

113. E. M. Hale, H. W. Fuller, and H. Rubinstein, *J. Appl. Phys., 30:* 189 (1959).

114. H. W. Fuller and M. E. Hale, *J. Appl. Phys., 31:* 238(1960).

115. H. W. Fuller and M. E. Hale, *J. Appl. Phys., 31:* 1699(1960).

116. J. Fidler and P. Skalicky, *Appl. Phys. Letters, 39*(7): 573(1981).

117. R. Levi-Setti, in *Scanning Electron Microscopy/1974,* O. Johari and I. Corvin (eds.), IIT Research Inst., Chicago, 1974, p. 125.

118. W. H. Escovitz, T. R. Fox, and R. Levi-Setti, *Proc. Nat. Acad. Sci. USA, 72*(5): 1826(1975).

119. R. Levi-Setti, in *Advances in Electronics and Electron Physics, Supplement 13A, Applied Charged Particle Optics,* A. Septier (ed.), Academic Press, New York, 1980, p. 261.

120. L. W. Swanson, G. A. Schwind, A. E. Bell, and J. E. Brady, *J. Vac. Sci. Technol. 16:* 1864(1979).

121. G. S. Cargill III, *Nature, 286:* 691(1980).

122. L. E. Murr, *Microelectronics Journal, 10:* 12(1979).

123. D. Gabor, *Proc. Roy. Soc., A197:* 454 (1949).

124. A. Tonomura, *J. Electron Microsc., 33(2):* 101 (1984); *38:* 543 (1989).

125. R. Castaing and L. Henry, *Compt. Rend.* (Paris), *255:* 76 (1962).

126. R. M. Henkelman and F. P. Ottensmeyer, *J. Microscopy, 102:* 79 (1974).

127. L. Reimer, I. Fromm, and R. Rennekamp, *Ultramicroscopy, 24:* 339 (1988); see also L. Reimer, et al., *EMSA Bulletin, 20(1):* 73 (1990).

128. C. W. Chu, et al., *Phys. Rev. Lett., 58:* 405 (1987).

129. G. Van Tendaloo, H. W. Zandbergen, and S. Amelinckx, *Solid St. Comm., 63:* 389 (1987).

130. J. G. Bednorz and K. A. Müller, *Z. Phys., B64:* 189 (1986).

131. L. E. Murr, et al., *Solid St. Comm., 73(10):* 695 (1990).

132. L. E. Murr, et al., *Appl. Phys. Lett., 55(15):* 1575 (1989).

133. Z. Z. Sheng and A. M. Hermann, *Nature, 332:* 138 (1988).

134. M. Maeda, et al., *Jpn. J. Appl. Phys., 27:* L209 (1988).

135. L.E. Murr and C. S. Niou, *J. Mater. Sci. Lett., 9:* 1103 (1990).

136. M. A. Subramanian, et al., *Science, 239:* 1015 (1988).

137. C. C. Torardi, et al., *Science, 240:* 633 (1988).

138. L. D. Marks Materials Research Center, Northwestern University, U.S.A. or Hitachi Scientific Instruments, Mountain View, CA.

139. See for example: P. A. Stadelmann, *Ultramicroscopy, 21:* 131 (1987); M. A. O'Keefe, et al., *Nature, 274:* 322 (1978); W. Krakow, *Ultramicroscopy, 18:* 197 (1985).

140. J. M. Cowley and A. F. Moodie, *Acta Cryst., 10:* 609 (1957).

141. P. Goodman and A. F. Moodie, *Acta Cryst., A30:* 280 (1974).

142. S. Horiuchi, et al., *Jpn. J. Appl. Phys., 27:* L1172 (1988).

143. R. Kilaas, *Proc. Electron Micro. Soc. Amer. 45th,* G. W. Bailey (ed.), San Francisco, CA, 1987, p. 66.

144. M. M. Al-Jassim, M. Hockley, and G. R. Booker, in *Defects in Semiconductors,* J. Narayan and T. Y. Tan (eds.), North-Holland Publishing Co., Amsterdam, 1981, p. 521.

145. T. T. Sheng and R. B. Marcus, *J. Electrochem. Soc., 127:* 737(1980).

146. J. C. Bravman and R. Sinclair, *J. Electron Microscopy Technique, 1:* 53(1984).

147. S. N. G. Chu and T. T. Sheng, *J. Electrochem. Soc.*, *131(11)*: 2665 (1984).

148. A. K. Head, et al., *Computed Electron Micrographs and Defect Identification*, North-Holland Publishing Co., Amsterdam, 1973.

149. P. W. Hawkes (eds.), *Computer Processing of Electron Microscope Images*, Topics in Current Physics, Springer-Verlag, Berlin, 1980.

150. W. O. Powers, et al., *Phil. Mag.*, *60(2)*: 227(1989).

151. W. L. Bell, *J. Appl. Phys.*, *47*: 1676(1976).

152. R. Sinclair, G. M. Michal, and T. Yamashita, *Met. Trans.*, *12A*: 1503 (1981).

153. L. E. Murr and M. C. Inman, *Phys. Status Solidi*, *10*: 441(1965).

154. E. W. Müller, *J. Appl. Phys.*, *28*: 6(1957).

SUGGESTED SUPPLEMENTARY READING

Amelinckx, S.: *Direct Observation of Dislocations*, Solid State Supplement no. 6, Academic Press Inc., New York, 1964.

Bethge, H. and J. Heydenreich (eds.): *Electron Microscopy in Solid State Physics*, Materials Science Monographs no. 40, Elsevier Publishers, New York, 1987.

Buseck, P., et al. (eds.): *High Resolution Transmission Electron Microscopy and Associated Techniques*, Oxford University Press, New York, 1989.

Drits, V. A.: *Electron Diffraction and High Resolution Electron Microscopy of Mineral Structures*, Springer-Verlag, New York, 1987.

Edington, J. W.: *Practical Electron Microscopy in Materials Science*, Van Nostrand Reinhold, New York, 1976.

Egerton, R. F.: *Electron-Energy Loss Spectrometry in the Electron Microscope*, Plenum Publishing Corp., New York, 1989.

Heidenreich, R. D.: *Fundamentals of Transmission Electron Microscopy*, Interscience Publishers Inc., New York, 1964.

Hirsch, P. B., et al.: *Electron Microscopy of Thin Crystals*, Butterworth Scientific Publications, London, 1965 (new edition, 1977).

Hirth, J. P., and Jens Lothe: *Theory of Dislocations*, McGraw-Hill Book Company, New York, 1968.

Johnson, J. E., and P. B. Hirsch (eds.): *High Resolution and High Voltage Electron Microscopy*, Wiley Interscience, New York, 1987.

Kay, D.: *Techniques for Electron Microscopy*, Blackwell Scientific Publications Ltd., Oxford, 1961.

Reimer, L.: *Transmission Electron Microscopy*, Springer-Verlag, New York, 1989.

Smallman, R. E., and K. H. G. Ashbee: *Modern Metallography*, Pergamon Press, New York, 1966.

Spence, J. C. H.: *Experimental High Resolution Electron Microscopy*, Clarendon Press, Oxford, 1981.

Thomas, G., and M. J. Goringe: *Transmission Electron Microscopy of Materials*, John Wiley, New York, 1979.

von Heimendahl, M. (translated by U. E. Wolff): *Electron Microscopy of Materials*, Academic Press, New York, 1980.

Weertman, Johannes, and Julia Weertman: *Elementary Dislocation Theory*, The Macmillan Company, New York, 1964.

8
HIGH-VOLTAGE
ELECTRON MICROSCOPY

8.1 INTRODUCTION

We have observed in the previous chapters that the penetration of solid
matter is enhanced by increasing the electron energy. Further, you will
recall from Chap. 3 that resolution increases with accelerating potential,
while spherical and chromatic aberrations are decreased. As a consequence,
high-voltage electron microscopy has a number of very obvious advantages
in the study of thin crystals, particularly where thicker samples can be
observed at greatly increased contrast than at standard voltages of 100
kV.

High-voltage electron microscopes - those operating at accelerating
potentials of 200 kV and above were developed concurrently with the more
conventional electron microscopes in the years just before World War II
[1-3]. These were not successful primarily because of limitations in design know-how, the lack of electronic stabilization at high voltages, poor
vacuum systems, and a host of related engineering problems in addition to
the prohibitive costs involved in construction. Two decades later, however, a number of electron microscopes had been constructed that proved
extremely useful and notably contributed to the present enthusiasm for

high-voltage electron microscopy [4-7]. Perhaps the most famous facility
is that of G. Dupouy and coworkers at Toulouse, France [6].

As we have already noted, the main advantages of high voltages in
transmission electron microscopy are increased specimen penetration and
greatly enhanced image contrast. Despite these features, there are certain
difficulties involved, especially where radiation damage becomes prominent,
and in situations where diffraction is characterized by many beams. We
will attempt to clarify these points in this chapter, in addition to demonstrating the dramatic clarity of high-voltage electron transmission
images; and the applications to which the high-voltage technique is particularly well suited.

8.2 DESIGN FEATURES OF THE HIGH-VOLTAGE ELECTRON MICROSCOPE

While a treatment of design features is not necessarily germane to our understanding of the applications of high-voltage electron microscopy (HVEM),
there are several features that, on clarification, may enhance your appreciation of the high-voltage electron microscope. For the most part, the
design features of most contemporary high-voltage electron microscopes are
fundamentally the same as those for more conventional instruments. In
other words, most HVEM designs involve scaled-up CTEM designs. We will
recall from Chap. 3 [Eq. (3.5) or (3.9)] that the lens focal length is
proportional to the accelerating potential and inversely proportional to
the square of the lens ampereturns as it relates to field strength. As a
consequence of this basic property, in order to maintain reasonable focal
lengths at higher voltages, it is necessary to enlarge the lens design,
that is, to include more windings or increase the current.

Enlargement of the lenses sometimes necessitates the enlargement of
the pole pieces, with the result that the spherical aberration coefficients
for the lenses increase. Of particular importance is the increased objective-lens aberration, which must be minimized by a large ampere-turn (NI)
requirement.

A somewhat bulky design is required in critical portions of the instrument to minimize radiation hazards. This is an especially important
feature of the viewing section and photographic chamber, where up to more
than 10 cm of shielded glass may be required in addition to mobile lead
shields. Since x-ray production is also severe in the gun and condenser
sections, heavy shielding is also required in this area. The fact that

x-ray excitation is strongly dependent on the beam current also places a limitation on the tolerable beam current, and in this respect an upper limit in brightness may be imposed. However new electron optical concepts may eventually be utilized in HVEM lens design which, along with field-emission guns can have a significant impact on these limitations.

8.2.1 HIGH-VOLTAGE POWER SUPPLIES

The really unique feature of the high-voltage microscope design is the high-voltage power source; not necessarily because of its circuit design, but because of its overall engineering. Historically there are two prominent dc high-voltage sources, namely, the Cockroft-Walfton and Van de Graaff generator. Both have been successfully applied in the design of high-voltage instruments; however, the Van de Graaff is usually found to be somewhat lacking in high-voltage stability. The essential features of the Cockroft-Walton generator have in fact been incorporated in approximately 90 percent of the instruments operating at 300 kV or above built since 1940. Most conventional microscope power supplies also employ the Cockroft-Walton circuit in the high-voltage supply, with the exclusion of the accelerator section attached to the electron gun.

In its simplest form the Cockroft-Walton generator is a voltage-multiplier circuit having the general features shown schematically in Fig. 8.1. This system consists of an arrangement of diodes or half-wave rectifiers in series connecting a ladder network of capacitors. Each pair of diodes in Fig. 8.1 operates so as to allow current to flow on alternate half-cycles specific to the driving frequency. Each even-subscripted capacitor is thus charged to a potential twice the previous one, so that successive stages act as voltage doublers. The potential across C_2 is therefore twice the primary voltage, and this voltage is again doubled across C_4, etc. The output or final voltage for the multiplier system is then given approximately by

$$V_o = 2^{N_s} V_p \qquad (8.1)$$

where N_s is the number of stages, and V_p is the initial or primary voltage.

In the simple voltage-multiplier system of Fig. 8.1, the ripple voltage ΔV_o is given by

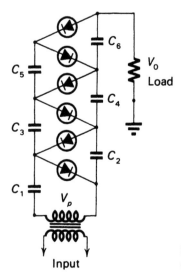

FIG. 8.1 *Cockroft-Walton voltage-multiplier circuit.*

$$\Delta V_o = \frac{I_L N_s (N_s + 1)}{2Cf} \tag{8.2}$$

where I_L = load current

C = capacitance

f = driving frequency

It is obvious from Eq. (8.2) that the higher the input frequency, the lower the output ripple. Ripple is also approximately proportional to N_s^2 for a large number of stages.

The circuit characteristics of Fig. 8.1 are directly employed in most conventional microscope power supplies. However, for operation at voltages in excess of 300 kV, two such circuits are often connected in a parallel or symmetrical arrangement as shown in the high-voltage generating circuit of Fig. 8.2. In the symmetrical Cockroft-Walton circuit, the output ripple is given simply by

$$\Delta V_o = \frac{I N_s}{2Cf} \tag{8.3}$$

and we observe that the stability of the symmetrical case does not rapidly decrease where a large number of stages is necessitated.

The Cockroft-Walton circuit is usually driven by a radio frequency source ($f \simeq 30$ kHz). The output, as shown in Fig. 8.2, is then connected to the accelerator tube of up to 30 stages of electrodes. Each electrode stage is supplied with a divided voltage supplied through the bleeder resistors from the output of the Cockroft-Walton generator as shown in Fig. 8.3. The physical makeup of the accelerator tube consists of a stack of

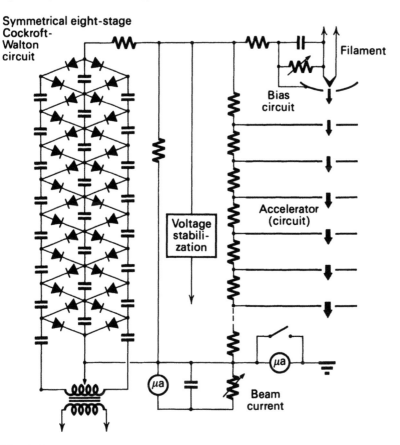

Symmetrical eight-stage Cockroft-Walton circuit

Filament

Bias circuit

Voltage stabili-zation

Accelerator (circuit)

Beam current

FIG. 8.2 *High-voltage generating circuit showing a symmetrical Cockroft-Walton voltage multiplier coupled with voltage cascade-type accelerator (schematic).*

porcelain insulators sealed together by metal flanges and o rings; and the accelerating electrodes themselves are made of highly polished metal. The regular spacing of the accelerating electrodes in such a stacked array function to linearly accelerate in stages the electrons emitted from the V filament at an initial potential of from 10 to 50 kV[†].

In the commercially designed high voltage electron microscopes such as those of A.E.I. (Kratos), Hitachi, and J.E.O.L., the standard V-type filament is sometimes incorporated into a turret design allowing filament

---

[†]*You might profit by glancing at Chap. 3 in A. E. S. Green, "Nuclear Physics," McGraw-Hill Book Company, New York, 1955; or by reading the original article by J. D. Cockcroft and E. Walton, Proc. Roy. Soc. (London), A136: 609(1932).*

changes to be made remotely and automatically, or a manual exchange device
similar to that incorporated in the exchange of specimens is employed.
Electromagnetic beam deflection in the gun section can facilitate alignment.
The dark-field microscopy beam deflector is located below the condenser
lens.

8.2.2 PHYSICAL DESIGN OF HIGH-VOLTAGE INSTRUMENTS

We emphasized previously that x-ray production in the high-voltage electron
microscopes presents a formidable design problem since adequate shielding
is a necessity. Radiation levels are generally kept below about 0.5 mr/h
during operation by using lead shielding. The area of particular concern
is the condenser section. Additional protection is also accommodated at
open slot sections in the column such as the aperture holders, etc., and
the camera area.

   The Hitachi high-voltage electron microscope is provided with a diffraction lens coupled between the intermediate and projector lenses. This
lens serves to increase the effective camera length L for selected-area
electron diffraction.

   The mechanical design of the high-voltage electron microscopes is of
necessity considerably more complicated than that of conventional instruments. Modifications must also be made insofar as vacuum systems are concerned because of the high-voltage system and related features.

   Since it is not possible to deal adequately with the details of the
design of high-voltage electron microscopes in this chapter, the reader
who is particularly interested in a more detailed discussion is referred
to the data sheets available from individual microscope manufacturers, or
the recent proceedings of electron microscopy symposia [8,9]. The design
features of most high-voltage instruments are clearly illustrated in Fig.
8.3, which shows an overall view of the AEI(Kratos)-EM7 high-voltage electron microscope. The enlargement of basic design features is perhaps obvious on comparison with the more conventional microscope depicted in Fig.
5.1.

8.3 HIGH-VOLTAGE ELECTRON DIFFRACTION

One of the most apparent phenomena occurring at high accelerating potentials in the electron microscope is the enlargement of the Ewald sphere
as indicated in Fig. 8.4. Since we have already noted that diffraction

FIG. 8.3 *High-voltage transmission electron microscope with a 1.3 million volt accelerating potential to achieve a point resolution of 1 Angstrom (0.1nm). (Photograph courtesy of Hitachi, Ltd.).*

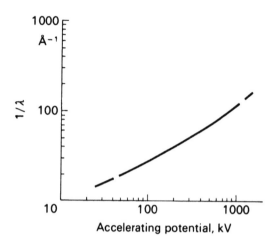

FIG. 8.4 *Variation of radius of reflection (Ewald) sphere with accelerating potential.*

intensities evolve phenomenologically by the intersection of the interference regions at the reciprocal-lattice points with the reflection sphere (Fig. 6.6), it becomes clear that despite a reduction in the extensions of these interference regions (rel rods) because of the increased specimen thickness, the number of excited reflections increases considerably at high voltages. This situation is of course conducive to exciting systematic reflections, many having intensities essentially equivalent to that of the primary beam. The diffraction, because thicker foils are employed, must of necessity be considered dynamical. However, because of the excitation of numerous systematic reflections, two-beam dynamical theory as discussed in Chap. 6 is no longer strictly applicable; and for the most part, if the diffraction effects are to be rigorously interpreted, we require a multiple-beam dynamical treatment. The systematic excitation of many Bragg reflections is illustrated in the diffraction contrast image of bend contours in tungsten observed at 1 MeV as shown in Fig. 8.5. The increase in the sphere of reflection will become quite obvious on comparison with Fig. 7.6b.

8.3.1 MANY-BEAM DYNAMICAL DIFFRACTION

The 2-beam dynamical theory of diffraction and image contrast (Chaps. 6 and 7) is applicable for HVEM images on a phenomenological basis, but in principle the two-beam approximation will break down and n-beam many-beam dynamical theory has to be used for a rigorous treatment. The methods of solution of n-beam cases are essentially calculating devices, and when 3 or more strong beams operate, difference equation methods must normally be used for a solution. These have been outlined by Howie [10], Howie and

FIG. 8.5 *Bend Contours in tungsten showing extended reflection surface in reciprocal space at 1000 kV. Bar represents 1 μm. (Courtesy Hitachi, Ltd., Tokyo.)*

Basinski [11], and Hashimoto [12] to name but a few, and you are referred
to these references for more details since our treatment here will be su-
perficial at best.

The electron waves passing through a perfect crystal foil can be expressed by a linear combination of so-called Bloch waves having a wave vector $\underline{K}^j$ [see Eq. (1.44) or Eq. (6.30)]. At voltages around 1MV, relativistic effects must be taken into account and the electron mass must be replaced by the relativistic mass, and the crystal potential [Eq. (6.28)]
must also be altered as already described in Eq. (7.35). In effect, we
must have

$$|\underline{K}_0| = \frac{1}{\lambda} = \frac{1}{h} \sqrt{2meV_0 + \left(\frac{eV_0}{c}\right)^2}$$

and consider this in Eq. (6.33). The generalized wave function will then
be expressed by [13,14]

$$\psi(r) = \sum_j A_j \sum_g c_g^{(j)} |\underline{K}^{(j)}| \exp[2\pi i (\underline{K}^{(j)} + g) \cdot \underline{r}] \tag{8.4}$$

The amplitude $c_g^{(j)} |\underline{K}^{(j)}|$ must satisfy the set of equations

$$\left[\chi_0^2 - (\underline{K}_0^{(j)} + g)^2\right] c_j^{(j)} |\underline{K}^{(j)}| + \sum_\ell U_\ell c_{g-\ell}^{(j)} |\underline{K}^{(j)}| = 0 \tag{8.5}$$

where $\chi_0^2 = 2meE/h^2 + U_0$. Here both ℓ and g denote reciprocal lattice
vectors. This is essentially the so-called dispersion equation obtained
in the original paper by Bethe [15] (which develops the dynamical theory
of electron diffraction) and expresses a general relationship between amplitude coefficients, crystal potential, and the total electron energy, E
(as determined by the accelerating potential V_0). Although it is not obvious when n plane-wave components of the Bloch waves of Eq. (8.4) are
considered, Eq. (8.5) consists of a set of n equations each having n terms.

For example, in the 2-beam approximation there are two equations each
consisting of two terms as follows:

$$(\chi_0^2 - |\underline{K}^{(j)}|^2) \, c_0^{(j)} + U_{-g} \, c_g^{(j)} = 0$$
$$U_g \, c_0^{(j)} + \chi_0^2 - (\underline{K}^{(j)} + \underline{g})^2 \, c_g^{(j)} = 0 \tag{8.6}$$

To expand this to three or more beams (see Fig. 6.18) it is convenient to use the notation of matrix algebra outlined briefly in Chap. 7 utilizing the generalized form

$$\underline{A}^{(j)} \left\{ c_g^{(j)} \right\} = 0 \tag{8.7}$$

where $\left\{ c_g^{(j)} \right\}$ is a column vector with amplitudes $c_g^{(j)}$, and $\underline{A}^{(j)}$ is a matrix with diagonal elements

$$a_{gg} = \chi_0^2 - (\underline{K}^{(j)} + \underline{g})^2 \tag{8.8}$$

and with off-diagonal elements

$$a_{g\ell} = U_{g-\ell} \tag{8.9}$$

where g refers to the row and ℓ refers to the column.

If we now consider three beams, for example -g, 0, +g, Eq. (8.7) appears as

$$\begin{pmatrix} a_{-g-g} & U_{-g} & U_{-g-g} \\ U_g & a_{00} & U_{-g} \\ U_{gg} & U_g & a_{gg} \end{pmatrix} \tag{8.10}$$

Thus, in the n-beam case, $\underline{A}^{(j)}$ [Eq. (8.10)] is n x n, and this concept can be applied to solutions of the contrast equations implicit in Eq. (7.42) for example, where there would be one transmitted beam in each crystal part, but more than one diffracted beam to be considered in evaluating the contrast at a phase-shift plane separating two crystal portions.

Fortunately, for most applications of the HVEM involving materials analysis or microstructural diagnosis, a rigorous solution of n-beam equations is usually not necessary. However, detailed examination of the n-beam equations, and especially the dispersion equation [Eq. (8.5)], can provide some unique insight into the crystal structure because, after all, it is a representation of the interaction of the electron (Bloch) waves with the crystal potential. This depends on the crystal (atomic) geometry and atomic structure. We can even give a geometrical interpretation to the dispersion equation called the *dispersion surface,* which for any diffraction vector is a *locus* it sweeps out in $\underline{K}$-space. You might recall from Chap. 2 that for free electrons this surface in a crystal unit cell

is a sphere (the Fermi surface). However in real crystals these surfaces
of constant energy are characterized by Brillouin zone boundaries as illustrated in Fig. 6.5. Indeed Eq. (8.5) includes χ_0 which is the energy function in the dispersion relation. The essence of all of this is that since
these phenomena are intrinsically related to the image, detailed examination of even simple images of interference phenomena as shown in Fig. 8.5
can yield some very detailed information about the Brillouin zone structure
and other subtle crystal structure because the detail in Fig. 8.5 is very
sensitive to the strength and shape of the crystal potential and the accelerating voltage. You should notice in particular that the arms radiating
out from the large pole figure are due to electrons essentially incident
parallel to the planes of atoms in tungsten, and the fine structure of the
intensity oscillations across these arms arises from Bloch wave interactions with the crystal potential. There are other subtle effects which
occur at high voltages and we will discuss their occurrence and applications in Sec. 8.7.

Electron diffraction at high voltages also results in increased accuracy of the diffraction pattern because of the reduction in the Bragg
angle. Calibration of high-voltage electron diffraction patterns following
the same basic guidelines established in Sec. 6.4.3 is accomplished with
increased accuracy; and the addition of a diffraction lens between the intermediate and projector, as indicated in Sec. 8.2.1, allows for an adjustment of the pattern size to be recorded. These features allow selected-
area electron diffraction patterns to be obtained from considerably smaller
areas at 1000 kV than at 100 kV. This is a particularly useful feature in
studies of precipitation and related analytical problems.

8.4 IMAGE CONTRAST AT HIGH VOLTAGES

8.4.1 *SPECIMEN TRANSMISSIBILITY*

The most obvious advantages of high-voltage electron microscopy over conventional microscopy at 100 kV is the ability of higher-energy electrons
to penetrate greater thicknesses of material. H. Hashimoto [16] measured
the energy dependence of absorption in Al, and related the penetration
power as transmissive power, defined as the reciprocal of the absorption
length δ_g. It has been generally established that the transparency of
specimens is dependent on the absorption coefficients as predicted theoretically [17,18]. Figure 8.6 illustrates the mean (theoretical) transmis-

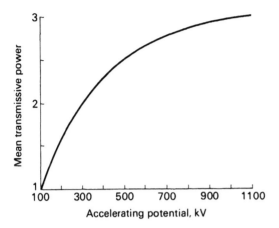

FIG. 8.6 *Theoretical penetration power
with increasing accelerating potential.*

sive power of high-energy electron waves as a function of accelerating
potential based on unity transmission at 100 kV.

We observe from Fig. 8.6 that penetration power increases rapidly between 100 and about 900 kV, and then levels off. Even at 125 kV, a 20 percent increase in specimen penetration is expected to occur, with penetration effectively doubling in specimen materials observed at 300 kV as compared to observations at 100 kV.

While in Fig. 8.6 we can gain some feel for the enhanced specimen penetration at high voltages, the nature of the factors limiting the maximum
usable specimen thickness for different materials is not clearly understood,
especially where the difference in specimen absorption is considered.
R. Uyeda and M. Nonayama [19] investigated the thickness limits in MoS_2
specimens by utilizing the criterion that the useful thickness limit corresponds to diffraction patterns containing only Kikuchi lines. G. Thomas
[20] measured the thickness limits of Si and stainless steel (materials
differing markedly in mean absorption) by adopting the criterion that the
thickness limit be defined as the condition where fringe contrast at twin
or stacking-fault interfaces was effectively destroyed (presumably as a
result of electron absorption) using a systematic <111> reflection. Figure 8.7 reproduces these data, which are observed not to be proportional
to $(v/c)^2$ as one is led to conclude by considering dependence strictly in
terms of the mean absorption parameter. The contrast features in high-
voltage electron microscope images are therefore presumably due in part to
contributions of inelastically scattered electrons.

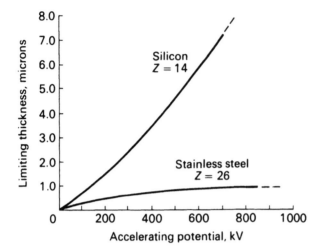

FIG. 8.7 *Thickness limits for Si and stainless steel as determined experimentally by G. Thomas, Phil. Mag., 17: 1097 (1968). Thickness limits are characterized by zero-fringe contrast.*

It must be cautioned that the data recorded in Fig. 8.7 are not necessarily absolute, and should serve only as a relative guide. This is particularly true at lower voltages since conditions do exist where contrast is sharp above or below the curves shown. A look at Fig. 7.28 in retrospect will testify to this statement. This figure should also be compared with essentially identical twin-boundary fringe images observed at 1 MeV as shown in Fig. 8.8. You might compare the thickness (see Prob. 8.2) of the stainless steel specimen shown in Fig. 8.8 with the limiting thickness illustrated in the stainless steel curve of Fig. 8.7. Values of the relativistic extinction distances for stainless steel and other materials appear in Appendix D (Table D.4).

Figure 8.7 shows generally that the value of the thickness limit decreases for materials with large atomic number Z determined by image contrast as opposed to brightness. The image formation, as shown in Fig. 8.8, is also ideally described by the two-beam dynamical theory so long as two-beam diffraction effects predominate. This condition can be assured by observing the selected-area electron diffraction pattern, and by utilizing the techniques outlined in Chaps. 6 and 7. Diffraction contrast in thick specimens has been observed to vanish when the Kikuchi diffraction pattern contains only Kikuchi lines [19].

FIG. 8.8 *Bright-field image of twin
boundary containing incoherent steps in
stainless steel foil observed at 1000 kV
in Hitachi Perkin-Elmer 1000 electron
microscope. Note electron absorption
effects (type 3). (Courtesy Hitachi,
Ltd., Tokyo.)*

8.4.2 DIFFRACTION CONTRAST AND IMAGE RESOLUTION AT HIGH VOLTAGE

On referring back to Fig. 3.15, you will recall that theoretical resolution in the electron microscope will effectively double at 1 MeV as compared with that at 100 kV. This occurs because at high voltages, the aperture angle α is reduced, causing a decrease in the magnitude of spherical
aberration [Eq. (3.11)] and the diffraction error. These reduced aberrations, coupled with a concomitant decrease in the chromatic aberration
[Eq. (3.14)] give rise to the enhanced resolution. Reduction of spherical
aberration at high voltages manifests itself in marked image sharpness of
focus.

Figure 8.9 illustrates the typical appearance of dislocation images
observed in an Al-Mg-Si alloy at 1000 kV in the Hitachi 1000 electron microscope. It may be instructive, in referring to Fig. 8.9, to work Prob.
8.3, which illustrates the sharpness of high-voltage images of defects,
essentially free of distortion.

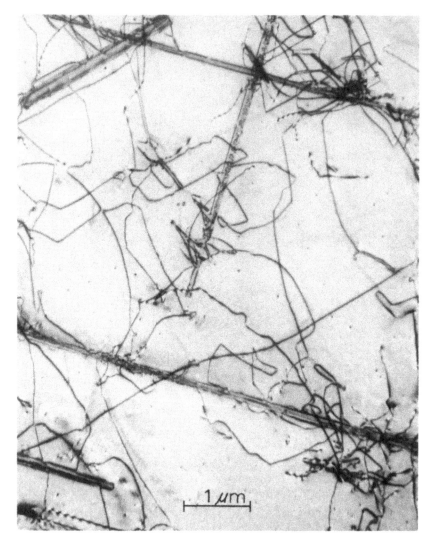

FIG. 8.9 *Dislocations and precipitates in Al-Mg-Si alloy observed at 1000 kV. (Courtesy Hitachi, Ltd., Tokyo.)*

Resolution and particularly contrast in high-voltage electron microscope images are also enhanced markedly by a reduction in chromatic image aberration. This is particularly noticeable in the reduction in the energy loss from thicker specimens observed at high accelerating potentials, in addition to the fact that $\Delta V/V_0$ [Eq. (3.14)] is inherently very small because of the large V_0.

Where the spread of energy loss of very thin specimens is given by nΔV (n being the number of inelastic-scattering acts incurring an energy loss ΔV), the corresponding loss in thick specimens is simply ΔV. Therefore chromatic aberration does not increase in proportion to the specimen thickness.

The poor resolution and lack of sharpness usually occurring in dark-field images in the more conventional electron microscopes operating at approximately 100 kV (which is particularly noticeable on comparison with the corresponding bright-field image) results primarily from displacements by chromatic aberration of additionally focused images formed by elastically scattered electrons that have suffered an appreciable energy loss (see section 3.3.2). In many cases multiple images of the object features result, each image recognizably displaced from the other by approximately [21].

$$\Delta(i) \simeq C_c \alpha f \frac{\Delta V}{V_o} \tag{8.11}$$

where f is the focal length.

Since spherical aberration due to the divergence of the electron beam is small, aperture dark-field images, being limited chiefly by chromatic aberration, are very drastically sharpened because in Eq. (8.11) both α and $\Delta V/V_o$ are appreciably reduced in comparison to observation at 100 kV. Figure 8.10 illustrates typically the marked clarity in aperture dark-field images at high voltages as compared with that attainable at more conventional accelerating potentials. It will also be instructive to observe briefly in retrospect the dark-field images reproduced in Figs. 7.25 and 7.37. Note in particular on comparison of these figures with Fig. 8.10d that the chromatic dispersion (and streaking along an edge of the image due to spherical aberration) effectively disappears at high voltage.

Although the diffraction pattern superimposed in Fig. 8.10d is not well defined in the central region, it illustrates nonetheless the fact that the Bragg angle at high voltage is extremely small. Consequently, the aperture need not to be moved very far from the optic axis to form dark-field images. Consequently, aperture dark-field images can be very clear in the HVEM. In addition, since the Bragg angle at high voltage (~1MV) is almost the same as the optimum aperture angle for high-resolution images, it is not necessary to use a small aperture to exclude diffracted beams at high voltages. This so-called multi-beam imaging method [12]

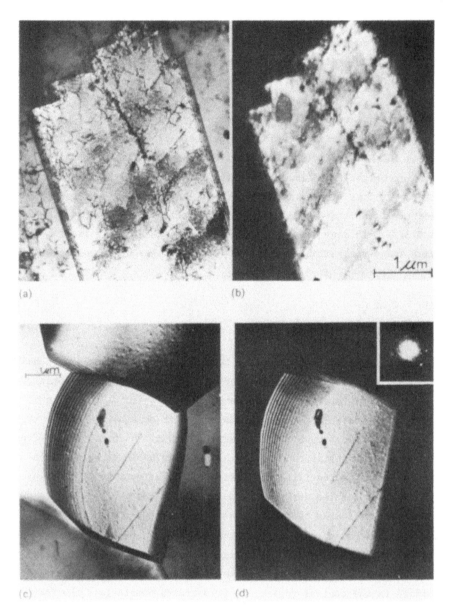

FIG. 8.10 *Enhanced image clarity at high voltage resulting from reduced chromatic aberration. (a) and (b) are bright- and dark-field images of annealing twin in deformed Inconel alloy observed at 125 kV. (c) and (d) show corresponding bright- and dark-field images, respectively, for small grain in stainless steel using 1000 kV accelerating potential. Dark-field images are aperture dark field. (c and d) are courtesy of Japan Electron Optics Laboratory Co., Ltd., Tokyo.)*

provides high contrast images which can be rapidly recorded because of the additional signal intensity. This situation also facilitates high-resolution lattice imaging in the HVEM.

In general, small objects near the entrance surface such as precipitates or other inclusions in a thick film observed in the HVEM will be somewhat blurred by comparison with the images of the same features near the exit surface. By comparison, Hashimoto [12] has observed that the fringe contrast of a stacking fault is higher near the entrance surface the exit surface. By comparison, Nonoyama, et al [22] have observed that the fringe contrast of a stacking fault is higher near the entrance surface than near the exit surface. This occurs mainly because the stacking fault image is formed by Bloch wave interference which is influenced by absorption. Of course the image features are also strongly influenced by absorption as influenced by the diffraction conditions in adjacent crystals such as at a grain or twin boundary as described in Chap. 7. These effects are observed in the HVEM images shown in Figs. 8.8 and 8.10 c and d.

8.5 ELECTRON IRRADIATION EFFECTS IN SPECIMENS OBSERVED AT HIGH VOLTAGE

The study of organic materials, polymers, and ionic solids appears to be particularly attractive at high voltages because critical total dose of electron irradiation, defined as the product of the current density and the exposure time, is linearly proportional to the accelerating potential [7]. Such materials, observed to degrade appreciably at 100 kV, will be maintained for longer periods because of the reduction in ionization rate. This reduction in ionization rate is accompanied by a loss in image brightness, which may be compensated for in some applications through the use of image intensification (Sec. 3.5).

In many materials, however, the ionization potential is lower than the energy of the electron beam. Electrons that strike the atoms of the specimen can thereby incur a displacement resulting in the creation of a vacancy - interstitial pair. We can consider the maximum energy transferred to a solid atom by an electron is given by

$$E_m = \frac{2V_d(V_d + 2mc^2)}{Mc^2} \tag{8.12}$$

where V_d = accelerating potential at which atomic displacement will occur
m = electron mass

M = atomic mass (or atomic weight of specimen material)

c = velocity light

Since the creation of such a defect pair requires an average energy of approximately 25 eV, we can compute the accelerating potential at which such phenomena will be expected to occur. Figure 8.11 is a plot of M versus V_d, assuming E_m in Eq. (8.12) to be 25 eV.

Note in Fig. 8.11 that the production of point defects by electron irradiation damage will occur in most solids observed at 1 MeV accelerating potential. The rate of formation will of course vary with the material, and the noticeable clustering of such defects will ultimately depend strongly on the local diffusion coefficients in the thin specimens.

The production of visible radiation damage in solids (resolvable as point-defect clusters) can be both beneficial and detrimental. In the first instance this phenomenon presents a simple means for studying radiation damage in solids. In the second, where point-defect generation by external parameters is to be investigated, it will become impossible to differentiate those formed by selected experimental means and those generated by the action of the electron beam. Figure 8.12 illustrates the degree of visible point-defect production that can routinely occur in observations of metal samples.

The number of secondary defects formed as a result of radiation damage in the HVEM is also strongly dependent upon crystallography, and in some instances the concentration of secondary defects tends to saturate at some accelerating potential [22]. Consequently, very high voltages may not necessarily produce proportionately higher radiation-induced defects.

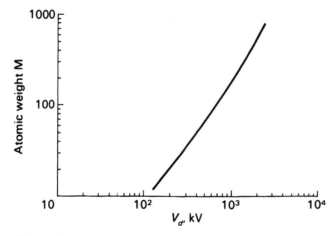

FIG. 8.11 *Threshold potentials of solids as function of atomic weight.*

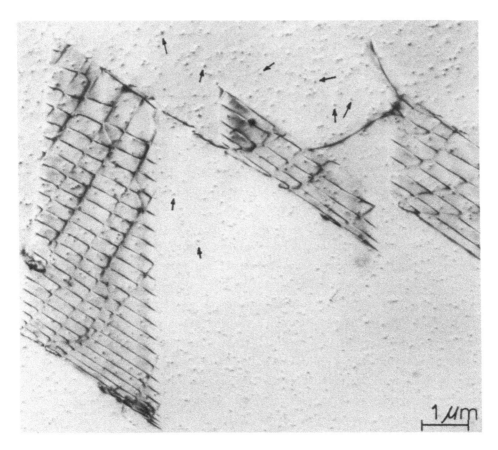

FIG. 8.12 *Dislocation networks lying in (111) planes in stainless steel observed in bright field at 1000 kV. Point-defect clusters are quite prominent in background as black dots or resolvable in terms of their characteristic contrast features (Sec. 7.3.2). (Courtesy Hitachi, Ltd., Tokyo.)*

8.6 IN-SITU DEVICES AND EXPERIMENTS

One of the most significant features of the HVEM is the ability to perform a variety of *in-situ* experiments on representative specimens. This is due in part because of the penetration power of the HVEM and the large size of contemporary instruments which facilitate a variety of rather complicated stages and experimental devices. Figure 8.3 is indeed an example of a facility designed primarily for the *in-situ* study of ion beam interactions with solid films, and this instrument can also accommodate a variety of environmental stages for the study of chemical reactions, hydrated-state phenomena, straining, and even combinations such as straining while being

bombarded by high-energy ions, stress corrosion experiments, etc. One particular advantage of the HVEM in observing chemical reactions *in situ* is
the higher penetrating power which allows the products to be observed in
association with the substrate or unreacted material.

Environmental cells for observing reactions in the HVEM have been
described by several investigators [23-25], and experiments have included
hydriding reactions, oxidation, reduction, and related reactions. These
cells permit differential pumping which allows uninterrupted observations
while a reaction gas is admitted to the specimen area. Specimens can also
be simultaneously heated in many facilities, and this can allow for a wide
range of meaningful observations.

Heating-cooling-straining stages for HVEM facilities can provide for a
wide range of *in-situ* experiments involving observations of precipitation,
deformation, transformations and the like, and indeed many observations of
these features have been reported in the literature [26]†. Even though *in-
situ* experiments involving deformation can be performed in the HVEM, it is
sometimes difficult to record events sequentially without the aid of a cine
camera or video recording system. These systems sometimes suffer resolution losses which make fine details difficult to discern. In addition,
experiments involving foil straining in the HVEM are difficult to execute
because it is very hard to make uniformly thin specimens of nearly any
material, and specimens which contain holes or irregularities fail unpre-
dictably, or make predictions of dynamic events at specific film locations
impossible. Indeed, it is generally impossible to choose specific areas
where predictable dislocation activity might occur. Figure 8.13 shows one
method of preparing thin stainless steel foils for *in-situ* straining in the
Kratos microscope illustrated in Fig. 8.3. In this technique, a modification of the pointed-electrode method described in Appendix B, only the foil
edges were electron transparent and irregular. Observations of dislocation
emission from grain boundaries have been made utilizing this technique as
illustrated in Fig. 8.14. Figure 8.15 shows a low-magnification sequence
illustrating the propagation of cracks in thin stainless steel foils shown
in Fig. 8.13, and the formation of voids ahead of the advancing crack tips.
Details of crack propagation phenomena have been of particular interest in

---

†*See the discussion of straining stages by U. Valdre in* High Voltage Electron Microscopy, *P. R. Swann, C. J. Humphreys, and M. J. Goringe (eds.),
Academic Press, New York, 1974, p. 124.*

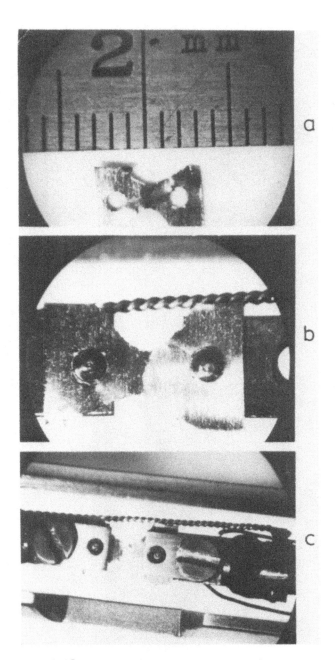

FIG. 8.13 *Preparation of type 304 stainless steel
specimens for* in-situ *HVEM straining experiments.
(a) as-prepared (electropolished) microtensile specimen; (b) microtensile specimen placed on crosshead pins of* in-situ *straining stage (single tilt);
(c) microtensile specimen of (b) strained to failure
showing fully assembled gripping arrangement of the
straining stage. Crosshead displacement occurs by
relaxing spring-loaded mechanism by heating wire
shown.* [After L. E. Murr, Mater. Sci. Engr., 51:
71(1981)].

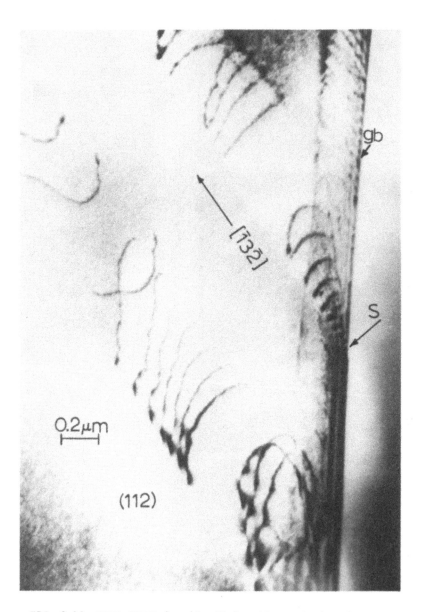

FIG. 8.14 *HVEM (1MV)* in-situ *dislocation emission from a*
grain boundary in a stainless steel sample strained in the
arrangement shown in Fig. 8.13(c). Wedge-type source in the
grain boundary (gb) is denoted S. While the emission was
actually observed, the event was too rapid to record sequen-
tially. [*From L. E. Murr*, Mater. Sci. Engr., 51: 71(1981).]

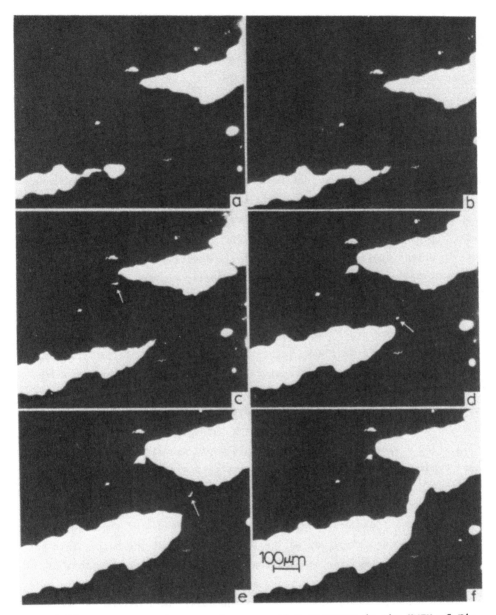

FIG. 8.15 *Room temperature crack propagation sequence in the HVEM of Fig.
8.3 (1MV). (a) Thin film region following straining by crosshead displacement of 200 μm showing crack tips approaching one another. (b) Advance of
lower crack by void coalescence after displacement of 233 μm. (c) Upper
crack producing void (microcrack) ahead of the crack tip (arrows) and advance of lower crack after 278 μm displacement. (d) Crack tip blunting and
void formation at arrow after 314 μm displacement. (e) Coalescence and
growth of lower crack and void production (arrow) after 360 μm displacement.
(f) Crack-tip coalescence after 375 μm displacement* [see L. E. Murr, Thin
Solid Films, 84: *131(1981)*].

of particular interest in HVEM *in-situ* experiments, and Fig. 8.16 shows the
details of the plastic zone structure at the tip of an advancing crack in a
thin MgO crystal strained in the HVEM [for more details see F. Appel and U.
Messerschmidt Mikuna, *Physica Status Solidi (a)*, 55: 529 (1979)].

8.7 APPLICATIONS OF HIGH-VOLTAGE ELECTRON MICROSCOPY

At this point it should be only too obvious that the applications of high-
voltage electron microscopy are largely identical to those outlined in
Chap. 7. The high-voltage instrument has the advantage of increased acura-
cy of electron diffraction, sharpness of detail in aperture dark-field images, and perhaps most importantly the ability to penetrate materials of
truly representative thicknesses. The latter feature is perhaps one of the
most important applications of high-voltage electron microscopy. In effect
many materials incapable of being satisfactorily thinned for transmission
observations in conventional (100 kV) instruments can be examined at voltages in the neighborhood of 1 MeV. We can observe from Fig. 8.7 that
while for the more dense materials, such as stainless steel, the limiting
thickness levels off at 1 MeV, lighter materials, particularly ceramics,
gain in limiting thickness at accelerating potentials considerably above
1 MeV. Extrapolations for Si in Fig. 8.7 in fact indicate no appreciable
leveling off even at 3 MeV, with the possibility of effectively penetrating
nearly 0.001 in. of Si at an accelerating potential of 3 MeV. Entire
solid-state devices and device components could be directly observed under
these conditions. In addition the details of semiconductor junctions could
be directly observed.

A great deal of the electron microscopy over the past two decades has
been subject to some question from time to time because of the argument
that thin-film structures may not be representative of the bulk. This has
been particularly true for dislocations, and the direct observations of
grain growth, phase transformations, and related dynamic phenomena in thin
specimens has been shown to differ from the bulk. H. Fujita et al [27]
have shown that for recrystallization or martensitic transformations to
occur representatively in electron microscope specimens requires a foil
thickness greater than 1 micron.

L. E. Murr [28] observed that grain growth of Er_2O_3 in the electron
microscope in foils approximately 0.08 micron thick (shown in Fig. 7.48)
appears to be influenced by the surfaces as indicated in the inclination

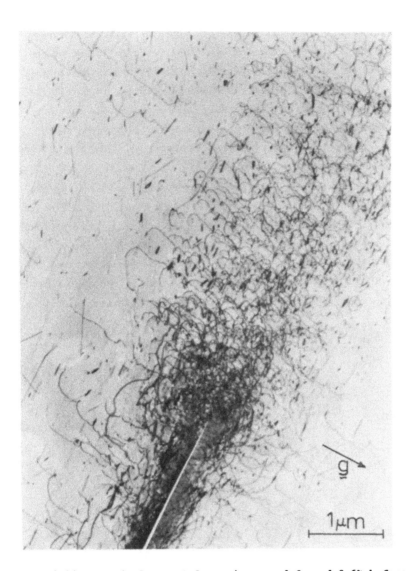

FIG. 8.16 *MgO single crystal specimen predeformed 0.6% before chemical thinning in hot phosphoric acid, then strained in the HVEM to produce the crack shown. The tensile axis was along <100> while the crack orientation is <010>. The dislocation structure at the crack tip and far removed from the crack tip is particularly dramatic in this bright-field electron micrograph. (1 MV accelerating potential) (Courtesy Drs. F. Appel and U. Messerschmidt, Akademie der Wissenschaften der DDR.)*

of the equilibrated grain-boundary planes. Grain-boundary inclinations of
grains equilibrated by the furnace annealing of thin sheet (bulk), and
electropolished uniformly from both sides, are representative of a random
slice of the bulk [28]. Observations of electropolished foils and of phenomena occurring in thin foils during their formation, appear therefore to
be representative of the bulk. Dynamic phenomena occurring in thin-films
directly within the microscope are presumably influenced by the section
thickness below some critical value. This feature is illustrated in the
curves of Fig. 8.17a that show the mean grain-boundary inclincation in electropolished stainless steel to be approximately 61° as compared with 71°
for the vapor-deposited Er films oxidized and recrystallized directly within the electron microscope at 100 kV. The statistical mean inclination for
a random section of bulk material is about 63° [29] and this feature prompts
the conclusions drawn above. In addition, the determination of the variation in Er_2O_3 boundary inclination with foil thickness, as illustrated in
Fig. 8.13b, indicates that perhaps bulk characteristics may be approximated
by thicknesses in excess of about 2000 Å. High-voltage electron microscopy

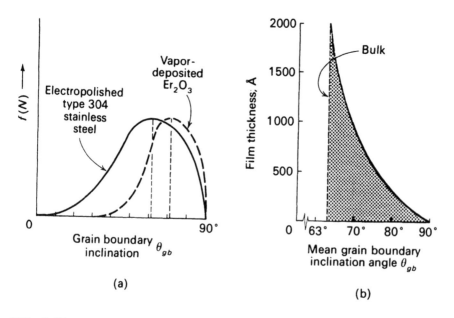

(a)

(b)

FIG. 8.17 *Comparison of equilibration processes in thin foils*
and bulk materials. (a) Distributions of equilibrium grain-
boundary inclinations in bulk stainless steel thinned to foil,
and vapor-deposited and oxidized Er foils (Er_2O_3). (b) Varia-
tion of mean grain-boundary inclination with thickness of Er_2O_3
foils having undergone grain growth and equilibration in elec-
tron microscope. (From L. E. Murr, [32].)

is therefore primarily an extension of contemporary materials science in the investigation of dynamic (in-situ) phenomena, and may not significantly alter the accuracy or conclusions of materials research performed in the range of 100 kV as described in Chap. 7.

Dupouy and Perrier [30.31] have demonstrated the remarkable clarity of the dark-field images of metals and alloys, particularly precipitates. We must be aware of the fact that in many cases the dimensions of precipitates or inclusions could exceed the foil thickness, and lead to some misinterpretations or false image characteristics. The ability of high-voltage electron microscopy to faithfully and clearly reveal solid-state precipitation is convincingly demonstrated in the bright-field image reproduced in Fig. 8.18.

It is also possible to observe the three-dimensional nature of dislocation cell structure in thicker foils at high voltages, in addition to being able to resolve the characteristic geometries of a three-dimensional polycrystalline solid. The geometrical nature of thin sections, whose image detail is projected two-dimensionally in the electron microscope (Sec. 7.4) is of course amenable to careful analysis using the high-voltage techniques discussed. However, this feature may lose some significance in very thick specimens where diffraction effects arising at overlapping-crystal defects, precipitates, etc., will require considerable interpretation. A good deal of the confusion in thicker specimen can be avoided through the use of stereo techniques. Stereo microscopy can also provide information relevant to the position of defects with respect to the specimen surfaces. Stereo pairs of electron micrographs are usually taken by tilting the specimen to obtain two views showing the same image features at the same magnification. As a general rule, a total tilt angle of 12° is optimum for a 1 μm thick specimen at 10,000X magnification. This angle must increase somewhat proportionately with decreasing magnification or thickness [33].

Bright-field pole figures, typically illustrated in Fig. 8.5 (and in Figs. 6.34 and 6.35) can also have many interesting applications in high-voltage electron microscopy as discussed in the work of Steeds and Eades [34] dealing with the development of the electron diffraction technique known as real space crystallography. Such pole figures can be used to determine point- and space- groups and are sensitively dependent upon the accelerating potential. As we mentioned earlier, these patterns are also projections of the electron wave interactions with the electronic structure of the specimen, and contain information about the Brillouin zone boundaries and the dispersion surfaces.

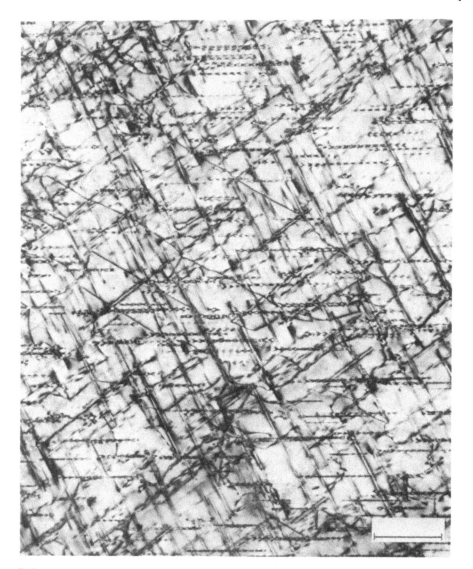

FIG. 8.18 *Dense precipitation in Al—Mg—Si alloy observed in bright field at 1 MeV in the Hitachi Perkin-Elmer 1000 electron microscope. Bar represents 1 μm. (Courtesy Hitachi, Ltd., Tokyo.)*

8.7.1 THE CRITICAL VOLTAGE EFFECT

When viewing the changes which occur in bend contour images such as Fig. 8.5 with accelerating voltage, certain contours are observed to disappear at some critical voltage, V_c. This occurs because pairs of Bloch waves which may be equal in amplitude are exactly out of phase. At this point, two branches of the dispersion surface make contact with a Brillouin zone boundary. This effect can also be observed as an extinction of specific Kikuchi lines or a change in the Kikuchi-line symmetry in the associated electron diffraction (diffuse-scattering) pattern. Consequently V_c can be determined experimentally for any material by systematically changing the accelerating potential and observing the disappearance voltage for second-order Kikuchi lines or second-order Bragg maxima on bend-contour images. Critical voltages can also be determined exactly from theory by examining the relevant Bloch wave excitation coefficients ($C_0^{(j)}$).

It has been observed that the critical voltage of alloys is affected by physical parameters which do not pertain to pure metals. Consequently, V_c depends on short-range order, and can provide some information on disordered solid solutions, ordered alloys, and the like as discussed by Fisher and Shirley [35].

It is apparent in retrospect from Chap. 7 (in particular Sec. 7.10.3) that high-voltage electron microscopy (at voltages of 300kV and above) has become a standard for high-resolution and atomic resolution imaging. The establishment of the ARM (atomic resolution microscope) facility at the University of California, Berkeley and other facilities throughout the world in the decade of the 1980's has been a testimonial to this phenomenon. Many commercial electron microscopes operating above 200kV are scattered around the world. High-voltage electron microscopes operating in the 1 - 3 MeV range are accessible to scientists world-wide, especially in Europe, Japan, and the United States. Voltages as high as 15MV are attainable in high-voltage electron microscopes in Japan. Indeed as high-resolution electron microscopy becomes more routine, high-voltage electron microscopes will also become more routine. Recent advances in modularizing high-voltage power supplies and in developing other high-voltage hardware will also allow microscopes operating in the range 300 - 400kV to be more routine and commercially available. As this happens, distinctions between high-voltage electron microscopy and high-resolution electron microscopy will effectively disappear.

PROBLEMS

8.1 Sketch the power circuit diagram for a high-voltage electron microscope intended for operation at 1.2 MeV if the input voltage is 30 kV at 3×10^4 hertz. The ratio of accelerator stages to generator stages is to be maintained at 3:1. What ideally will be the magnitude of electron acceleration between successive stages in the accelerator tube?

8.2 The operating reflection for diffraction contrast in Fig. 8.8 is $[\bar{1}11]$. Calculate the specimen thickness and the angle of inclination of the twin plane with the surfaces. What is the likely orientation for the grain surface? What percentage of the limiting thickness does the specimen represent?

8.3 Refer to Fig. 8.9 and consider the operating reflection for a two-beam dynamical situation giving rise to the dislocation contrast to $\underline{g} = [\bar{1}11]$. Calculate the expected dislocation width and compare this value with the mean dislocation width measured directly from the micrograph. Plot the dislocation widths as a histogram and comment on the statistical spread of the measurements. (The alloy content of the specimen of Fig. 8.9 is Mg, 17 percent; Si, 9 percent; balance Al.) Assume n = 2 for the dislocations.

8.4 Estimate the density of defect clusters that occur in the image of Fig. 8.12. Plot a histogram of their sizes as measured directly from the micrograph and find the mean image size. Explain the differences in the image size observed. Would you expect the movement of dislocations in the specimen area of Fig. 8.12 to differ markedly from that of Fig. 8.8? Explain your conclusions.

8.5 Devise an experiment to determine the effect of the specimen surfaces on the mobility of grain boundaries in a polycrystalline material. Outline your procedure and the expected significance of your observations. Sketch the potential distribution for two approaching surfaces.

8.6 If the precipitates in Fig. 8.18 lie in the {111} planes of the fcc alloy, determine the possible surface orientations and sketch the details of the corresponding selected-area electron diffraction patterns. Assuming the effective camera length to remain constant, by how much will these patterns change in size if the area is examined at 100 kV? Will the patterns at 100 kV be larger or smaller than those at 1000 kV, that is, will there be more spots or less in the diffraction pattern?

8.7 In the dark-field image of small grains in vapor-deposited Al observed at 60 kV (Fig. P8.7), the effective aperture was found to be

0.003 cm^{-1}. If the objective focal length is maintained constant at 5 mm, and assuming $C_c = 0.3$ f_{ob}, find the associated energy loss. Assuming the energy loss, chromatic aberration coefficient, and focal length to remain constant, calculate the image displacement for the grains shown above at 1 MeV if $\alpha = 0.0001$ cm^{-1}. If the resolving power of the high-voltage instrument is 10 Å, will the image displacement be observable?

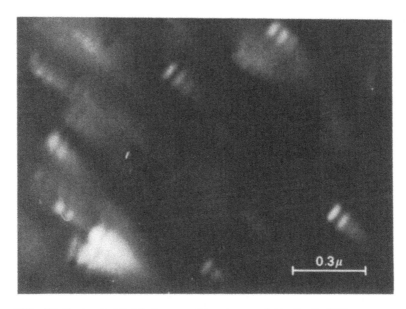

FIG. P8.7 *Dark-field image. (Courtesy of J. S. Lally)*

8.8 Suppose that in a high-voltage electron microscope operated at 1 MeV, 2.5-microns-thick Inconel is observed after cold work and a very short anneal. The average grain size is noted to be roughly 1 micron. Assuming the minimum field for selected-area electron diffraction to be 0.1 micron, indicate in sketches the appearance of a selected-area electron diffraction pattern. Assuming the size of precipitates in this alloy to be unchanged from those observed at 100 kV shown in 7.44, comment on the ability to analyze these precipitates in detail. How might the images of grain boundaries appear assuming mean inclination angles with the specimen surfaces of 63°?

8.9 We want to investigate radiation damage to Ti and a chrome-nickel-thoria dispersion-hardened alloy (Ni, 20 percent Cr, 2 percent ThO$_2$)

if samples of these materials are irradiated in a nuclear reactor for a total dose of 10^{20} nvt. Discuss any special handling required to examine these specimens, waiting time, degree of radioactivity, etc. What would be the optimum accelerating voltage to be employed in a high-voltage electron microscope for direct observation of these materials, keeping in mind that electron irradiation-induced defect clusters are unwanted in the sample?

REFERENCES

1. V. K. Zworykin, J. Hillier, and A. W. Vance, *J. Appl. Phys., 12:* 738(1941).

2. H. O. Muller and E. Ruska, *Kolloid-Z., 95:* 21(1941).

3. M. von Ardenne, *z. Physik, 117:* 657(1941).

4. B. Tadano et al., *J. Electron Microscopy, 4:* 5(1956).

5. N. M. Popov, *Bull. Sci. U.S.S.R., 23:* 436(1959).

6. G. Dupouy, F. Perrier, and R. Fabre, *Compt. Rend., 252:* 627(1961).

7. K. Kobayashi et al., *Japan. J. Appl. Phys., 2:* 47(1963).

8. C. J. Arceneaux (ed.), *Proc. Electron Microscopy Soc. Assn.,* Claitor's Publishing Division, Baton Rouge, La., 1967, 1968.

9. *Proc. Sixth Intern. Conf. Electron Microscopy (Kyoto),* 1(1966).

10. A. Howie, *Phil. Mag., 14:* 223(1966).

11. A. Howie and Z. S. Basinski, *Phi. Mag., 17:* 1039(1968).

12. H. Hashimoto, *Jernkont. Ann., 155:* 479(1971).

13. H. Hashimoto in *High Voltage Electron Microscopy,* P. R. Swann, C. J. Humphreys, and M. J. Goringe (eds.), Academic Press, Inc., New York, 1074, p. 9.

14. C. J. Humphreys, *Phil. Mag., 25:* 1459(1972).

15. H. A. Bethe, *Ann. d. Physik, 87:* 55(1928).

16. H. Hashimoto, *J. Appl. Phys., 35:* 277(1964).

17. H. Hashimoto et al., *J. Phys. Soc. Japan, Suppl. B-11, 17:* 170(1962).

18. P. B. Hirsch, *J. Phys. Soc. Japan, Suppl. B-11, 17:* 143(1962).

19. R. Uyeda and M. Nonoyama, *Japan, J. Appl. Phys., 6:* 557(1967).

20. G. Thomas, *Phil. Mag., 17:* 1097(1968).

21. J. S. Lally et al., *Proc. Electron Microscopy Soc. Am.,* p. 246, Claitor's Publishing Division, Baton Rouge, La., 1967.

22. H. Fujita and N. Sumida, *J. Phys. Soc. Japan, 35:* 224(1973).

23. D. L. Allinson, *J. Microsc., 97:* 209(1973).

24. P. R. Swann and N. J. Tighe, *Jernkont. Ann, 155:* 497(1971).

25. P. R. Swann, G. Thomas, and N. J. Tighe, *J. Microsc., 97:* 249(1973).

26. P. Haasen, et al. (eds.), *Strength of Metals and Alloys,* vol. I - II, Pergamon Press, London, 1979.

27. H. Fujita, et al, *Japan J. Appl. Phys., 6:* 214(1967).

28. L. E. Murr, *Phys. Stat. Sol., 24:* 135(1967).

29. L. E. Murr, *J. Appl. Phys., 39:* 5557(1968).

30. G. Dupouy and F. Perrier, *J. Microscopie, 3:* 233(1964).

31. G. Dupouy and F. Perrier, *Ann. Phys., 8:* 251(1963).

32. L. E. Murr, *Scripta Met., 3:* 167(1969).

33. B. Hudson and M. J. Makin, *J. Phys. E. Scient. Instrum., 3:* 311(1970).

34. J. W. Steeds and J. A. Eades, *Surface Sci., 38:* 187(1973).

35. R. M. Fisher and C. G. Shirley, *J. Metals, 33 (3):* 26(1981).

SUGGESTED SUPPLEMENTARY READING

Arceneaux, C. J. (ed.): *Proc. Electron Microscopy Society of America,* Claitor's Publishing Division, Baton Rouge, La., 94-111(1969).

Bailey, G. W. (ed.): *Proc. Electron Microscopy Society of America,* Claitor's Publishing Division, Baton Rouge, La., 2-30(1980).

Bailey, G. W. (ed.): *Proc. 48th Annual Meeting of the Electron Microscopy Society of America,* San Francisco Press, Box 6800, San Francisco, CA, 1990.

Brederoo, P., and J. Van Landuyt (eds.): *Electron Microscopy 1980,* vol. 4: High Voltage, North-Holland Publishing Co., New York (1980).

Buseck, P., et al. (eds.): *High-Resolution Transmission Electron Microscopy and Associated Techniques,* Oxford University Press, New York, 1989.

Cosslett, V. E.: The High Voltage Electron Microscope, *Contemp. Phys.,* vol. 9, no. 4, July 1968.

Johnson, J. E. and P. B. Hirsch (eds.): *High Resolution and High Voltage Electron Microscopy,* Wiley, New York, 1987.

Makin, M. J., and J. V. Sharp: An Introduction to High-Voltage Electron Microscopy, *J. Mater. Sci.,* vol. 3, p. 360, 1968.

Reimer, L.: *Transmission Electron Microscopy,* 2nd edition, Springer-Verlag, New York, 1989.

Swann, P. R., C. J. Humphreys and M. J. Goringe (eds.): *High Voltage Electron Microscopy,* Academic Press, New York, 1974.

Appendix A
PRINCIPLES OF
CRYSTALLOGRAPHY
AND SOME USEFUL
CRYSTALLOGRAPHIC
AND DIFFRACTION DATA

A.1 INTRODUCTION

While you are assumed to already be fairly familiar with crystal structures, structure notation, and basic crystallographic data, it is certainly
worthwhile to have, during the course of studying the various topics related in one way or another to basic crystallographic principles, a readily
available source of the most useful information. At the same time, the
manipulation of crystallographic equations and notation, especially where
such is required in the analysis of a diffraction pattern, surface markings, etc., is readily forgotten if not frequently used. Consequently a
brief review of some first principles can be very enlightening. It is
with all these shortcomings in mind that the presentation of the material
here has been formulated. You are reminded, perhaps unnecessarily, that
the presentation is by no means complete. At the expense of completeness,
we have really striven for usefulness, perhaps at the expense of clarity
at some points. For a more extensive treatment of much of the material
presented, you are referred to the International Tables for X-ray Crystallography [1] and the books by M. J. Buerger [2], R. W. G. Wyckoff [3],
C. S. Barrett and T. B. Massalski [4], J. B. Cohen [5], and W. B. Pearson
[6].

A.2   UNIT CELLS OF CRYSTAL SPACE LATTICES

Since most of our dealings with materials is concerned with the three-dimensional solid nature of these materials, the two-dimensional crystal and the associated symmetry properties will be ignored here, and we will briefly treat the three-dimensional crystalline lattice.   In effect, three-dimensional lattices evolve simply by stacking two-dimensional layers.   If these layers are considered as arrays of atoms, the smallest consistent symmetry unit, or unit cell, viewed as a space lattice can consist of low-symmetry arrays where the unit cell contains neither orthogonal features nor any equivalence of unit-cell dimensions (the triclinic); or it can consist of perfectly symmetric arrangements of atoms, as in the cubic space lattice.   If we adhere to the notation shown in Fig. A.1, the 7-space lattice unit cells that result form 14 Bravais lattices depicting the possible arrangements of atoms in a solid, as shown in Fig. A.2.   The space lattices of Fig. A.2 are necessarily somewhat exaggerated since the arrangement of the atoms is not very real.   Ideally, in a real solid, the atoms, as balls, are considered to touch; and in reality they compress one another slightly due to nuclear and electron attraction coupled with mutual repulsion.

A.3   THE IDENTIFICATION OF PLANES AND DIRECTIONS IN A CRYSTAL-
CRYSTAL GEOMETRY

A.3.1   CRYSTAL PLANE NOTATION USING MILLER INDICES

The Miller indices are used as a standard notation for the various faces or planes of a crystal referred to in unit-cell convention.   In this notation, the plane formed by the intersection of the axes a, b, and c of Fig. A.1 is designated by taking the reciprocals of the intercepts and reducing these to the integers having a lowest common denominator.   The denominator is then ignored, and the resulting integers correspond to the Miller indices

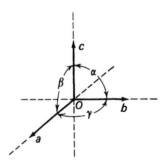

FIG. A.1 *Unit-cell convention.*

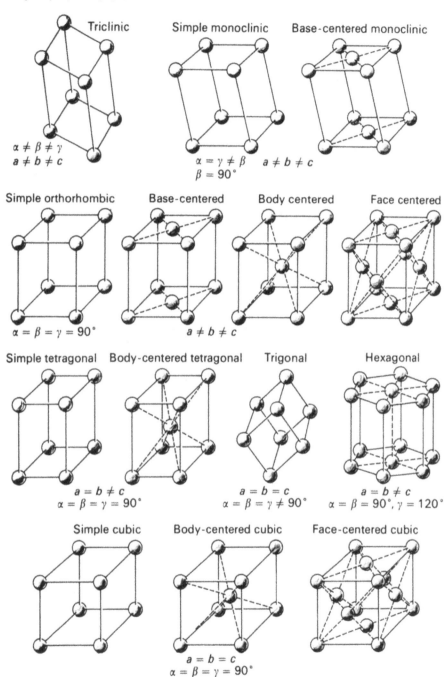

FIG. A.2 *Fourteen Bravais crystal lattices composed of seven crystal systems.*

(hkℓ), the parentheses indicating a plane hkℓ. Figure A.3 illustrates the intercept notation for several crystal planes. Note that the method of reciprocals simply allows all intercepts to be related to the unit cell specified by edge lengths a,[†] b, and c.[*] Consequently a plane with intercepts a/4, b/4, c/8 has reciprocals 4, 4, 8, and Miller indices (448), and is parallel to the (112) plane formed by intercepts a, b, and c/2. All planes having intercepts outside the unit cell are related to the lowest order plane within the unit cell [for example, intercepts at 2a, 2b, c reduce to reciprocals 1/2, 1/2, 1 or (112)], while parallel planes having intercepts within the unit cell become higher-order planes of the form {112}. The braces significant of planes of the same indices indicate equivalent planes of a form in a crystal. Thus the multiplicity of the {112} form is 34 planes or

$$\{112\} = \begin{array}{cccccc} (112), & (121), & (211), & (\bar{1}12), & (\bar{1}21), & (\bar{2}11) \\[4pt] (1\bar{1}2), & (1\bar{2}1), & (2\bar{1}1), & (11\bar{2}), & (12\bar{1}), & (21\bar{1}) \\[4pt] (\bar{1}\bar{1}2), & (\bar{1}\bar{2}1), & (\bar{2}\bar{1}1), & (1\bar{1}\bar{2}), & (1\bar{2}\bar{1}), & (2\bar{1}\bar{1}) \\[4pt] (\bar{1}1\bar{2}), & (\bar{1}2\bar{1}), & (2\bar{1}\bar{1}), & (\bar{1}\bar{1}\bar{2}), & (\bar{2}\bar{2}\bar{1}), & (\bar{2}\bar{1}\bar{1}) \end{array}$$

where indices with a bar above indicate negative intercepts. For example, observe that in Fig. A.3, the intercepts -a, -b, -c result in ($\bar{1}\bar{1}\bar{1}$), which again must be viewed within the unit cell as shown. It is sometimes confusing in this respect to realize that a plane having negative intercepts should, if possible, be translated into the bounds of the unit cell. In effect the negative-index convention simply allows planes of a form as they relate to the symmetry properties of a crystal lattice to be identified on the basis of the unit cell coordinate convention specified (as in Fig. A.1). It is useful in working with planes of a crystal to utilize the eight-cube indexing system shown in Fig. A.3 to relate negatively indexed planes to the unit cell.

In an attempt to clarify this point still further, let us observe the several crystal plane conventions illustrated in Fig. A.4. Here the positive-index convention (100), (010), (001) is readily observed. The negatively indexed planes ($\bar{1}$00), (0$\bar{1}$0), and (00$\bar{1}$) are brought into coincidence with the bounds of the unit cell by the translation scheme shown in Fig.

[†]*Note that a in the text is the same as a in the figures.*
[*]*Note that in determining the indices of a plane, a, b, and c are treated as unit dimensions.*

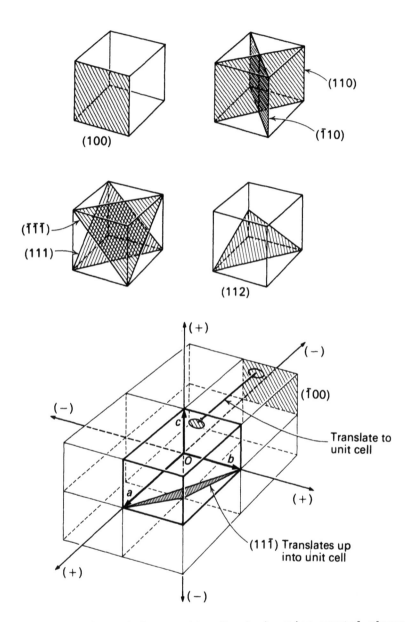

FIG. A.3 *Miller index notation for designating crystal planes and the eight-cube-reference translation scheme for indicating negative index planes.*

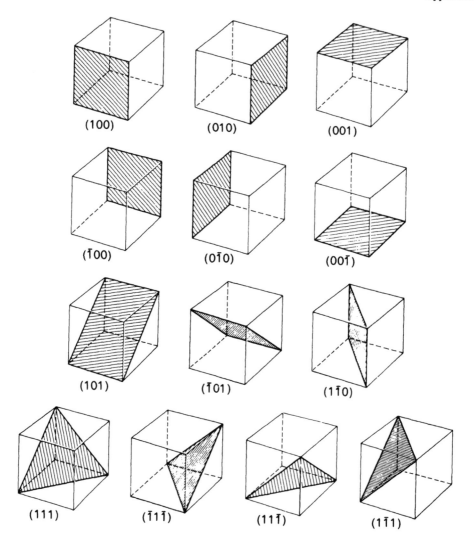

FIG. A.4 *Some important crystal planes characterized by negative Miller indices.*

A.3, and form opposite and equivalent crystal faces. In the case of (101), ($\bar{1}$01), and ($\bar{1}$10) shown, the negative indices specify distinct planes within the unit cell; ($\bar{1}\bar{1}$0) is again an equivalent opposite plane with respect to (110), but not describable within the bounds of the unit cell. This feature isn't really important since we are only interested in the kind of plane in the crystal lattice. A similar scheme evolves for (111), ($\bar{1}$1$\bar{1}$), and (11$\bar{1}$), which represent distinct planes, while (1$\bar{1}$1) is describable within the unit cell as an equivalent parallel plane, the negative of ($\bar{1}$1$\bar{1}$).

A.3.2 INTERPLANAR SPACING

The spacing between equivalent atomic planes in an orthogonal crystal, such as $(1\bar{1}1)$ and $(\bar{1}1\bar{1})$ in Fig. A.4, is given by

$$d_{\{hk\ell\}} = \frac{1}{\sqrt{(h/a)^2 + (k/b)^2 + (\ell/c)^2}} \tag{A.1}$$

For hexagonal crystals, however, the interplanar spacings for planes designated by Miller indices $(hk\ell)$ are given by

$$d_{\{hk\ell\}} = \frac{1}{\sqrt{4/3[(h^2 + hk + k^2)/a^2] + (\ell/c)^2}} \tag{A.2}$$

A.3.3 INDEX NOTATION FOR HEXAGONAL CRYSTAL PLANES

The use of the Miller index notation when dealing with hexagonal crystals has the disadvantage that equivalent hexagonal lattice planes do not have similar indices. In view of this difficulty a notation called the Miller-Bravais indices is generally employed to designate a hexagonal crystal plane $(hk\ell)$, where $i = -(h + k)$. Consequently, $(\bar{1}10)$ and (100) in the hexagonal system are observed to be equivalent as $(\bar{1}100)$ and $(10\bar{1}0)$ of the form $\{1\bar{1}00\}$. Figure A.5 illustrates a number of hexagonal crystal planes in terms of the (3) Miller and (4) Miller-Bravais indices.

A.3.4 ANGLE BETWEEN CRYSTAL PLANES

It is also interesting if not necessary to be able to designate the angle one plane of a crystal makes with another. This is especially important in investigations of surface crystallographies and surface orientations, as well as the characterization of slip traces, specimen thickness, etc., in thin-film sections observed by transmission electron microscopy. If θ is the dihedral angle a plane $(hk\ell)$ makes with a reference plane or surface plane (HKL), then for an orthogonal crystal system (that is, orthorhombic, tetragonal, cubic) the general equation is

$$\text{Orthogonal crystals} \quad \cos \theta =$$

$$\frac{H/h/a^2 + Kk/b^2 + L\ell/c^2}{\sqrt{(H/a)^2 + (K/b)^2 + (L/c)^2} \sqrt{(h/a)^2 + (k/b)^2 + (\ell/c)^2}} \tag{A.3}$$

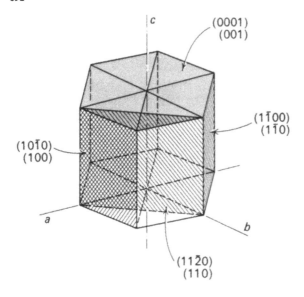

FIG. A.5 *Crystal planes in hexagonal system.*

For the hexagonal crystal lattice, the relationship takes the form

$$\left|\begin{matrix}\text{Hexagonal} \\ \text{crystals}\end{matrix}\right| \ \cos\ \theta\ =$$

$$\frac{Hh\ +\ Kk\ +\ (1/2)(Hk\ +\ Hh)\ +\ (3/4)(a/c)^2L\ell}{\sqrt{[H^2\ +\ K^2\ +\ HK\ +\ (3/4)(a/c)^2L^2][h^2\ +\ k^2\ +\ hk\ +\ (3/4)(a/c)^2\ell^2]}} \qquad (A.4)$$

A.3.5 DIRECTIONS IN A CRYSTAL

Since we have fairly rigorously specified the notation for planes in a crystal, we require now a notation to specify directions in a crystal as well. The notation [uvw] is generally adopted for this purpose; the specification of a direction is [uvw] and directions are indicated by carets <uvw>. uvw simply specifies the coordinates of a point from an origin (Fig. A.1), while [uvw] denotes the direction of a line from the origin to the point. In the designation of crystal directions, the smallest integers, u v, w are employed, and fractional coordinates are generally avoided. The notation is therefore rather standard insofar as the crystal directions constitute simple vector notation. In specific cases, negative indices also appear in crystal direction notation, for example, the direction perpendicular to a negatively indexed plane.

The [uvw] direction notation suffices for orthogonal crystals and crystals in general, with the exception of the hexagonal crystal lattice. A notation widely used in the hexagonal system consists in writing [UVW] or [uvtw] where

$$U = u - t \qquad V = v - t \qquad W = w \qquad t = -(u + v)$$

The transformation from [UVW] to [uvtw] therefore requires

$$u = (1/3)(2U - V) \qquad v = (1/3)(2V - U) \qquad t = -(u + v) \qquad w = W$$

A.3.6 ANGLE BETWEEN CRYSTAL DIRECTIONS

The angle θ' between directions $[u_1 v_1 w_1]$ and $[u_2 v_2 w_2]$ in a general crystal system is given by

$$\cos \theta' = \frac{a^2 u_1 u_2 + b^2 v_1 v_2 + c^2 w_1 w_2 + A\,bc + B\,ac + C\,ab}{I_1 I_2} \tag{A.5}$$

where

$$A = (v_1 w_2 + w_1 v_2) \cos \alpha$$

$$B = (u_1 w_2 + w_1 u_2) \cos \beta$$

$$C = (v_1 u_2 + v_2 u_1) \cos \gamma$$

and

$$I_1 = \sqrt{a^2 u_1^2 + b^2 v_1^2 + c^2 w_1^2 + 2bc v_1 w_1 \cos \alpha + 2ca w_1 u_1 \cos \beta + 2ab u_1 v_1 \cos \gamma}$$

$$I_2 = \sqrt{a^2 u_2^2 + b^2 v_2^2 + c^2 w_2^2 + 2bc v_2 w_2 \cos \alpha + 2ca w_2 u_2 \cos \beta + 2ab u_2 v_2 \cos \gamma}$$

In the cubic system, Eqs. (A.3) and (A.5) are identical if θ is the angle between planes (HKL) and (hkℓ) and θ' is the angle between directions [HKL] and [hkℓ], which in turn are perpendicular to (HKL) and (hkℓ), respectively. We can then write

$$\theta = \theta' = \cos^{-1} \frac{Hh + Kk + L\ell}{\sqrt{H^2 + K^2 + L^2}\,\sqrt{h^2 + k^2 + \ell^2}} \tag{A.6}$$

for the angle between planes or directions in the cubic system. It should be cautioned that the direction [HKL] is necessarily perpendicular to the plane (HKL) only in the cubic system. For a general treatment of the condition for perpendicularity between a direction [uvw] and plane (hkℓ),

you are urged to consult *International Tables for X-ray Crystallography*,
Vol. II, p. 106 [1]. The positive first-quadrant angles between numerous
low-index planes (HKL) and (hkℓ) or directions [HKL] and [hkℓ] in cubic
crystals computed from Eq. (A.6) are listed in Table A.1.

A.3.7 *THE CRYSTALLOGRAPHIC ZONE AXIS*

Since our particular concern in this text is electron optical analysis, a
property of considerable use in electron diffraction analysis is the zone
axis [uvw] of a crystal. Crystal planes parallel to a zone axis [uvw] are
regarded as planes of the zone. Since $\underset{\sim}{r}^{*}_{hk\ell}$, the reciprocal lattice vector,
is normal to the plane (hkℓ), $\underset{\sim}{r}^{*} \cdot \underset{\sim}{r} = 0$ expresses the condition that (hkℓ)
belongs to the zone [uvw], or

$$hu + kv + \ell w = 0 \tag{A.7}$$

For the cubic system, we can specify the zone axis by [HKL], which is
normal to a crystallographic surface (HKL). Consequently, for the incident
electron beam parallel to [HKL] and normal to (HKL), the corresponding
reflections from allowed {hkℓ} planes are found from

$$Hh + Kk + L\ell = 0 \tag{A.8}$$

A.4 DETERMINATION OF CRYSTALLOGRAPHIC ORIENTATIONS

A.4.1 *SINGLE-CRYSTAL ELECTRON DIFFRACTION PATTERNS*

Since the Bragg angles in electron diffraction are very small, with respect to x-ray diffraction, the use of Eq. (A.8) allows the expected reflections {hkℓ} (all allowed reflections determined by the corresponding
crystal structure factors as discussed in Chap. 6) to be determined for
the electron beam along [HKL]. As we have mentioned, for cubic crystals,
and in the hexagonal system for the condition

$$\frac{1}{h}\left(u - \frac{v}{2}\right) = \frac{1}{k}\left(v - \frac{u}{2}\right) = \frac{1}{\ell}w\left(\frac{c}{a}\right)^{2}$$

[HKL] is perpendicular to (HKL). It is therefore possible, considering
the allowed reflections for a particular crystal system (see Table 6.1),
to construct single-crystal diffraction spot patterns representing a specific crystalline surface orientation (HKL) of the reciprocal lattice.

For example, Fig. A.6 illustrates a composite diffraction pattern indicating a (100) surface orientation for a diamond cubic, bcc, and fcc crystal system. Thus, with [100] oriented parallel to the electron beam, the reflections observed in a selected-area electron diffraction pattern will be described by the corresponding diffraction nets of Fig. A.6. Let us, as a simple exercise, consider [100] in Eq. (A.8). The first reflections to be expected for diamond cubic will be 022, $0\bar{2}2$, and $02\bar{2}$ (brackets or parentheses being deleted to indicate both here and in Fig. A.6 that the indices represent crystal directions with respect to the central beam or zero-order reflection 000, or reflections from corresponding planes {hkℓ}. The next reflecting planes will be 004 and $00\bar{4}$; and we observe, in fact, all reflections of an allowed form {0hk} for the diamond cubic, bcc, and fcc crystal systems depicted in Fig. A.6. Similarly, for [HKL] = [110] in Fig. A.7, the conditions for expected reflections are governed by {hhℓ} or <H$\bar{\text{H}}$L>, where h or H can be zero.

Numerous commonly encountered crystallographic orientations in electron studies for the diamond cubic, bcc, and fcc systems are included in

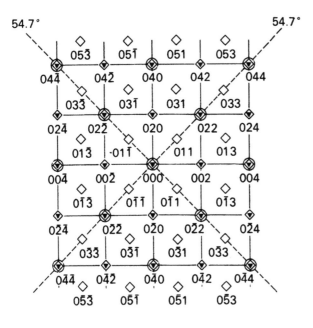

FIG. A.6 *(100) diffraction patterns.* ◯ *is diamond cubic,* ◇ *is bcc,* ▼ *is fcc. Dashed lines are traces of* {111}.

TABLE A.1 *Angles between planes* {*HKL*} *and* {*hkℓ*} *in the cubic crystal system*[†]

{HKL}	{hkℓ}	Angles between HKL and hkℓ, in degrees					
100	100	0.0	90.0	——	——	——	——	——
	110	45.0	90.0	——	——	——	——	——
	111	54.7	——	——	——	——	——	——
	210	26.6	63.4	90.0	——	——	——	——
	211	35.3	65.9	——	——	——	——	——
	221	48.2	70.5	——	——	——	——	——
	310	18.4	71.6	90.0	——	——	——	——
	311	25.2	72.5	——	——	——	——	——
	320	33.7	56.3	90.0	——	——	——	——
	321	36.7	57.7	74.5	——	——	——	——
110	110	0.0	60.0	90.0	——	——	——	——
	111	35.3	90.0	——	——	——	——	——
	210	18.4	50.3	71.6	——	——	——	——
	211	30.0	54.7	73.2	90.0	——	——	——
	221	19.5	45.0	76.4	90.0	——	——	——
	310	26.6	47.9	63.4	77.1	——	——	——
	311	31.5	64.8	90.0	——	——	——	——
	320	11.3	54.0	66.9	78.7	——	——	——
	321	19.1	40.9	55.5	67.8	79.1	——	——
111	111	0.0	70.5	——	——	——	——	——
	210	29.2	75.0	——	——	——	——	——
	211	19.5	61.9	90.0	——	——	——	——
	221	15.8	54.7	78.9	——	——	——	——
	310	43.1	68.6	——	——	——	——	——
	311	29.5	58.5	80.0	——	——	——	——
	320	36.8	80.8	——	——	——	——	——
	321	22.2	51.9	72.0	90.0	——	——	——
210	210	0.0	36.9	53.1	66.4	78.5	90.0	——
	211	24.1	43.1	56.8	79.5	90.0	——	——
	221	26.6	41.8	53.4	63.4	72.7	90.0	——
	310	8.1	32.0	45.0	64.9	73.6	81.9	——
	311	19.3	47.6	66.1	82.3	——	——	——
	320	7.1	29.7	41.9	60.3	68.2	75.6	82.9
	321	17.0	33.2	53.3	61.4	69.0	83.1	90.0
211	211	0.0	33.6	48.1	60.0	70.5	80.4	——
	221	17.7	35.3	47.1	65.9	74.2	82.2	——
	310	25.4	40.2	58.9	75.0	82.6	——	——
	311	10.0	42.4	60.5	75.8	90.0	——	——
	320	25.1	37.6	55.5	63.1	83.5	——	——
	321	10.9	29.2	40.2	49.1	56.9	70.9	77.4
		83.7	90.0	——	——	——	——	——
221	221	0.0	27.3	38.9	63.6	83.6	90.0	——
	310	32.5	42.5	58.2	65.1	84.0	——	——
	311	25.2	45.3	59.8	72.5	84.2	——	——
	320	22.4	42.3	49.7	68.3	79.3	84.7	——
	321	11.5	27.0	36.7	57.7	63.6	74.5	79.7
		84.9	——	——	——	——	——	——

TABLE A.1 *Angles between planes {HKL} and {hkℓ} in the cubic crystal system (Continued)*[†]

{HKL}	{hkℓ}	Angles between HKL and hkℓ, in degrees						
310	310	0.0	25.8	36.9	53.1	72.5	84.3	——
	311	17.6	40.3	55.1	67.6	79.0	90.0	——
	320	15.3	37.9	52.1	58.3	79.7	79.9	——
	321	21.6	32.3	40.5	47.5	53.7	59.5	6.50
		75.3	85.2	90.0	——	——	——	——
311	311	0.0	35.1	50.5	63.0	84.8	——	——
	320	23.1	41.2	54.2	65.3	75.5	85.2	——
	321	14.8	36.3	49.9	61.1	71.2	80.7	——
320	320	0.0	22.6	46.2	62.5	67.4	72.1	——
	321	15.5	27.2	35.4	48.2	53.6	58.7	68.2
		72.8	77.2	85.8	90.0	——	——	——
321	321	0.0	21.8	31.0	38.2	44.4	50.0	64.6
		69.1	73.4	85.9	——	——	——	——

[†]*Actually it should be pointed out that the angles are also the angles between directions <HKL> and <hkℓ>; and that either HKL or hkℓ can be designated the reference or surface plane or zone axis in studies involving crystallographic orientation, etc.*

the diffraction nets of Figs. A.6 to A.14, and several common hexagonal-pattern orientations are also included as Figs. A.15 to A.17.

Figures A.6 to A.17 have been constructed to an arbitrary scale that maintains dimensions between reflections in such a way that measuring directions directly from the nets and taking the ratios will produce values consistent with taking the actual ratio of the indices as outlined in Chap. 6. It is therefore possible to magnify or demagnify the patterns of Figs. A.6 to A.17 consistent with the particular camera constant of a particular electron microscope, in order to obtain diffraction pattern "overlays" from which direct crystallographic information may be obtained as outlined in Chap. 6.

You should observe that the {111} traces sketched in the diffraction nets of Figs. A.6 to A.14 represent slip traces, stacking-fault planes, twin planes, etc., for fcc and diamond cubic materials. The angles associated with these traces indicate the angle the particular (111) plane makes with the crystallographic surface plane represented by the diffraction net. Crystallographic information relating to these features can be readily identified by comparing appropriate nets with the corresponding electron optical image data.

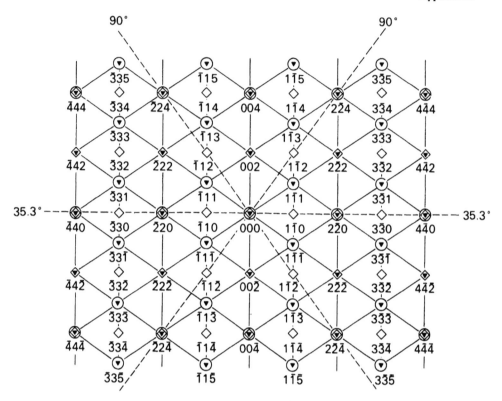

FIG. A.7 *(110) diffraction patterns.* ◯ *is diamond cubic,* ◇ *is bcc,* ▼ *is fcc. Dashed lines are traces of* {1̄11}.

A.4.2 CRYSTAL STEREOGRAPHY AND STEREOGRAPHIC PROJECTIONS

In dealing with the geometrical features of solid crystals as outlined
above, it is sometimes very difficult, if not impossible, to simultaneously
visualize the angular relationships between the corresponding planes and
directions, normals to planes (poles), and zone axes. It is also difficult
to sketch in projection many angular and/or geometric features of a solid
crystal. A convenient means for simultaneously performing such manipulations is found in crystal stereography and crystal stereographic projections. This scheme evolves as follows: consider a unit cube of a solid
(a cube solid) to be enclosed in a transparent sphere, and to be positioned
at the center of the sphere. If we now expose the various crystallographic
planes of the cube we can "project" each plane onto the transparent "reference sphere" by defining a normal, symmetrically disposed with respect to
the dimensions of the "exposed" planes, as a crystallographic "pole".

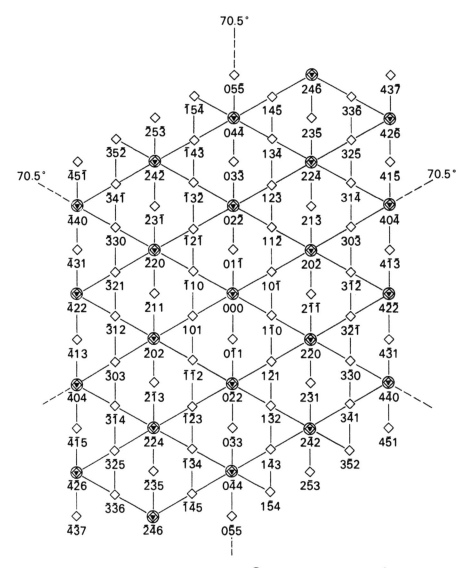

FIG. A.8 *(111) diffraction patterns.* ◯ *is diamond cubic,* ◇ *is bcc,* ▼ *is fcc. Dashed lines are traces of* {111}.

Figure A.18 illustrates this concept. A reference circle that cuts a number of poles, whose normals lie parallel to and in the circular plane formed by the reference circle, represents a crystallographic zone since by definition (for the cubic system) a zone axis is characterized by all planes parallel to the axis. Consequently we are able, with regard to Fig. A.18, to project the hemispherical surface portion of the reference

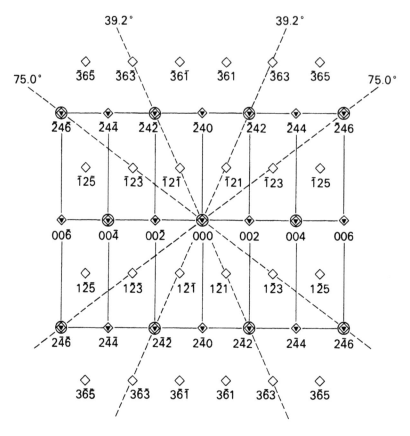

FIG. A.9   *(210) diffraction patterns.* ◯ *is diamond cubic,* ◇
is bcc, ▼ *is fcc.   Dashed lines are traces of* {111}.

sphere onto any designated plane by looking along a zone axis or pole normal to this plane. Furthermore, if the reference sphere is ruled as a
globe in degrees longitude and degrees latitude, then poles lying on common great circles as shown in Fig. A.18 are uniquely representative of the
angle one plane makes with another. In effect, we have only to measure
the angular displacement from one pole to another on a common great circle
or meridian projection to find the angle between crystal planes of a zone.
If, in Fig. A.18, we designate 001 as north, 00Ī as south, 0Ī0 as west,
and 010 as east, and rule the globe accordingly, the projection of such a
ruled globe would appear as shown in Fig. A.19. Such a meridian projection is called a Wulff net or simply a stereographic net, the net being
projected onto a reference plane normal to 100. Of course, the designation of the reference poles of the globe or reference sphere of Fig. A.18

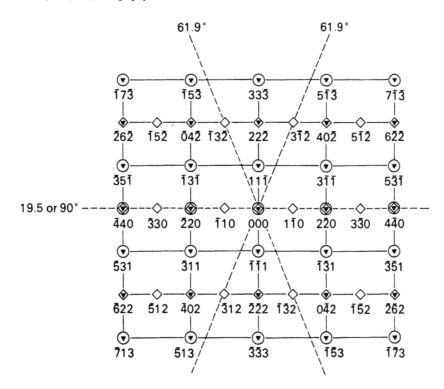

FIG. A.10 *(112) diffraction patterns.* ○ *is diamond cubic,* ◇ *is bcc,* ▼ *is fcc. Dashed lines are traces of* {111}.

as N, S, E, and W is quite arbitrary, and the stereographic net of Fig. A.19 can be used as a standard for a projection of the reference sphere normal to any plane.

In Figure A.18, a projection of the reference sphere (or hemisphere) onto the 001 plane (shown shaded) along the [001] zone axis is indicated by the arrow. The stereographic projection observed on the shaded plane would then appear as shown in Fig. A.20. A similar projection along [110] as shown by the corresponding arrow in Fig. A.18 would also result in the stereographic projection shown in Fig. A.21.[†]

In the stereographic reference net of Fig. A.19, the meridians (longitude lines) extend from top to bottom, while the latitude lines extend from left to right across the projection. Such nets can be ruled to a desired graduation so that measurement errors are not greater than a degree. The measurement of angles between poles in the projections of Figs.

[†]*Some useful stereographic data can be found in* Acta Cryst., 5: 294(1952).

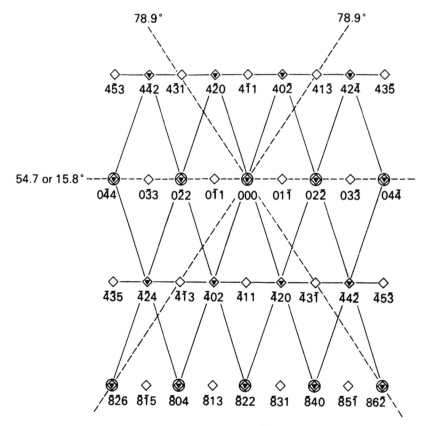

FIG. A.11 *(122) diffraction patterns.* ◯ *is diamond cubic,* ◇ *is bcc,* ▼ *is fcc. Dashed lines are traces of* {111}.

A.20 and A.21 is performed by superposing the projection over Fig. A.19 in an overlay scheme. The coincidence of any two poles on the same meridian by rotation of the projection over the stereographic net then allows the angle between the corresponding crystal planes to be accurately determined by simply counting the latitude lines between the poles along the meridian. The manipulation of stereographic projections is very useful when the projections of Figs. A.20 and A.21 or similar projections are imprinted on transparencies, with the circumference of the great circle coincident with that of a ruled stereographic net.[†]

[†]*Stereographic nets are available commercially from the Hydrographic Office, Navy Department, Washington, D.C.; N. P. Nies, Laguna Beach, Calif.; and Polycrystal Book Service, Box 11567, Pittsburgh, Penn. The latter two sources also supply standard stereographic projections of sizes commensurate with the nets, and in transparent forms.*

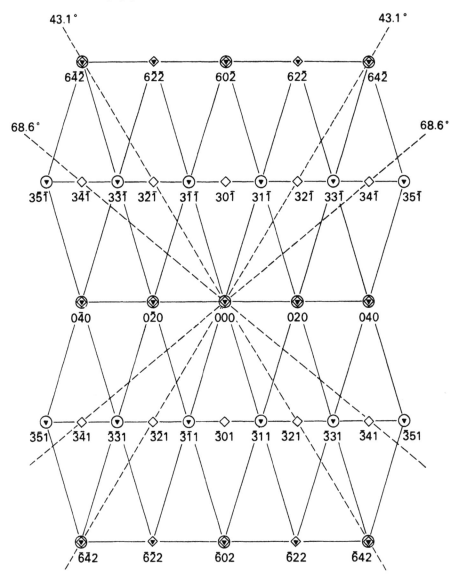

FIG. A.12 *(103) diffraction patterns.* ◯ *is diamond cubic,* ◇ *is bcc,* ▼ *is fcc. Dashed lines are traces of* {111}.

Crystallographic analysis with stereographic projections One obvious property of standard projections such as those shown in Figs. A.20 and A.21 is that they contain a graphical reference for rapidly finding the angles between planes of a crystal. And, because in the cubic crystal system (simple cubic, bcc, fcc, diamond cubic), directions are normal to the cor-

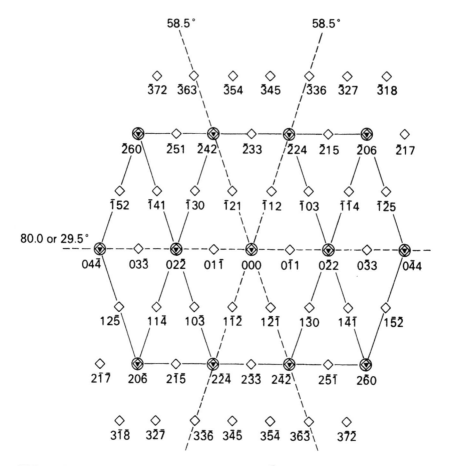

FIG. A.13 *(311) diffraction patterns. ◯ is diamond cubic, ◇ is bcc, ▼ is fcc. Dashed lines are traces of {111}.*

responding planes of the crystal, it is possible to analyze surface crystallographic orientations where two definable traces or trace directions are observed. This is particularly useful in transmission electron microscopy, where an identifiable selected-area electron diffraction pattern is not available to establish the crystallographic surface of a corresponding transmission image that contains two or more identifiable traces. Of particular importance, and as an adequate example, we can consider an electron transmission microscope image of an fcc film containing two traces. We can at the outset assume they are traces of {111} planes. The technique then involves drawing a great circle on tracing paper representing the unknown specimen surface. Diameters are then drawn on this circle normal

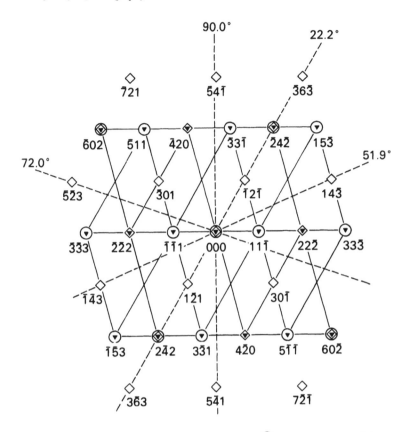

FIG. A.14 *(123) diffraction patterns.* ◯ *is diamond cubic,* ◇ *is bcc,* ▼ *is fcc.* *Dashed lines are traces of* {111}.

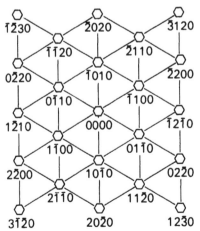

FIG. A.15 *(0001) diffraction pattern; hcp.*

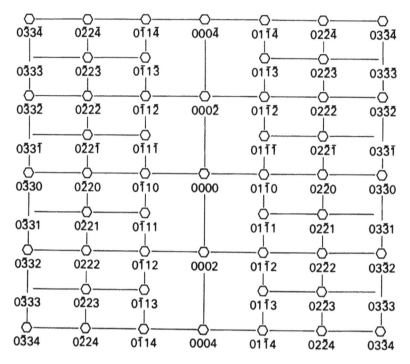

FIG. A.16   *(10Ī0) diffraction pattern: hcp*

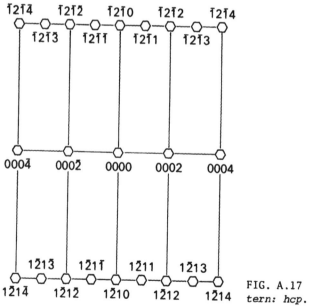

FIG. A.17   *(11Ī0) diffraction pattern: hcp.*

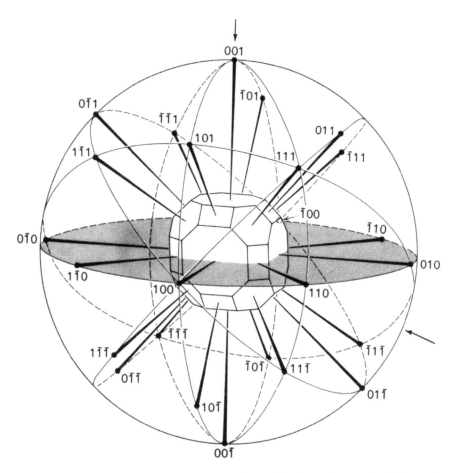

FIG. A.18 *Stereographic reference sphere showing pole projections of cubic crystal planes.*

to the traces observed. A standard projection containing all poles of {111}, such as Fig. A.20, is then superposed as a transparency over this tracing with a stereographic net of the same size great circle as an underlay. These three sheets are pinned through the projection circle centers so that they can be independently rotated with respect to one another. An attempt is then made to find a situation where a pole of the standard projection can be rotated into coincidence with the diameters on the tracing sheet by an equivalent rotation about the stereographic-net axis. The position of the {111} poles following such an operation describes a possible specimen crystal orientation of a form {hkℓ}. Unfortunately a specific plane (hkℓ) cannot unambiguously be described unless a third nonparallel trace is present.

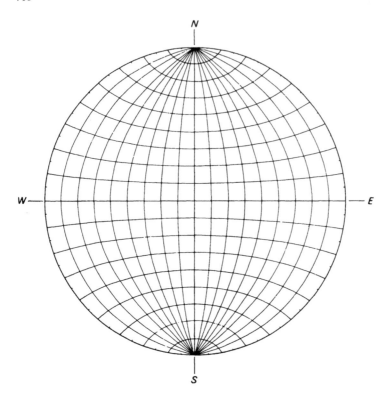

FIG. A.19 *Meridian stereographic net with 10° graduations.*

Some refinements as a result of detailed investigations of the determination of surface crystallography by stereographic projections have been
made by M. P. Drazin and H. M. Otte [7] who have also devised tables for
the orientation of specimens from measurements of {111} traces, as in an
electron transmission microscope image [8]. You may find additional help
in orientation problems by consulting E. A. Wood, *Crystal Orientation
Manual,* Columbia University Press, New York, 1963, or B. D. Cullity, *Elements of X-ray Diffraction,* Addison-Wesley Publishing Company, Inc.,
Reading, Mass., 1956.

As noted in Chap. 2, the projection of a metal whisker in a field
electron or ion microscope is in reality an orthographic projection of
ions representing zone or crystallographic net plane atoms, or those atoms
representing the edges or ledges of crystal planes accommodating the pseudohemispherical end form. Superposition of a stereographic net on a field-
ion micrograph of exactly the same size showing the complete great-circle
plane of projection can allow the crystal zones to be fairly rigorously

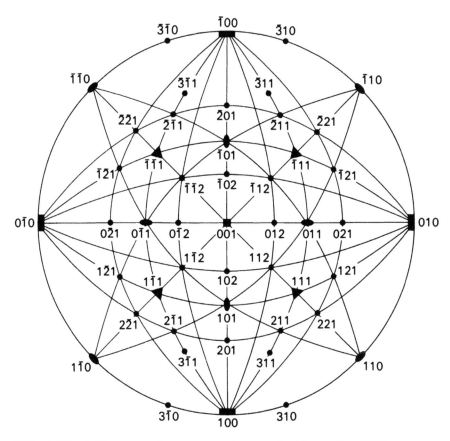

FIG. A.20 *Cubic crystal stereographic projection onto (001).*

described, and the crystallographic features of the end form to be studied
in some detail. For whiskers with known growth axes, standard stereographic projections along these axes can be most useful in the identification
of the tip crystallography.

Stereography of noncubic crystals While we have been concerned, in the
preceding presentation of stereographic projections, with cubic crystals,
stereographic projections can be devised for any crystal system. The basic proposition that poles are normal to planes in the system is adhered
to, and the angular properties of poles can be accurately measured on projections using the stereographic net and calculating the angles from equations such as Eq. (A.4) for hexagonal crystals for example, or Eq. (A.3)
for other orthogonal systems. An excellent reference for constructing
such projections is to be found in F. C. Phillips, *An Introduction to
Crystallography,* 3rd, ed., Longmans, Green & Co., Ltd., London, 1964.

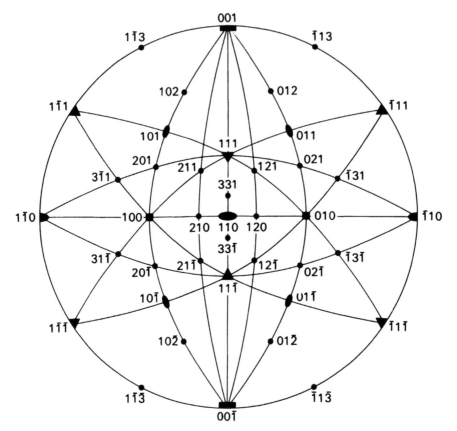

FIG. A.21 *Cubic crystal stereographic projection onto (110). Only a few prominent zones are shown.*

REFERENCES

1. *International Tables for X-ray Crystallography*, vols. I, II, and III, Kynoch Press, Birmingham, England, 1962.

2. M. J. Buerger, *Elementary Crystallography*, John Wiley & Sons, Inc., New York, 1956.

3. R. W. G. Wyckoff, *Crystal Structures*, vols. I and II, Interscience Publishers, a division of John Wiley & Sons, Inc., New York, 1963.

4. C. S. Barrett and T. B. Massalski, *Structure of Metals*, 3d ed., McGraw-Hill Book Company, New York, 1966.

5. J. B. Cohen, *Diffraction Methods in Materials Science*, The Macmillan Company, New York, 1966.

6. W. B. Pearson, *A Handbook of Lattice Spacings and Structures of Metals and Alloys*, Pergamon Press, New York, 1958.

7. M. P. Drazin and H. M. Otte, *Phys. Status Solidi, 3*: 814(1963).

8. M. P. Drazin and H. M. Otte, *Tables for Determining Cubic Crystal Orientations from Surface Traces of Octahedral Planes*, Harrod Company, Baltimore, 1964.

Appendix B
SPECIMEN PREPARATION

B.1   INTRODUCTION

This appendix is intended as a convenience to those readers actively en-
gaged in the investigation of materials by any of the electron optical and
electron or ion optically oriented techniques outlined in the text. We
will attempt to demonstrate briefly the techniques employed in preparing
specimens for the various modes of analysis, in addition to tabulating the
more reliable formulas and recipes. It must be cautioned at the outset
that this presentation is not exhaustive - it is intended only as an introduction to the techniques of specimen preparation[†].

Because of the large number of techniques in use for preparing specimens, it has been necessary in the present treatment to minimize the details and to present, especially in the preparation of thin films for electron microscopy, only those descriptions and formulas that require the
least amount of effort and experience. In most cases where appropriate,

---

[†]*You might wish to consult L. S. Brammar and M. A. P. Dewey,* Specimen Preparation for Electron Metallography, *Elsevier Publishing Company, Amsterdam, 1966, for a more rigorous treatment of specimen preparation.*

the reference to the original work is quoted as a convenience to the reader. The references are listed consecutively by number and collected at the conclusion of this Appendix.

B.2 SPECIMEN PREPARATION FOR THERMIONIC EMISSION MICROSCOPY

As discussed in Sec. 2.4.2, the specimen surface is responsible for the image features in the thermionic emission microscope. The surface to be observed - metallic or nonmetallic - is prepared by any of the conventional metallographic techniques employing chemical polishing or surface (mechanical) polishing (see Tables B.3 and B.4). Specimen size is limited only by the instrument dimensions. In most cases, emission must be enhanced by the application of a suitable activator to the prepared surface to reduce the effective work function. This can be accomplished simply by "painting" the surface, or by surface-tension spreading of a fluid activator, or by vapor depositing the activator onto the prepared emission surface. Successful images then depend on the operating potential and the specimen temperature.

Table B.1 lists a number of materials and activators that show good emission characteristics. This table also gives related details of the operating potential and the sample temperature. For details of vapor deposition, consult Sec. B.6.

B.3 *PREPARATION OF SPECIMENS FOR FIELD EMISSION AND SCANNING TUNNELING MICROSCOPIES*

The preparation of suitable emission tips for surface observations and for tunneling and atomic force microscopes is, like the preparation of electron transmission specimens described briefly in Sec. B.6, somewhat of an art. Wires of the material of interest welded (spot welded) onto a suitable loop of wire (such as Mo about 0.010 in. ($\sim$ 0.025cm) in diameter or less) can be etched to a point by dipping the end into a suitable etchant, or the polishing can be aided by the application of an electric current. As illustrated in Fig. B.1, the manual dipping methods, as they are called, require little more than patience. The ultimate production of a satisfactory emission tip will, however, usually require some particular electrolyte or etchant composition; and the manner of dipping or electrification will also invariably influence the tip character. Figure B.2 illustrates the three fundamental tip shapes, each influenced somewhat by the manner of dipping and the conditions of etching or electropolishing, as well as the viscosity of the electrolyte. The most desirable tip shape is the concave shape of Fig. B.2c because in this case the tip radius does not

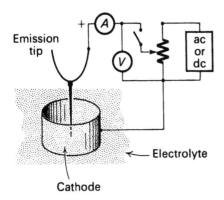

FIG. B.1 *Manual dipping method for polishing emission tips.*

change rapidly with field evaporation. Figure B.2b also satisfies this condition if the polishing takes place uniformly along a considerable portion of the tip length. The tip shape can also be altered by varying the cathode shape during electropolishing as shown in Fig. B.1. In some cases the cathode can be a thin plate lying flat in the electrolyte bath.

Y. Yashiro [4] has also perfected a thin-layer condenser discharge-electrolytic polishing method that preferentially produces characteristic concave tips somewhat automatically. In this method, illustrated schematically in Fig. B.3, the electrolyte, having a specific gravity less than that of the insulating layer, floats above this layer and does not mix with it. Generally a high-viscosity liquid is desirable as the electrolyte. The insulating layer is usually CCl_4 or some similar fluid. By adjusting the electrolyte layer thickness, it is possible to control the tip shape to some extent; and the application of pulse current by the condenser discharge promotes surface smoothness and circularity of the tip cross section. Polishing is continued until the wire separates in the concave cross section. Near the moment of separation, the voltage is kept low to minimize stresses induced by an impulsive separation.

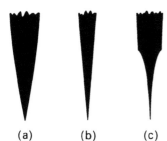

(a)   (b)   (c)

FIG. B.2 *Characteristic emission-tip shapes.* *(a) Convex; (b) straight; (c) concave.*

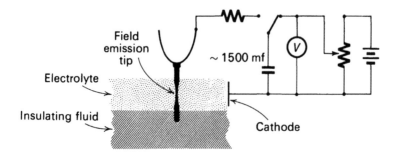

FIG. B.3 *Thin-layer condenser-discharge (electrolytic) polishing of emission tips.* [*After Y. Yashiro,* Oyobutsuri
(Applied Physics, Japan), 33: 912(1964).]

Final polishing of tips to hemispherical perfection is normally accomplished directly within the field emission microscope by the application of
a field of sufficient intensity to cause desorption of surface protrusions,
by heating the tip (high-temperature anneal), or a combination of both.

During the polishing of certain materials either by the dipping method of Fig. B.1 or the discharge-layer method of Fig. B.3, the production
of heat and gas bubbles at the tip surface may disturb the process. Certain additives can be incorporated in the electrolyte to reduce the bubble
problem.

Table B.2 outlines the methods for polishing a number of field emission tips. This table is not necessarily complete; and in some cases you
will profit by consulting specific texts treating field emission microscopy[†]. It must also be cautioned that in many instances of successful
preparations of emission tips reported in the literature, the critical
steps in producing the tip are excluded. In frequent cases the fine details for successfully producing certain emission samples appear to be
eliminated in the description.

B.4 SPECIMEN PREPARATION FOR ELECTRON PROBE MICROANALYSIS AND SCANNING ELECTRON MICROSCOPY

You will recall from Chaps. 4 and 5 that in the examination of specimen
composition in the electron microprobe, the specimen cannot be altered.
The requisites are that in the surface be conducive so that surface charges do not accumulate, and that the sample be of a sufficient size to accommodate analysis.

[†]*Consult the Supplementary Reading List of Chap. 2.*

TABLE B.1 *Sample activation for thermionic emission*

Sample material	Activator	Mode of activation	Cathode (sample) temp., °C	kV	Comments	Reference
Co	Ba or Sr formate	Paint surface with solution	<1000	20	Images also formed at 800°C without activation using higher potentials and low vacuum	1
Cr	Sr	Vapor deposition	<1000	20	Ba activation will give slightly poorer image contrast	1
Cu	Ba formate	Paint surface with formate solution	700–800	20	Sr-Ba formate activator can enhance emission above 850°C	1
Fe	Ba	Vapor deposition *in situ*	1000	20	When vapor depositing Ba, vacuum should be <10^{-4} torr; purge with pure Ar	1,2
	Ba-Co formate	Paint surface	<1000			
Mo	Ba	Vapor deposition	1000	25	Purge system with Ar-image in high vacuum	2
Ni and Ni alloys	Ba	Vapor deposition	>750	20–25	Emission reversals may occur at different temperatures	2,3
Si	Ba-Co formate	Solution painting of specimen surface	800–900	25	Images can be obtained without activation	1
Steels (carbon)	Cs	Vapor deposition	650–750	25–30	Emission reversal occurs between two activators	2
	Ba	Vapor deposition	>1000	25		
Steels (stainless) (Ni, Cr, Fe alloys)	Ba	Vapor deposition	>1000	25	System must be purged with Ar. Images also of good quality without activation	
Ta	Ba-Co formate	Solution painted	<1000	25	Add Co formate to Ba formate in 1 : 2 ratio	1
Ti	Ba	Vapor deposition	1000	25–30	Image quality improved with high vacuum	2
W	Ba	Vapor deposition	>1000	25	Image quality improved with high vacuum	2
/r	Ba	Vapor deposition	>1000	25	Image quality improved with high vacuum	2

TABLE B.2 *Preparation of field emission tips*

Material	Polishing Method	Electrolyte	Comments	Reference
Ag	Manual dipping (Fig. B.1)	40% NHO_3 solution or 15% KCN solution	Use Ag cathode and apply about 20 volts (dc). Technique applies to 4- to 10-mil wires	
Au	Condenser discharge (Fig. B.3)	15% KCN solution (2 mm layer)	Use Au or Pt electrode (cathode) and apply approximately 10 to 15 volts (ac). CCl_4 serves as insulating layer	
	Manual dipping (Fig. B.1)	50% HNO_3 50% HF	Use about 10 volts (ac)	—
Au-Pt	Manual dipping (Fig. B.1)	1N KCN	Apply 105 volts (ac) for 2-mil wires	14
Be	Manual dipping (Fig. B.1)	50% conc. HCl 50% H_3PO_4	———— ————	6
Co	Manual dipping (Fig. B.1)	Conc. HCl	Apply 1 to 3 volts (ac) for 8-mil wires. Use Co or stainless steel cathode	5
Co-Pt	Manual dipping (Fig. B.1)	1N KCN	Apply 1 to 10 volts (ac) for 2-mil wires. Pt cathode	14
Cu	Manual dipping (Fig. B.1)	40% NHO_3 60% CH_3OH Conc. H_3PO_4	Apply approximately 10 volts (dc) for 5- to 10-mil wires. Use Cu cathode and maintain solution temp. below about −20°C. Use ac for H_3PO_4	—
Fe	Manual dipping (Fig. B.1)	10% HCl	Final tip cleaning and shaping must be done by field evaporation at 21°K	—
Fe	Condenser discharge (Fig. B.3)	KOH solution (40%)	CCl_4 insulating layer 0 to 100 volts (dc)	4
	As above	1 part HCl, 1 part HNO_3, 2 parts H_2O	Apply 2 to 3 volts (dc)	—
Fe-C Steels	As above	As above	As above	—
Ir	Manual dipping (Fig. B.1)	Molten $KOH-NaNO_3$ solution	No voltage applied for preparing tips from 5- to 10-mil wires. Image at 78°K after 15 sec anneal at 1000°C	7
	As above	2N KCN	Apply 5 to 10 volts (ac). Use stainless steel cathode	54
Mo	As above	As above	As above	4
Mo-Re	As above	10% KOH solution	4- to 8-mil wires requires 0 to 70 volts dc or ac (pulsed by manual switch)	—
Nb	Manual dipping (Fig. B.1)	40% HF 60% HNO_3	Maintain solution temp. near 0°C applying about 10 to 15 volts for 5- to 10-mil wires	—
Ni	Condenser discharge (Fig. B.3)	10 to 20% KOH solution	For 4-mil wires use 0 to 70 volts dc. NH_4OH solution can be added to reduce bubbles formed. CCl_4 provides insulating fluid layer	4
	Manual dipping (Fig. B.1)	30% HCl saturated with $KClO_3$	Use 1 volt (ac)	—
Ni-Cr	Manual dipping (Fig. B.1)	12 N H_2SO_4	Apply 6 to 9 volts (dc). Use stainless steel cathode	—
Ni-Pt	Manual dipping (Fig. B.1)	1N KCN	Apply 1 to 10 volts (ac) for 2-mil wires	14

TABLE B.2 *Preparation of field emission tips (Continued)*

Material	Polishing method	Electrolyte	Comments	Reference
Pd	Condenser discharge (Fig. B.3)	10 to 20% KOH solution	Apply 0 to 70 volts (dc) for 4-mil wires	—
	Manual dipping (Fig. B.1)	70% HNO_3 30% HCl	Apply 2 to 4 volts (ac) for 5- to 10-mil wires	—
Pd-Pt	Manual dipping (Fig. B.1)	1N KCN	Apply 1 to 5 volts (ac) for 2-mil wires. Use Pt cathode	14
Pt	Condenser discharge (Fig. B.3)	2 mm layer of 20% KCN	0 to 10 volts ac using Pt electrode. CCl_4 insulating fluid	8
	Manual dipping (Fig. B.1)	10 N KCN	Apply 2 to 7 volts (ac)	54
Pt-Re	As above	As above	As above	54
Re	As above	2 mm layer of 20% KOH solution	0 to 70 volts dc using CCl_4 insulating fluid	4
Re-W	Manual dipping (Fig. B.1)	1N NaOH	Apply about 1 volt (ac) for 2-mil wires	14
Si	Manual dipping (Fig. B.1)	46% HNO_3 37% HF 15% CH_3COOH 2% Br	No voltage applied	—
Ta	Manual dipping (Fig. B.1)	4 parts 48% HF 2 parts conc. H_2SO_4 2 parts H_3PO_4 1 part CH_3COOH	Apply 15 volts dc using Pt cathode for 4-mil wires	9
V	Manual dipping (Fig. B.1)	20% H_2SO_4 80% CH_3OH	Apply 10 to 20 volts. Ni cathode recommended	—
W	Manual dipping (Fig. B.1)	5% NaOH solution or conc. KOH	Maintain electrolyte at about 20°C employing about 5 volts (dc or ac). W or stainless steel cathode recommended	—
	As above	8 parts NH_4OH 1 part KOH 2 parts H_2O	Apply 4 volts (ac)	—
W	Condenser discharge (Fig. B.3)	10% KOH (2 mm layer)	4-mil wires can be polished using 0 to 70 volts (dc). Slight addition of NH_4OH will reduce bubble formation	4
W-Mo	As above	As above	As above	
W-Pt	Manual dipping (Fig. B.1)	1N KCN	Apply about 1 volt (ac) for 2-mil wires	14
W-Re	As above	As above	For 4- to 10-mil wires use 0 to 100 volts (dc) and 2000 mfd condenser (Fig. B.3). Wires having up to 35% Re have been prepared	8
Zn	Manual dipping (Fig. B.1)	30% NHO_3 70% CH_3OH or conc. KOH	Employ Zn or stainless steel cathode and apply 5 to 10 volts (dc)	—

Normally, specimens for quantitative examination in the electron microprobe are cut and polished by conventional metallurgical techniques using the various string saws, diamond or cutting wheels, and related techniques followed by chemical or abrasive polishing. Specimens that are nonconductive in their final state of preparation are then coated with a few hundred angstrom units of carbon vapor deposited onto the surface to be investigated using a carbon arc in vacuum ($<10^{-3}$ mm Hg).

Naturally, when specimen morphology and related composition are to be investigated, the specimen is neither polished nor distorted. Only its shape is altered to fit the instrument observation chamber. If the specimen is nonconductive, carbon is vapor deposited onto its surface in vacuum. For samples having irregular fine shapes, facets, etc., rotation of the specimen and a change in the angle of shadow may be necessary to ensure a uniform conducting layer.

Observations of surface morphology in the scanning electron microscope are enhanced somewhat by the addition of a conductive layer capable of generating an abundance of secondary electrons in cases where composition is unimportant. In such cases, a layer of Au is vapor deposited onto the sample in place of the carbon. This is followed by a second vapor-deposited layer of Pd. Each layer should be a few hundred angstrom units thick to produce a total coating of less than 1000 Å. This process is particularly necessary for observations of ceramic and glassy materials. It has become common practice to evaporate the Au-Pd simultaneously from an alloy-wire source.

As indicated above, the application of the Au-Pd vapor deposit is advisedly performed with the specimen rotating, especially where the surface to be observed is irregular. Variations in the layer thickness and the mode of evaporation may contribute markedly to the image quality.

B.5 PREPARATION OF SURFACE REPLICAS FOR THE ELECTRON MICROSCOPE

The basic features of surface replication have been outlined in Chap. 5. We will attempt here to list some of the chemical preparations and solvents required in forming suitable replicas, in addition to indicating the particular surface characteristic to be replicated. Table B.3 lists a number of the more reliable replicas and indicates several prominent features that can be studied in various materials. You should realize that plastic replicas or replicas involving plastic stages that must be dissolved away are treated in a vapor-reflux apparatus designed for this specific "washing"

TABLE B.3 *Preparation of surface replicas*

Replica type	Plastic/solvent	Shadow angle	Features observed	Comments
Extraction (single stage)	Parlodion or collodion/amyl acetate	——	Carbide extracts; $M_{23}C_6$, M_6C, M_7C_3, etc.	Etch surface with 30% HCl, especially satisfactory for steels. Use 5% parlodion in amyl acetate or 10% collodion in amyl acetate
			Precipitates such as Ni_3 (AlTi) in Ni-Cr alloys, etc.	Etch surface with 35% H_3PO_4
Extraction (single-stage carbon)	——	90°	Precipitates and inclusions in steels, high-temp. alloys	Etch surface with 10% H_2SO_4 and dry-strip carbon vapor deposited to a thickness of 400–800 Å. Use Scotch tape soluble in acetone
Extraction (two-stage plastic-Ge)	Parlodion/amyl acetate	C or Ge at 90° Cr at 60°	Inclusions in steels or high-temp. alloys	Use etching procedures indicated above. Grid may be placed in wet plastic
Single-stage plastic	Parlodion/ collodion/amyl acetate	——	Phase structure in steels such as pearlite, ferrite, etc.	Use 5% parlodion or 10% collodion solution (in amyl acetate). Strip with Scotch tape—acetone or water soluble
	Formvar/dioxane	——	Same as above in addition to some surface features such as cleavage steps, etc.	Use 0.5% Formvar in dioxane. If grid is placed in wet plastic, cut hole in stripping tape
Cr/SiO or alternate forms Pt/SiO, etc.	——	Cr at 45° SiO at 90°	Surface pits and cleavage steps; related artifacts on most surfaces	Vapor deposition performed from 2 separate sources simultaneously. Replica is dry-stripped with soluble tape
Pt-C (two-stage) plastic)	Collodion or parlodion/amyl acetate	C at 90° Pt at 10 to 45°	General surface features including slip lines, fracture, etc.	Strip plastic using soluble tape. Shadowing can be done simultaneously in vacuum of 10^{-5} torr
Pt-Pd-C (two-stage plastic)	Collodion/amyl acetate	C at 90° Pt at 15 to 30° Pd at 45°	General fractography	Strip using soluble tape. Technique gives reliable detail and high contrast
Cr-C (two-stage plastic)	Parlodion/amyl acetate	C at 90° Cr at 45°	General fractography	Use 5% parlodion in amyl acetate solution
Pt-C (two-stage selected area)	Collodion/amyl acetate	C at 90° Pt at 10 to 30°	Any cumulative surface feature associated with stress, corrosion, etc.	Use 3 to 5% collodion in amyl acetate. Apply grid to wet plastic. Strip with soluble tape, cut out around support grid

operation. It is a simple matter to construct such a vapor washer in the laboratory by using a machined holder with inserts the size of the grid holder. This can be placed in a petri dish suspended above the solvent and covered. The dish is then gently heated so that the vapor engulfs the plastic replica and condenses on the lid. If the operation is conducted for a sufficient time, the replica is properly washed, that is, the plastic will have been dissolved. The various washing solvents are listed in Table B.3.

It should be noted that the surface treatments to highlight grain boundaries, precipitates, etc., are identical to those employed in classical (optical) metallography [12,13].

B.6 SPECIMEN PREPARATION FOR TRANSMISSION ELECTRON MICROSCOPY

Specimens for examination by transmission electron microscopy must ultimately meet only two requirements: (1) They must be thin enough for electrons to penetrate without excessive energy loss. (2) They must be representative of the sample in structure and composition; that is, reaction products in the form of artifacts and contamination as well as secondary deformation and heat treatment are to be avoided. A variety of methods have been devised over the past decades that involve the reduction of bulk solids to thin sheet and the subsequent polishing to foil, or the direct fabrication of solid films by vapor deposition in a vacuum or chemical vapor deposition onto a suitable substrate. Many crystals can be cleaved directly to thicknesses sufficiently thin to allow electron beam transmission. Mica will perhaps strike you as the most obvious example of this type.

The preparation of thin sections from thick-bulk samples is accomplished by several methods. These involve mechanical sawing by which slices as thin as 0.5 mm can be obtained if the sawing is performed under a coolant; chemical string sawing as shown in Fig. B.4a, electrolytic jet machining of thick sections to thinner sections as illustrated in Fig. B.4b; and the spark-erosion cutters manufactured commercially.[†]

Many thicker sections can be thinned appreciably by simple dissolution processes in chemical solutions that rapidly reduce the sample but these do not for the most part culminate in a suitably polished or thinned foil. Table B.4 details the methods and solutions for the reduction of

---

[†]*Cambridge Instruments Company, 420 Lexington, Avenue, New York.*

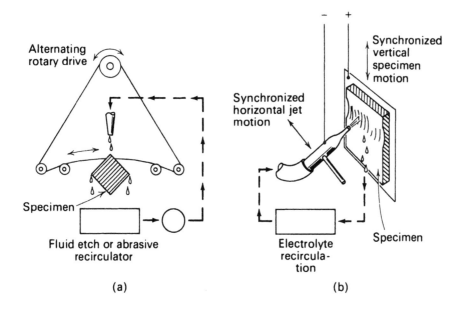

FIG. B.4 *Two stages in preparation of thin sections from bulk samples. (a) Chemical string saw. (b) Electrolytic jet machining, method for reducing thick slices to thin section.*

bulk samples, including cleavage. Ceramic materials and some metals can also be reduced by ion bombardment as indicated in Table B.4.

Samples for observation by transmission electron microscopy must be finally reduced somewhat uniformly and without etching or pitting. This normally involves a controlled dissolution process or electrolytic polishing. Samples to be thinned to final electron transparency must be no thicker than about 0.10 in.; however, many exceptions are found in certain instances and for particular electrolytes.

Two somewhat standard techniques have evolved for electropolishing thin foils. These are the *window method,* and the *Bollman method,*[†] each of which have numerous modifications. The window method is also called the *electrolytic cell method,* and the Bollmann method is also known by other names.

We shall treat the two basic electropolishing techniques in our discussion as the window method and the *pointed-electrode method,* these methods are illustrated in Fig. B.5a and b, respectively. In the window

_____

[†]*W. Bollmann,* Phys. Rev., 103: *1588(1956).*

TABLE B.4 *Preparation of samples from bulk*

Material	Method	Chemical solution or electrolyte	Final sample thickness	Comments	Reference
Ag	Electrolytic jet	20% HNO_3	<1 micron	Use about 20 volts at <1 amp for mechanically sawed sections	—
Ag alloys	Chemical dissolution	50% HNO_3 in H_2O	Can be thinned to electron transparency	Sections can be mechanically sawed from stock	—
Al and Al alloys	As above	20 g NaOH 100 cc H_2O	<1 micron	Use solution temp. of ~70°C	15
Al_2O_3	Ion bombardment	—	Can be thinned to electron transparency	Technique requires several hours for starting thicknesses of ~ 1 mm	—
	Liquid jet	85% H_3PO_4	Can be thinned to electron transparency	Requires ~500°C bath temp. Samples can be dipped	16
$BaTiO_3$	As above	Hot H_3PO_4	As above	—	17
Be	Chemical dissolution	60% H_3PO_4	<1 micron	Agitate vigorously	—
Bi_2Se_3	Cleavage	—	Can be cleaved along basal plane to electron transparency	—	—

Material	Method	Solution	Thinning	Notes	Ref.
Bi_2Te_3	Chemical dissolution	Dilute aqua regia	<1 micron	—	13
$CaFe_2$	As above	Conc. H_2SO_4	Can be thinned to electron transparency	Begin dissolution at 130°C and lower to 20°C for final thinning	18
Cu and Cu alloys	Chemical dissolution	80% HNO_3 20% H_2O	~1 micron	Agitate vigorously Electrolytic jet using 20 volts can also be used	19
Cu-Al	Chemical dissolution	40% HNO 50% H_3PO_4	~1 micron	—	20
	Electrolytic jet	75% H_3PO_4	<1 micron	Use about 100 volts (dc)	—
Diamond	Oxidation	Aqua regia	Can be thinned to electron transparency	Oxidize at 1350°C in CO_2, then boil in solution to remove carbon	21
Fe and Alloy steels	Chemical dissolution	30% HNO_3 15% HCl 10% HF 45% H_2O	~1 micron	Keep solution temp. hot (~70°C)	22
Fe-Mn	As above	75% HCl 25% H_2O	~1 micron	Vigorous agitation	—
Ge	As above	25cc HNO_3 15cc HF 15cc CH_3COOH 0.3cc Br	Can be thinned to electron transparency	Agitate solution. Thin specimen will float in solution	23

TABLE B.4 *Preparation of samples from bulk (Continued)*

Material	Method	Chemical solution or electrolyte	Final sample thickness	Comments	Reference
GaAs	Chemical dissolution and electrolytic jet	Chemical polish with 15% Br in CH_3OH; electrolytically in 25% $HClO_4$ 75% CH_3COOH	Can be thinned to electron transparency	Use 42 volts electrolytically and wash in conc. HF to remove oxide. Rinse in distilled H_2O	53
Inconel (all alloys)	Electrolytic jet	42% H_3PO_4 34% H_2SO_4 24% H_2O	Can be thinned to electron transparency	Use total current of roughly 1 amp or greater	——
KCl	Cleavage	H_2O	Can be dissolved to electron transparency	Place cleaved section between glass slides with a few drops of H_2O and repeat until section thins	——
Mg	Chemical dissolution	5% HCl 95% H_2O	<1 micron	Thinning is controlled by agitation	——
Mg alloys	Chemical dissolution	2% HCl 98% C_2H_5OH	<1 micron	Samples should be initially thinned in 2% HNO_3 in ethyl alcohol	24
Mg-Al, Zn	Chemical dissolution	15% HNO_3 85% H_2O	<1 micron	Agitate continuously	25
MgO	Chemical dissolution	15% H_3PO_4	Can be thinned to electron transparency	Agitate continuously and maintain solution at ~100°C	26

Material	Method	Solution	Capability	Notes	Ref.
Mica	Cleavage	—	Can be cleaved to electron transparency	Use sticky tape to cleave single layers and dissolve tape in solvent	—
Nb	Chemical dissolution	70% HNO_3 30% HF	Can be thinned to electron transparency	When sample becomes very thin, add HCl to bring conc. to about 40%, and lower temp. to 0°C	27
Ni and Ni alloys	Electrolytic jet	—	—	Agitate in hot solution	23
NiO	Liquid jet	Conc. H_3PO_4	Can be thinned to electron transparency	Mask specimen when thickness becomes <1 micron using black wax soluble in ethylene dichloride or allow H_2O to float. Stop action by flooding solution with deionized H_2O	28
Si	Chemical dissolution	95% HNO_3 5% HF	As above	Allow samples to float on solution. Stop action by flooding with deionized H_2O. Use as a sequence to previous technique for final thinning of small pieces	29
SiO_2	As above	50% HNO_3 50% HF	Can be thinned to electron transparency	Agitate vigorously	30
Ta	As above	As above	As above	As above	—

TABLE B.4 *Preparation of samples from bulk (Continued)*

Material	Method	Chemical solution or electrolyte	Final sample thickness	Comments	Reference
Ti and Ti alloys	*Electrolytic jet*	*50% H_2SO_4*	*<1 micron*	*Use approximately 30 volts (dc)*	—
U and U alloys	*Chemical dissolution*	*50% HCl*	*~1 micron*	*Agitate solution*	*31*
UC	*Electrolytic jet*	*50% CH_3OH 50% H_3PO_4*	*As above*	*Use 220 volts (dc) at start and reduce to about 80 volts for final thinning*	*33*
UO_2	*Liquid jet*	*Conc. H_3PO_4*	*Can be thinned to electron transparency*	*Solution temp. should be about 100°C*	*32*
V	*Electrolytic jet*	*70% CH_3OH*	*<1 micron*	*Use 30 to 50 volts (dc) and maintain solution <30°C*	—
W	*As above*	*80% solution of NaOH*	*Can be thinned to electron transparency*	*Agitate jet and use about 10 to 20 volts (dc). Final polish may require addition of H_2O to electrolyte*	—
An	*Chemical dissolution*	*10% HCl 5% HNO_3 10% CH_3OH 75% H_2O*	*<1 micron*	*Agitates speimen and maintain temp. <30°C*	—

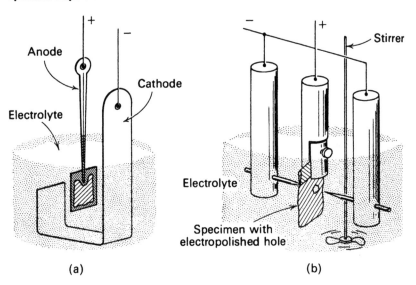

FIG. B.5 *Electrolytic polishing methods. (a) Window method (electrolytic cell); (b) pointed-electrode (Bollmann) method.*

method, the sample, painted at its edges with a protective lacquer[†] to prevent attack, is suspended in an electrolyte contained in a small vessel of roughly 200 ml capacity. The cathode consists generally of the same material[§] as the sample formed in a U-strip electrode having a width of about 1 in. The lacquered specimen (as the anode) is then suspended in the electrolytic cell and a direct current is applied. In certain situations a very large electrolyte bath is used with the cathode formed into a ring around the inner circumference of the vessel. The specimen is then suspended in the electrolyte at the center of the vessel and agitated. With the current and voltage properly adjusted, holes will form in the *window* of the foil and , if they are allowed to come together, they will contain electron transparent sections at the edges. A random array of holes is facilitated by rotating the specimen in the anode holder. The correct current-voltage setting for good polishing is the plateau of the voltage-current plot obtained for the system as shown in Fig. B.6.

[†]*For most applications, Microstop[(C)], an electroplating stop-off manufactured by Michigan Chrome & Chemical Company, 3615 Grinnel Avenue, Detroit, Mich., (order from Tobler Div., 220 W. Fifth St. Hope, Arkansas 71801) will be most satisfactory. This lacquer is soluble in acetone.*

[§]*Stainless steel will also perform satisfactorily in most cases.*

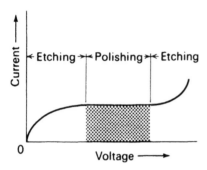

FIG. B.6 *Polishing plateau.*

In the pointed-electrode method (Fig. B.5b), the anode (specimen holder) and cathode holding the pointed electrodes are stainless steel. Generally the specimen, measuring 0.5 X 1.0 in., is suspended from the anode with the pointed electrodes spaced about 0.5 mm from the surfaces near the center. After polishing a hole, the specimen is shifted so that the electrodes are between the nearest specimen edge and the hole; they are withdrawn to a distance of 1 to 2 mm from the surfaces. Polishing is then continued at the same current as initially used until the edges come together as shown in Fig. B.7. In some applications it is desirable to stop the process before the edges polish through; the foil section between the edges can then be cut and will have the final appearance of a dog bone. This is useful in producing microtensile specimens for in-situ straining studies as illustrated in Fig. 8.13. Continued electropolishing will then produce two nodes as shown in Fig. B.7.

In most cases where specimens are to be examined somewhat routinely in the TEM, and where the starting material is in the form of a thin sheet roughly 10 mil (0.025 cm) thick, discs having the diameter of the standard specimen holder (~ 3 mm) can be punched and specimens prepared by jet electropolishing from both sides in an automated device having a light sensor indicating when a hole has been electropolished. Such automatic electropolishing apparatus, described earlier by Schoone and Fischione [55] are now available commercially. It might be interesting to note that some of the very first specimens for TEM were made by jet electropolishing [see R. D. Heidenreich, *J. Appl. Phys., 20:* (1949)]. Figure B.8 illustrates schematically the principles of operation of the automatic disc electropolishing units.

Just as in the preparation of emission tips outlined in Sec. B.3, the successful preparation of thin foils is somewhat more of an art than a

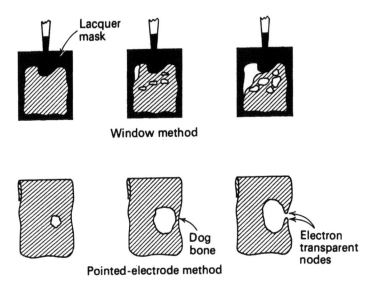

Window method

Dog bone

Electron transparent nodes

Pointed-electrode method

FIG. B.7 *Development of electron transparent sections during electropolishing.*

science. The success of an attempt to polish a material will all too frequently depend on circumstances usually not reported by an investigator either because of his short-sightedness, or because of a deliberate attempt to conceal his success. In the outline of techniques and electrolyte formulas in Table B.5, an attempt has been made to ensure that those listed can be successful. Numerous techniques are included that have been successfully and routinely used by the author. You are reminded that the voltage and current settings indicated refer to the methods outlined above, and are specific to the dimensions discussed and illustrated in Fig. B.5. Stirring of the solution or other means to prevent bubble formation, pitting, etc., are also indicated.

We should also mention that numerous thin-foil specimens, especially ceramic and related inorganic samples, can be prepared by ion bombardment. The details of this method have been described by Castaing [10] and Graf and Genty [11]. It works well for some inorganic materials but it produces ion damage in metals and is not a good method for preparing metals.

The alternative method for producing thin-foil specimens for transmission electron microscopy is by evaporation onto a suitable substrate in vacuum using a thermally heated source, an electron gun focused onto the source, or a sputtering arrangement where the source material is ionized by an electric field. A suitable substrate is normally one that allows

TABLE B.5 *Preparation of electron transmission samples*

Material	Technique	Electrolyte	Temp °C	Voltage (dc volts)	Current amp	Comments	Reference
Ag and Ag alloys	Window or pointed electrode or jet polish	36g each of AgCN, KCN, and KCO$_3$ in 1 ℓ H$_2$O	0	Variable	0.2	Voltage will depend on technique. Conditions are critical to eliminate anodic deposit. Wash in ethyl alcohol. Use stainless steel cathode	—
		1.5% KCN solution	0	6-8	<0.1	Use stainless steel cathode	
Ag–Al	Window	20% HClO$_4$ 10% C$_3$H$_5$(OH)$_3$ 70% C$_2$H$_5$OH	30	20-40	>1	Use stainless steel cathode	15
Al and Al alloys	Window or jet polish	20% HClO$_4$ 80% C$_2$H$_5$OH	20 -60	15-20 80	0.5 0.1-1	Use stainless steel cathode Use 5mm jet height	34
Al–Cu	Pointed electrode	617 cc H$_3$PO$_4$ 240 cc H$_2$O 134 cc H$_2$SO$_4$ 156 g H$_2$CrO$_4$	60	9-12	0.1-0.2	Oxide layer may be removed by soaking in 35 cc H$_3$PO$_4$, 65 cc H$_2$O with 16 g H$_2$CrO$_4$ added. Wash in H$_2$O before and after soaking.	35
Al$_2$O$_3$	See Table B.4	—	—	—	—	——————	—
Au and Cu–Au	Window	34 g KCN 7.5 g K$_3$Fe(CN)$_6$ 7.5 g KNaC$_4$H$_4$O$_6$·4H$_2$O 10 cc H$_3$PO$_4$ 2 cc NH$_4$OH Add 500 cc H$_2$O	60	6-8	1	Use stainless steel cathode and fresh solutions	36

Material	Technique	Solution	Temp	Voltage	Current	Remarks	Ref
Be	Window	60% H_3PO_4 35% $C_3H_5(OH)_3$ 2.5% H_2O 2.5% conc. H_2CrO_4 solution	60	30	—	Foils must be washed in hot distilled H_2O and alcohol and cleaned ultrasonically in alcohol (ethyl)	37
Be	Jet polish	25 mL HNO_3 5 mL H_2SO_4 5 mL HCl 335 mL C_2H_5OH 165 mL ethylene glycol	0	85	0.7	Current control is important	
Cd	Jet polish	100 mL $HClO_4$ 50 mL $C_3H_5(OH)_3$ 150 mL CH_3OH 200 mL C_2H_5OH	0	20	0.05	Set current below point where brown film forms	57
Co	Window	77% CH_3COOH	20	22	2	Use stainless steel cathode	34
Co–Cr–C	Jet polish	15% $HClO_4$ 85% CH_3COOH	10	20	—	wash in ethyl alcohol (ethanol)	58
Cr–Fe	Pointed electrode	As above	<10	12	2	Use stainless steel cathode	38
Cu and Cu alloys	Window	67% CH_3OH 33% HNO_3	-30	10–20	0.1	Use Cu ring or plane electrode and agitate specimen or use brush	—
Cu–Al	As above	50% H_3PO_4	0	3–5	0.5–1.0	to remove bubbles periodically. Wash in cold methanol or deionized water	
Cu–Ni	As above	55% CH_3COOH 34% HNO_3 11% H_3PO_4	20	3–4	0.1	As above	

TABLE B.5   *Preparation of electron transmission samples (Continued)*

Material	Technique	Electrolyte	Temp., °C	Voltage, dc volts	Current, amp	Comments	Reference
Fe–B	Jet Polish	30 mL HCl 660 mL CH_3OH 100 mL butyl cellosolve	-60	100	0.03	Amorphous/Metglass alloys use ~ 5 mm Jet height	—
Fe, carbon steels	Pointed electrode	20 parts CH_3COOH 1 part HClO	15	40	~1	Maintain voltage constant. Wash sample in ethyl alcohol. Where possible specimen should be withdrawn from electrolyte in N_2. Stir solution at 200 rpm	—
	Jet polish	87% C_2H_5OH 13% HCl	-60	75	0.03		
Fe–Mn	As above or window	200 cc H_2O 150 g H_2CrO_4	60	6-8	1-3		
Fe–Ni	As above	42% H_3PO_4	30	8-9	9-10	Use stainless steel cathode, rinse in running H_2O and ethyl alcohol	—
Fe–Ni–Cr alloys (stainless steel)	As above	As above	30-60	8-9	9-10	As above; thicknesses up to 0.30 in. can be thinned directly. Solution must be stirred at 200 to 300 rpm. Rinse in H_2O and ethyl alcohol	39
	Jet polish	20% $HClO_4$ 80% C_2H_5OH	-50	60-90	0.05		
Ge	See Table B.4	—	—	—			—
Inconel	Pointed electrode	42% H_3PO_4	30-60	8-9	9-10	Starting material having thicknesses <0.004 in. will thin rapidly. Rinse in running H_2O and ethyl alcohol. Stir solution at 100 to 300 rpm	—
	Jet polish	15% $HClO_4$	0 to -40	80-160	0.9 - 1		

Material	Method	Electrolyte	Temp (°C)	Voltage	Current	Comments	Ref.
Mg and Mg alloys (Mg-Al, Mg-Zn)	Window	33% HNO_3	0	9-10	1-2	Use stainless steel cathode and rinse first in phosphoric acid-ethanol solution (1:5) and finally in pure ethyl alcohol	25,34 40
Mo	Window	87% CH_3OH 13% H_2SO_4	0	5	1-2	Agitate specimen or periodically brush surface to remove bubbles.	—
Mo	Jet polish	As above	0	22	0.08	Remove oxide film by immersion in NH_4OH	—
Mo-Re	Window	Conc. H_2SO_4	20	10	1	Use stainless steel cathode. Rinse first in ammonia, then in ethyl alcohol	—
Nb	As above (see also Table B.4)	85% HNO_3	50	8	~1	Use Pt cathode and wash in ethyl alcohol	41
Nb	Jet polish	saturated NH_4F in CH_3OH	-10	300	0.02	Use 1 mm jet diameter	56
Nb-Zr	As above	93% CH_3OH	-60	25	~1	Use Pt cathode	42
Ni,Ni-Cr Ni-Cr-Fe (Ni alloys) TD-Ni	Pointed electrode	42% H_3PO_4 34% H_2SO_4 24% H_2O	30-60	9-10	9-10	Rinse in running water and ethyl alcohol	—
TD-Ni and TD-Ni-Cr (approximately 2% THO_2)	Jet polish	20% $HClO_4$ 80% C_2H_5OH or CH_3OH	0 to -60	50-100	0.03-0.05	Some alloys thin better in electrolytes previously used for thinning stainless steel, thus solution should appear deep green. Stir solution at ~300 rpm. Add H_2O if sample etches	

TABLE B.5 *Preparation of electron transmission samples (Continued)*

Material	Technique	Electrolyte	Temp., °C	Voltage, dc volts	Current, amp	Comments	Reference
Pb	Jet polish	20% $HClO_4$	-40	100	0.09	Use 3.5 mm jet height 1.5 mm diameter	—
Pt	Window	Eutectic mixture of Na and KCl	500	—	<1	Wash specimen in boiling aqua regia	43
Pu, alloys	Jet polish	133 mL CH_3COOH 35 g CrO_3 25 mL H_2O	10	10	0.02	Wash specimens in cold methanol	59
Ta	As above	94% CH_3OH 5% H_2SO_4 1% HF	0	50–70	8	Wash specimens in methanol	15
	Jet polish	add butyl cellosolve to above	-50	40	-0.02	Use 1.5 mm jet	—
Ti, Ti-Al Ti-Al-V, Ti alloys	Window	65% CH_3COOH 30% C_4H_9OH 5% H_3PO_4	15	6–8	1	Stir solution at approximately 300 rpm. Rinse specimens in butanol and ethyl alcohol	—
Ti	Jet polishing	0.5 mol/L Mg $(ClO_4)_2$ in CH_3OH	-30	100	0.04	Use 1.5 mm jet method reduces hydride formation	56
TiC	As above	50% HNO_3 25% HNO_3 25% CH_3COOH	20	5–10	1	Use stainless steel cathode	—
U	Pointed electrode and jet polish	133 cc CH_3COOH 33 cc H_2O 39 g H_2CrO_4	>30	20	1	Use stainless steel cathode and clean in solution of 75% H_2SO_4, 18% glycerol, 7% H_2O	44

Material	Technique	Electrolyte	Temp	Voltage	Current	Remarks	Ref.
UC	See Table B.4	—	—	—	—	—	—
UO_2	See Table B.4	—	—	—	—	—	—
V	Pointed electrode / Jet Polish	80% CH_3OH 20% H_2SO_4 / As above	20 / 20	12-20 / 20	2-3 / 0.03	Use Ni or stainless steel cathode. Rinse in H_2O and ethyl alcohol	—
W	As above	1% NaOH solution	30-60	10	1-2	Use stainless steel cathode and wash specimens in deionized water. Stir solution at 100 to 200 rpm	—
W	Window	94% CH_3OH 5% H_2SO_4 1% HF	0	50-70	8	Agitate specimen	15
Zn	As above	63% C_2H_5OH	<20	2-4	0.1	Use stainless steel cathode and wash specimens in H_2O	45
Zr and Zr alloys	Pointed electrode	80% C_2H_5OH 20% $HClO_4$	0	25	0.5-1	Use stainless steel cathode and stir solution at 100 to 300 rpm	46

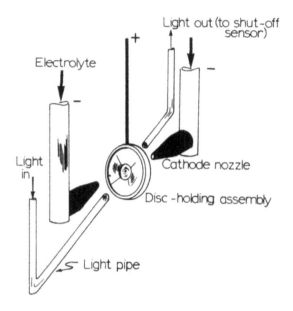

FIG. B.8 *Schematic view of a submerged jet elec-*
tropolishing unit for the automatic preparation
of perforated disc specimens for TEM. The light
pipes shown enable automatic shutoff when the
disc is perforated.

the foil to be separated from it without damage to the foil. This can be
accomplished either by preferentially dissolving away the substrate, or by
loosening the foil so that it becomes free. For most purposes, NaCl is a
perfect substrate since it is easily prepared, cheap, and foils can be
floated off by submersion in water.

Normally an arrangement as shown in Fig. B.9 will be satisfactory for
vapor depositing most pure metals. The substrate, rigidly attached to a
Cu block, can be heated, or a cooling system can be built into the vacuum
system for flushing liquid nitrogen, etc., over the Cu block. The evaporation source is usually tungsten or another very high-temperature material
that can be easily fabricated from a 0.001-in.-thick sheet cut into 0.25-
in. widths, and crimped with a needle-noise pliers at the center to form
a V-boat with the ends folded for mounting. Tungsten is the simplest material to use, but in many cases it reacts with the evaporant and forms
compounds that either destroy the boat or contaminate the vapor deposit.

Thin foils can be prepared with many residual structures, depending
on the evaporation rate, the substrate temperature and structure, and the
degree of vacuum. These features have been outlined previously to some

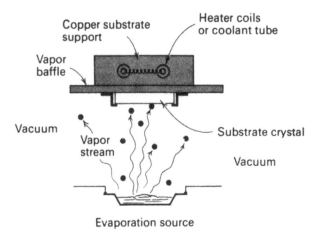

Vapor baffle

Copper substrate support

Heater coils or coolant tube

Vacuum

Vapor stream

Substrate crystal

Vacuum

Evaporation source

FIG. B.9 *Physical arrangement of components for vapor deposition of thin foils.*

extent in Chap. 7. [See L. E. Murr and M. C. Inman, *Phil Mag.*, *14:* 135 (1966).] Evaporation of materials in low vacuum (10^{-4} torr) or onto crystalline substrates at room temperature usually results in polycrystalline foils. Post-annealing of the foils on the substrate can, however, increase the grain size. On raising the substrate tmperature or cooling it to some critical level in very high vacuum ($< 10^{-7}$ torr), epitaxial single-crystal films can be prepared.

Table B.6 outlines the preparation of foils by physical vapor deposition (PVD). Thin films are also routinely made by other means as well, including chemical vapor deposition (CVD), sputter deposition, and electron beam heating of the evaporation sources instead of resistively heating a "boat" as illustrated in Fig. B.9. The reader interested in reviewing a variety of these techniques should refer to the following: L. Holland, *The Vacuum Deposition of Thin Films,* J. Wiley & Sons, Inc., New York, 1956; C. F. Powell, J. H. Oxley, and J. Blocher, *Vapor Deposition,* J. Wiley & Sons, Inc., New York, 1966, and R. F. Bunshah, et al., *Deposition Technologies for Films and Coatings,* Noyes Publications, Park Ridge, N.J., 1982.

Selected-area TEM and build-up techniques It is often necessary to isolate a region or volume element in a material system for TEM analysis. While we have described selected-area replication techniques in Chap. 5, there are some novel approaches to selected-area or selected-volume isolation for TEM. Figure B.10 shows a simple example of this technique

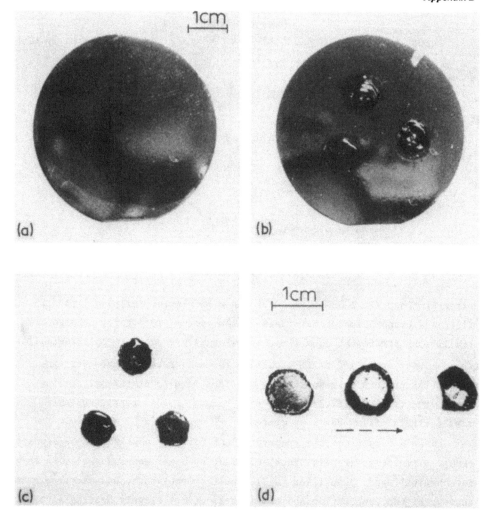

FIG. B.10 *Preparation and isolation of selected areas (and volumes) on silicon wafers for TEM. (a) Mechanically polished Si wafer (0.04mm thick). (b) Selected areas on surface or bulk of Si wafer to be isolated using black-wax buttons over the areas coincidentally masking both surfaces. (c) Wax buttons isolated by dissolution of Si wafer matrix in 95% HNO_3 + 5% HF. The desired area or volume is sandwiched between the buttons. (d) Sequence showing the thinning stages of isolated Si section for TEM. The wax serves as a surface preservative for one surface (where desired) and a support. (From Reference [60]).*

which involves the isolation of three buttons from selected regions in a small diameter silicon wafer. This is accomplished by placing wax beads on the surfaces to include the volume of interest, dissolving away the remaining matrix (Fig. B.10(c)) (Fig. B.10(b)), removing one layer of wax to expose the volume element, and then chemically polishing the remaining element until it is electron transparent (Fig. B.10(d)). As shown in Fig. B.10(d), this technique also preserves one surface, and when the final thin section is prepared, the volume microstructure can be related unambiguously with surface detail. If a uniform volume element without surface correspondence is desired, then all the wax would be dissolved in Fig. B.10(d), and the remaining silicon section dissolved to produce a thin electron transparent section. The rate of dissolution in Fig. B.10(d) can be controlled so that the reduction in area of the specimen and the rate of reducing the thickness will result in a roughly 2 to 3mm disc area, with the thickness electron transparent. The ability to control this thinning will also allow an area chosen near the center of the beads to be preserved as a thin area which can be viewed in the TEM.

In utilizing a technique illustrated by Fig. B.10, it is important to use a wax mask which can be completely and easily dissolved so that no organic residues are left on the electron transparent thin section. Organics left on the specimen surface will interact with the electron beam and contribute variously to forming contamination layers or other artifacts which confuse the observations of microstructures or related crystallographic information. This feature cannot be stressed enough. For example washing TEM specimens with benzene, acetone (which often contains traces of benzene) or other organic solvents is bound to create problems because these ring compounds are polymerized in the electron beam and produce contamination layers which can seriously compromise the analysis. Samples should always be washed in pure ethanol - not denatured or methanol. These contain organic compounds, including benzene, which will leave a residue and create compromising polymer contaminants on the specimen surface.

In contrast to isolating volume regions to create a selected thin section from a materials system, it is often necessary to build up a very tiny volume of material to allow it to be sliced, polished, and electropolished or ion milled to electron transparency. It is often desirable to work with a reasonably standard specimen disc of about 3mm; and to produce a working disc of 3mm from a specimen particle having a maximum dimension of 0.5 to 1mm would require some kind of build-up. In addition, to examine transverse slices of a very tiny wire (such as an electrical contact wire in a micro-chip (Fig. 5.43)) would also require the wire to be built-up to a sufficient diameter to allow a slice to be cut which approximates a standard (3mm) TEM specimen disc.

There are a number of ways to achieve these features. For example a tiny wire could be repeatedly dipped in a suitable material such as epoxy or mounted in a rigid epoxy to allow a slice to be made which can be ion milled to produce an electron transparent thin section of the wire cross-section. An alternative would also involve electrodeposition or other chemical or vapor deposition of the same or similar metal or material uniformly distributed over the surface to achieve sufficient build up.

A novel application of this approach is illustrated in the sequence of schematics and TEM image shown in Fig. B.11. In this sequence, a copper jet fragment created explosively as a shaped charge is detonated is soft-

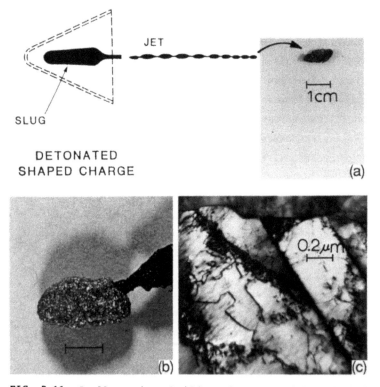

FIG. B.11 *Small specimen build-up for TEM. (a) Schematic showing the soft recovery of a copper jet fragment in a shaped charge jet. The photo-graphic insert shows the recovered fragment. (b) Electrodeposited (Cu) particle in (a) prior to slicing transverse section and punching a 3mm disc. (c) TEM bright-field image showing jet fragment microstructure after jet polishing disc cut from (b). (Courtesy Alan Gurevitch [61]).*

recovered in "shaving cream" (Fig. B.11(a) and (b)), wrapped with a thin
copper wire and electrodeposited with copper (Fig. B.11(c)); and finally
sliced to punch a 3mm TEM disc which was electropolished as shown in Fig.
B.9 to produce an electron-transparent thin section shown typically in Fig.
B.11(d) [61].

Ion-milling and XTEM specimen preparation In recent years specimen preparation by ion milling has become popular especially in dealing with composite systems and organic or inorganic regimes where other approaches to
forming thin electron transparent sections are untenable. Of course any
method for forming an electron transparent thin section which does not
alter the regime nor create artifacts contributing to observations of the
regime is acceptable. This includes smashing ceramics into flakes or
chards or stripping thin layers from bulk specimens, cleaving small pieces,
or other approaches. But to observe the interface between a ceramic fiber
in a metal matrix or an insulating layer on a semiconductor normally pre-
cludes these approaches as well as any kind of electropolishing.

The principles of ion sources and the ability to focus and direct ion
beams has been described in various chapters throughout this book, especially Chaps. 3, 4, and 5. Commercial units for ion erosion or ion milling
(Fig. B.12) have been available since about 1980. Focused ion beams of

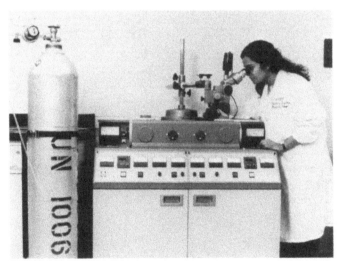

FIG. B.12 *Commercial (Gatan, Inc.) ion mill. The unit consists of an oil-
diffusion pumped vacuum system, ion sources, high-voltage accelerating
system, and a convenient specimen exchange system which functions without
disturbing the vacuum or milling conditions.*

argon accelerated typically at voltages ranging from 4 to 10kV provide
milling rates of 10 to 100 microns/hour in a variety of materials including
metals, ceramics, semiconductors, and superconductors. Figure B.13 shows
some typical data for comparison for the commercial ion mill shown in Fig.
B.12.   In Fig. B.13(b) the beam angle relative to the specimen surface is
observed to have an important effect on the specimen current which in turn
influences the rate of material removal. Optimum beam angles (shown for
copper in Fig. B.13(b)) often range between 15 to 20 degrees, and these low
beam angles also assure a much larger thin area because the milling is done
at a shallow angle which "defocuses" the ion beam and enlarges the milling
area. In addition, the specimen can also be rotated and this can also enlarge the thinned area and produce a more uniformly thinned area.

At higher beam voltages and at elevated temperatures, many materials
can be ion damaged. This damage can take the form of defects and defect
clusters created by ions knocking out lattice atoms. These effects can be

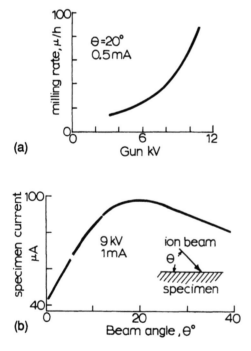

FIG. B.13 *Typical argon ion milling rate versus ion source potential (a)
and specimen current versus argon ion beam angle (b) for copper. (Based
on data from Gatan, Inc.).*

reduced by reducing the accelerating voltage (thereby reducing the milling rate as shown in Fig. B.13(a)) and by cooling the specimen. Ion milling of sensitive specimens is therefore often done at liquid nitrogen temperature.

Ion milling can be accomplished using one ion gun to mill one side of the specimen or a dual gun system where the same area (volume element) is milled from two sides simultaneously. Thinning is also facilitated by dimpling specimens from one or both sides in the area to be milled, and this can involve local grinding and polishing in a small area prior to milling. This is especially important for thicker specimens in order to minimize the milling time and the prospects for specimen heating and ion beam damage.

While the conditions shown in Fig. B.13 will change for different materials, the milling conditions for many composites are sufficiently similar to allow thin sections to be made of ceramics in metals, various metal phases, and ceramic/semiconductor or related layered materials which can be effectively thinned to examine the cross-sections and the details of the interface between the layers. These considerations are important in ion milling of superconductor oxides and related materials illustrated typically in Figs. 7.85 to 7.89.

Cross-sectional transmission electron microscopy (XTEM) was introduced somewhat in parallel with the introduction of commercial ion milling systems, and is the epitome of a regime in which ion milling is an essential and unique approach to producing electron transparent thin sections [62-65]. Cross-sectional TEM specimens are important for determining the properties of a variety of layered materials, including integrated circuit components and especially interfaces between materials regimes as illustrated, for example, in Fig. 7.90. The essence of this technique is illustrated in Fig. B.14 which shows the development of a sufficiently thick sandwich array from which to make a slice large enough to cut a standard (3mm) TEM disc from which to create a thin section which can be dimpled (mechanically thinned) and ion-milled to electron transparency. The thinnest area, located in the center of the 3mm disc after dimpling, will allow the electron-transparent region to be optimized at the interface between regimes (2) and (3). By rotating the specimen and controlling the ion beam angle during thinning, it is possible to create very large electron-transparent areas which can allow many interfaces to

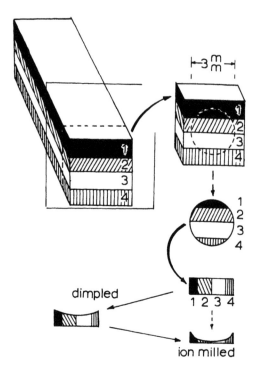

dimpled

ion milled

FIG. B.14 *Schematic representation of composite material system and preparation of electron-transparent cross-section for the study of interface structure. Natural composite section or sections glued together to form thick, built-up composite from which slices can be cut as shown by the dotted line. The section is then ground to form a 3mm disc and the disc is mechanically dimpled to reduce the cross-sectional thickness to shorten ion milling time. The dimpled disc is then ion milled to electron transparency which allows the regions (2) and (3) to be viewed along with the interface which separates these two regions.*

be examined in complex, multi-layer composites. Figure B.15 illustrates this feature for two multi-layer Si/SiO_2 regimes which were glued together using a fast-setting epoxy. The insert shows a lattice image of a precipitate at the junction region between an SiO_2 insulator layer and a poly-Si grain boundary [64]. Epoxies which can be used to efficiently and effectively bond composites together are described in [63-65]. Superglues and similar fast setting resins are readily available.

It might be useful in summarizing the tables in this appendix to point out that they are not by any means complete or exhaustive, and are intended only as a useful guide. More exhaustive data can be found in the

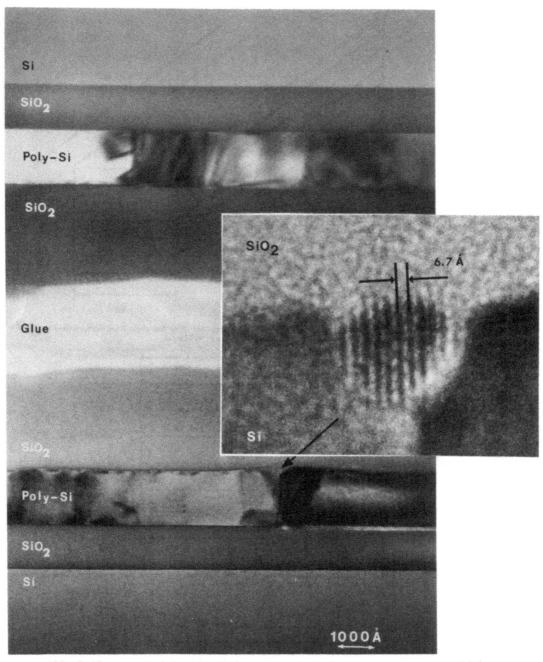

FIG. B.15 *TEM (bright-field) image showing electron-transparent Si/SiO$_2$ composite prepared as illustrated conceptually in Fig. B.14. The insert also shows a lattice image of a precipitate at the intersection of a poly-Si grain boundary with the SiO$_2$ layer. (Courtesy Dr. John Bravman, Stanford University).*

literature cited as well as a great deal of the literature not cited. In
addition, texts which deal specifically with electron microscope specimen
preparation continue to be written and updated, and the reader should
refer, for example, to: P. Goodhew, *Specimen Preparation for Transmission
Electron Microscopy,* Oxford University Press, England, 1984.

Finally, Fig. B.16 provides a schematic summary of the transmission
electron microscope techniques which are useful in preparing a host of
electron-transparent thin sections and other specimen configurations for
analytical electron microscopy. While we have demonstrated some extremely
versatile uses for XTEM in elucidating precipitates at interfacial regimes
in micro-circuits and integrated circuit chips (Fig. B.15), older techniques such as replication electron microscopy will also find useful
applications well into the new frontiers in materials research which will
require analysis by electron microscopy and microanalysis. One example of
this has recently involved the examination of materials surfaces exposed
to space in a low-earth orbit experimental satellite placed in orbit in
1986 and recovered in 1990. Preliminary SEM analysis of alloy surfaces
showed micrometeorite and other space particle impact damage, but in
addition, many growths and crystalline-looking materials were found on the
surfaces. To examine these effectively required they be isolated from the
surface by a replica-based lift-off technique illustrated conceptually in
Fig. B.16(b) [66].

This appendix is intended as a useful guide in developing and utilizing certain skills required for specimen preparation in the context of the
importance of specimen preparation to various types of electron microscopy,
and particularly transmission electron microscopy. These techniques are
of course useful in preparing specimens for various situations and analytical techniques which have been discussed and illustrated throughout
this book.

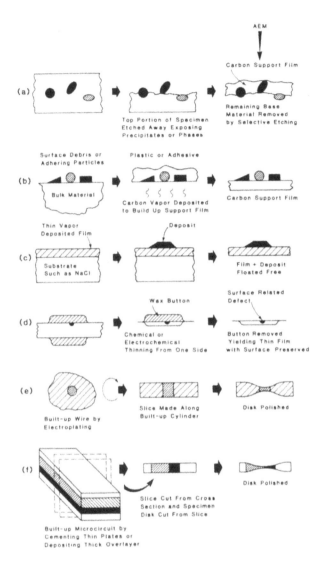

FIG. B.16 *Summary of techniques for thin specimen preparation for AEM.*
(a)   Extraction of precipitates or specific phase from a bulk material or
film.   (b)   Isolation of wafer section or surface preservation and correla-
tion of defects, etc. with a surface feature or region.   The wax button
can be a lacquer coating or other masking material.   (c)   Stripping of
debris or support of atomized powders (or other fine particles) for obser-
vation and microanalysis.   This can include the deposition of a carbon
film onto the surface of NaCl onto which nuclei, etc. have been vapor
deposited.   The nuclei will be maintained in their spatial regime and
carbon film containing them can be floated on water.   (d)   Preparation of a
metal support film onto which a second phase is deposited.   The support
film can also serve as growth substrate or reaction substrate.   In the
example shown, the support film and overgrowth/deposit are removed from
the NaCl substrate crystal by floating in water and picking up the floating
film using a small mesh grid.   (e)   A thin edge or fine wire can be built
up by electrodeposition to its surfaces until it is sufficiently thick to
cut a slice parallel to the surface to be viewed and examined.   In the
schematic the slice is thinned by dual jet electropolishing to produce a
thin dimple in the center of the disc, which is placed in the AEM specimen
holder.   (f)   XTEM/composite section preparation.   Sections can be glued
together, sliced, rounded, polished, dimpled, and ion milled to
electron transparency to view interfaces in particular.

TABLE B.6 *Preparation of thin foils by vacuum evaporation*

Material	Sub-strate	Source heater	Optimum substrate temp., °C	Evaporation rate, Å/s	Grain size at 10^{-5} torr, Å	Grain size at 10^{-9} torr, Å	Comments	Reference
Ag	NaCl	W-boat	250-300	$10^2 - 10^3$	10^4	Single crystal	Orientation is (100) strip by immersion of substrate in H_2O	47
Al	NaCl	W-boat	450	$10^2 - 10^3$	10^4	2×10^4	(111) orientation for thicker foils. Large grain can be grown on annealing at substrate temp. for 1 to 2 hours.	47
Au	NaCl	W-boat	275	10^3	2×10^3	Single crystal	(100) orientation with high-fault density, can be reduced by substrate anneal at 300°C	48
Bi	NaCl	W-boat	150	10	10	—	—	—
Cd	NaCl	W-boat	100	10	—	—	—	—
Co	NaCl	W-boat	300	$10^2 - 10^3$	10^4	10^6	(100) orientation, fcc to hcp transformation occurs at elevated temp.	—
Cr	NaCl	W-boat	250	10^3	$<10^2$	10^2 to 10^3	Grain size increased by anneal at 350°C. Orientation is (100).	—

Cu	NaCl	W-boat	300	10^3	10^2 to 10^3	10^3 to 10^4	Large grain (100) structure results when foils are annealed on substrate at 400°C for 2 hr	48
Cu–Au	NaCl	W-boat	270	$10^2 - 10^3$	$10^4 - 10^5$	—	Evaporate Ag followed by Au. Float films free in H_2O and dissolve Ag in HNO_3. Evaporate Cu and anneal for 1 hr at 350 C to homogenize	—
Cu–Ni	NaCl	W-boat	300	10^3	10^2	$10^5 - 10^6$	Use a homogenized power or wire section. Films loose homogeneity for thickness >0.5µ	—
Fe	NaCl	W-boat	380	$10^2 - 10^3$	$10^2 - 10^3$	Single crystal	Evaporate as powder	—
Ge	NaCl	W-boat	450–700	$10 - 10^2$	—	—	Strip from NaCl substrates by immersion in H_2O and from MgO by immersion in 35% HCl. (111) orientation.	50
In	NaCl	W-boat	130	10	3×10^3	$10^4 - 10^6$	Grain size not increased by annealing	51
InP	NaCl	W-boat	50	$>10^3$	—	—	Strip by immersion in ethyl alcohol and H_2O solution	—

TABLE B.6 Preparation of thin foils by vacuum evaporation (Continued)

Material	Sub-strate	Source heater	Optimum substrate temp., °C	Evaporation rate, Å/s	Grain size at 10^{-5} torr, Å	Grain size at 10^{-9} torr, Å	Comments	Reference
Ni	NaCl	W-boat	300	10^3	1.5×10^3	5×10^4 single crystal	Use heavy boat or Al_2O_3 boat for thick foils. Evaporate Ni in powder form	—
Ni-Cr	NaCl	W-boat	300	10^3	$10^2 - 10^3$	10^5	Large grain sizes occur on annealing either in situ or in electron microscope	—
Ni-Fe	NaCl	W-boat	300	10^3	$10^2 - 10^3$	$10^4 - 10^6$	Increase grain size by heating at 400°C for 4 hr	—
Pb	NaCl	W-boat	130	$10^2 - 10^3$	4×10^3	$10^4 - 10^5$	Increase grain size by annealing for several hours at 150°C	—
Pd	NaCl	W-boat	325 250	10^3 3×10^3	4×10^4 ──	$10^5 - 10^6$ Single Crystal	Can be grown single crystalline with (100) orientation by annealing at 400°C for several hours.	—
Pt	NaCl	W-boat	275	10^3	$10^2 - 10^3$	10^5	Strip by immersion in H_2O	—

Rh	NaCl	W-boat	275	10^3	$<10^2$	10^4	Strip by immersion in H_2O	—
Si	Mica	W-boat	300	$10^2 - 10^3$	$10^2 - 10^3$	$10^5 - 10^6$	(111) orientation	—
Sn	NaCl	W-boat	-196	10^3	$10^4 - 10^5$	Single crystal or large grain	(100) orientation Strip by immersion in H_2O	52
V	NaCl	W-boat	250	10	$<10^2$	$10^3 - 10^4$	Larger grains result on annealing at 350°C	51
Zn	NaCl	W-boat	100	$10^2 - 10^3$	10^2	—	Large grains result on annealing at 250°C for 2 hr. Strip film by immersion in H_2O	—

REFERENCES

1. R. D. Heidenreich, *J. Appl. Phys., 26:* 757(1955).

2. W. L. Grube and S. R. Rouze in H. I. Aaronson and G. S. Ansell (eds.), *High Temperature, High-Resolution Metallography,* p. 316, Gordon and Breach, Science Publishers, Inc., New York, 1965.

3. A. Sandor, *J. Electron Controls, 15:* 101, 111 (1963).

4. Y. Yashiro, *Oyobutsuri* (Applied Physics, Japan), *33:* 912(1964).

5. O. Nishikawa and E. W. Müller, *J. Appl. Phys., 38:* 3159(1967).

6. W. Bonfield and C. H. Li, *Acta Met., 11:* 585(1963).

7. K. Rendulic and E. W. Müller, *J. Appl. Phys., 37:* 2593(1966).

8. B. Ralph and D. G. Brandon, *Phil. Mag.,* 8: 919(1963).

9. S. Nakamura and E. W. Müller, *J. Appl. Phys., 36:* 2535(1965).

10. R. Castaing, *Rev. Met., 52:* 669(1955).

11. R. Graf and B. Genty in D. Kay (ed.), *Techniques for Electron Microscopy,* chap. 10, Blackwell Scientific Publishers, Ltd., Oxford, 1961.

12. P. A. Jaquet, *The Mechanism of Electrolytic Polishing of Metals,* Pergamon Press, New York, 1956.

13. W. J. McG. Tegart, *The Electrolytic and Chemical Polishing of Metals,* Pergamon Press, New York, 1956.

14. W. Dubroff and E. S. Machlin, *Acta Met., 16:* 1313(1968).

15. G. Thomas, *Transmission Electron Microscopy of Metals,* John Wiley & Sons, Inc., New York, 1962.

16. N. J. Tighe, *Rev. Sci. Instr., 35:* 520(1964).

17. H. B. Kirkpatrick and S. Amelinckx, *Rev. Sci. Instr., 33:* 488(1962).

18. E. Schuller and S. Amelinckx, *Naturwissenschaften, 47:* 491(1960).

19. H. Ormerod and W. J. McG. Tegart, *J. Inst. Met., 92:* 297(1963-64).

20. P. R. Swann and H. Warlimont, *Acta Met., 11:* 511(1963).

21. T. Evans and C. Phaal, *Phil. Mag., 7:* 843(1962).

22. S. R. Keown and F. B. Pickering, *J. Iron Steel Inst., 200:* 757(1963).

23. B. A. Irving, *Brit. J. Appl. Phys., 12:* 92(1961).

24. P. Lelong and J. Herenquet, *Rev. Met., 58:* 587(1961).

25. R. Phillips, in D. Kay (ed.), *Techniques for Electron Microscopy,* Blackwell Scientific Publishers, Ltd., Oxford, 1961.

26. R. J. Stokes and D. J. Sauve, *Rev. Sci. Instr., 35:* 1363(1964).

27. L. I. Van Torne and G. Thomas, *Acta Met., 11:* 881(1963).

28. J. E. Lawrence and H. Koehler, *J. Sci. Instr., 42:* 270(1965).

29. R. F. Finch et al., *J. Appl. Phys., 34:* 406(1963).

30. S. Weissman et al., *J. Phys. Soc. Japan, 18* (Suppl. S111): 179(1963).

31. D. W. Dawe and M. A. P. Dewey, *Aeon Lab. Rept. No. 288* (1963).

32. H. Blank and S. Amelinckx, *J. Appl. Phys., 34:* 2200(1963).

33. B. L. Eyre and M. J. Sole, *Phil. Mag., 9:* 545(1964).

34. H. M. Tomlinson, *Phil.Mag., 3:* 867(1958).

35. R. B. Nicholson, G. Thomas, J. Nutting, *Brit. J. Appl. Phys., 9:* 25(1958).

36. J. Silcox and P. B. Hirsch, *Phil. Mag., 4:* 72(1959).

37. G. P. Walters and W. C. Fuller, *AERE Rept. No. R4319*(1963).

38. S. Nenno et al., *J. Phys. Soc. Japan, 15:* 1409(1960).

39. L. E. Murr, *Appl. Mater. Res., 3:* 153(1964).

40. R. G. Davies and N. S. Stoloff, *Trans. AIME, 230:* 390(1964).

41. A. Fourdeaux and A. Wronski, *J. Less-common Metals, 6:* 11(1964).

42. M. S. Walker, R. Sticker, and F. E. Werner, *Z. Metallk. 54:* 331(1964).

43. P. M. Kelly and J. Nutting, *Laboratory Workers Handbook,* pp. 4-84, Newnes, London, 1964.

44. P. Feltham and P. Ryder, *Less-common Metals, 7:* 144(1964).

45. A. Berghezan, A. Fourdeaux, and S. Amelinckx, *Acta Met., 9:* 464(1961).

46. L. M. Howe, J. L. Whitton, and J. F. McGurn, *Acta Met., 10:* 773(1962).

47. L. E. Murr and M. C. Inman, *Phil Mag., 14:* 135(1966).

48. L. E. Murr, *Brit. J. Appl. Phys., 15:* 1511(1964).

49. D. W. Pashley and A. E. B. Presland, *J. Inst. Metals, 87:* 419(1958-(1959).

50. B. W. Sloope and C. O. Tiller, *J. Appl. Phys., 33:* 3458(1962).

51. L. E. Murr, *Proc. Electron Microscopy Soc. Am.* (1967).

52. R. W. Vook, *J. Appl. Phys., 32:* 1557(1961).

53. E. S. Meieran, *J. Appl. Phys., 36:* 2544(1965).

54. Osman T. Inal, private communication.

55. R. O. Schoone and E. A. Fischione, *Rev. Sci. Instr., 37:* 1351(1966).

56. T. Schober and D. G. Westlake, *Metallography, 14:* 359(1981).

57. H. O. Kirchner, *J. I. M., 97:* 256(1969).

58. E. R. Thompson and F. D. Lemkey, *Met. Trans., 1:* 2799(1970).

59. Dana Rohr and K. P. Staudhammer, Los Alamos Scientific Laboratory, private communication (1982).

60. W. A. Szilva, L. E. Murr, and M. L. Sattler, *J. Mater. Sci. Lett., 9:* 859(1974).

61. A. Gurevitch, et al., "Comparative Studies of Shaped Charge Component Microstructures," in, *Shock Wave and High-Strain-Rate Phenomena in Materials,* M. A. Meyers, L. E. Murr, and K. P. Staudhammer (eds.), Marcel Dekker, New York, 1991.

62. T. T. Sheng and R. B. Marcus, *J. Electrochem. Soc., 127:* 737(1980).

63. M. M. Al-Jassim, M. Hockley, and G. R. Booker in *Defects in Semiconductors,* J. Narayan and T. Y. Tan (eds.), North-Holland, Amsterdam, 1981, p. 521.

64. J. C. Bravman and R. Sinclair, *J. Electron Microscopy Technique, 1:*
     53(1984).

65. S. H. G. Chu and T. T. Sheng, *J. Electrochem. Soc., 131(11):* 2665
     (1984).

66. C. Miglionico, C. Stein, and L. E. Murr, *J. Mater. Sci.,* in press
     (1991).

Appendix C
COMPUTER PROGRAMS
FOR ELECTRON DIFFRACTION
ANALYSIS

Two computer programs are included here, along with tabulated examples of
output data, which are devised to aid the experimentalist in the rapid calculation of diffraction pattern data. These programs, while useful in
their present form, also serve to illustrate the use of the computer in
the rapid solution and handling of diffraction data. They will serve to
illustrate the general format for calculating diffraction ring or spot
vector magnitudes for any crystal system by making the proper adjustments
in the appropriate functions. The first program is written in simple Fortran using a file system for index and material organization. This can be
easily reprogrammed for more convenient use in small computers having
graphic display or paper data display systems. Typical data display for
an fcc material is shown in Table C.1.

Program C.1 is intended simply as an illustration of the use of the
computer to produce data for analyzing electron diffraction patterns. The
program format is cumbersome in some respects since the index library can
simply be replaced by the structure factor in trigonometric form which can
be solved to produce the diffraction ring or $|g|$ values.

Program C.2 which follows produces scaled plots of reciprocal lattice
zones or diffraction nets utilizing a program modified from the original

plotting routine described by R. J. DeAngelis [in *Metallography, 6:* 439
(1973)]. The program reproduced here was written by Rob Aiken at Los
Alamos National Laboratory based upon assignments given in an electron
microscope course utilizing this book in its original form (*Electron Optical Applications in Materials Science,* McGraw-Hill Book Co., 1970). Similar computer-generated transmission electron diffraction pattern routines
have also been described by R. A. Ploc and G. H. Keech in *J. Appl. Cryst.,*
5: 244(1972), and Ploc and co-workers might be contacted at Atomic Energy
of Canada Ltd., Chalk River, Ontario, Canada for other computer projection
routines, including stereographic projections by computer.

Program C.2 was written specifically for an HP9835A computer with an
HP9872A plotter. The program must be carefully examined in order to understand its components and routine structure, but is generally applicable to
any of the Bravais lattices. Figure C.1 illustrates a typical computer
plot of an fcc nickel (110) diffraction net. Obviously this program format can be modified to display diffraction nets on a CRT graphic display.
A novel application of this technique is to utilize a flat, horizontal CRT
display upon which diffraction patterns can be laid for comparison. The
CRT display pattern can then be adjusted to the actual SAD camera constant
size using an appropriate scale factor for comparison.

You should also be alerted to the fact that many organizations have
available standard computer programs for numerous analytical routines involved in x-ray diffraction, and these can be utilized in electron diffraction work as well. The interested reader is urged to consult the following:

American Crystallographic Association
335 E. 45th Street
New York, New York   10017

American Society for Testing and Materials (ASTM)
1916 Race Street
Philadelphia, Pennsylvania   19103

International Centre for Diffraction Data
1601 Park Lane
Swarthmore, Pennsylvania   19081

International Union of Crystallography
Five Abbey Square
Chester CH1 2HU
ENGLAND

A recently described technique (M. J. Carr, *Analytical Electron Micros-copy - 1981,* San Francisco Press, 1981, p. 139) has made possible the real time acquisition and reduction of electron diffraction data with an analytical electron microscope. M. J. Carr and W. F. Chambers (*Proc. Electron Microscopy Soc. of America,* G. W. Bailey, ed., Claitor's Publishing Div., Baton Rouge, 1982, p. 744) have outlined a comprehensive software system called "RAD" designed to facilitate both electron diffraction analysis and EDX analysis in real time on an AEM. The controlling software is a set of Flextran programs developed by Tracor Northern Corporation. The interested reader might contact Tracor Northern or M. J. Carr, Rockwell International, Rocky Flats Plant, Golden, Colorado for details.

A recently described technique (M. J. Carr, *Analytical Electron Microscopy - 1981,* San Francisco Press, 1981, p. 139) has made possible the real time acquisition and reduction of electron diffraction data with an analytical electron microscope. M. J. Carr and W. F. Chambers (*Proc. Electron Microscopy Soc. of America,* G. W. Bailey, ed., Claitor's Publishing Div., Baton Rouge, 1982, p. 744) have outlined a comprehensive software system called "RAD" designed to facilitate both electron diffraction analysis and EDX analysis in real time on an AEM. The controlling software is a set of Flextran programs developed by Tracor Northern Corporation. The interested reader might contact Tracor Northern or M. J. Carr, Sandia National Laboratories, Albuquerque, New Mexico for details.

Many new and useful software packages for electron diffraction pattern analysis and comparison are also available through the Electron Microscopy Society of America, and Dr. Nestor Zaluzec, Argonne National Laboratory, Argonne, Illinois has also been instrumental in developing and documenting a variety of software routines. The Bulletin of the Electron Microscopy Society of America (*Bulletin of EMSA*) can also be consulted for help in locating suitable software or in contacting individuals who will be able to provide educational tools, including software, which can be helpful in electron microscopy analysis.

PROGRAM C.1   *Diffraction data calculations for bcc and fcc materials*

```
C       CALCULATION OF DIFFRACTION RADII FOR METAL CRYSTAL SAMPLES
C       IN ELECTRON MICROSCOPY
C       IN THIS PROGRAM, MP DENOTES MELTING POINT—BP DENOTES BOILING POINT—
C       IN DEGREES CENTIGRADE
C       LP DENOTES THE CRYSTAL LATTICE PARAMETER IN ANGSTROM UNITS
C       ABC NOTATION IMPLIES MILLER INDICE NOTATION HKL
C       DHKL IS THE INTERPLANAR SPACING IN ANGSTROM UNITS
C       RECIPD = 1/DHKL
C       R IMPLIES DIFFRACTION RADII IN CENTIMETERS
C       AS MEASURED IN THE ELECTRON MICROSCOPE DIFFRACTION IMAGE
        DIMENSION A1(25), B1(25), C1(25), A2(25), B2(25), C2(25)
        DIMENSION METAL (3)
        DRIFT F(X, Y, Z) = ALP/SQRTF (X*X+Y*Y+Z*Z)
1       READ 2, ALL, KV
2       FORMAT (1F10.5, 110)
4       READ 6, (A1(I), B1(I), C1(I), I = 1,20)
5       READ 6, (A2(I), B2(I), C2(I), I = 1,20)
6       FORMAT (20F4.1/20F4.1/20F4.1)
7       READ 8, METAL, SYSTEM, SY, MP, BP, ALP
8       FORMAT (3A5, 1A5, F4.1, 15, F6.1, F10.5)
81      IF (WY) 9, 26, 9
9       PRINT 10
10      FORMAT (1H1//////, 12X, 5HMETAL, 10X, 6HSYSTEM, 10X, 2HMP, 10X, 2HBP, 10X, 2H 2LP)
11      PRINT 12, METAL, SYSTEM, MP, BP, ALP
12      FORMAT (1H, 2X, 3A5, 12A, 1A5, 9X, 15, 7X, F6.1, 8X, F6.3//)
112     PRINT 212
212     FORMAT (1H, 12X, 5HLAMDL, 11X, 6HKVOLTS)
312     PRINT 412, ALL, KV
412     FORMAT (1H, 11X, F6.3, 10X, 16)
13      PRINT 14
14      FORMAT (///1H, 20X, 3HHKL, 20X, 4HDHKL, 20X, 6HRECIPD, 20X, 1HR//)
15      DO 24 I = 1,20
16      IF (SY)17, 26, 18
```

PROGRAM C.1   *(Continued)*

```
17    A = A1(I)
      B = B1(I)
      C = C1(I)
      GO TO 19
18    A = A2(I)
      B = B2(I)
      C = C2(I)
19    D2 = DRIFTF (A, B, C)
20    RECIPD - 1.0/D2
21    R = RECIPD*ALL
      IA = A
      IB = B
      IC = C
22    PRINT 23, IA, IB, IC, D2, RECIPD, R
23    FORMAT (1H, 20X, 3I1, 18X, F8.5, 17X, F8.5, 16X, F8.5)
24    CONTINUE
25    GO TO 7
26    STOP
27    END
```

```
2.19000          100                    (Camera constant and operating voltage entry)
1.0 0.0 2.0 0.0 0.0 2.0 1.0 2.0 2.0 0.0 1.0 3.0 0.0 2.0 2.0 3.0 2.0
1.0 4.0 0.0 1.0 3.0 1.0 3.0 4.0 2.0 3.0 3.0 2.0 4.0 2.0 4.0 2.0 4.0
3.0 1.0 5.0 2.0 1.0 4.0 4.0 0.0 5.0 0.0 3.0 3.0 3.0 3.0 6.0 0.0 0.0
1.0 1.0 0.0 2.0 0.0 3.0 0.0 1.0 1.0 4.0 2.0 0.0 4.0 0.0 0.0 0.0 3.0
1.0 4.0 2.0 2.0 5.0 1.0 0.0 3.0 3.0 2.0 4.0 0.0 5.0 3.0 1.0 6.0
0.0 0.0 4.0 6.0 2.0 0.0 5.0 0.0 6.0 2.0 2.0 4.0 4.0 4.0 7.0 1.0
ALPHA BRASS    FCC 1.0   0000000.0    3.6840
ALUMINIUM      FCC 1.0   12202060.0   4.0496
CESIUM         BCC-1.0   28 690.0     6.0790
CHROMIUM       BCC-1.0   18902500.0   2.8850
```

Examples of data/data base structures. Third column shows melting and boiling points in Celsius.

TABLE C.1 *Example output data for bcc and fcc diffraction pattern analysis*

STAINLESS STEEL	METAL	SYSTEM FCC	MP 1450	BP 0.0	LP 3.560

	LAMDL 2.190	KVOLTS 100		

HKL	DHKL	RECIPD	R
111	2.05537	0.48654	1.06550
200	1.78000	0.56180	1.23034
220	1.25865	0.79450	1.73996
311	1.07338	0.93164	2.04028
222	1.02768	0.97306	2.13101
400	0.89000	1.12360	2.46067
331	8.81672	1.22441	2.68146
420	0.79604	1.25622	2.75112
422	0.72668	1.37612	3.01370
511	0.68512	1.45959	3.19651
333	0.68512	1.45959	3.19651
440	0.62933	1.58900	3.49772
531	0.60175	1.66182	3.63939
600	0.59333	1.68539	3.69101
442	0.59333	1.68539	3.69101
620	0.56289	1.77656	3.89067
533	0.54289	1.84198	4.03393
622	0.53669	1.86327	4.08057
444	0.51384	1.94612	4.26201
711	0.49350	2.00602	4.39318

PROGRAM C.2 Generation and plotting of electron diffraction patterns (diffraction nets)

```
80    ! This program consists of two distinct parts. First the interactive
90    ! main program which acts as a driver for the routine PLOT_NET, which
100   ! in turn is the core of the program.
110   !
120   !
130   PLOTTER IS 7,5, "9872A"          ! Define plotting device.
140   OPTION BASE 0                    ! Set initial index of arrays to be zero.
150   DEG                              ! Set so trig functions are computed in deg's.
160   DIM Screen$(19)[72]
170   COM REAL A,B,C,Alpha,Beta,Gamma,C_const,C_mult,P(3),S(3,3),Vol,INTEGER U,V,
      W,L_zone,C_sys,S_lat,Limit,Header$(1)[72]
180   !
190   ! Data for Menu display
200   !
210   S_data: DATA " MENU "                          ! Screen line 0
220   DATA ""                                        ! 1
230   DATA "Variables"                               ! 2
240   DATA " 1    Header"                            ! 3
250   DATA " 2    Zone axis        "                 ! 4
260   DATA " 3    Crystallographic system"           ! 5
270   DATA " 4    a=      b=      c=      "           ! 6
280   DATA " 5    alpha=    beta=    gamma=    "      ! 7
290   DATA " 6    Index limit       "                ! 8
300   DATA " 7    Camera constent   "                ! 9
310   DATA " 8    Chart multiplier  "                ! 10
320   DATA " 9    Paper size        "                ! 11
330   DATA " 10   Csize:net,header  "                ! 12
340   DATA ""                                        ! 13
350   DATA "Commands"                                ! 14
360   DATA " 11   Print diffraction net"             ! 15
370   DATA " 12   Set for new crystal"               ! 16
380   DATA " 13   Exit"                              ! 17
390   DATA "","",                                    ! 18-19
400   !
```

PROGRAM C.2 Generation and plotting of electron diffraction patterns (diffraction nets) (Continued)

```
410  ! Initialized data - Default values
420  !
430  I_data:  DATA 1.369                    ! Camera const      (angstrom-cm)
440  DATA 1                                 ! Chart multipication factor
450  DATA 200,270                           ! Paper size (mm)
460  DATA 2,3                               ! Character size
470  DATA 3                                 ! Indicies Limit
480  !
490  ! Initialize
500  !
510  MAT READ Screen$
520  READ C_const,C_mult,P(0),P(1),P(2),P(3),Limit
530  Screen$(8)=Screen$(8)&VAL$(Limit)
540  Screen$(9)=Screen$(9)&VAL$(C_const)
550  Screen$(10)=Screen$(10)&VAL$(C_mult)
560  Screen$(11)=Screen$(11)&VAL$(P(0))&" x "&VAL$(P(1))
570  Screen$(12)=Screen$(12)&VAL$(P(2))&", "&VAL$(P(3))
580  !
590  ! Print screen then preform the desired command by branching.
600  !
610  Menu: CALL P_screen(Screen$(*))
620  Input: New_xal=0
630  INPUT "Enter number (1-14)",Com
640  IF (Com>=1) AND (Com<=13) THEN Doit
650  CALL Error("value out of range",1500)
660  GOTO Input
670  Doit: ON Com GOTO One,Two,Three,Four,Five,Six,Seven,Eight,Nine,Ten,Eleven,Tw
elve,Thirteen
680  !
690  !
700  ! Because of the "set for New Crystal" command the order of the label
710  ! blocks to follow are critical.
720  !
730  Two: INPUT "Enter zone axis: (U,V,W)",U,V,W
740        Screen$(4)="     2   Zone axis ["&VAL$(U)&", "&VAL$(V)&", "&VAL$(W)&"]"
```

PROGRAM C.2 *Generation and plotting of electron diffraction patterns (diffraction nets) (Continued)*

```
750        GOTO Menu
760    !
770    Twelve: New_xal=1              ! This drops though.
780    !
790    One:  INPUT "Enter header:",Header$(0)
800        Screen$(3)=" 1    Heading-- "&Header$(0)
810        IF New_xal<>1 THEN Menu
820    !
830    Three: CALL Get_c_sys(C_sys,S_lat,Screen$(5),Header$(1))
840        IF New_xal<>1 THEN Menu
850        PRINT CHR$(12)
860    !
870    ! If the xal system is cubic only ask for "a" else get "a","b","c".
880    !
890    Four: IF C_sys<>1 THEN Skip1
900        INPUT "Enter lattice constent a: (angstroms)",A
910        B=A
920        C=A
930        GOTO Skip2
940    Skip1: INPUT "Enter lattice constent a,b,c: (angstroms)",A,B,C
950    Skip2: Screen$(6)=" 4    a="&VAL$(A)&"    b="&VAL$(B)&"    c="&VAL$(C)
960        IF New_xal<>1 THEN Menu
970    !
980    ! If xal system is cubic then set angles to 90 deg, else ask for angles.
990    !
1000   Five:IF C_sys=1 THEN Alpha=Beta=Gamma=90
1010       IF (C_sys=0) OR (C_sys>=4) THEN INPUT "Enter lattice angles Alpha,Beta,
Gamma: (rad)",Alpha,Beta,Gamma
1020       Screen$(7)=" 5    Alpha="&VAL$(Alpha)&"    Beta="&VAL$(Beta)&"    Ga
mma="&VAL$(Gamma)
1030       GOTO Menu
1040   !
1050   Six: INPUT "Enter index limit:",Limit
1060       Screen$(8)=" 6    Index Limit    "&VAL$(Limit)
1070       GOTO Menu
1080   !
```

PROGRAM C.2 Generation and plotting of electron diffraction patterns (diffraction nets) (Continued)

```
1090 Seven: INPUT "Enter camera constant: (angstrom-cm)",C_const
1100      Screen$(9)="   7    Camera constant  "&VAL$(C_const)
1110      GOTO Menu
1120 !
1130 Eight: INPUT "Enter chart multiplier:",C_mult
1140      Screen$(10)="   8    Chart multiplier "&VAL$(C_mult)
1150      GOTO Menu
1160 !
1170 Nine: INPUT "Enter paper size: (mm)",P(0),P(1)
1180      Screen$(11)="   9    Paper size    "&VAL$(P(0))&" x "&VAL$(P(1))
1190      GOTO Menu
1200 !
1210 Ten: INPUT "Enter character size:(net,heading)",C_size_c,C_size_h
1220      Screen$(12)="  10    Csize:net,header "&VAL$(C_size_c)&", "&VAL$(C_size_h)
1230      GOTO Menu
1240 !
1250 Eleven: IF NOT ((A=0) OR (B=0) OR (C=0) OR (Alpha=0) OR (Beta=0) OR (Gamma=0) OR (C_sys=0) OR (S_lat=0) OR (U=0) AND (V=0) AND (W=0)) THEN P_net
1260      CALL Error("variables not set:",1500)
1270      GOTO Menu
1280 P_net: CALL Plot_net
1290      GOTO Menu
1300 !
1310 Thirteen: PEN 0
1320      PRINT CHR$(12)
1330      STOP
1340 !
1350 END
1360 !
1370 !
1380 ! Calculate the angle between two planes described by
1390 ! (H1,K1,L1) and (H2,K2,L2).
1400 !
1410 SUB Angle_calc(Theta,INTEGER H1,K1,L1,H2,K2,L2)
```

PROGRAM C.2 *Generation and plotting of electron diffraction patterns (diffraction nets)* *(Continued,*

```
1420      COM REAL A,B,C,Alpha,Beta,Gamma,C_const,C_mult,P(*),S(*),Vol,INTEGER U,
V,W,L_zone,C_sys,S_lat,Limit,Header$(*)
1430      DEG
1440      !
1450      ! Check for special conditions.
1460      !
1470      IF (H1=0) AND (K1=0) AND (L1=0) THEN Zero
1480      IF (H2=0) AND (K2=0) AND (L2=0) THEN Zero
1490      FOR J=1 TO Limit
1500        FOR I=1 TO Limit
1510          IF (H1/I=H2/J) AND (K1/I=K2/J) AND (L1/I=L2/J) THEN Zero
1520          IF (H1/I=-H2/J) AND (K1/I=-K2/J) AND (L1/I=-L2/J) THEN Pi
1530        NEXT I
1540      NEXT J
1550      !
1560      ON C_sys GOTO Cubic,Tetra,Ortho,Hex,Rhomb,Mono,Tric
1570      !
1580 Cubic: Theta=ACS((H1*H2+K1*K2+L1*L2)/SQR((H1^2+K1^2+L1^2)*(H2^2+K2^2+L2^2))
)
1590      SUBEXIT
1600 Tetra: Theta=ACS((H1*H2+K1*K2/A^2+L1*L2/C^2)/SQR(((H1^2+K1^2)/A^2+L1^2/C^2)
*((H2^2+K2^2)/C^2+L2^2/C^2)))
1610      SUBEXIT
1620 Ortho: Theta=ACS((H1*H2/A^2+K1*K2/B^2+L1*L2/C^2)/SQR((H1^2/A^2+K1^2/B^2+L1^
2/C^2)*(H2^2/A^2+K2^2/B^2+L2^2/C^2)))
1630      SUBEXIT
1640 Hex: Theta=ACS((H1*H2+K1*K2+.5*(H1*K2+H2*K1)+3*A^2*L1*L2/(4*C^2))/SQR((H1^2
+K1^2+H1*K1+3*A^2*L1^2/(4*C^2))*(H2^2+K2^2+H2*K2+3*A^2*L2^2/(4*C^2))))
1650      SUBEXIT
1660 Rhomb: CALL D_calc(A,B,C,Alpha,Beta,Gamma,D1,C_sys,H1,K1,L1)
1670      CALL D_calc(A,B,C,Alpha,Beta,Gamma,D2,C_sys,H2,K2,L2)
1680      Vol=A^3*SQR(1-3*COS(Alpha)^2+2*COS(Alpha)^3)
1690      Theta=ACS(A^4*D1*D2/Vol^2*(SIN(Alpha)^2*(H1*H2+K1*K2+L1*L2)+(COS(Alpha)
^2-COS(Alpha))*(K1*L2+K2*L1+L1*H2+L2*H1+H1*K2+H2*K1)))
1700      SUBEXIT
1710 Mono: CALL D_calc(A,B,C,Alpha,Beta,Gamma,D1,C_sys,H1,K1,L1)
```

PROGRAM C.2 Generation and plotting of electron diffraction patterns (diffraction nets) (Continued)

```
1720        CALL D_calc(A,B,C,Alpha,Beta,Gamma,D2,C_sys,H2,K2,L2)
1730        Theta=ACS(D1*D2/SIN(Beta)^2*(H1*H2/A^2+K1*K2*SIN(Beta)^2/B^2+L1*L2/C^2-
(L1*H2+L2*H1)*COS(Beta)/(A*C)))
1740        SUBEXIT
1750 Tric: CALL D_calc(A,B,C,Alpha,Beta,Gamma,D1,C_sys,H1,K1,L1)
1760        CALL D_calc(A,B,C,Alpha,Beta,Gamma,D2,C_sys,H2,K2,L2)
1770        Theta=ACS(D1*D2/Vol*(S(1,1)*H1*H2+S(2,2)*K1*K2+S(3,3)*L1*L2+S(2,3)*(K1*
L2+K2*L1)+S(1,3)*(L1*H2+L2*H1)+S(1,2)*(H1*K2+H2*K1)))
1780        SUBEXIT
1790 Zero: Theta=0
1800        SUBEXIT
1810 Pi:  Theta=180
1820        SUBEXIT
1830 SUBEND
1840 !
1850 !
1860 ! Calculate the "D" spacing between given (H,K,L) planes.
1870 !
1880 SUB D_calc(D,INTEGER H,K,L)
1890     COM REAL A,B,C,Alpha,Beta,Gamma,C_const,C_mult,P(*),S(*),Vol,INTEGER U,
V,W,L_zone,C_sys,S_lat,Limit,Header$(*)
1900        DEG
1910        IF (H=0) AND (K=0) AND (L=0) THEN GOTO Zero
1920        ON C_sys GOTO Cubic,Tetra,Ortho,Hex,Rhomb,Mono,Tric
1930        !
1940 Cubic: D=SQR(A^2/(H^2+K^2+L^2))
1950        SUBEXIT
1960 Tetra: D=SQR(1/((H^2+K^2)/A^2+L^2/C^2))
1970        SUBEXIT
1980 Ortho: D=SQR(1/(H^2/A^2+K^2/B^2+L^2/C^2))
1990        SUBEXIT
2000 Hex:  D=SQR(1/(4*(H^2+H*K+K^2)/(3*A^2)+L^2/C^2))
2010        SUBEXIT
2020 Rhomb: D=SQR(A^2*(1-3*COS(Alpha)^2+-2*COS(Alpha)^3)/((H^2+K^2+L^2)*SIN(Alph
a)^2+2*(H*K+K*L+H*L)*(COS(Alpha)^2-COS(Alpha))))
2030        SUBEXIT
```

PROGRAM C.2 *Generation and plotting of electron diffraction patterns (diffraction nets) (Continued)*

```
2040 Mono: D=SQR(SIN(Beta)^2/(H^2/A^2+K^2*SIN(Beta)^2/B^2+L^2/C^2-2*H*L*COS(Beta
     )/(A*C)))
2050     SUBEXIT
2060 Tric: D=SQR(Vol^2/(S(1,1)*H^2+S(2,2)*K^2+S(3,3)*L^2+2*S(1,2)*H*K+2*S(2,3)*K
     *L+2*S(1,3)*H*L))
2070     SUBEXIT
2080 Zero: D=0
2090 SUBEND
2100 !
2110 !
2120 SUB Error(S$,N)
2130     BEEP
2140     DISP S$
2150     IF N<0 THEN SUBEXIT
2160     WAIT N
2170 SUBEND
2180 !
2190 !
2200 ! Check if the given diffraction spot (H,K,L), is forbidden. If so
2210 ! RETURN 1 else RETURN 0.
2220 !
2230 DEF FNForbidden(INTEGER H,K,L,C_sys,S_lat)
2240     ON S_lat GOTO Simple,Body,Face,Base
2250 Simple: RETURN 0
2260 Body: IF (H+K+L) MOD 2=1 THEN RETURN 1
2270     RETURN 0
2280 Face: IF (H+K) MOD 2+(H+L) MOD 2+(K+L) MOD 2=2 THEN RETURN 1
2290     RETURN 0
2300 Base: IF (H+K) MOD 2=1 THEN RETURN 1
2310     RETURN 0
2320 FNEND
2330 !
2340 !
2350 SUB Get_c_sys(INTEGER C_sys,S_lat,S$,L$)
2360     DIM Sys$(7),Lat$(4)
```

PROGRAM C.2   *Generation and plotting of electron diffraction patterns (diffraction nets)   (Continued)*

```
2370      DATA "", "Cubic", "Tetragonal", "Orthorhombic", "Hexagonal", "Rhombohedral",
"Monoclinic", "Triclinic"
2380      DATA "", "Simple", "Body-Centered", "Face-Centered", "Base-Centered"
2390      MAT READ Sys$
2400      MAT READ Lat$
2410      !
2420      PRINT CHR$(12);CHR$(129);" Set Crystal System ";CHR$(128)
2430      PRINT ""
2440      FOR I=1 TO 7
2450          PRINT "      ";I;"      ";Sys$(I)
2460      NEXT I
2470      INPUT "Enter number: (1-7)",C_sys
2480      !
2490      PRINT CHR$(12);CHR$(129);" Set Space Lattice Type ";CHR$(128)
2500      PRINT ""
2510      FOR I=1 TO 4
2520          PRINT "      ";I;"      ";Lat$(I)
2530      NEXT I
2540      INPUT "Enter number: (1-4)",S_lat
2550      !
2560      S$="      3      "&Lat$(S_lat)&"  "&Sys$(C_sys)
2570      L$=Lat$(S_lat)&"  "&Sys$(C_sys)
2580 SUBEND
2590 !
2600 !
2610 ! Print the Heading for a diffraction pattern.
2620 !
2630 SUB Heading(X,Y)
2640      COM REAL A,B,C,Alpha,Beta,Gamma,C_const,C_mult,P(*),S(*),Vol,INTEGER U,
V,W,L_zone,C_sys,S_lat,Limit,Header$(*)
2650      CSIZE P(3)
2660      MSCALE 0,Y
2670      R_limit=X-10
2680      L_limit=10
2690      Y_inc=(Y-X)/6
2700      IF Y_inc(C_size*X/100 THEN Y_inc=(C_size+.1)*X/100
```

PROGRAM C.2 Generation and plotting of electron diffraction patterns (diffraction nets) (Continued)

```
2710    LORG 5
2720    MOVE X/2,-Y_inc
2730    LABEL Header$(0)
2740    MOVE X/2,-2*Y_inc
2750    LABEL Header$(1)
2760    LORG 2
2770    MOVE L_limit,-3*Y_inc
2780    LABEL "[UVW]="&VAL$(U)&VAL$(V)&VAL$(W)
2790    MOVE L_limit,-4*Y_inc
2800    IF (A=B) AND (B=C) THEN LABEL "A=B=C="&VAL$(A)
2810    IF (A<>B) OR (B<>C) THEN LABEL "A="&VAL$(A)&" B="&VAL$(B)&" C="&VAL$(C)
2820    MOVE L_limit,-5*Y_inc
2830    IF (Alpha=Beta) AND (Beta=Gamma) THEN LABEL "Alpha=Beta=Gamma="&VAL$(Al
pha)
2840    IF (Alpha<>Beta) OR (Beta<>Gamma) THEN LABEL "Alpha="&VAL$(Alpha)&" Bet
a="&VAL$(Beta)&" Gamma="&VAL$(Gamma)
2850    MOVE L_limit,-6*Y_inc
2860    IF C_mult=1 THEN LABEL "Camera const="&VAL$(C_const)
2870    IF C_mult<>1 THEN LABEL "Camera const="&VAL$(C_const)&" ("&VAL$(C_mult)
&"x)"
2880    IF L_zone=0 THEN SUBEXIT
2890    LORG 8
2900    MOVE R_limit,-3*Y_inc
2910    LABEL "Laue Zone="&VAL$(L_zone)
2920    SUBEND
2930    !
2940    !
2950    ! This routine is the heart of the program. It goes through an iteration
2960    ! of possible diffraction indicies. Each one is then checked to see if
2970    ! it satisfies the equation, UH + VK + WL = 0. It is then checked to see
2980    ! if it is a forbidden reflection. If all is well, the angular relationship
2990    ! and D values are calculated and the spot is plotted using the two points
3000    ! (Hp,Kp,Lp) and (Hpp,Kpp,Lpp) as reference. Two reference points are
3010    ! needed since ARC-COSINE yields values between 0-180 deg's.
3020    !
```

PROGRAM C.2   *Generation and plotting of electron diffraction patterns (diffraction nets)*   *(Continued)*

```
3030  SUB Plot_net
3040    COM REAL A,B,C,Alpha,Beta,Gamma,C_const,C_mult,P(*),S(*),Vol,INTEGER U,
V,W,L_zone,C_sys,S_lat,Limit,Header$(*)
3050    INTEGER Zone,H,K,L,Hp,Lp,Kp,Hpp,Kpp,Lpp
3060    DEG
3070    OVERLAP
3080    LIMIT 0,P(0),0,P(1)                              ! Set physical limits
3090    PRINT CHR$(12)                                   ! Clear CRT
3100    INPUT "Print heading? (Y/N)",Tmp$
3110    IF (Tmp$[1]="Y") OR (Tmp$[1]="y") THEN CALL Heading(P(0),P(1))
3120    IF C_sys=7 THEN CALL Triclinic_calc
3130    CALL Set_net_area(P(0),P(2))
3140    !
3150    FOR Zone=-L_zone TO L_zone
3160      FOR L=Limit TO -Limit STEP -1
3170        FOR K=Limit TO -Limit STEP -1
3180          FOR H=Limit TO -Limit STEP -1
3190            Ip=U*H+V*K+W*L
3200            IF Ip<>Zone THEN Next_h
3210            IF FNForbidden(H,K,L,C_sys,S_lat)=1 THEN Next_h
3220            IF ((Hp<>0) OR (Kp<>0) OR (Lp<>0)) AND (Hpp=0) AND (Kpp
=0) AND (Lpp=0) THEN GOSUB Set_pp
3230            IF (Hp=0) AND (Kp=0) AND (Lp=0) THEN GOSUB Set_p
3240            CALL D_calc(D,H,K,L)
3250            IF D<>0 THEN R=C_const*C_mult*10/D
3260            IF D=0 THEN R=0
3270            CALL Angle_calc(Theta,H,K,L,Hp,Kp,Lp)
3280            CALL Angle_calc(Thetapp,H,K,L,Hpp,Kpp,Lpp)
3290            IF PROUND(Theta,-2)<>PROUND(Thetap+Thetapp,-2) THEN The
ta=-Theta
3300            X_point=R*COS(Theta)
3310            Y_point=R*SIN(Theta)
3320            IF (ABS(2*X_point)>P(0)) OR (ABS(2*Y_point)>P(0)) THEN
Next_h
3330            CALL Plot_pt(X_point,Y_point,H,K,L)
```

PROGRAM C.2 *Generation and plotting of electron diffraction patterns (diffraction nets) (Continued)*

```
3340 Next_h:          NEXT H
3350             NEXT K
3360         NEXT L
3370     NEXT Zone
3380     SERIAL
3390     PEN 0
3400     SUBEXIT
3410     !
3420     ! Set reference points.
3430     !
3440 Set_p: Hp=H
3450     Kp=K
3460     Lp=L
3470     PRINT "Reference points are: ";Hp;Kp;Lp
3480     RETURN
3490     !
3500 Set_pp: Hpp=H
3510     Kpp=K
3520     Lpp=L
3530     PRINT "        ";Hpp;Kpp;Lpp
3540     CALL Angle_calc(Thetap,Hp,Kp,Lp,Hpp,Kpp,Lpp)
3550     RETURN
3560 SUBEND
3570     !
3580     ! Plot a "o" at the point X,Y then label the corresponding indicies.
3590     !
3600     !
3610 SUB Plot_pt(X,Y,INTEGER H,K,L)
3620     LORG 5
3630     MOVE X,Y
3640     LABEL "o"
3650     LABEL VAL$(H)&VAL$(K)&VAL$(L)
3660 SUBEND
3670     !
3680     !
```

PROGRAM C.2 *Generation and plotting of electron diffraction patterns (diffraction nets) (Continued)*

```
3690  SUB P_screen(S$(*))
3700     PRINT CHR$(12);S$(0)
3710     FOR I=1 TO 19
3720        PRINT S$(I)
3730     NEXT I
3740  SUBEND
3750  !
3760  !
3770  ! Set the size of the plotting area and also define the plotting
3780  ! units to be millimeters.
3790  !
3800  SUB Set_net_area(X,C)
3810     MSCALE X/2,X/2
3820     SETUU
3830     CSIZE C
3840  SUBEND
3850  !
3860  !
3870  SUB Triclinic_calc
3880     COM REAL A,B,C,Alpha,Beta,Gamma,C_const,C_mult,P(*),S(*),Vol,INTEGER U,
       V,W,L_zone,C_sys,S_lat,Limit,Header$(*)
3890     Cosa=COS(Alpha)
3900     Cosb=COS(Beta)
3910     Cosg=COS(Gamma)
3920     S(1,1)=A^2*C^2*SIN(Alpha)^2
3930     S(2,2)=A^2*B^2*SIN(Beta)^2
3940     S(3,3)=A^2*B^2*SIN(Gamma)^2
3950     S(1,2)=A*B*C^2*(Cosa*Cosb-Cosg)
3960     S(2,3)=A^2*B*C*(Cosb*Cosg-Cosa)
3970     S(1,3)=A*B^2*C*(Cosg*Cosa-Cosb)
3980     Vol=A*B*C*SQR(1-Cosa^2-Cosb^2-Cosg^2+2*Cosa*Cosb*Cosg)
3990  SUBEND
4000  !
4010  ! the last line.
```

	$\bar{5}\bar{5}9$ +	$\bar{3}\bar{3}9$ +	$\bar{1}\bar{1}9$ +	$\bar{1}19$ +	$3\bar{3}9$ +	$5\bar{5}9$ +	
$\bar{6}\bar{6}8$ +	$\bar{4}\bar{4}8$ +	$\bar{2}\bar{2}8$ +	$00\bar{8}$ +	$2\bar{2}8$ +	$4\bar{4}8$ +	$6\bar{6}8$ +
	$\bar{5}\bar{5}7$ +	$\bar{3}\bar{3}7$ +	$\bar{1}\bar{1}7$ +	$\bar{1}17$ +	$3\bar{3}7$ +	$5\bar{5}7$ +	
$\bar{6}\bar{6}6$ +	$\bar{4}\bar{4}6$ +	$\bar{2}\bar{2}6$ +	$00\bar{6}$ +	$2\bar{2}6$ +	$4\bar{4}6$ +	$6\bar{6}6$ +
	$\bar{5}\bar{5}5$ +	$\bar{3}\bar{3}5$ +	$\bar{1}\bar{1}5$ +	$\bar{1}15$ +	$3\bar{3}5$ +	$5\bar{5}5$ +	
$\bar{6}\bar{6}4$ +	$\bar{4}\bar{4}4$ +	$\bar{2}\bar{2}4$ +	$00\bar{4}$ +	$2\bar{2}4$ +	$4\bar{4}4$ +	$6\bar{6}4$ +
	$\bar{5}\bar{5}3$ +	$\bar{3}\bar{3}3$ +	$\bar{1}\bar{1}3$ +	$\bar{1}13$ +	$3\bar{3}3$ +	$5\bar{5}3$ +	
$\bar{6}\bar{6}2$ +	$\bar{4}\bar{4}2$ +	$\bar{2}\bar{2}2$ +	$00\bar{2}$ +	$2\bar{2}2$ +	$4\bar{4}2$ +	$6\bar{6}2$ +
	$\bar{5}\bar{5}1$ +	$\bar{3}\bar{3}1$ +	$\bar{1}\bar{1}1$ +	$\bar{1}11$ +	$3\bar{3}1$ +	$5\bar{5}1$ +	
$\bar{6}\bar{6}0$ +	$\bar{4}\bar{4}0$ +	$\bar{2}\bar{2}0$ +	000 +	$2\bar{2}0$ +	$4\bar{4}0$ +	$6\bar{6}0$ +
	$\bar{5}51$ +	$\bar{3}31$ +	$\bar{1}11$ +	111 +	331 +	551 +	
$\bar{6}62$ +	$\bar{4}42$ +	$\bar{2}22$ +	002 +	222 +	442 +	662 +
	$\bar{5}53$ +	$\bar{3}33$ +	$\bar{1}13$ +	113 +	333 +	553 +	
$\bar{6}64$ +	$\bar{4}44$ +	$\bar{2}24$ +	004 +	224 +	444 +	664 +
	$\bar{5}55$ +	$\bar{3}35$ +	$\bar{1}15$ +	115 +	335 +	555 +	
$\bar{6}65$ +	$\bar{4}46$ +	$\bar{2}26$ +	006 +	226 +	446 +	666 +
	$\bar{5}57$ +	$\bar{3}37$ +	$\bar{1}17$ +	117 +	337 +	557 +	
$\bar{6}68$ +	$\bar{4}48$ +	$\bar{2}28$ +	008 +	228 +	448 +	668 +
	$\bar{5}59$ +	$\bar{3}39$ +	$\bar{1}19$ +	119 +	339 +	559 +	

FIG. C.1 *(110) fcc diffraction pattern generated by the computer program listed in Table C.2.*

Appendix D
TABULATED MATERIALS
CONSTANTS AND USEFUL
PHYSICAL PROPERTIES

This Appendix is simply a collection of important and useful physical constants and related physical properties of the elements and other materials.
In Table D.1, the elements and their physical properties are listed periodically (increasing Z from 1 to 100). The data composing this table have
been obtained from numerous sources over the years, with corrections being
made as they became available. The major sources for initial reference
purposes have been the periodic chart No. TER-C (1964) by Samuel Ruben,
H. W. Sams and Company, Inc., and C. S. Barrett and T. B. Massalski, *Structure of Metals*, McGraw-Hill Book Company, New York, 1966. This latter
source has been revised and republished (Pergamon Press, New York, 1980).

Table D.2 gives the necessary correction data for electrons accelerated by increasing potentials from 0.1 kV to 1000 kV and includes the velocity, wavelength, wavelength reciprocal or Ewald-sphere radius ($1/\lambda$), and
the relativistic mass correction m/m_0) or $[1 - (v/c)^2]^{-\frac{1}{2}}$.

In Table D.3 the amplitudes of the electron scattering factors are
listed for specific values of $\sin \theta/\lambda$ or ($1/d_{hk\ell}$) for each element from
$Z = 1$ to $Z = 100$. These values are essentially those given by J. A. Ibers
and B. K. Vainshtein in Vol. III of the *International Tables for X-ray
Crystallography* [Tables 3.33 A(1) and A(2)]. These values are assumed to

to be approximately correct. More recent values of $f_{el}(\mu)$ calculated are observed to be approximately 10 percent larger [see, for example, P. A. Doyle and P. S. Turner, *Acta Cryst.*, *A24*: 390(1968)].

Tables D.4, D.5, and D.6 utilize the data of Tables D.2 and D.3 in the calculation of the dynamical extinction distances for some fcc (including the diamond cubics Si and Ge), bcc, and hcp materials, respectively, for several of the low-order reflections ($g_{hk\ell}$). The values tabulated are computer solutions of Eq. (7.6), with the structure factor for the hcp materials expressed as

$$
\begin{aligned}
F_{hcp}(hk\ell) &= \left[F(real)^2 + F(imaginary)^2 \right]^{\frac{1}{2}} \\
F(real) &= f_{el}(\mu) \left[1 + \cos 2\pi \left(\frac{2h}{3} + \frac{k}{3} + \frac{\ell}{2} \right) \right] \\
F(imaginary) &= f_{el}(\mu) \left[1 + \sin 2\pi \left(\frac{2h}{3} + \frac{k}{3} + \frac{\ell}{2} \right) \right]
\end{aligned}
$$

These values are of course only approximations, with an accuracy contingent on the faith one puts in the values of $f_{el}(\mu)$ interpolated from Table D.3 for the appropriate reflection ($g_{hk\ell}$).

Values of the commonly used physical constants, derived constants, and some useful conversion factors are included in Table D.7, which concludes this Appendix. These have been obtained from numerous sources over the years and appropriately corrected.

TABLE D.1 Periodic table of the elements and some of their physical properties

Atomic No. Z	Element	Symbol	Atomic weight, M	Density (ρ at 20°C), g/cm³	Crystal structure	Lattice constants, Å a	b	c	Temp., °C	Melting point, °C	Boiling point, °C
1	Hydrogen	H	1.008	0.084 × 10⁻³	Hexagonal	3.750	—	6.120	-271	-259.4	-252.7
2	Helium	He	4.003	0.166 × 10⁻³	HCP	3.570	—	5.830	-271.5	-271.4	-268.9
3	Lithium	Li	6.940	0.53	BCC	3.509	—	—	20	186	1370
4	Beryllium	Be	9.02	1.84	HCP	2.285	—	3.584	20	1300	2770
5	Boron	B	10.81	2.34	Rhombic	9.450	—	23.800	20	2280	2470
6	Carbon (diamond)	C	12.01	3.51	Diamond cubic	3.568	—	—	20	—	—
6	Carbon	C	12.01	2.26	Hexagonal	2.461	—	6.701	20	3750	4830
	Carbon (amorphous)	C	12.01	1.95	—	—	—	—	—	—	—
7	Nitrogen (α)	N	14.01	1.165 × 10⁻³	Cubic	5.661	—	—	-253	-210.1	-195.8
8	Oxygen	O	16.00	1.332 × 10⁻³	Cubic	6.83	—	—	-225	-218.8	-183
9	Fluorine	F	19.00	—	—	—	—	—	—	-230	-188
10	Neon	Ne	20.18	0.839 × 10⁻³	FCC	4.52	—	—	-268	-249	-246
11	Sodium	Na	22.99	0.97	BCC	4.29	—	—	20	98	892
12	Magnesium	Mg	24.31	1.74	HCP	3.209	—	5.21	20	650	1110
13	Aluminum	Al	26.98	2.702	FCC	4.05	—	—	20	660	2060
14	Silicon	Si	28.09	2.33	Diamond cubic	5.4308	—	—	20	1440	2300
15	Phosphorus	P	30.98	1.82	Cubic	7.71	—	—	-35	44	280
16	Sulfur	S	32.06	2.07	Orthorhombic	10.51	12.94	24.60	20	119	445
17	Chlorine	Cl	35.46	—	Tetragonal	8.56	—	6.12	20	-101	-35
18	Argon	A	39.94	1.663 × 10⁻³	FCC	5.42	—	—	-185	-189	-186
19	Potassium	K	39.10	0.86	BCC	5.247	—	—	-233	6.3	770
20	Calcium (α)	Ca	40.08	1.55	FCC	5.582	—	—	20	860	1400
20	Calcium (γ)	Ca	40.08	1.55	BCC	4.486	—	—	500	—	—
21	Scandium (α)	Sc	45.10	3.0	FCC	4.541	—	—	20	1200	2380
21	Scandium (β)	Sc	45.10	3.0	HCP	3.31	—	5.273	20	—	—
22	Titanium (α)	Ti	47.90	4.51	HCP	2.951	—	4.683	25	1825	3030
22	Titanium (β)	Ti	47.90	4.51	BCC	3.306	—	—	900	—	—
23	Vanadium	V	50.94	6.1	BCC	3.028	—	—	30	1745	3400
24	Chromium	Cr	52.00	7.19	BCC	2.885	—	—	20	1895	2500
25	Manganese (α)	Mn	54.94	7.43	Cubic‡	8.914	—	—	20	1245	2150
	Manganese (β)	Mn	54.94	7.43	Cubic	6.314	—	—	20	—	—
	Manganese (γ)	Mn	54.94	7.43	FCC	3.862	—	—	1100	—	—
26	Iron (α)	Fe	55.84	7.86	BCC	2.866	—	—	20	1540	2740
	Iron (γ)	Fe	55.84	7.86	FCC	3.571	—	—	20	—	—
	Iron (γ')	Fe	55.84	7.86	FCC	3.647	—	—	900	—	—
27	Cobalt (α)	Co	58.93	8.9	HCP	2.505	—	4.089	20	1495	2900
27	Cobalt (β)	Co	58.93	8.9	FCC	3.544	—	—	—	—	—
28	Nickel	Ni	58.7	8.9	FCC	3.524	—	—	20	1455	2730
29	Copper	Cu	63.57	8.96	FCC	3.615	—	—	20	1083	2600

TABLE D.1 *Periodic table of the elements and some of their physical properties (Continued)*

Atomic No. Z	Element	Symbol	Atomic weight M	Density (ρ at 20°C), g/cm³	Crystal structure	Lattice constants, Å a	b	c	Temp., °C	Melting point, °C	Boiling point, °C
30	Zinc	Zn	65.37	6.49	HCP	2.665	—	4.947	20	420	906
31	Gallium	Ga	69.72	5.91	FC orthorhombic	4.52	7.66	4.526	20	30	2070
32	Germanium	Ge	72.59	5.32	Diamond cubic	5.656	—	—	20	912	2690
33	Arsenic	As	74.92	5.727	Rhombic	4.131	—	—	20	612	820
34	Selenium	Se	78.96	4.79	Hexagonal	4.366	—	4.959	20	220	680
35	Bromine	Br	79.91	3.15	Orthorhombic	4.49	6.65	8.74	20	-72	19.2
36	Krypton	Kr	83.70	3.488 × 10⁻³	FCC	5.68	—	—	-150	-157	-152
37	Rubidium	Rb	85.47	1.53	BCC	5.70	—	—	-191	40	680
38	Strontium	Sr	87.62	2.6	FCC	6.085	—	—	20	775	1380
39	Yttrium	Y	88.91	4.47	HCP	3.647	—	5.731	20	1500	2515
40	Zirconium (α)	Zr	91.22	6.49	HCP	3.231	—	5.148	20	1870	2910
40	Zirconium (β)	Zr	91.22	6.49	BCC	3.609	—	—	20		
41	Niobium	Nb	92.91	8.4	BCC	3.301	—	—	20	2495	3710
41	Columbium	Cb	92.91	8.4	BCC	3.301	—	—	20	2495	3710
42	Molybdenum	Mo	95.94	10.2	BCC	3.147	—	—	20	2750	4800
43	Technetium	Tc	99.0	11.5	HCP	2.735	—	4.388	20	2130	
44	Ruthenium	Ru	101.7	12.2	HCP	2.698	—	4.273	20	2550	4900
45	Rhodium	Rh	102.91	12.4	FCC	3.803	—	—	20	1970	4500
46	Palladium	Pd	106.4	12.0	FCC	3.891	—	—	20	1554	4000
47	Silver	Ag	107.87	10.5	FCC	4.089	—	—	20	960	2210
48	Cadmium	Cd	112.4	8.65	HCP	2.979	—	5.617	20	321	765
49	Indium	In	114.82	7.31	FC tetragonal§	4.598	—	4.4947	20	156	2000
50	Tin (α)	Sn	118.64	7.30	Diamond cubic	6.892	—	—	20	232	2270
50	Tin (β)	Sn	118.64	7.30	Tetragonal	5.831	—	3.181	20	232	2270
51	Antimony	Sb	121.75	6.62	Rhombic	4.507	—	—	20	631	1440
52	Tellurium	Te	127.6	6.24	Hexagonal	4.457	—	5.927	20	455	1390
53	Iodine	I	126.9	4.94	Orthorhombic	4.787	7.266	9.793	20	114	183
54	Xenon	Xe	131.3	5.495 × 10⁻³	FCC	6.24	—	—	-185	-112	-108
55	Cesium	Cs	132.91	1.90	BCC	6.079	—	—	-173	28	690
56	Barium	Ba	137.34	3.5	BCC	5.019	—	—	20	715	1640
57	Lanthanum (α)	La	138.9	6.17	HCP	3.770	—	12.16	20	830	3465
57	Lanthanum (β)	La	138.9	6.17	FCC	5.296	—	—	20		
58	Cerium (α)	Ce	140.12	1.90	FCC	5.161	—	—	20	805	3415
58	Cerium (β)	Ce	140.12	1.90	HCP	3.62	—	5.99			
59	Praseodymium (α)	Pr	140.91	6.77	HCP	3.673	—	5.92	20	930	3025
59	Praseodymium (β)	Pr	140.91	6.77	FCC	5.151	—	—	20		
60	Neodymium	Nd	144.24	7.00	HCP	3.658	—	11.8	20	860	3160

Atomic number	Element	Symbol	Atomic weight	Density		Crystal structure	a	b	c	T (°C)	Melting point	Boiling point
61	Illinium	Il	—	—	—	—	—	—	—	—	—	—
	Promethium	Pr	147.0	—	—	—	—	—	—	—	—	—
62	Samarium	Sm	150.42	7.7	—	Rhombic	8.996	—	—	20	1300	1645
63	Europium	Eu	152.0	5.26	—	BCC	4.606	—	—	20	825	1500
64	Gadolinium	Gd	157.25	7.89	—	HCP	3.636	—	5.783	20	1340	2690
65	Terbium	Tb	158.92	8.27	—	HCP	3.601	—	5.694	20	1330	2510
66	Dysprosium	Dy	162.50	8.54	—	HCP	3.590	—	6.648	20	1390	2330
67	Holmium	Ho	164.93	8.80	—	HCP	3.577	—	5.616	20	1410	2310
68	Erbium	Er	167.26	9.05	—	HCP	3.559	—	5.59	20	1400	2320
69	Thulium	Tm	168.93	9.33	—	HCP	3.54	—	5.555	20	1510	1710
70	Ytterbium	Yb	173.04	6.98	—	FCC	5.486	—	—	20	860	1430
71	Lutetium	Lu	174.97	9.84	—	HCP	3.503	—	5.55	20	1610	1690
72	Hafnium	Hf	178.49	13.1	—	HCP	3.194	—	5.051	20	1700	3210
73	Tantalum	Ta	180.95	16.6	—	BCC	3.298	—	—	20	3010	4110
74	Tungsten	W	183.85	19.3	—	BCC	3.165	—	—	20	3425	5930
75	Rhenium	Re	186.2	21.0	—	HCP	2.761	—	4.458	20	3130	5880
76	Osmium	Os	190.2	22.6	—	HCP	2.735	—	4.319	20	2800	5500
77	Iridium	Ir	192.20	22.5	—	FCC	3.839	—	—	20	2455	5300
78	Platinum	Pt	195.09	21.4	—	FCC	3.924	—	—	20	1774	4410
79	Gold	Au	196.97	19.3	—	FCC	4.079	—	—	20	1063	2970
80	Mercury	Hg	200.59	13.6	—	Rhombic	3.005	—	—	20	-39	357
81	Thallium	Tl	204.39	11.85	—	HCP	3.457	—	5.525	20	302	1460
82	Lead	Pb	207.19	11.4	—	FCC	4.95	—	—	20	327	1740
83	Bismuth	Bi	208.98	9.8	—	Rhombic	4.746	—	—	20	271	1420
84	Polonium (α)	Po	210.0	9.31	—	Simple cubic	3.345	—	—	20	600	965
84	Polonium (β)	Po	210.0	9.31	—	Rhombic	3.359	—	—	20	—	—
85	Astatine	At	211.0	—	—		—	—	—	—	—	—
86	Radon	Rn	222.0	4.4	—		—	—	—	—	-71	-62
87	Francium	Fa	223.0	—	—		—	—	—	—	—	—
88	Radium	Ra	226.05	5.0	—		—	—	—	—	690	1150
89	Actinium	Ac	227.05	—	—		—	—	—	—	1120	—
90	Thorium	Th	232.04	11.7	—	FCC	5.084	—	—	20	1900	4500
91	Protactinium	Pa	231.0	15.37	—	BC tetragonal§	3.925	—	3.238	20	3000	—
92	Uranium	U	238.04	19.07	—	Orthorhombic	2.858	5.877	4.955	20	1130	3810
93	Neptunium (α)	Np	237.0	20.45	—	Orthorhombic	4.2723	4.887	6.663	20	630	—
94	Plutonium (α)	Pu	239.0	19.81	—	Monoclinic	6.1835	4.824	10.973	20	630	3255
95	Americium	Am	241.0	11.87	—	Hexagonal	3.642	—	11.76	20	855	—
96	Curium	Cm	247	—	—		—	—	—	—	—	—
97	Berkelium	Be	247	—	—		—	—	—	—	—	—
98	Californium	Cf	249	—	—		—	—	—	—	—	—
99	Einsteinium	Es	254	—	—		—	—	—	—	—	—
100	Fermium	Fm	253	—	—		—	—	—	—	—	—

† Melting and boiling points correspond to standard pressure (760 mm Hg (torr)).

‡ Stable form.

§ Body-centered and face-centered tetragonal are the same except $a(\text{bct}) = a/\sqrt{2}$ (fct).

TABLE D.2 *Relativistic velocity v, wavelength* λ*, Ewald sphere radius* $(1/\lambda)$*, and relativistic correction factor* $\left[1 - (v/c)^2\right]^{-1/2}$ *for electrons with increasing accelerating potential* V_O

V_O, kV	$v(\times 10^{10}$ cm/sec)	λ, Å	λ^{-1}, Å^{-1}	$\left[1 - (v/c^2\right]^{-1/2}$
0.1	0.059	1.226	0.82	1.002
0.2	0.084	0.867	1.15	1.004
0.3	0.103	0.708	1.40	1.006
0.4	0.119	0.613	1.63	1.008
0.5	0.133	0.548	1.82	1.0010
0.6	0.145	0.500	2.00	1.0012
0.7	0.157	0.463	2.15	1.0014
0.8	0.168	0.433	2.31	1.0016
0.9	0.178	0.408	2.45	1.0018
1.0	0.187	0.388	2.68	1.0020
2.0	0.256	0.274	3.65	1.0040
3.0	0.323	0.224	4.47	1.0059
4.0	0.373	0.194	5.17	1.0078
5.0	0.417	0.173	5.77	1.0098
6.0	0.456	0.158	6.33	1.0117
7.0	0.491	0.146	6.83	1.0137
8.0	0.525	0.137	7.30	1.0157
9.0	0.555	0.129	7.76	1.0176
10.0	0.585	0.122	8.20	1.0196
20.0	0.815	0.086	10.65	1.0391
30.0	0.985	0.070	14.32	1.0587
40.0	1.122	0.060	16.68	1.0783
50.0	1.237	0.054	18.71	1.0978
60.0	1.337	0.049	20.53	1.1174
70.0	1.427	0.045	22.25	1.1370
80.0	1.506	0.042	23.95	1.1565
90.0	1.578	0.039	25.48	1.1761
100.0	1.644	0.037	27.05	1.1957
200.0	2.084	0.0251	39.9	1.3915
300.0	2.329	0.0197	50.6	1.5873
400.0	2.482	0.0164	60.8	1.7830
500.0	2.588	0.0142	70.3	1.9785
600.0	2.663	0.0126	79.6	2.1740
700.0	2.718	0.0113	88.5	2.3700
800.0	2.761	0.0103	97.3	2.5650
900.0	2.795	0.0094	106.0	2.7610
1000.0	2.822	0.0087	115.0	2.9570

TABLE D.3   Atomic scattering factors for electrons $f_{el}(\mu)$ in Å†

Element	Z	0.00	0.05	0.10	0.15	0.20	0.25	0.30	0.35	0.40	0.50	0.60	0.70	0.80	0.90	1.00
											sin θ/λ					
H	1	0.53	0.51	0.45	0.38	0.31	0.25	0.20	0.16	0.13	0.09	0.06	0.05	0.04	0.03	0.02
He	2	0.45	0.43	0.40	0.37	0.33	0.29	0.25	0.22	0.19	0.14	0.11	0.09	0.07	0.06	0.05
Li	3	3.31	2.78	1.88	1.17	0.75	0.53	0.40	0.31	0.26	0.19	0.14	0.11	0.09	0.08	0.06
Be	4	3.09	2.82	2.23	1.63	1.16	0.83	0.61	0.47	0.37	0.25	0.19	0.15	0.12	0.10	0.08
B	5	2.82	2.62	2.24	1.78	1.37	1.04	0.80	0.62	0.50	0.33	0.24	0.18	0.14	0.12	0.10
C	6	2.45	2.26	2.09	1.74	1.43	1.15	0.92	0.74	0.60	0.41	0.30	0.22	0.18	0.14	0.12
N	7	2.20	2.10	1.91	1.68	1.44	1.20	1.00	0.83	0.69	0.48	0.35	0.27	0.21	0.17	0.14
O	8	2.01	1.95	1.80	1.62	1.42	1.22	1.04	0.88	0.75	0.54	0.40	0.31	0.24	0.19	0.16
F	9	1.84	1.77	1.69	1.53	1.38	1.20	1.05	0.91	0.78	0.59	0.44	0.35	0.27	0.22	0.18
Ne	10	1.66	1.59	1.53	1.43	1.30	1.17	1.04	0.92	0.80	0.62	0.48	0.38	0.30	0.24	0.20
Na	11	4.89	4.21	2.97	2.11	1.59	1.29	1.09	0.95	0.83	0.64	0.51	0.40	0.33	0.27	0.22
Mg	12	5.01	4.60	3.59	2.63	1.95	1.50	1.21	1.01	0.87	0.67	0.53	0.43	0.35	0.29	0.24
Al	13	6.10	5.36	4.24	3.13	2.30	1.73	1.36	1.11	0.93	0.70	0.55	0.45	0.36	0.30	0.25
Si	14	6.00	5.26	4.40	3.41	2.59	1.97	1.54	1.23	1.02	0.74	0.58	0.47	0.38	0.32	0.27
P	15	5.40	5.07	4.38	3.55	2.79	2.17	1.70	1.36	1.12	0.80	0.61	0.49	0.40	0.33	0.28
S	16	4.70	4.40	4.00	3.46	2.87	2.32	1.86	1.50	1.22	0.86	0.64	0.51	0.42	0.35	0.30
Cl	17	4.60	4.31	4.00	3.53	2.99	2.47	2.01	1.63	1.34	0.93	0.69	0.54	0.44	0.37	0.31
A	18	4.71	4.40	4.07	3.56	3.03	2.52	2.07	1.71	1.42	1.00	0.74	0.58	0.46	0.38	0.32
K	19	9.00	7.00	5.43	4.10	3.15	2.60	2.14	1.90	1.49	1.07	0.79	0.61	0.49	0.40	0.34
Ca	20	5.40	5.08	4.57	3.85	3.13	2.52	2.06	1.72	1.45	1.07	0.82	0.65	0.53	0.44	0.37
Sc	21	5.60	5.27	4.72	3.98	3.24	2.61	2.14	1.78	1.51	1.12	0.86	0.68	0.55	0.45	0.38
Ti	22	5.80	5.46	4.88	4.12	3.35	2.70	2.21	1.85	1.57	1.16	0.89	0.71	0.57	0.47	0.40
V	23	5.91	5.65	5.03	4.24	3.45	2.79	2.29	1.91	1.62	1.20	0.93	0.74	0.60	0.49	0.41
Cr	24	6.10	5.84	5.17	4.37	3.56	2.88	2.36	1.98	1.68	1.25	0.96	0.76	0.62	0.51	0.43
Mn	25	6.19	5.93	5.34	4.49	3.66	2.97	2.43	2.04	1.73	1.29	0.99	0.79	0.64	0.53	0.45
Fe	26	6.38	6.13	5.48	4.62	3.76	3.05	2.51	2.10	1.79	1.33	1.03	0.82	0.66	0.55	0.46
Co	27	6.51	6.32	5.62	4.73	3.87	3.14	2.58	2.16	1.84	1.37	1.06	0.84	0.69	0.57	0.48
Ni	28	6.70	6.41	5.74	4.85	3.97	3.22	2.65	2.23	1.89	1.41	1.09	0.87	0.71	0.59	0.49
Cu	29	6.80	6.61	5.89	4.97	4.06	3.30	2.72	2.29	1.95	1.45	1.13	0.90	0.73	0.60	0.51
Zn	30	7.00	6.70	6.03	5.08	4.16	3.38	2.79	2.35	2.00	1.49	1.16	0.92	0.75	0.62	0.52
Ga	31	7.21	6.89	6.15	5.20	4.25	3.46	2.86	2.41	2.05	1.53	1.19	0.95	0.77	0.64	0.54
Ge	32	7.30	7.09	6.29	5.32	4.35	3.54	2.93	2.46	2.10	1.57	1.22	0.97	0.79	0.66	0.56
As	33	7.50	7.18	6.41	5.43	4.44	3.62	2.99	2.52	2.15	1.61	1.25	1.00	0.82	0.68	0.57
Se	34	7.60	7.37	6.56	5.53	4.54	3.70	3.06	2.58	2.20	1.65	1.28	1.02	0.84	0.70	0.59
Br	35	7.80	7.47	6.68	5.63	4.63	3.78	3.13	2.64	2.25	1.69	1.32	1.05	0.86	0.71	0.60
Kr	36	7.88	7.56	6.80	5.74	4.71	3.85	3.19	2.69	2.31	1.73	1.35	1.08	0.88	0.73	0.62
Rb	37	8.00	7.75	6.92	5.85	4.80	3.93	3.26	2.75	2.35	1.77	1.38	1.10	0.90	0.75	0.63
Sr	38	8.22	7.85	7.04	5.96	4.89	4.00	3.32	2.80	2.40	1.80	1.41	1.13	0.92	0.77	0.65

TABLE D.3 *Atomic scattering for electrons* $f_{el}(\mu)$ Å† *(Continued)*

Element	Z	\(sin\,\theta/\lambda\)														
		0.00	0.05	0.10	0.15	0.20	0.25	0.30	0.35	0.40	0.50	0.60	0.70	0.80	0.90	1.00
Y	39	8.30	8.04	7.16	6.06	4.98	4.07	3.38	2.86	2.45	1.84	1.44	1.15	0.94	0.78	0.66
Zr	40	8.49	8.14	7.28	6.16	5.06	4.15	3.45	2.91	2.50	1.88	1.47	1.17	0.96	0.80	0.68
Nb	41	8.6	8.23	7.40	6.27	5.15	4.22	3.51	2.97	2.54	1.92	1.50	1.20	0.98	0.82	0.69
Mo	42	8.7	8.42	7.52	6.36	5.24	4.29	3.57	3.02	2.59	1.95	1.53	1.22	1.00	0.84	0.71
Tc	43	8.9	8.52	7.63	6.47	5.31	4.36	3.63	3.08	2.64	1.99	1.56	1.25	1.02	0.85	0.72
Ru	44	9.0	8.62	7.75	6.56	5.40	4.43	3.69	3.13	2.68	2.03	1.58	1.27	1.04	0.87	0.74
Rh	45	9.1	8.81	7.85	6.66	5.48	4.50	3.75	3.18	2.73	2.06	1.61	1.30	1.06	0.89	0.75
Pd	46	9.3	8.90	7.97	6.75	5.56	4.57	3.81	3.23	2.77	2.10	1.64	1.32	1.08	0.90	0.77
Ag	47	9.4	9.00	8.07	6.85	5.64	4.64	3.87	3.28	2.82	2.13	1.67	1.34	1.10	0.92	0.78
Cd	48	9.5	9.19	8.19	6.95	5.72	4.71	3.93	3.34	2.86	2.17	1.71	1.37	1.12	0.94	0.79
In	49	9.6	9.29	8.31	7.03	5.80	4.78	3.99	3.39	2.91	2.20	1.73	1.39	1.14	0.95	0.81
Sn	50	9.8	9.38	8.40	7.13	5.88	4.84	4.05	3.44	2.95	2.24	1.76	1.41	1.16	0.97	0.82
Sb	51	9.9	9.48	8.50	7.22	5.95	4.91	4.10	3.49	3.00	2.27	1.79	1.44	1.18	0.99	0.84
Te	52	10.0	9.57	8.62	7.31	6.03	4.97	4.16	3.54	3.04	2.31	1.81	1.46	1.20	1.00	0.85
I	53	10.1	9.77	8.71	7.39	6.11	5.04	4.22	3.59	3.08	2.34	1.84	1.48	1.22	1.02	0.87
Xe	54	10.2	9.86	8.81	7.49	6.19	5.10	4.27	3.64	3.13	2.38	1.87	1.51	1.24	1.04	0.88
Cs	55	10.4	9.96	8.93	7.57	6.26	5.17	4.33	3.68	3.17	2.41	1.90	1.53	1.26	1.05	0.89
Ba	56	10.5	10.05	9.02	7.66	6.34	5.23	4.39	3.73	3.21	2.45	1.93	1.55	1.28	1.07	0.91
La	57	10.6	10.15	9.12	7.75	6.40	5.30	4.44	3.78	3.26	2.48	1.95	1.57	1.30	1.09	0.92
Ce	58	10.7	10.24	9.21	7.84	6.49	5.36	4.50	3.83	3.30	2.51	1.98	1.60	1.32	1.10	0.94
Pr	59	10.8	10.44	9.31	7.92	6.56	5.42	4.55	3.88	3.34	2.55	2.01	1.62	1.33	1.12	0.95
Nd	60	10.9	10.53	9.41	8.01	6.63	5.48	4.60	3.93	3.38	2.58	2.03	1.64	1.35	1.13	0.96
Pm	61	11.0	10.63	9.53	8.10	6.70	5.55	4.66	3.97	3.43	2.61	2.06	1.66	1.37	1.15	0.98
Sm	62	11.1	10.72	9.62	8.17	6.77	5.61	4.71	4.02	3.47	2.65	2.09	1.69	1.39	1.17	0.99
Eu	63	11.2	10.82	9.72	8.25	6.85	5.67	4.77	4.07	3.51	2.68	2.11	1.71	1.41	1.18	1.00
Gd	64	11.4	10.92	9.79	8.34	6.91	5.73	4.82	4.11	3.55	2.71	2.14	1.73	1.43	1.20	1.02
Tb	65	11.5	11.01	9.88	8.42	6.98	5.79	4.87	4.16	3.59	2.74	2.17	1.75	1.45	1.21	1.03
Dy	66	11.6	11.11	9.98	8.50	7.05	5.85	4.92	4.20	3.63	2.78	2.19	1.77	1.47	1.23	1.05
Ho	67	11.7	11.20	10.08	8.58	7.12	5.91	4.98	4.25	3.67	2.81	2.22	1.80	1.48	1.25	1.06
Er	68	11.8	11.30	10.17	8.66	7.19	5.97	5.03	4.30	3.71	2.84	2.25	1.82	1.50	1.26	1.07
Tm	69	11.9	11.49	10.27	8.74	7.26	6.03	5.08	4.34	3.75	2.87	2.27	1.84	1.52	1.28	1.09
Yb	70	12.0	11.59	10.36	8.82	7.33	6.09	5.13	4.39	3.79	2.91	2.30	1.86	1.54	1.29	1.10
Lu	71	12.1	11.68	10.44	8.90	7.40	6.15	5.18	4.43	3.83	2.94	2.32	1.88	1.56	1.31	1.11
Hf	72	12.2	11.78	10.53	8.98	7.46	6.20	5.23	4.48	3.87	2.97	2.35	1.90	1.58	1.32	1.13
Ta	73	12.3	11.87	10.63	9.05	7.53	6.26	5.28	4.52	3.91	3.00	2.38	1.93	1.59	1.34	1.14
W	74	12.4	11.97	10.72	9.13	7.59	6.32	5.33	4.56	3.95	3.03	2.40	1.95	1.61	1.35	1.15

TABLE D.3   *Atomic scattering for electrons* $f_{el}(\mu)$ Å† *(Continued)*

	Z														
Re	75	12.5	12.06	10.79	9.21	7.66	6.38	5.38	4.61	3.99	3.06	2.43	1.97	1.63	1.37	1.17
Os	76	12.6	12.16	10.89	9.29	7.72	6.43	5.43	4.65	4.03	3.09	2.45	1.99	1.65	1.38	1.18
Ir	77	12.7	12.26	10.96	9.36	7.79	6.49	5.48	4.70	4.07	3.12	2.48	2.01	1.66	1.40	1.19
Pt	78	12.8	12.35	11.06	9.44	7.86	6.55	5.53	4.74	4.11	3.16	2.50	2.03	1.68	1.42	1.21
Au	79	12.9	12.45	11.13	9.51	7.92	6.60	5.58	4.78	4.14	3.19	2.53	2.05	1.70	1.43	1.22
Hg	80	13.0	12.54	11.23	9.58	7.98	6.66	5.63	4.83	4.18	3.22	2.55	2.07	1.72	1.45	1.23
Tl	81	13.1	12.64	11.32	9.66	8.05	6.71	5.68	4.87	4.22	3.25	2.58	2.10	1.74	1.46	1.25
Pb	82	13.2	12.69	11.39	9.74	8.11	6.77	5.72	4.91	4.26	3.28	2.60	2.12	1.75	1.48	1.26
Bi	83	13.2	12.75	11.49	9.81	8.18	6.82	5.77	4.95	4.30	3.31	2.63	2.14	1.77	1.49	1.27
Po	84	13.3	12.83	11.56	9.87	8.24	6.88	5.82	4.99	4.33	3.34	2.65	2.16	1.79	1.51	1.28
At	85	13.4	12.93	11.66	9.95	8.30	6.93	5.87	5.04	4.37	3.37	2.68	2.18	1.81	1.52	1.30
Rn	86	13.5	13.02	11.73	10.02	8.36	6.98	5.92	5.08	4.41	3.40	2.70	2.20	1.82	1.54	1.31
Fr	87	13.6	13.12	11.80	10.10	8.42	7.04	5.96	5.12	4.44	3.43	2.73	2.22	1.84	1.55	1.32
Ra	88	13.7	13.22	11.90	10.16	8.49	7.09	6.01	5.16	4.48	3.46	2.75	2.24	1.86	1.56	1.34
Ac	89	13.8	13.31	11.97	10.24	8.55	7.14	6.06	5.20	4.52	3.49	2.78	2.27	1.87	1.58	1.35
Th	90	13.9	13.41	12.04	10.30	8.61	7.20	6.10	5.24	4.55	3.52	2.80	2.29	1.89	1.59	1.36
Pa	91	14.0	13.50	12.14	10.37	8.67	7.25	6.15	5.28	4.59	3.55	2.82	2.31	1.91	1.61	1.37
U	92	14.1	13.60	12.21	10.45	8.73	7.31	6.19	5.33	4.63	3.58	2.85	2.33	1.93	1.62	1.39
Np	93	14.2	13.69	12.28	10.51	8.79	7.35	6.24	5.37	4.66	3.61	2.87	2.35	1.94	1.64	1.40
Pu	94	14.3	13.77	12.38	10.59	8.85	7.41	6.28	5.41	4.70	3.63	2.90	2.37	1.96	1.65	1.41
Am	95	14.4	13.83	12.45	10.65	8.91	7.46	6.33	5.45	4.74	3.66	2.92	2.39	1.98	1.67	1.43
Cm	96	14.4	13.90	12.52	10.71	8.97	7.51	6.38	5.49	4.77	3.69	2.94	2.41	1.99	1.68	1.44
Bk	97	14.5	13.98	12.59	10.79	9.03	7.56	6.42	5.53	4.81	3.72	2.97	2.43	2.01	1.70	1.45
Cf	98	14.6	14.08	12.69	10.85	9.09	7.61	6.47	5.57	4.84	3.75	2.99	2.45	2.03	1.71	1.46
Es	99	14.7	14.17	12.76	10.92	9.14	7.67	6.51	5.61	4.88	3.78	3.01	2.47	2.04	1.73	1.48
Fm	100	14.8	14.27	12.83	10.99	9.20	7.72	6.56	5.65	4.91	3.81	3.04	2.49	2.06	1.74	1.49

†Thomas-Fermi-Dirac statistical calculations. For relativistic considerations, values must be multiplied by $[1 - (v/c)^2]^{-\frac{1}{2}}$ (see Table D.2).

TABLE D.4 *Dynamical extinction distances (Å) for some fcc materials—ξ_g (fcc)* †

Accelerating potential, kV	Al (fcc), Z = 13.0			γ-Fe (fcc), Z = 26.0			Inconel (fcc), Z = 27.2		
	[111]	[200]	[220]	[111]	[200]	[220]	[111]	[200]	[220]
75	489	591	939	229	264	398	219	253	382
100	548	662	1052	256	296	446	245	284	428
125	595	719	1142	278	321	484	267	308	465
150	634	766	1217	296	342	516	284	328	495
200	695	839	1334	325	375	565	311	360	543
300	776	937	1489	363	418	631	347	402	606
400	827	999	1587	387	446	673	370	428	646
500	862	1041	1654	403	465	701	386	446	673
600	887	1071	1702	415	478	722	397	459	693
700	905	1094	1738	423	488	737	406	469	707
800	920	1111	1765	430	496	748	412	476	718
900	931	1124	1787	435	502	757	417	482	727
1000	940	1135	1804	439	507	765	421	487	734

Accelerating potential, kV	Stainless steel (fcc), Z = 25.8			β-Co (fcc), Z = 27.0			Ni (fcc), Z = 28.0		
	[111]	[200]	[220]	[111]	[200]	[220]	[111]	[200]	[220]
75	228	264	397	218	266	383	211	244	369
100	256	295	444	244	285	429	236	273	413
125	278	321	483	265	309	466	256	297	448
150	296	342	514	283	330	496	273	316	478
200	324	374	564	310	361	544	299	346	523
300	362	418	629	346	403	607	334	387	584
400	386	446	671	369	430	647	356	412	623
500	402	464	699	384	448	675	371	430	649
600	414	478	719	396	461	694	382	442	668
700	422	488	735	404	471	709	390	452	682
800	429	496	746	410	478	720	396	459	693
900	434	502	755	415	484	729	401	464	701
1000	438	506	762	419	489	736	405	469	708

Accelerating potential, kV	Cu (fcc), Z = 29.0			Rh (fcc), Z = 45.0			Ag (fcc), Z = 47.0		
	[111]	[200]	[220]	[111]	[200]	[220]	[111]	[200]	[220]
75	216	249	372	176	202	291	199	227	324
100	241	279	417	197	226	326	223	254	363
125	262	303	453	214	245	354	242	276	394
150	279	323	482	228	261	377	258	294	420
200	306	354	528	250	286	414	283	322	461
300	342	395	590	279	320	462	316	360	514
400	364	421	629	297	341	492	337	384	548
500	380	439	656	310	355	513	351	400	571
600	391	452	675	319	365	528	361	411	588
700	399	461	689	326	373	539	369	420	600
800	405	469	700	331	379	548	375	427	610
900	410	474	708	335	384	554	379	432	617
1000	414	479	715	338	387	559	383	436	623

Accelerating potential, kV	α-Brass (fcc), Z = 29.2			Pd (fcc), Z = 46.0			Ir (fcc), Z = 77.0		
	[111]	[200]	[220]	[111]	[200]	[220]	[111]	[200]	[220]
75	223	257	381	182	209	300	125	142	200
100	250	288	426	204	234	336	141	160	224
125	271	313	463	221	254	364	153	173	244
150	289	333	493	236	271	388	163	185	260
200	317	365	541	259	297	426	178	202	285
300	354	408	604	289	331	475	199	226	318
400	377	434	643	308	353	507	212	241	339
500	393	453	671	321	368	528	221	251	353
600	404	466	690	330	379	543	227	258	363
700	413	476	704	337	387	555	232	264	371
800	419	483	715	342	393	563	236	268	377
900	424	489	724	346	398	570	239	271	381
1000	428	494	731	350	401	576	241	274	385

†The values tabulated here are relativistic, but should be considered approximate since the accuracy is contingent on the accuracy of the atomic scattering factor for electrons used in the calculations.

TABLE D.4 *Dynamical extinction distances (Å) for some fcc (diamond cubic) materials—ξ$_g$ (diamond cubic) (Continued)*

Accelerating potential, kV	[111]	[220]		[111]	[220]
	Si (Diamond), Z = 14.0			Ge (Diamond), Z = 32.0	
75	778	674		545	405
100	871	755		610	454
125	946	820		662	492
150	1007	874		706	524
200	1104	958		773	575
300	1233	1069		864	642
400	1314	1140		921	684
500	1370	1188		959	713
600	1409	1222		987	734
700	1439	1243		1008	749
800	1462	1267		1024	762
900	1479	1283		1036	770
1000	1493	1295		1046	777

TABLE D.5 *Dynamical extinction distances (Å) for some bcc materials— ξ$_g$ (bcc)*

Accelerating potential, kV	[110]	[200]	[211]	[110]	[200]	[211]	[110]	[200]	[211]
	β-Ti (bcc), Z = 22.0			α-Fe (bcc), Z = 26.0			β-Zr (bcc), Z = 40.0		
75	360	520	658	240	352	446	289	395	493
100	403	582	736	268	394	500	324	442	552
125	438	633	800	292	428	543	352	480	599
150	466	674	852	311	456	578	375	512	639
200	511	739	934	340	499	634	410	561	700
300	571	825	1043	380	558	708	458	626	782
400	608	879	1112	405	594	755	489	667	833
500	634	916	1158	422	619	786	509	696	868
600	652	943	1192	435	637	809	524	716	894
700	666	963	1217	444	651	826	535	731	912
800	676	978	1236	451	661	839	543	742	927
900	685	990	1251	456	669	849	550	751	938
1000	691	999	1263	460	675	857	555	758	947
	V (bcc), Z = 23.0			Rb (bcc), Z = 37.0			Nb (bcc), Z = 41.0		
75	292	427	549	915	1100	1286	231	325	406
100	327	478	615	1025	1232	1448	259	364	454
125	355	519	668	1113	1338	1564	281	396	493
150	378	553	712	1185	1426	1666	300	421	526
200	415	606	780	1299	1563	1826	329	462	476
300	463	677	871	1451	1745	2039	367	516	643
400	494	721	929	1547	1860	2174	391	550	686

TABLE D.6 *Dynamical extinction distances (A) for some hcp materials—ξ$_g$ (hcp)*

Accelerating potential, kV	[10$\bar{1}$0]	[0002]	[10$\bar{1}$1]	[10$\bar{1}$0]	[0002]	[10$\bar{1}$1]	[10$\bar{1}$0]	[0002]	[10$\bar{1}$1]
	Be (hcp), Z = 4.0			α-Co (hcp), Z = 27.0			α-Zr (hcp), Z = 40.0		
75	1277	733	1037	418	219	311	540	285	392
100	1431	821	1161	469	245	348	604	319	439
125	1552	892	1261	509	267	378	656	347	477
150	1654	950	1343	542	284	402	700	369	508
200	1813	1041	1472	595	311	441	766	405	557
300	2025	1163	1644	664	348	493	856	452	622
400	2209	1239	1752	707	371	525	912	482	663
500	2250	1291	1826	738	386	547	951	502	691
600	2314	1329	1879	759	397	563	979	517	711
700	2363	1357	1919	775	406	575	999	528	726
800	2400	1378	1949	786	412	584	1015	536	737
900	2429	1395	1972	796	417	591	1027	542	746
1000	2453	1408	1991	804	421	597	1037	548	753
	α-Ti (hcp), Z - 22.0			Zn (hcp), Z = 30.0			Cd (hcp), Z = 48.0		
75	665	353	489	505	241	362	475	218	335
100	745	395	547	566	270	405	532	245	375
125	809	429	595	615	293	440	578	266	407
150	862	457	633	655	312	469	616	283	443
200	944	501	694	718	342	514	675	310	476
300	1055	559	775	801	382	574	953	346	531
400	1124	596	826	854	407	613	803	369	566
500	1171	662	861	890	424	637	837	385	590
600	1206	640	886	916	437	656	861	396	607
700	1231	643	905	935	446	670	879	404	620
800	1250	663	919	950	453	680	893	410	630
900	1265	671	930	961	458	688	904	415	638
1000	1278	678	939	971	463	695	913	419	644
	Nd (hcp), Z = 60.0			Er (hcp), Z = 68.0			Re (hcp), Z = 75.0		
75	1106	444	755	476	249	339	256	132	185
100	1238	497	845	534	278	380	287	148	208
125	1345	540	918	580	302	413	311	161	225
150	1432	575	978	618	322	440	331	171	240
200	1570	631	1072	677	353	482	364	188	263
300	1753	704	1197	756	394	538	405	210	294
400	1869	751	1276	806	420	574	432	223	313
500	1948	782	1329	840	438	598	450	233	326
600	2004	805	1368	864	451	615	464	240	336
700	2046	822	1397	882	460	628	473	245	343
800	2079	835	1419	896	467	638	480	248	348
900	2103	845	1436	907	473	646	487	252	352
1000	2123	853	1449	916	478	652	491	254	356

TABLE D.7 *Common physical constants, conversion factors, derived constants, and related data*

Avogadro's number $\qquad N_A = 6.0226 \times 10^{23} \ (g \ mole)^{-1}$
Bohr radius $\qquad (h^2/4\pi^2 m_0 e^2) = 0.53 \ \mathring{A}$
Boltzmann's constant $\qquad K = 1.3805 \times 10^{-16} \ erg/deg$
Compton wavelength $\qquad \lambda_C = h/m_0 c = 0.0243 \ \mathring{A}$
Electron charge $\qquad e = 4.8029 \times 10^{-10} \ esu$
$\qquad\qquad = 1.59 \times 10^{-19} \ coulomb$

Electron mass (rest) $\qquad m_0 = 9.1083 \times 10^{-28} \ g$
$\quad e/m_0 \qquad\qquad = 5.273 \times 10^{17} \ esu/g$
Gas constant $\qquad R = 8.317 \times 10^7 \ erg/deg \ mole$
Light velocity $\qquad C = 2.998 \times 10^{10} \ cm/sec$
Permittivity of free space $\qquad \epsilon = 8.854 \times 10^{-12} \ farad/m$
Planck's constant $\qquad h = 6.626 \times 10^{-27} \ erg \ sec \qquad \hbar = h/2\pi$
$\qquad\qquad = 4.1 \times 10^{-15} \ volt \ sec$
Proton rest mass $\qquad m(P) = 1.672 \times 10^{-27} \ kg$

$1 \ eV = 1.602 \times 10^{-12} \ erg$
$1 \ volt = 1/300 \ esu = 10^8 \ emu$
$1 \ esu \ (charge) = 2.998 \times 10^{10} \ emu$
$1 \ cal \ (15°C) = 4.186 \ joules$
$1 \ kcal = 4.186 \times 10^{10} \ ergs$
$1 \ psi = 0.070307 \ kg/cm^2$
$1 \ kbar = 0.1 \ GPa = 10^9 \ dynes/cm^2 = 986.92 \ atmospheres = 14,504 \ psi$
$1 \ rad = 57.296°$
$1 \ micron \ (\mu) = 10,000 \ \mathring{A} = 10^{-3} \ mm$
$1 \ angstrom \ (\mathring{A}) = 10^{-8} \ cm = 10^{-4} \ \mu m$
Natural logarithm base $(e) = 2.718$
Electron rest energy $(m_0 c^2) = 51.5 \ kV$
$1 \ kV \ (kilovolt) = 1000 \ volts$
$1 \ eV/molecule = 23.05 \ kcal/mol$
$1 \ gauss = 10^{-4} \ weber/m^2 \qquad m = meter$
$1 \ oersted = 7.96 \ amp/m$
To convert susceptibility in cgs units to mks units, multiply by 4π

Appendix E
PROBLEM SOLUTIONS
AND DISCUSSION

The problems associated with each chapter are intended to broaden specific topics and to provide some direction in the application of a concept or formulation. This intended exercise loses some significance if the solution is consulted too early. On the other hand, it is helpful for self-study if solutions are given or the methodology and approach are discussed. Therefore, it is hoped that a well disciplined effort be made to solve a problem before consulting this section. You must also note that in some problems there is really not a single or unique solution, and these situations will be discussed as appropriate. Problems should be an important part of the learning process and to that extent this section seeks to continue this philosophy through additional discussion, and in pointing the reader to additional references, etc. In some cases the final solution is still left to the reader. You should look carefully at your solution in cases where actual values are not given and determine whether the magnitude or the units really make sense. If they do, the chances are the problem has been satisfactorily solved.

Chapter 1:

1.1 Recall from the definition of the coulomb, 1 ampere = 1 coulomb/second. Consequently, if a current of 10^{-4} amperes impinges on a thin foil (a beam of that current) then the electrons are passing a cross section at a rate of 6.24×10^{14} electrons/second. The electrical energy lost can be calculated to be 36% of the beam energy or 0.43 Cal./s.

1.2 Consider the total energy $E = eV_o + m_o C^2$ and $E^2 = p^2 c^2 + m_o^2 E^4$ and solve for P.

1.3 Use Eq. (1.8) to find $\lambda = 0.037$ A for 100 kV electrons, and substituting into Eq. (1.9) for n = 2, $\alpha/\sqrt{3}$ [$\alpha \equiv d_{hk\ell}$ in Eq. (A.1)] we find $\theta = \sin^{-1}(0.064/a)$; where a is the lattice parameter of any fcc material.

1.4 This is basically a potential well problem. We set the potential energy, U, inside the wrapper equal to zero. Thus, inside the foil wrapper we have from Eq. (1.16) $d^2\psi/dz + 2mE\psi/\hbar^2 = 0$ whose solution is $\psi(z) = A\sin n\pi z/t$, where t is the thickness of the wrapper. We now consider Eq. (1.18) in one dimension (z-direction) and substitute for $\psi(z)$ and integrate, i.e.

$$\int_0^t A^2 \sin^2 \frac{n\pi}{t} z \, dz$$

to find that $A = \sqrt{2/t}$. Since $n\pi/t = \sqrt{2mE/\hbar^2}$ we substitute into $\psi(z)$ to obtain the desired form of the wave function. The corresponding wave functions are plotted as sine functions between zero and t. Some additional details of the solution are given in 1.5 below.

1.5 Substitute U = 0 into Eq. (1.16) and solve $\nabla^2 + \gamma^2 = 0$, where $\gamma^2 = (\gamma_x^2 + \gamma_y^2 + \gamma_z^2) = 2mE/\hbar^2$. Solutions are found in the form $\psi(x) = A\sin\gamma_x x$, $\psi(y) = A\sin\gamma_y y$, $\psi(z) = A\sin\gamma_z z$ where $\gamma_x = n_x\pi/a$, $\gamma_y = n_x\pi/a$, $\gamma_y = n_y\pi/a$, $\gamma_z = n_z\pi/a$ when boundary conditions are substituted, i.e. $\psi(x) = 0$ at x = a so $\sin\gamma_x a = 0$, etc. We then solve for E as required.

1.6 Equation 1.39 will serve as a starting point, and when the terms of the energy equation are evaluated for hydrogen, with hc/λ substituted for E, we obtain a general form for the wavelength of the emission lines as $\lambda = 912(1/4 - 1/n)$Å where n = 3, 4, 5, 6 (see Fig. 1.5). Consequently, λ = 7015Å, 4877Å, 4342Å, and 4108Å.

1.7 For continuity of Ω at z = 0, A + B = C + D.

For continuity of $\partial\Omega/\partial z$ at z = 0 we obtain

$$A(-i|K_o| + i\alpha) + B(-i|K_o| - i\alpha) = C(-i|K_o| + \beta) + D(-i|K_o| - \beta)$$

For periodicity of Ω,

$$Ae^{-i(|K_o| + \alpha)R} + Be^{-i(|K_o| + \alpha)R} = Ce^{i(|K_o| - \beta)b} + De^{i(|K_o| + \beta)b}$$

For periodicity of $\partial\Omega/\partial z$,

$$A(-i|K_o| + i\alpha)e^{-i(|K_o| - \alpha)R} + B(-i|K_o| - i\alpha)e^{-i(|K_o| + \alpha)R} =$$

$$C(-i|K_o| + \beta)e^{i(|K_o| - \beta)b} + D(-i|K_o| - \beta)e^{i(|K_o| + \beta)b}$$

Stipulating that the determinant of the coefficients of A, B, C, and D must be zero then results in the form of the equation shown. Invoking the limit property discussed on page 25 then results in Eq. 1.49.

1.8 Substitute for A and B and differentiate as required to obtain the expression of Eq. (1.22).

1.9 Evaluate α and β from Eqs. (1.47) and (1.48) respectively with U_b = 10 eV and E = 8 eV. Substituting for α, β, and b in the transmission equation we find T $\cong$ .02 (or 2 percent transmission). Thus, if each barrier only transmits 2 percent, we find after the first barrier the electrons which have penetrated to number $0.02 \times 10^4 = 2 \times 10^2$ and after the next barrier only 4 electrons would remain, then less than 1 after the third barrier. Consequently no electrons would penetrate the 10-barrier analog.

1.10 We assume in this problem that the blue-green activation is a characteristic of the electron wavelength. Therefore the wavelength might ideally be around 4000 Å. If the electrons are indeed accelerated through a potential difference to attain this wavelength, then we use Eq. (1.8) to solve for V_o = 9.5 μV.

1.11 We let $<P(x)> = \int \psi^* \frac{h}{2\pi i} \frac{\partial\psi}{\partial x}$ dx and find $\psi^* = C^* e^{-i\gamma(x - \omega t/\gamma)}$. Note that $\gamma = n\pi/x$, and $<P(x)> = \frac{h\gamma}{2\pi} \int C^*Cdx$. From Eq. (1.18), $\int C^*C = 1$, thus $<P(x)> = nh/2x$; $\Delta q = x/2\pi n$. So if $<P(x)> = \Delta P$, then $\Delta P \cdot \Delta q = h/4\pi$.

1.12   Simply write out Eq. (1.35) with N = 3 (for Z = 3) and let j = 1, 2, 3, k = 1, 2, 3, etc. The important feature to note is the complexity, as a result of the electron-electron and electron-nuclear interactions as compared to the simple hydrogen atom where only one electron-nuclear interaction is included.

1.13   Use eV - $Mv^2/2$ and solve for V. For mass M expressed in amu and B in Gauss, the radius in cm is given by $144\sqrt{MV/B}$, and on substituting we find R = 181 cm.

1.14   Solve Eq. (1.7) substituting 1900 m_o for m_o for a hydrogen ion (proton: H^+). The wavelength associated with 100 kV electrons is 0.037 Å [on solving Eq. (1.8)]. So we set λ in Eq. (1.7) equal to 0.037 Å, substitute again 1900 m_o for m_o, and solve for V_o. You must pay attention to the units.

Chapter 2:

2.1   Since $E_W = hc/\lambda_o$, we can determine the threshold wavelength, λ_o by substituting E_W = 4.62 eV. We find that λ_o = 2680 Å (~0.27 μm). Consequently, since the U.V. source wavelength is below this value, no electrons are emitted. If $h\nu_o - E_W > 0$, electrons would be emitted with kinetic energy $mv^2/2$. So we could then set $mv^2/2 = h\nu_o - E_W$ and solve for v.

2.2   Again consider $E_W = hc/\lambda_o$. For Ag, you need to observe from Table 2.3 that E_W = 4.78 eV. From Eq. (2.7), note that $E_B = E_F + E_W$, and that E_F can be calculated at T = 0°K from Eq. (2.6); with $N_o = N/a^3$. As an example, N_o for Ag is 5.8 x 10^{28} electrons/m³. Consequently, E_F E_F = 5.51 eV. Consequently for Ag the barrier potential is 10.29 eV (4.78 eV + 5.51 eV).

2.3   Note initially that the potential difference or field intensity (E_z) is V_o/z = 1 kV/0.1 m = 10 kV/m. The thermal energy required (in eV) is then obtained by solving Eq. (2.11), substituting 5.30 eV for E_W, and 10 kV/m for E_z. At T = 0°K we assume that electrons are emitted only by a high-field condition so we could set E_W = 0 in Eq. (2.11) and solve for E_z, then $V_o = zE_z$. Again you will need to pay attention to the units and should not be alarmed at the high voltage required.

2.4   It is sufficient to simply note the change in the average work func-
        tion if we assume a high-field condition. For example, initially
        E_W = 0.8(4.89 eV) + 0.2(5.94 eV) = 5.09 eV. Then after recrystalli-
        zation we have E_W = 0.5(4.89 eV) + 0.5(5.94 eV) = 5.42 eV. The
        values of Cu work functions for the [111] and [100] orientations have
        been taken from Table 2.1. With these work function values, ΔJ can
        be calculated from Eq. (2.15).

2.5   From Table 2.1 we observe that $E_W[001]$ for Ag is 4.81 eV. Substitut-
        ing into Eq. (2.16) then results in $E_W' \cong 3$ eV. Here again you must
        be careful to use consistent units. Note that $n_{A(+)} \cong 3.5 \times 10^{19}/\text{m}^2$
        and $\delta = 3.4 \times 10^{-10}$ m.

2.6   It might be possible to compare intensities for specific orientations
        under constant imaging conditions if the orientations are large
        enough to be separately identified by x-ray diffraction. For very
        small-grain metals this would prove difficult.

2.7   Calibration of any type requires a known standard. If a feature such
        as a thin wire is observed which does not change shape (size) then
        lens parameters can be plotted against the feature size measured. If
        the temperature is not high enough to change features in the thermi-
        onic electron emission microscope, then the same features could be
        compared by optical microscopy.

2.8   The magnification expected from Eq. (2.22) would be 450,000 X. The
        field intensity from Eq. (2.21) is 10^{10} V/m. The magnification of
        the image in Fig. P2.8(d) is obtained by recognizing that a unit cell
        of iridium is imaged in the central (001) plane, and that the unit
        cell dimension (from Table D.1) is 3.839 Å. Therefore when you
        measure the side of the unit cell directly from the figure in the
        text, the magnification will be this dimension divided by the iridium
        lattice parameter. You should get a magnification of about 6 million
        times. Notice that the atoms in the edges of this unit cell face are
        slightly elliptical (egg-shaped rather than round). This is a high-
        field effect. The face-centered atom is not recognizable because the
        field is very weak by comparison to the corner atoms. You can prove
        it is there by measuring the diagonal distances and the unit cell
        edge lengths for example.

2.9   From Table 2.1, E_W for W is 4.9 eV along [112]. If we assume the
      operating conditions are as given in 2.8 above, then $E_z = 10^{10}$ V/m.
      The Fermi energy can be estimated if we assume the temperature is
      near absolute zero from Eq. (2.6), where $N_o = N/a^3$; $a = 3.165$ Å.
      Substitute into Eq. (2.20) for example to obtain current density,
      and multiply by the emission surface area which can be determined
      from the geometry in 2.8 above to yield total current flow.

2.10 You must assume that the operating condition (voltage) is the same as
      in Prob. 2.8. However the geometrical features will give a magnifi-
      cation of 833,000 X. Since the operating voltage is the same, the
      field strength, $E_z = 1.7 \times 10^{10}$ V/m. The optimum resolution is then
      obtained from Eq. (2.26) (with appropriate units). You need to
      compare this value with the interatomic spacing in the (112) plane.
      If it is smaller, then atoms are observable. The emission current
      will of course depend upon E_z and the work function for [112] W (from
      Table 2.1).

2.11 The percent vacancy concentration can be estimated from the equation
      on p. 75 when the vacancy activation energy, E_V is known. To prove
      this concentration, a Ta wire could be quenched from $2300°$K, "freez-
      ing in" the vacancies present, and then observing the wire in the
      FIM by field evaporation as in the sequence of Fig. 2.25 to actually
      count the concentration in a known volume as a percent.

2.12 The surface tension presumably can be estimated from Eq. (2.25) but
      the field intensity must be approximated by assuming it is equivalent
      to just 6 percent of the modulus of elasticity of Pt in tension. The
      surface tension so determined would then presumably correspond to
      $20°$K so in order to determine the temperature at $1000°$C, we would
      assume a linear temperature coefficient of energy of -1.1 ergs/(cm^2)
      ($°$C) which means that we subtract this quantity times the temperature
      difference from the value obtained in Eq. (2.25). To compare the
      evaporation field consult Table 2.4.

2.13 One possibility might be to fatigue small wires in intermittant ten-
      sion until failure occurs, then produce emission end forms near the
      fatigued end for observation in the FIM. Utilizing field evaporation,
      vacancies could be counted in a volume of wire as in Fig. 2.25.
      These measurements would then be compared to unfatigued wire prepared

and observed in the same way. Significantly more vacancies in the fatigued wire might then support this suggestion.

2.14 A, C, E, G, are the 102 poles; B, D, F, and <111> poles; vacancies occur at 1, 4, 8; interstitial possibilities occur at 3, 5, 6, 7; a dislocation occurs at 2 (follow the rings). H was somehow missed, J is [$\bar{1}$01], and I is [$\bar{1}$10]. You can ponder K. (Figure A.20 will be helpful but you might also consult the book by Müller and Tsong listed in the Suggested Supplementary Reading list which follows).

2.15 A ruby laser will have a wavelength of 6860 Å. Consequently the mass (M) or equivalent pressure exerted by such a beam shining on an area will be $Mc^2 = hc/\lambda$ (quantum energy or Plank's relation) or $M = h/c\lambda$, where c is the velocity of light. This is an approximation of the photon mass and is so much smaller than the applied force that it has no effect whatever.

Chapter 3:

3.1 $M_1 = 50$, $M_2 = 100$, therefore $M = M_1 M_2 = 5000$. Substitute to find dM/M.

3.2 Substituting for H in Eq. (3.5) can allow for the form of Eq. (3.8) to be obtained readily if we simply let

$$\zeta_c = 1/Ad_f^4 \int \frac{dz}{(z^2 + d_f^2)^3}$$

where A is a constant. It is recognized that indeed ζ_c is a geometrical parameter.

3.3 Since the focal length is to be maintained constant, and the form factor does not change we have $V_o/(NI)^2 = V_o'/(NI^1)^2$. We assume the number of turns will also not change so the new current at 125 kV must be 1.58 times the current at 50 kV. At high voltages where limitations on current-carrying capacity limit the necessary current increases, then N must increase. This requires the lens become larger in physical size. In effect the entire column will normally become larger at high voltages. You should compare Fig. 8.3 with Fig. 5.1.

3.4 Use Eq. (3.6) in the general form $4\pi NI/10 - H\Delta z$, where Δz is the gap length of the pole piece. With the proper units, we find I = 239 mA.

From Eq. (3.10), with P = 1.7 μΩ·cm we find the power dissipated to be 19.5 watts. From the I^2R relationship the equivalent lens resistance, R, is 342 ohms and from Ohm's law, the lens voltage is 82 volts.

3.5   In a standard electron microscope, there are three lenses contributing to image magnification: the objective lens the intermediate lens, and the projector lens. The total magnification is the product of these individual lens magnifications. Since only the intermediate is unknown, we calculate it from M = f/p -f); where p = 0.65 cm, and f is determined from Eq. (3.9) with the intermediate parameters to be 0.625 cm. Consequently the total magnification is 15 x 25 x 6 = 2250 X.

3.6   Note that γ_o and $\gamma_o{}'$ refer to the constants in Eqs. (3.13) and (3.15) respectively and solve for the corresponding values of C_s and C_c. Then evaluate Eq. (3.18) after having evaluated δ_{LL}, δ_{Sp}, and δ_{Cr}. Note that ΔV and $\Delta I = 10^{-5}$. Compare solution to Eq. (3.17) (7 Å) with Eq. (3.18) (7 Å). There is essentially no error because at this resolution, the design features of the lenses do not contribute significantly.

3.7   Consider Eq. (3.12). The resolution improves, if we consider only the line resolution in Eq. (3.17) from 7 Å at 5×10^5 volts to 4.35 Å at 10^6 volts accelerating potential. The point resolution from Eq. (3.20) is 5.22 Å.

3.8   The most effective electron spot size, δ_f is about 10.44 Å so $D_f \cong$ 1 μm. This means that for the conditions in Prob. 3.7, focus would be maintained even for a sample as thick as 1 μm if the beam were to fully penetrate this thickness. For electrons in the SEM originating beyond the effective D_f, the image of that region would become unfocused.

3.9   We are assuming here that the effective electron wavelength is 12.3 Å. Consequently from Eq. (3.17) we observe the resolution to be about 1230 Å. That would be just below the best resolution in the light microscope.

3.10 From Eq. (3.17) we find the resolution to be 25 Å since the electron wavelength is 38.9 Å.

3.11 You need only draw a complex line spectrum and drawing a line through the smaller peaks illustrate the elimination of background by raising the cut-off level.

3.12 Consider Eq. (3.24) and let $(n^+/n_o) = 10^{-16}$, then solve for T. Note that E_W and E_i must be expressed in volts in this equation, not electron volts.

3.13 You should look at Fig. 3.21 and then solve Eq. (3.26) after finding R in Eq. (3.27) to be 19.6 m.

Chapter 4:

4.1 You can use Eq. (4.4), but a simplified form would be

$$\frac{1}{\lambda'} = \left(\frac{2\pi^2 mZ^2 e^4}{ch^3}\right)\left(\frac{1}{n_2{}^2} - \frac{1}{n_1{}^2}\right)$$

from which λ' (Kα) for Er is 0.253 Å. The K(β) wavelength is 0.221 Å so the difference, $\Delta\alpha$, is 0.032 Å. If a sample contains Tm, its Kα wavelength is about 0.243 Å. Comparing this with Er you can observe that they are very close and discrimination could be difficult. In fact if the signal is detected by a recording device, these two peaks could essentially overlap.

4.2 For Ag, λ(Kα) $\cong$ 0.564 Å (ignore the screening constant). Consequently using an NaCl crystal oriented with a cube face as the spectrometer, $\theta = \sin^{-1}(\lambda/a)$, where a = 5.64 Å. Consequently, $\theta \cong 5.7°$.

4.3 Refer to Fig. 4.15 to obtain an effective Z for each ratio I_B/I_A. Calculate C_2 from Eq. (4.24) and find $C_1 = 1 - C_2$.

4.4 See Sec. 4.4.2

4.5 Al segregation could be observed in electron backscatter by a change in the I_B/I_A ratio, which would decrease. The excitation potentials for Ti and Al (Kα) are obtained directly from Table 4.2. V_K(Al) = 1.55 kV, V_K(Ti) = 4.95 kV. To determine the detector angle you must determine the bragg angle, θ, by first determining the characteristic wavelengths (as in 4.1 above) and then considering 2d for mica from Table 4.1.

4.6 Calculate the characteristic wavelengths as in 4.1 and then consider λ' (a). E$'$(kV) - 12.4 (evaluated form of Planck's equation). Solving

for E' we obtain for Cu for example 8 and 0.9 kV or keV for K(α) and
L(α) characteristic energies. The excitation energy must therefore
exceed these values (perhaps by 10%) and we then estimate 8.8 and
1 kV potentials for Cu respectively. Compare these estimates with
the values presented in Table 4.2.

4.7 You need to recognize that the concentration of Cu in the sample is
0.7, and find the ratio $I_A/I_{A'}$ as 0.07, 0.11, 0.16, 0.23, 0.33, and
0.46, and dividing these ratios by 0.7, find g(E) which is plotted
against the corresponding values of ϕ_t given. (Compare to Fig. 4.23.)

4.8 Refer to 4.6 and consult some of the references cited in the Suggested
Supplementary Reading list in Chap. 4.

4.9 The advantages in using higher voltages include the excitation of
higher-order spectra. This can be important when two elements having
closely-spaced emission spectra occur. In some cases, higher voltages
are required to excite x-ray emission. This is true for heavy elements (large Z).

4.10 The absence of matter is the only logical explanation for image contrast differences which do not simultaneously produce x-ray excitation for a characteristic elemental species.

4.11 We might fracture the Cu and assume the P embrittlement to be segregated near the fracture surfaces. Either microprobe analysis or
Auger analysis could be utilized. Auger spectrometry would be accurate. Ion microprobe techniques could provide a depth profile from
the fracture surface, etc. See Table 4.3 for comparison of sensitivities.

4.12 Solve Eq. (4.40) for d_2 = 0.63 μm. You should consult Ref. 43 in
Chap. 4.

4.13 From Fig. 4.49 the flight time for Si^+ is about 584 ns so V = 25.6
cm/μs. The total voltage (V_{dc} + V_p) which must be applied is obtained
from Eq. (4.46).

4.14 The beam broadening is calculated from Eq. (4.36) by considering the
initial broadening, b, of a 300 Å thick carbon film, and the broadening after an increase in thickness of 50 Å of essentially the same
material, expressed as a percent increase or fraction increase b'/b =
$(t'/t)^{3/2}$ = 1.26 or roughly 26% increase in beam broadening.

Chapter 5:

5.1 The critical thickness is given by Eq. (5.6) and a value could be estimated from Table 5.1 at 65 kV by determining an effective Z to be essentially the same as Al_2O_3; $t_c \cong 600$ Å at 65 kV. If we now stipulate that for this thickness 60% of the incident beam must penetrate, we can solve this condition as a transmission problem (0.6T).

5.2 See Sec. 5.2.3. With $Q_c t_c = Q_s t < 1$ we observe $\Delta I/I \cong K t_c/t$ as in Eq. (5.7a).

5.3 Plotting ρ and t_c in Table 5.1 will produce a poor relationship for some elements. Assuming t_c for Ti at 65 kV $\cong 200$ Å, we can estimate from Eq. (5.14a) that (Δt) min $\cong 20$ Å. You need to recognize that ρ and t_c are not well correlated for some elements.

5.4 From Eq. (5.13) we find α changes from 0.005 to 0.001. You might approximate changes in $(\rho t)_c$ from Fig. 5.4 assuning the response for carbon and pllystyrene are the same. Then substitute for ρ from Table 5.1 and find Λ and $Q = 1/\Lambda$. We obtain

$$\Delta I = I_0 (e^{-t/\Lambda'} - e^{-t/\Lambda})$$

where t = 1000 A. Find the minimum detectable thickness variation for each case Λ and Λ' from Eq. (5.14a).

5.5 The thickness will be given approximately by Eq. (5.18). Assuming a point source, note that the distance $d_s = 5$ cm normal to the substrate, but this distance increases to $\sqrt{d_s^2 + 1.27^2}$ cm at the edges of the 1 inch square substrate (1.27 cm center-to-edge). So the thickness change (in percent) is 15% (you really do not have to consider the actual thickness evaporated). The contrast at the scratch varies but is 100 percent maximum.

5.6 Discrepancies will occur in the calculations of film thickness using Eq. (5.18) and comparing the experimental values because the geometry of the source were not the same. Note that the W-source was fashioned into a boat having a $30°$ angle. As a consequence of this, the source cannot be treated as a simple point source and the change is considered to be evaporated onto a $30°$ surface section.

5.7 Use Eq. (5.4) and assume for simplicity $I_0 = 1$. For the carbon granules (assumed to be spheres) we can estimate the contrast from Eq.

(5.12). From Eq. (5.23) we estimate n by considering D_{min}/δ_{opt} = 100 Å/20 Å = 5. (D_{min} is assumed to be the granule size given).

5.8   You will need to measure the image features (black regions) consider-
      ing the micrographs in the book represent a magnification of approxi-
      mately 5480 X (by plotting the calibration curve implicit in Fig.
      5.9).

5.9   Landing gear and other similar structures can be checked for micro-
      cracks by carefully cleaning a surface area and stripping a plastic
      replica from this surface. The same area could in fact be examined
      repeatedly by placing a plastic protector over a cleaned area after
      each replica is made, then stripping the patch, making a new replica
      after a certain service time, etc. The replica could be examined in
      the TEM or SEM as a non-destructive testing (diagnostic) method.

5.10 Determine the magnification as in 5.9 above and multiply by 2.3 to
      obtain the actual print magnifications in the book. The shadow angle
      is calculated as in Fig. 5.12 [Eq. (5.22)] assuming the features in
      Fig. 5.6 to be hemispherical (h_p = diameter of feature or maximum
      shadow width). The black rim represents the thickness of contamina-
      tion in Fig. 5.6 so it is divided by 100 Å/min.

5.11 In Fig. 5.6(c) the contrast is sufficient to clearly see the inter-
      sections of slip traces. Simply measure the angle of intersection
      and assume that these markings represent the intersection of slip
      traces at 90° to the specimen surface. [This will allow you to con-
      clude the orientation might be (110)]. You might conclude this with-
      out the assumption made. See Table A.1 and the diffraction nets in
      Appendix A.

5.12 Solve Eq. (5.30).

5.13 One could simply mask the surface of interest and electropolish the
      section cut from a test specimen from the back side. Removing the
      lacquer mask would then allow the surface to be viewed in an analyti-
      cal electron microscope (See Sec. 7.7) where the surface could be
      examined by SEM and the dislocation structure in a specific area
      could be observed by TEM or STEM.

5.14 You need to consider Sec. 5.5.2 and Eq. (4.10). Contrast will result
      by differences in the secondary electron emission at the points A, B,
      C, as a result of the different geometries.

5.15 Use Eq. (5.37) and consider Fig. 5.35 to be applicable. The fracture shown is a shear failure. The view shown in Fig. 5.30 illustrates elongated ductile dimples characteristic of shear (in the elongation direction).

5.16 D_m = 4 mm.

5.17 See L. E. Murr, F. L. Williams, D. M. Smith, P. Predecki, and S-H. Wang, *Proc. 3rd IEEE International Pulsed Power Conference*, June, 1981. Notice the fact that ESCA can provide information on the oxidation states and the actual oxide compositions while AES can only elucidate the fact that oxygen is present. You should consult some of the Suggested Supplementary Reading list at the conclusion of Chap. 5.

Chapter 6:

6.1 This is a somewhat classical problem in analytical geometry. You might consider a plane intersecting a cartesian coordinate system or unit cell system a, b, c at points A, B, C along these respective axes. Then we define

$$\underline{r}^* = \underline{g}_{hk\ell} = \underline{a}^*h + \underline{b}^*k + \underline{c}^*\ell$$

which extends from the unit cell origin. We can observe from the unit cell origin. We can observe from this geometry that

$$\overline{AC} = -\frac{\underline{a}}{h} + \frac{\underline{c}}{\ell} \text{ and } \overline{AB} = -\frac{\underline{a}}{h} + \frac{\underline{b}}{k}$$

and these two directions define the plane (hkℓ). The proof is then that $\underline{r}^* \cdot \overline{AC} = 0$ and $\underline{r}^* \cdot \overline{AB} = 0$, since if $\underline{r}^*$ is perpendicular to two directions in the plane it is perpendicular to the plane.

6.2 For the diamond cubic unit cell, atom positions are defined by 000, $\frac{1}{2}\frac{1}{2}0$, $\frac{1}{2}0\frac{1}{2}$, $0\frac{1}{2}\frac{1}{2}$, $\frac{1}{4}\frac{1}{4}\frac{1}{4}$, $\frac{3}{4}\frac{3}{4}\frac{1}{4}$, $\frac{3}{4}\frac{1}{4}\frac{3}{4}$, $\frac{1}{4}\frac{3}{4}\frac{3}{4}$, but basic cell coordinates can be considered to be simply 000 and $\frac{1}{4}\frac{1}{4}\frac{1}{4}$. Since there are ideally 4 such cells composing the diamond unit cell, the structure factor is $4(1 + e^{\pi i(h + k + \ell)/2})$ where the factor 4 is the fcc structure factor.

6.3   For the base-centered orthorhombic cell, we consider (as shown in
      Fig. A.2) $a \neq b \neq c$ and that the cell can be described by atom posi-
      tions 000, $\frac{1}{2}\frac{1}{2}0$ in the base and then turned. Reflections occur for
      h + k even so extinctions occur for h + k odd. Therefore all planes
      diffract except (100).

6.4   MgO has the NaCl structure which is visualized as two interpenetrat-
      ing fcc unit cells. The important feature is that this is a unit
      cell of mixed species and the situation is as described in Sec. 6.3.2.

6.5   For a bcc lattice which you can sketch the basis vectors are given by

$$\underset{\sim}{a} = a\underset{\sim}{i}_x$$

$$\underset{\sim}{b} = a\underset{\sim}{i}_y$$

$$\underset{\sim}{c} = \frac{a}{2}(\underset{\sim}{i}_x + \underset{\sim}{i}_y + \underset{\sim}{i}_z)$$

Reciprocal lattice vectors can be evaluated as

$$\underset{\sim}{a}^* = \frac{1}{\underset{\sim}{a}} = \frac{-\underset{\sim}{i}_z + \underset{\sim}{i}_x}{a}$$

$$\underset{\sim}{b}^* = \frac{1}{\underset{\sim}{b}} = \frac{-\underset{\sim}{i}_z + \underset{\sim}{i}_y}{a}$$

$$\underset{\sim}{c}^* = \frac{1}{\underset{\sim}{c}} = \frac{2\underset{\sim}{i}_z}{a}$$

Then considering

$$\underset{\sim}{r}^* = h\underset{\sim}{a}^* + k\underset{\sim}{b}^* + \ell\underset{\sim}{c}^*$$

we find

$$\underset{\sim}{r}^* = \frac{1}{a}[h\underset{\sim}{i}_x + k\underset{\sim}{i}_y + (-h - k + 2\ell)\underset{\sim}{i}_z]$$

and from this you can observe that the reciprocal lattice points are
those with coordinates of the form

$$\frac{1}{a}[h, k, (-h - k = 2\ell)]$$

when the bcc reciprocal lattice is constructed from these reciprocal
lattice coordinates, the resulting structure is fcc. You might find

it convenient to construct a table of corresponding values and then plot out the values.

6.6 You should refer to Reference 5 in Chapter 6.

6.7 We choose any arbitrary circle and allow this to represent the incident beam intensity, I_o. Then using this circle (with area $\pi r^2 =$ 1 unit) we need to draw the other spots in the "hypothetical" diffraction pattern having intensities I_g. Since this is a kinematical case, $I_g < I_o$ and we evaluate I_g at 100 kV for $t = 100$ Å to be much less than 1, and construct circles of area $\pi r'^2$, where $r' < r$. We evaluate V_C and $F(hk\ell)$ for Ni. If the Ewald sphere passes through each reciprocal lattice point (an unlikely situation) all diffraction circles would be the same.

6.8 Be careful to distinguish the rings in P6.8(a). The camera constant is determined from Eq. (6.23a) and Fig. 6.11. Note that by inspection of the ring sequences and pattern appearance that (b) and (c) are bcc and fcc respectively in Fig. P6.8. Careful measurement of ring radii should convince you that (b) is V.

6.9 In Fig. P6.9 (a) is (110), (b) is (103) and (c) is (112). (b) shows two spots close together where one is a twin spot (near the lower left corner). From Fig. A.12, recognize that the deviation from $[31\bar{1}]$ is negative and by measuring the distance from this reflection and the bright Kikuchi line (same as R in Fig. 6.21), $\Delta\theta$ can be determined from Eq. (6.51). Calculate λL from any of the patterns. For example in (a) the {111} diffraction spots lie on a common circle of diameter R so $\lambda L = 3.56 R/\sqrt{3}$. You need only measure R from the pattern.

6.10 You should use the computer program of Appendix C to check your work. Refer to Sec. 6.8.4. The (101) pattern will look identical to the (110) pattern except the indices of the spots will be different.

6.11 Construction of bct diffraction nets simply requires the use of the "appropriate" structure factor. This is an exercise similar to that in 6.10.

6.12 Since the pattern and brigh-field image are already rotated into coincidence, the rotation information can be ignored. Note that the orientation is (110) and the linear defects extend along the $[\bar{1}12]$

direction. Extra spots at <111>/3 and streaks normal to the defects and through the extra spots are unambiguous evidence for deformation twins. Compare with Fig. 7.59.

6.13 The pattern in Fig. 6.21 is fcc and can therefore be compared to Fig. 6.34 by reducing the scale significantly on a Xerox machine and making a Xerox transparency to superimpose upon a transparency of Fig. 6.34. You can also index the numerous Kikuchi line pairs in Fig. 6.21 since the transparency comparison is cumbersome. Note in Fig. 6.34 that [001] is at the right, [111] at the upper-left of the triangle. The bright zone which looks almost like a square in the pattern is [125]. The orientation is actually to the right of this zone (near [33$\bar{1}$]).

6.14 You will need to first determine from inspection if possible the pattern unit cell or crystal lattice. This can be done by trial and error as well. You must measure the ring radii and then determine values of $d_{hk\ell}$ from Eq. (6.23a) by knowing λL. You must calibrate the instrument using a known standard to determine λL as illustrated in Fig. 6.11. If the pattern is a spot pattern not all possible reflections might occur for any particular crystal system. You can check the $d_{hk\ell}$ values you measure with possibilities you could think of from the ASTM x-ray card files for example. If a spot pattern only occurs you could of course conclude that the flake is a single crystal [provided only one set of spots (or one distinct pattern)] occurs. If only Kikuchi lines are present you can essentially measure corresponding values of "R" in the pattern and similarly measure $d_{hk\ell}$ values.

6.15 The spacing of the surface atoms can be determined by using Eq. (6.54a) with $\theta = 44.4°$ and $\lambda = 1.226$ Å to obtain d' = 2.49 Å. This assumes a simple linear array (100) as in Fig. 6.24(d) for example. Since the atoms in a simple linear sketch might be made along the <110> direction in the surface where the bulk atoms would be separated by 2.49 Å also. This shows a virtual coincidence of the surface atoms positions with the bulk.

6.16 The critical dynamical thickness is given by Eq. (6.48). Substitute corresponding values for λ and corresponding values for F(hkℓ). To examine a film kinematically could conceptually require the actual thickness to be less than t_{dyn} for a particular operating reflection, $\underset{\sim}{g}$. This is done by tilting the foil to excite a specific reflection $\underset{\sim}{g}$.

6.17 You need to measure the first [(111)] diffraction ring width in each pattern and measure the average grain size using Eq. (6.56). The grain sizes in Fig. P6.17b and c are 25 Å and 8 Å respectively.

6.18 Note the three major spots in Fig. 6.44 making angles of 6° and 16°. This corresponds to three overlapping gratings - two rotated by 6° and the third rotated with respect to the reference for the other two by 16°. The exercise using transparencies as in Fig. 6.43 should produce a pattern which looks similar to Fig. 6.44.

6.19 You should note that the diffraction patterns go through systematic changes with increasing reduction (strain). The changes progress from single-crystal patterns to distorted polycrystalline patterns. Note too that the diffraction spots begin to "split" at lower strain values and are separated by larger angles with higher strains. Consequently, you could actually measure the approximate angles of misorientation of crystallites (actually sub-grains created by reduction) and plot these angular misorientations against true strain values. This is indicative of microstructural change with rolling reduction.

Chapter 7:

7.1 Use Eq. (7.6) and make relativistic corrections as necessary. Find the atomic scattering factor from Table D.3. You can check the values to be calculated in Table D.4.

7.2 Use Eq. (7.6). You can determine the effective Z in order to evaluate the atomic scattering factor from Table D.3.

7.3 Fig. 7.9a is a stainless steel foil. From Table D.4 [or you can use Eq. (7.6)], you can find the extinction distance to be 256 Å (at 100 kV as stipulated in the caption of Fig. 7.9). From Eq. (7.14) $t \cong 1024$ Å since $N \cong 4$ (counting the fringes around the holes). In Fig. 7.10 the extinction distance is (from Table D.4) 820 Å. There are roughly 3 fringes associated with the etch pits so from Eq. (7.14) the thickness is 2460 Å. Since the pits are tetragonal and

in the (111) plane, the sides of the pits can be assumed to be parallel to {111}. Consequently you can measure the projected width of
the ramp side of each pit to be about 1000 Å. Since the geometrical
thickness is given by Eq. (7.13), and $\theta \cong 70°$, $t = w \tan \theta \cong 2750$ Å.
The difference is due to the uncertainty in the number of fringes to
some extent.

7.4   This is a "topo problem". Note that the thickness is changing (increasing from the hole and into the material. The twin is a solid
section even though the contrast conditions make it appear as though
the twin interior is a "vacuum". The thickness changes can be measured by considering Eq. (7.13). Picking any specific point, the
thickness can be calculated geometrically, and this thickness when
divided by the number of fringes at that location will give an approximate extinction distance. This is roughly 290 Å. From Table D.4 at
125 kV, the minimum extinction distance for stainless steel is 278
corresponding to $\underset{\sim}{g}$ = <111>.

7.5   Use a constant circle. Note that for simplicity you can assume the
faults are all equally separated and perhaps equally separated from
the surfaces. Then refer to Fig. 7.12 and draw a perfect crystal
reference circle. Move along this circle the arbitrary distance to
the first fault and establish a tangent line. Draw a fault circle
tangent line making an angle of 120° with this line and draw the
fault circle. You now have two circles. The next fault construction
will be a circle forming three equal circles displaced 120° with
centers joined by 60° angles. The final fault will cause a circle
coincident with the first reference (perfect crystal) circle. Consequently, the net phase shift will be zero since the perfect crystal
controls the final contrast. You can approximate the conditions in
Fig. 7.16b by sketching overlapping faults to be represented by lines
joining dislocations. Three overlapping lines correspond to zero
phase shift (or nearly so).

7.6   Measure the true radius of curvature from Eq. (7.23) and calculate
the stress from Eq. (7.24). In Fig. 7.23, the dislocations bow and
split near the arrow at the upper left portion (in b). To actually
"prove" that the dislocations have split you might consider that the
curvature of the two dislocations that actually appear to be split is
smaller than that of the dislocation which precedes them. We could

reason that if the stress is uniform in this slip plane, and certainly G is constant, then from Eq. (7.24) for b/R to be constant requires b to decrease when R decreases. Consequently, we would argue that b for these two dislocations is smaller than that for the preceding dislocation. This would happen if indeed the dislocation split into partials, i.e. $\frac{a}{2}$ <110> $\longrightarrow$ 2($\frac{a}{6}$ <112>).

7.7 Use Eq. (7.25) for the force and Eq. (7.26) for the total stress. The normal stress on any individual dislocation is obtained from Eq. (7.24).

7.8 This is a good opportunity to review your sophomore course in differential equations.

7.9 If region II is tilted away from any diffracting condition this means that $S_2 = 0$. So we simply duplicate the situation in I which is given in Eq. (7.48) for II. If neither I nor II are diffracting then there may not be much contrast at the interface. Ideally it could be "invisible".

7.10 The wave amplitude will be given by

$$\int_0^{t_1} e^{2\pi i s z} \, dz + e^{i\phi_1} \int_{t_1}^{t_1 + t_2} e^{2\pi i s z} \, dz +$$

$$e^{i(\phi_1 + \phi_2)} \int_{t_1 + t_2}^{t_1 + t_2 + t_3} e^{2\pi i s z} \, dz$$

where t_1 is the distance from the entrance surface to the first fault, and t_3 is the distance to the exit surface. Consequently $t = t_1 + t_2 + t_3$. The phase shifts of the faults are ϕ_1 and ϕ_2. Solving we find

$$e^{\pi i s t_1} \sin\pi s t_1 + e^{\pi i s (2t_1 + t_2)} e^{i\phi_1} \sin\pi s t_2 +$$

$$e^{\pi i s (2t_1 + 2t_2 + t_3)} e^{i(\phi_1 + \phi_2)} \sin\pi s t_3$$

It is apparent just from the form of this solution that there is a dependence on the fault spacing and their disposition in the foil. You need to consider this further then add a third fault and examine the equations which result.

7.11   The dislocation will be visible if $\underset{\sim}{g} \cdot \underset{\sim}{b} \neq 0$. We observe from the
parameters given that $\underset{\sim}{g} \cdot \underset{\sim}{b} = 1$. If, as a consequence of the cross-
slip, the Burgers vector changes as described, then $\underset{\sim}{g} \cdot \underset{\sim}{b} = 0$ and
the dislocation becomes invisible.

7.12   The faults in Fig. 7.34 are intrinsic. If the diffraction pattern
were reversed, the character would be extrinsic. You need simply use
Table 7.2 as a reference. The fault nature in Fig. P7.12 is also in-
trinsic. Note that $\underset{\sim}{g}$ in Fig. P7.12 is $\langle 111 \rangle$ and points to the right
on the page. From Table 7.3 the fault is class B. From Table 7.2
you should recognize the foil geometry and conclude that the upper
surface is on right side of the projected image.

7.13   You need to first determine the Burgers vector which results when the
two dislocations combine. This is $\frac{a}{2} [01\bar{1}]$. Then since this combina-
tion disappears we require $\underset{\sim}{g} \cdot \underset{\sim}{b} = 0$ and find $\underset{\sim}{g}$ to accomplish this to
be either $[111]$ or $[200]$. If this total dislocation splits into two
mobile Shockley partials, one of which is visible ($\underset{\sim}{g} \cdot \underset{\sim}{b} \neq 0$) and one
of which is invisible ($\underset{\sim}{g} \cdot \underset{\sim}{b} = 0$), we can have $\frac{a}{6} [\bar{1}12]$ and $\frac{a}{6} [12\bar{1}]$.
The first is essentially invisible while the second is visible. We
cannot uniquely have $\underset{\sim}{g} \cdot \underset{\sim}{b} = 0$ for this situation.

7.14   In Fig. 6.41b the twins are all inclined to the surface 55° [the surface orientation is (100)]. You need to measure the length of all
the twins in both directions and substitute the parameters into
Eq. (7.51). The area can be the area of the micrograph on the page.

7.15   In Fig. 7.20a you can actually count the individual dislocations in
the area if you exclude the crack region to the extreme left of the
image. Roughly 35 dislocations are evident. The area is approxi-
mately 48 x 10^{-10} cm^{-2}, so $\rho \cong 7$ x 10^9 cm^{-2}. In Fig. 7.20c the dislocations are more complicated and you must use Eq. (7.53). The
thickness can be approximated by counting dots on the dislocations
(corresponding to fringes) for those cases where individual disloca-
tions are clearly distinguishable (N $\cong$ 10) and assume $\underset{\sim}{g}$ = <111> for
the 100 kV operating condition.

7.16   Choose some convenient area. The volume in this area is then found
by multiplying by 1100 Å. Count the precipitates in this area.
Their size is certainly within the isostrain contours and you must

estimate it. Then determine the volume and determine the precipitate volume as a fraction of the total volume.

7.17 Measure the apparent geometries and determine the dihedral angles from equations in the form of Eq. (7.56). Insert the dihedral angles into Eq. (7.61) to obtain C_{AB} and $C_{T_A B}$.

7.18 There are 5 fringes in the high-angle grain boundary. Referring to reference 23 in Chap. 7 you will recognize that the accelerating potential was 125 kV. The thickness can be determined, allowing the calculations to be executed.

7.19 Refer to Reference 23 in Chap. 7 for the details.

7.20 You can measure the dislocation node radii directly from the image of Fig. 7.56b and recognize that since the nodes lie in a plane normal to the beam, the apparent radii are the true radii. The stacking fault free energy is then calculated readily from Eq. (7.68).

7.21 The dispersoids occur in Fig. 7.77b. Simply measure their sizes within some area you choose and plot a histogram (number of particles in a size range, N, versus the size range). Pick a small area and determine the associated volume by multiplication by 1000 Å. Then sum the volumes of all particles within this area and take the volume ratio. Dislocations in Fig. 7.77b can actually be counted or determined as in Eq. (7.53).

7.22 You might question why the Sb cannot be directly observed. If it segregates to the boundary as a monolayer (a Gibbs adsorption layer) this would certainly be true unless the boundary were normal to the surface in a very thin film and lattice imaging techniques could be reliably utilized. You might, however, in looking at the experiments suggested by Fig. 7.49 compare systematic changes in the dihedral angles of wire samples doped with Sb and tested for embrittlement. See Reference 23 in Chap. 7 for additional details.

7.23 Refer to Sec. 7.8.2.

7.24 Make a thin foil of the material and observe the changes in a speciman heating stage in the electron microscope. You might look at Sec. 8.6.

7.25 The critical question in this problem is whether the fringe reversal is representative of the special phase angle, or possibly due to the

fact that it is a microtwin and not a stacking fault. You also need
to consider that complete reversal will occur for kinematical diffraction and since only 4 fringes occur the film is indeed thin if
the true extinction distance is $\xi_g/2$. You should be aware of the
fact that when faults are characterized by a phase shift of $\phi = \pi$,
the number of fringes will be larger than one might expect. This is
generally rare in practice. You should consult the reference noted.

7.26   The problem is outlined in detail in the reference cited.

7.27   You must recognize in Fig. P7.27 that the Burgers vector is approximately a line normal to $\underline{g} = [\bar{2}3\bar{1}]$. By measuring the angle between
this line and individual dislocation lines, values of character angle
can be determined and plotted in a histogram. The average angle can
be determined from this plot.

7.28   The oxide particles in Fig. P7.28 are Cr_2O_3 particles determined by
calibrating the electron diffraction pattern (finding λL) using the
(111) Ni ring, and then indexing the spots in the pattern using the
ASTM x-ray card file. The angle between the two spots represents the
angle of misorientation of the individual crystals. This angle is
about $4°$.

7.29   The magnification is determined simply by identifying the planes (d_1
or d_2) and measuring the actual spacing (in mm or $Å$), then dividing
this measurement by the interplanar spacing (about 1.4 million times).

7.30   You can measure the fringe spacing in Fig. 7.62 (smaller fringes) and
use Eq. (7.72) to determine the rotation $\cdot$ $d_1 = d_2$. The diffraction
pattern would have a double diffraction spot and the main beam which
would be simultaneously admitted through the objective aperture (a
circle you draw).

7.31   There are 5 dislocations easily recognizable in the lattice image
if you sight down the lattice planes. If you consider the total
area of the micrograph (in cm) you enlarge or reproduce, the dislocation density will be 5/(area). You can calculate the area by
considering the lattice planes to be spaced $a/\sqrt{h^2 + k^2 + l^2}$ or
$3.99/2 = 1.99$ $Å$. The dislocation images could suggest that the
difference between experiments reporting cold fusion products
(heat, neutrons, etc.) and those not reporting these effects
results by vastly different deformation (dislocation) microstruc-

tures for the palladium electrodes. The cold fusion controversy
has gone on variously since the turn of the century and may well
go on into the next century.

Chapter 8:

8.1   In order to produce 1.2 MV output for 30 kV input we will have a
circuit like Fig. 8.1 with 6 stages (actually 5.3 stages are required).
If ideally the accelerator has only 2 stages as required, the acceleration between stages will be greater than 0.5 MV.

8.2   The accelerating potential for recording Fig. 8.8 was 1 MV and the
extinction distance from Table D.4 is 438 Å. Since there are roughly
21 fringes in the boundary projection, the approximate thickness is
0.92 μm.

8.3   Because the dislocations are in such high contrast and so sharp, this
is an interesting opportunity to make a comparison with dynamical
image width calculations. The image width $\Delta x = \xi_g/\pi$ or $\Delta x = 280$ Å.
The measured width averages about 300 Å when numerous measurements
are plotted as a histogram. The extinction distance was estimated
to be 880 Å by determining an effective Z for the alloy composition
given.

8.4   The mean image size is around 60 Å. The density can only be expressed in terms of the thickness of the sheet which cannot be determined since insufficient information exists to make such a determination. Dislocations would perhaps become blocked by these point
defect clusters and this might generally impede dislocation motion.

8.5   Certainly one interesting experiment you could try could involve
making different uniform thicknesses of the same foil having similar
grain sizes (see Fig. 7.75 for example). These could then be systematically heated in a high-voltage TEM and the grain boundary movement
directly observed. In the HVEM thick sections could be examined and
the dynamic events compared. This experiment could also be performed
in part outside the HVEM in a clean environment.

8.6   It is a simple matter to compare the directions in Fig. 8.18 with
patterns generated by using the computer program in Appendix C, or
by comparison with the "diffraction nets" in Appendix A. (201) is

a possibility by inspection. The surface could be tilted off (201).
Recalling from Eq. (6.23) that $R = \lambda L/d_{hk\ell}$, where R is a size feature of the pattern, we can observe that at 100 kV λ will increase
from 1000 kV, and as a consequence the pattern size will increase.

8.7 In the sense that chromatic aberration is really dependent upon electron energy loss, ΔV, in a specimen, it can be estimated by considering the image displacement to be represented by δ_{Cr} in Eq. (3.14).
Ignoring $\Delta I/I$ the value of ΔV at 60 kV can be estimated.
$\Delta V \cong V_0 \delta_{Cr}/2C_C\alpha \cong 0.53$ kV. For all parameters constant except α,
δ_{Cr} at 1 MV would be roughly 2 Å and would not be observable.

8.8 This is a situation which involves overlapping grains. For a 2.5 μm
thick section with 1 μm diameter grains, at least 2 grains would
overlap everywhere in the foil. Consequently we would expect multiple diffraction effects as described in Chap. 6. Precipitates would
be difficult to analyze from the electron diffraction pattern because
of the potentially complex arrangement of spots posing a challenge to
index. The projections of grain boundaries inclined 63° in each
grain would be roughly 0.5 μm.

8.9 You will want to consider radioactive characteristics and the actual
specimen sizes. Very tiny samples (very small volumes) could be
safely handled with a tweezer under the right conditions. The examination of radioactive samples in a TEM is indeed a serious problem.
In many cases a "hot" microscope is used; i.e., one dedicated to
handling radioactive specimens.

Appendix F
COURSE OUTLINES

Two suggested course outlines are attached here which have been utilized somewhat successfully in courses taught over a period of roughly a decade. These outlines have a standard semester format which has also been reorganized for teaching in the quarter system either by extending each to a two-quarter sequence, or by placing more emphasis and teaching time in each week of the quarter.

The first course represents an introductory course in electron microscopy which assumes laboratory exercises to be an integral and inseparable part of the course. This course has been successful at both the senior undergraduate level and as a first course at the graduate level. The second course is an extension of the introductory course emphasizing electron and ion microanalysis. This course has been expanded to a two-semester sequence or a three-quarter sequence, but can be given as a one-semester course or a two-quarter sequence. The second course would normally be a graduate course and the suggested format has also been utilized in several intensive short courses over a 4-day period.

These course outlines are intended to provide some ideas on the organization of the topics in this book into a formal instructional format. There are certainly many possible outlines which might be constructed and

effectively utilized depending upon the intent of the course and the specific interests or areas of expertise of the instructor, etc.

ELEMENTS OF ELECTRON MICROSCOPY

Suggested Course Outline (A First Course at the Senior or Graduate Level)

Lecture 1   a) Fundamental properties of electrons (Chap. 1)

            b) Electron optics and electron optical design (Chap. 3).
               Theory of electromagnetic lenses; image rotation.

Lab. 1   The electron microscope - Instrumentation and hardware, funda-
           mental techniques for specimen observation (Chap. 3, Sec. 8.2) -
           Demonstration. (Appendix B)

Lecture 2   Lens properties and resolution. Formation of high-resolution
           images in the electron microscope.

Lab. 2   Image rotation calibration - using MoO_3 crystal standards.

Lecture 3   Electron microscopy of surfaces (Chap. 5) - Mass - Thickness
           contrast in "non-crystalline" images. Replication techniques.
           (Biological specimen contrast).

Lab. 3   Image magnification calibration. Demonstration of replication
           techniques (Chap. 5).

Lecture 4   Electron diffraction (Chap. 6, Sec. 6.1 - 6.4).

Lab. 4 λL calibration (Diffraction camera constant determination).

Lecture 5   Diffraction intensities (Chap. 6, Sec. 6.5).

Lab. 5   Diffraction patterns: Interpretation, simulation (production
           of diffraction nets; Appendix A), indexing (Chap. 6, Sec. 6.8.4).
           Lecture - Demonstration. (Computer techniques; Appendix C)

Lecture 6   Kikuchi diffraction patterns (Chap. 6, Sec. 6.6).

Lab. 6   Demonstration of properties of kikuchi patterns. Continuation
           of indexing techniques. - Demonstration - lecture (Computer
           techniques; Appendix C)

Lecture 7   Special diffraction effects (Sec. 6.8.7).

Lab. 7   Selected-area techniques - twin spots. Demonstration - lecture
           (utilizing the TEM) (Sec. 6.8.7)

Lecture 8   Transmission electron microscopy of crystalline materials
           (Chap. 7, Sec. 7.1 - 7.2) Kinematical diffraction contrast.

Lab. 8   Demonstration of bright and dark-field techniques in thin metal
           films - preparation of thin metal films by electropolishing
           techniques. (Appendix B)

Lecture 9   Continuation of kinematical diffraction contrast theory.

Lab. 9   Demonstration of weak-beam kinematical images and applications
           of the weak-beam method of imaging (Sec. 7.2.2).

Lecture 10 Dynamical theory of diffraction contrast (Chap. 7, Sec. 7.3).

Lab. 10 Calculation of film thickness, determination of extinction distances, and the concept of operating reflections (Sec. 7.2.1, 7.4 7.4).

Lecture 11 Crystal imperfections (Chap. 7, Sec. 7.3.1).

Lab. 11 Grain boundaries and twin boundaries (crystal boundaries) - dark-field techniques. Demonstration - lecture (Sec. 7.3.1, 7.4).

Lecture 12 Continuation of imperfection contrast.

Lab. 12 Determination of dislocation density and burgers vectors.

Lecture 13 Structural geometry in thin films (Chap. 7, Sec. 7.4).

Lab. 13 Applications of structural geometry. Determination of dihedral angles (Sec. 7.4.1).

Lecture 14 Applications of the electron microscope in the physical sciences (Sec. 7.5, 7.10.

Lab. 14 Grain size measurements and particle distributions (Sec. 6.8.2).

Lecture 15 Applications of the electron microscope in metallurgical and mineral sciences (Chap. 7, Sec. 7.10).

Lab. 15 Application examples and special problems.

ELECTRON AND ION OPTICAL APPLICATIONS/ELECTRON AND ION MICROANALYSIS

Suggested Course Outline (A Second Course at the Graduate Level)

WEEK I Introduction
Thermionic emission (Chap. 2, Sec. 2.1 - 2.4)

WEEK II Field emission (Chap. 2, Sec. 2.5)
Field-ion microscopy, Atom probe field-ion microscopy
Atom probe field-ion spectroscopy (Chap. 2, Sec. 2.6.3; Chap. 4, Sec. 4.7)

WEEK III Ion microprobe analysis (Chap. 4, Sec. 4.6.1 - 4.6.4)
Electron probe microanalysis (Chap. 4, Sec. 4.1 - 4.5)
Scanning electron microscopy (Chap. 4, Sec. 4.2.1; Chap. 5, Sec. 5.5)

WEEK IV Scanning electron microscopy
Secondary electron emission from solids (Chap. 5, Sec. 5.5.1)

WEEK V Secondary electron images
X-ray spectrometry in the SEM (Energy dispersive analysis)
(Chap. 4, Sec. 4.2.1, Sec. 4.4.1; Chap. 5, Sec. 5.5.8)

WEEK VI Electron spectroscopy for chemical analysis (Chap. 4, Sec. 4.3.3)
Scanning auger microscopy and scanning auger electron
spectroscopy (Chap. 4, Sec. 4.3.3, Sec. 4.5.3)

WEEK VII Diffraction intensities (Chap. 6, Sec. 6.5)
Electron diffraction applications

WEEK VIII Review of kinematical theory (Chap. 7)
Review of dynamical theory (Chap. 7)

WEEK IX   Applications of 2-beam theory (Chap. 7)
             Weak beam dark-field applications (Chap. 7)
             High-resolution electron microscopy (Chap. 7, Sec. 7.5)

WEEK X   Lattice fringe imaging and applications (Chap. 7, Sec. 7.5.1)

WEEK XI   Scanning transmission electron microscopy (STEM) (Chap. 7,
             Sec. 7.6)
             STEM applications
             Advances in materials characterization

WEEK XII   Electron energy loss spectroscopy (Chap. 4, Sec. 4.3.3)
             Analytical electron microscopy and applications (Chap. 7,
             Sec. 7.7)

WEEK XIII   *In-situ* experiments in the electron microscope
             High-voltage electron microscopy (Chap. 8)

WEEK XIV   Many-beam theory (Chap. 8, Sec. 8.3.1)
             Image analysis and simulation (Chap. 7, Sec. 7.10.1)

WEEK XV   Applications of electron microscopy in metallurgy and the
             mineral and materials sciences (Chap. 7 and 8)

Author Index

Numbers in parentheses designate the page numbers on which a complete citation (usually a literature citation) appears. Other page citations indicate that an author's work is referred to although his or her name may not be cited in the text. Author citations are usually listed numerically in each chapter and the citations are collected in the reference section at the end of each chapter or appendix.

A

Abbe, E., (578)
Agar, W., 421, (471)
Aigeltinger, E. A., 446, (472)
Aiken, R., (756)
Akiyama, K., 463, (472)
Albrecht, T. R., 88, (100)
Alexander, L. E., 424, (471)
Alexander, S., 88, (100)
Al-Jassim, M. M., 626, (646), (753)
Allinson, D. L., 670, (682)
Amelinckx, S., 344, (360), 500, 519, 520, 526, 529-531, (535), (536), 541, 578, 603, (642-648), 723, 726, (752), (753)
Amy, J. H., 293, (358)
Anderson, C. D., 1
Anderson, P. W., (100)
Andrews, K. W., (473)
Appel, F., (674)
Arceneaux, C. J., (554), 654,

(682), (683)
Arnot, F. L., 228, (253)
Art, A., 519, 534, (535), 536, (643)
Ashbee, K. H. G., (487), (647)
Ashby, M. F., 557, 573, (644)
Auger, P., 180, (252)
Avery, C. H., (294)

B

Bahadur, K., 69, 70, (99)
Bailey, G. W., (472), (683)
Bailey, J., 550, (643)
Baird, D. L., (100)
Baker, R. F., 215, (253)
Barbour, J. P., 67, (99)
Barer, R., (44), (148), (259)
Barnes, R. S., 553, (644)
Barrett, C. S., 453, (470), 685, (710), (775)
Basinski, Z. S., 658, (682)

Subject Index

Milton Keynes UK
Ingram Content Group UK Ltd.
UKHW052032071024
449327UK00027B/2524

9 780367 402945